Library of
Davidson College

**PLEISTOCENE
AND
RECENT
ENVIRONMENTS
OF
THE
CENTRAL
GREAT
PLAINS**

PLEISTOCENE AND RECENT ENVIRONMENTS OF THE CENTRAL GREAT PLAINS

Edited by
Wakefield Dort, Jr., and J. Knox Jones, Jr.

Department of Geology, University of Kansas
Special Publication 3

THE UNIVERSITY PRESS OF KANSAS
Lawrence/Manhattan/Wichita

© Copyright 1970
by the University Press of Kansas

Standard Book Number 7006-0063-9
Library of Congress Catalog Card Number 79-629062

Printed in the United States of America
Designed by Fritz Reiber

EDITORIAL NOTE

The Great Plains of North America support one of the largest contiguous grasslands in the world. Together with the adjacent tall-grass prairie of the Central Lowlands to the east, these grasslands occupy a region stretching from southern Canada to northern Mexico and from the foothills of the Rocky Mountains in the west to beyond the Mississippi River in the east. The natural setting of this vast area long has interested natural and physical scientists; yet there have been few attempts to focus on the environmental interrelationships as a whole.

In the course of informal conversations several years ago, we discussed the possibility of organizing an interdisciplinary symposium on the natural and physical environments of the grasslands region—one that would stress the contributions of the various disciplines, with particular reference to the kinds of information useful to scientists with differing basic backgrounds. Because of our own interests and those of many of our colleagues at the University of Kansas, we decided that the emphasis in such a symposium ought to be on the central part of the Great Plains. Further discussions with other scientists at Kansas and elsewhere indicated strong interest in a symposium, and in the autumn of 1967 an ad hoc committee was formed to arrange for one, the principal aims of which, aside from bringing scientists together in one place to promote informal exchange of ideas, were to be: (1) to attempt to summarize, in a general way at least, what had been done in the plains region in each of the interrelated disciplines of anthropology, botany, geology, and zoology; (2) to call attention to the important concept of interdisciplinary studies and to place information from a wide variety of sources in the appropriate geographic and temporal setting; and

(3) to focus attention on areas where further investigations clearly are needed and to suggest means by which interdisciplinary approaches might be useful in considering many problems.

Besides ourselves, the organizational committee consisted of Ronald W. McGregor (later replaced by Philip V. Wells), Department of Botany, Alfred E. Johnson, Department of Anthropology, and Richard F. Treece, representing University Extension (Institutes and Conferences). The committee, meeting informally through the 1967-68 school year, agreed that the proposed symposium should be held in the autumn of 1968 despite the fact that financial support from outside the University of Kansas would not be available. Within the University, the departments of Anthropology, Botany, Geology, and Zoology, the Museum of Natural History, and the University's Committee on Systematic and Evolutionary Biology agreed to sponsor the symposium. The latter two groups, together with the Department of Geology (through the Shell Aid Fund) and the departments of Anthropology and Botany, generously contributed financial support to underwrite part of the venture, and the "Symposium on Pleistocene and Recent Environments of the Central Plains" was held in Lawrence, Kansas, on October 25 and 26, 1968. More than 500 people attended the opening session. Participants came from twenty states and four Canadian provinces.

John P. Dessauer, former Director of the University Press of Kansas, arranged for publication of the present volume. Manuscripts were called for at the end of October 1968. Some were received on time, but others arrived much later (the latest manuscript was submitted to the editors in October 1969). All authors had the opportunity to insert limited new material at the time galley proof was available to them early in 1970.

Our editorial changes were minimal in most papers, consistent with the style established for the volume. Virginia Seaver diligently and thoughtfully attended to editorial matters for the University Press of Kansas. Yvonne Willingham, presently Acting Director of the Press, handled production. We are most grateful to these two persons and to the University of Kansas Printing Service for their aid in seeing this volume through to publication.

Wakefield Dort, Jr.
J. Knox Jones, Jr.

CONTRIBUTORS

David A. Baerreis
Department of Anthropology, University of Wisconsin,
Madison, Wisconsin 53706

Reid A. Bryson
Center for Climatic Research, Department of Meteorology,
University of Wisconsin, Madison, Wisconsin 53706

Frank B. Cross
Museum of Natural History, University of Kansas,
Lawrence, Kansas 66044

Wakefield Dort, Jr.
Department of Geology, University of Kansas,
Lawrence, Kansas 66044

Vincent H. Dreeszen
Nebraska State Geological Survey, University of Nebraska,
Lincoln, Nebraska 68508

David R. Fisher
Department of Zoology, University of Kansas,
Lawrence, Kansas 66044

C. Vance Haynes
Department of Geological Science, Southern Methodist University,
Dallas, Texas 75222

Claude W. Hibbard
Museum of Paleontology, University of Michigan,
Ann Arbor, Michigan 48104

Robert S. Hoffmann
Museum of Natural History, University of Kansas,
Lawrence, Kansas 66044

G. K. Hulett
Division of Biological Sciences, Fort Hays Kansas State College,
Hays, Kansas 67601

J. Knox Jones, Jr.
Museum of Natural History, University of Kansas,
Lawrence, Kansas 66044

Ronald O. Kapp
Department of Biology, Alma College,
Alma, Michigan 48801

James E. King
Department of Geochronology, University of Arizona,
Tucson, Arizona 85721

Marvin F. Kivett
Nebraska State Historical Society,
1500 R Street, Lincoln, Nebraska 68508

Richard A. Krause
Department of Anthropology, University of Missouri,
Columbia, Missouri 65202

Donald J. Lehmer
Department of Anthropology, Dana College,
Blair, Nebraska 68008

Everett H. Lindsay
Department of Geology, University of Arizona,
Tucson, Arizona 85721

Larry D. Martin
University of Nebraska State Museum,
Lincoln, Nebraska 68508

Peter J. Mehringer, Jr.
Department of Geochronology, University of Arizona,
Tucson, Arizona 85721
(Present address: Department of Anthropology, University of Utah,
Salt Lake City, Utah 84112.)

Robert M. Mengel
Museum of Natural History, University of Kansas,
Lawrence, Kansas 66044

Herbert H. Ross
Illinois Natural History Survey, Urbana, Illinois 61801
(Present address: Department of Entomology, University of Georgia,
Athens, Georgia 30602.)

Robert V. Ruhe
U.S. Department of Agriculture, Soil Conservation Service,
Agronomy Building, Iowa State University, Ames, Iowa 50010

C. Bertrand Schultz
University of Nebraska State Museum,
Lincoln, Nebraska 68508

Gerald R. Smith
Museum of Natural History, University of Kansas,
Lawrence, Kansas 66044
(Present address: Museum of Zoology, University of Michigan,
Ann Arbor, Michigan 48104.)

G. W. Tomanek
Division of Biological Sciences, Fort Hays Kansas State College,
Hays, Kansas 67601

Waldo R. Wedel
Division of Anthropology, Smithsonian Institution,
Washington, D.C. 20560

Philip V. Wells
Department of Botany, University of Kansas,
Lawrence, Kansas 66044

Fred Wendorf
Anthropology Research Center, Southern Methodist University,
Dallas, Texas 75222

Wayne M. Wendland
Center for Climatic Research, Department of Meteorology,
University of Wisconsin, Madison, Wisconsin 53706

H. E. Wright, Jr.
Limnological Research Center, University of Minnesota,
Minneapolis, Minnesota 55455

CONTENTS

1

EARTH SCIENCES AND CLIMATE

Wakefield Dort, Jr.
Recurrent Climatic Stress on Pleistocene and Recent Environments 3

Vincent H. Dreeszen
The Stratigraphic Framework of Pleistocene Glacial and Periglacial Deposits in the Central Plains 9

Fred Wendorf
The Lubbock Subpluvial 23

Robert V. Ruhe
Soils, Paleosols, and Environment 37

Reid A. Bryson, David A. Baerreis, and *Wayne M. Wendland*
The Character of Late-Glacial and Post-Glacial Climatic Changes 53

2

ANTHROPOLOGY
(consulting editor, Alfred E. Johnson)

C. Vance Haynes
Geochronology of Man-Mammoth Sites and Their Bearing on the Origin of the Llano Complex 77

Marvin F. Kivett
Early Ceramic Environmental Adaptations 93

Richard A. Krause
Aspects of Adaptation Among Upper Republican Subsistence Cultivators 103

Donald J. Lehmer
Climate and Culture History in the Middle Missouri Valley 117

Waldo R. Wedel
Some Environmental and Historical Factors of the Great Bend Aspect .. 131

3

BOTANY
(consulting editor, Philip V. Wells)

Ronald O. Kapp
 Pollen Analysis of Pre-Wisconsin Sediments from the Great Plains 143

H. E. Wright, Jr.
 Vegetational History of the Central Plains 157

Peter J. Mehringer, Jr., James E. King, and *Everett H. Lindsay*
 A Record of Wisconsin-Age Vegetation and Fauna from the Ozarks of Western Missouri 173

Philip V. Wells
 Vegetational History of the Great Plains: A Post-Glacial Record of Coniferous Woodland in Southeastern Wyoming 185

G. W. Tomanek and *G. K. Hulett*
 Effects of Historical Droughts on Grassland Vegetation in the Central Great Plains 203

Philip V. Wells
 Historical Factors Controlling Vegetation Patterns and Floristic Distributions in the Central Plains Region of North America 211

4

ZOOLOGY

Herbert H. Ross
 The Ecological History of the Great Plains: Evidence from Grassland Insects 225

Frank B. Cross
 Fishes as Indicators of Pleistocene and Recent Environments in the Central Great Plains 241

Gerald R. Smith and *David R. Fisher*
 Factor Analysis of Distribution Patterns of Kansas Fishes 259

Robert M. Mengel
 The North American Central Plains as an Isolating Agent in Bird Speciation 279

C. Bertrand Schultz and *Larry D. Martin*
 Quarternary Mammalian Sequence in the Central Great Plains 341

Robert S. Hoffmann and *J. Knox Jones, Jr.*
 Influence of Late-Glacial and Post-Glacial Events on the Distribution of Recent Mammals on the Northern Great Plains 355

Claude W. Hibbard
 Pleistocene Mammalian Local Faunas from the Great Plains and Central Lowland Provinces of the United States 395

EARTH SCIENCES AND CLIMATE

Recurrent Climatic Stress on Pleistocene and Recent Environments

WAKEFIELD DORT, JR.

ABSTRACT

Fluctuations of glaciers are mainly the result of fluctuations of climate that can be classified as glaciations, stades, substades, and lesser pulses. The number of fluctuations that occurred during the Quaternary Period is unknown, but is undoubtedly large. Some changes may have taken place in cycles lasting only a few hundred years, and changes in adjacent areas were not necessarily either synchronous or of the same magnitude and direction. Each change in climate exerted a corresponding stress on the local environment that affected the existing population of plants, animals, and aboriginal man.

Study of the myriad events of Pleistocene and Recent time—the Quaternary Period—is without question the most complex field of the natural sciences because inevitably it involves the meshing of information from several of the physical and biological disciplines. When grappling with this complexity in Quaternary problems, investigators frequently seek to ease frustration by assuming simplicity wherever there is no compelling evidence to the contrary. Such assumptions of simplicity have been made most frequently in the area of past climate and climatic change and, of course, in regard to events controlled by climatic change. The purpose of the brief discussion that follows is to call attention to the multiplicity of climatic changes, at various levels of intensity, which characterizes all of Quaternary time.

To a degree much greater than for any other segment of earth history, the Quaternary time scale has been built on the basis of variations in climatic regime or, more usually, on the basis of evidence of erosion or deposition that was the direct consequence of variation in climate. A detailed and penetrating discussion of the principles and problems involved in compilation of the Quaternary time scale has been presented by Morrison (1968). He commented that although the traditional basis for division of Quaternary time and stratigraphic record has been the fluctuation of glaciers in the north-central United States and western Europe, much of the world was not subjected to glaciation at all. Therefore, the use of classifications based on a geographically distant glacial record requires forced correlations that may not be justifiable, at least within existing limits of knowledge.

Currently there exists a considerable degree of controversy and confusion regarding differentiation and proper usage of time-stratigraphic, rock-stratigraphic, and geologic-climate units for the Quaternary record. There is even less agreement, especially at the lower levels of stratigraphic nomenclature, about the numbers of such units that are applicable. The largest units of Quaternary classifications are, of course, based on the glaciations and intervening interglaciations. It seems to be universally accepted that, at least in the Northern Hemisphere, there were four glaciations, the familiar Nebraskan-Kansan-Illinoian-Wisconsin sequence of North America or the Gunz-Mindel-Riss-Wurm sequence of Europe. There is, however, the possibility that there was an even earlier glaciation. This has been suggested in Europe—the Donau episode—but the record as thus far deciphered is not clear. Nevertheless, the apparent absence of truly definitive field evidence is not sufficient reason, by itself, to state categorically that there were only four episodes of climatic cooling and ice formation of glaciation rank during the Pleistocene Epoch. Every project involving field study of supposed lowest Pleistocene glacial deposits must reassess the question on the basis of available evidence.

The 1933 Stratigraphic Code authorized use of the terms "stage" and "substage" for climatic subdivisions of the Quaternary Period. Such usage led to confusion with other specified use of these words in stratigraphic nomenclature. Therefore, the 1961 Stratigraphic Code specifically rejected use of the term "stage" in this meaning and established a category of geologic-climate units. According to current proper usage, a "glaciation" was a climatic episode during which extensive glaciers developed, attained a maximum extent, and receded. A "stade" was a climatic episode within a glaciation during which a secondary advance of glaciers took place. Stades were separated by "interstades." If the record is sufficiently detailed, a stade may be subdivided into "substades."

The extent to which a glaciation can be subdivided into stades and substades depends largely on the degree of preservation of the geological record, but also to some extent on the diligence with which that record has been studied. Evidence of fluctuations of the continental ice sheet during Wisconsin time is especially well preserved in the state of Illinois and adjacent portions of Iowa, Minnesota, Wisconsin, and Indiana. The Illinois Geological Survey has been particularly active in formulating a classification that, in effect, subdivides the Wisconsin glaciation into several stades (Valderan, Twocreekan, Woodfordian, Farmdalian, Altonian), and also recognizes a number of substades (see, for example, Frye et al., 1965).

Farther back in the Pleistocene record, the combined effects of irregular erosion and general burial of deposits have resulted in a progressive obscuring of detail so that few subdivisions are known. In fact, it is only recently that clear-cut evidence of stadial units of the Kansan and Nebraskan glaciations have been recognized (Bayne, 1969; Dort, 1965, 1966a, 1966b; Reed and Dreeszen, 1965; Reed et al., 1965).

Almost all variations in ice-front position, with consequent shifts in erosional and depositional processes, are the result of variations in the climatic regime either at the ice front or in the interior of the area covered by the ice sheet. A possible exception is the glacial surge or short-period advance that may result from occasional dynamic inequilibrium in a part of the glacier mass. Compilation of a record of ice-front fluctuations, therefore, indirectly but rather soundly establishes a parallel record of the controlling climatic fluctuations.

It is now pertinent to inquire how many fluctuations of climate have occurred and how many fluctuations of ice-front position have been recorded for North American continental ice sheets—how many glaciations, stades, substades, and even lesser pulses were there and how many have been recognized? Because there is nowhere a complete stratigraphic column for the entire Quaternary Period, information must be acquired in bits and pieces from many locations and put together by various means of correlation, procedures that sometimes rest on rather insecure foundations. For this reason, it is by no means certain how many valid subdivisions have already been recognized, and it is essentially impossible to arrive at a meaningful estimation of the number of fluctuations that actually did occur. It is almost surely true, however, that the number is far greater than students of Pleistocene history have yet realized.

Five stades of the Wisconsin glaciation are reasonably well established on the basis of field evidence present in the United States; other stades have been based on evidence found mainly in Canada. The existence of more stades has been postulated from indications found in limited areas. It may well be that additional fluctuations of stadial rank actually occurred, but evidence for them has yet to be recognized, and undoubtedly each stade was comprised of several, perhaps many, substadial variations.

Three stades of the Illinoian glaciation have been recognized by the Illinois Geological Survey. Others have been suggested from various localities. It is highly probable that more occurred than have yet been suspected, and that all had substadial variations.

Despite accumulating incontrovertible evidence of stadial and substadial fluctuations of the Wisconsin glaciation and, to a lesser degree, of the Illinoian glaciation, it was until recently believed that the Kansan and Nebraskan glaciations were simple, single-pulse events. It is now known that evidence does exist for stadial fluctuations of both of the glaciations, and there are indications in northeastern Kansas and eastern Nebraska of at least three major subdivisions of each. In all probability many more actually occurred and each had substadial variations as well.

It can thus be reasoned that a considerable number of climatically controlled fluctuations of the North American ice sheets took place during Pleistocene time. As the magnitude of the fluctuation decreases with additional subdivisions of the nomenclature, there is increasing opportunity for the insertion of purely local events, either of climatic fluctuation or of movement of the ice-front position. The front of the ice sheet was not a simple arc extending for hundreds of miles across the continent. It was composed of many coalescing lobes influenced by subglacial and proglacial topography and by localizations of nourishing precipitation. These lobes, even those directly contiguous, did not always advance or retreat simultaneously, hence a minor pulse recorded at one spot may not have occurred at some other place nearby.

Additional insight into the possible frequency, duration, and magnitude of climatic fluctuations that may have characterized at least the waxing and waning phases of each glaciation can be gained by studying events of that part of the Quaternary usually considered to be post-glacial. The most clearly detailed evidence comes from mountainous areas rather than the interior lowlands. Without engaging in essentially meaningless debate over the question of whether the present climate is interglacial or truly post-glacial, it can be noted that a major portion of so-called post-glacial time was a time of glacier regeneration—the Neoglaciation. This episode lasted about 4000 years, ending at most only a few centuries ago.

Of greater importance to the present discussion, the Neoglaciation can be subdivided into at least two stades, each of which is comprised of several substades (see Richmond, 1965). It is thus clear that under favorable conditions there may be preserved in the geologic record evidence of climatic fluctuations that lasted for only a few centuries yet were of sufficient magnitude to cause significant changes in sensitive indicators such as alpine glaciers.

An example illustrating that changes of this minor intensity affected other areas and other processes is an unusual record preserved in a lava cave on the Snake River Plain of Idaho (Butler, 1968; Dort, 1968). Here, excavations by archaeologists revealed a sequence of laminated silts that had been periodically augmented by windblown sediment and then deformed by frost heaving and ice-wedge formation. Radiocarbon dates indicate that each cycle of rather drastic climatic change in this localized environment lasted about 400 years.

Beyond the limits of glaciation, variations in other climatically controlled processes provide evidence of fluctuations of the climate itself. Valley filling, terrace cutting, increase or decrease in loess accumulation or sand dune activity, formation and movement of colluvium, rise and fall of lakes in closed basins, development and modification of soils, and even records of population changes for plants, animals, or aboriginal man all provide important clues. Less is known about fluctuations in these warmer areas, which actually constitute more than half of the land areas of the world, because their records have received less intensive investigation than those of the glaciated regions. Nevertheless, it appears reasonable to suppose that those land areas not covered by ice during a glaciation, and all land areas during the interglaciations, were subjected to numerous and sometimes marked variations in climate.

Each of the Quaternary climatic variations exerted a stress on the environment of the area affected. Obviously, the fluctuations of higher rank, the glaciations and stades, had greater effect than lesser pulses, but each change of climate undoubtedly caused some change in the viability of the total plant and animal population that had existed during the preceding interval. Depending on the direction of change, organisms for which the previous environment provided only marginal existence were able to expand or else were forced out completely, and the intensity of competition among other organisms shifted accordingly. But when climatic conditions reverted to the pattern of a former episode, it was not always possible for the organic population to revert to an equal degree. The timing of application of the climatic stress in relation to establishment or differentiation of the population is of considerable importance. Similar environments separated by time or space often had dissimilar inhabitants.

Because changes in climate were so frequent during the Quaternary Period, grossly misleading impressions can be given by categorizing in a few words the climate of a major interval of time. True, the climate of a glaciation was in general colder and wetter than that of an interglaciation, but to speak of "the climate of the Wisconsin" when referring to a large area is almost meaningless. The climate underwent major variations during the time of the Wisconsin glaciation, and the climate in one part of a state differed from that in another part at any one moment. Much of this record remains to be deciphered. Furthermore, it is known that the sequential climates of the four well-known glaciations were not the same in detail. During some intervals, the climate of the ice-free areas of North America showed as much diversification as exists now; at other times there was a marked difference.

Finally, one additional point deserves emphasis. A major shift in one aspect of the climate will cause shifts in other aspects, but these secondary changes may not be everywhere of the same intensity, or even in the same direction (see Bryson *et al.*, this volume). This fact introduces additional complications and makes generalizations, especially over large areas, even more hazardous.

LITERATURE CITED

Bayne, C. K.
 1969. Evidence of multiple stades in the Lower Pleistocene of northeastern Kansas. Trans. Kansas Acad. Sci., 71:340-349.

Butler, B. R.
 1968. An introduction to archaeological investigations in the Pioneer Basin locality of eastern Idaho. Tebiwa, Jour. Idaho State Univ. Mus., 11:1-26.

Dort, W., Jr.
 1965. Nearby and distant origins of glacier ice entering Kansas. Amer. Jour. Sci., 263:598-605.
 1966a. Multiple Early Pleistocene glacial stades, northeastern Kansas. Geol. Soc. Amer. Spec. Paper, 87:47.
 1966b. Nebraskan and Kansan stades: complexity and importance. Science, 154:771-772.
 1968. Paleoclimatic implications of soil structures at the Wasden Site (Owl Cave). Tebiwa, Jour. Idaho State Univ. Mus., 11:31-36.

Frye, J. C., H. B. Willman, and R. F. Black
 1965. Outline of glacial geology of Illinois and Wisconsin. Pp. 43-61, in The Quaternary of the United States (H. E. Wright, Jr., and D. G. Frey, eds.), Princeton Univ. Press, Princeton, New Jersey, x+922 pp.

Morrison, R. B.
 1968. Means of time-stratigraphic division and long-distance correlation of Quaternary succession. Pp. 1-113, in Means of correlation of Quaternary successions (R. B. Morrison and H. E. Wright, Jr., eds.), Univ. Utah Press, Salt Lake City, xi+631 pp.

Reed, E. C., and V. H. Dreeszen
 1965. Revision of the classification of the Pleistocene deposits of Nebraska. Bull. Nebraska Geol. Surv., 23:1-65.

Reed, E. C., V. H. Dreeszen, C. K. Bayne, and C. B. Schultz
 1965. The Pleistocene in Nebraska and Kansas. Pp. 187-202, in The Quaternary of the United States (H. E. Wright, Jr., and D. G. Frey, eds.), Princeton Univ. Press, Princeton, New Jersey, x+922 pp.

Richmond, G. M.
 1965. Glaciation of the Rocky Mountains. Pp. 217-230, in The Quaternary of the United States (H. E. Wright, Jr., and D. G. Frey, eds.), Princeton Univ. Press, Princeton, New Jersey, x+922 pp.

The Stratigraphic Framework of Pleistocene Glacial and Periglacial Deposits in the Central Plains

VINCENT H. DREESZEN

ABSTRACT

Classification of the Pleistocene deposits in the Central Plains appears to be in a process of evolution. Evidence that continental glaciation was complex and that each major glacial period included multiple advances is accumulating. Moreover, the concept of four major glaciations separated by distinct interglacial periods needs reexamination. Problems of correlation between deposits related to continental sheets of glacial ice and those related to streams, probably in part rejuvenated by crustal adjustment, remain to be solved. The precision of correlation of Pleistocene deposits in the Central Plains with those of other geographical areas diminishes beyond the reach of radiocarbon dating. Greater precision will require interdisciplinary action among Pleistocene scientists and a considerable attention to subsurface data.

INTRODUCTION

The Pleistocene Epoch spans geologic time from the end of the Tertiary to the beginning of the Holocene, or Recent. Figured roughly, it began with the first major advance of continental ice and ended with melting of the last. Curray (1965, p. 724) suggested that the boundary between the Pleistocene and Holocene be placed at approximately 7000 BP, based on evidence that by then the sea level had stabilized near its present level. Frye and Willman (1960, p. 4) inferred from radiocarbon dates that the division was at 5000 BP, whereas Richmond (1965, p. 226 and table 2) indicated that deglaciation was completed at about 6500 BP. General agreement on dating this boundary can be foreseen. On the other hand, agreement among Pleistocene geologists in dating the beginning of the Pleistocene still is far from being reached. For example, many geologists who are not directly concerned with the problem consider the Pleistocene to represent about one million years of time, whereas radiogenic dates published by Evernden et al. (1964) have caused Hibbard et al. (1965, p. 509) to assume the base of the Pleistocene, as they define it, to be two to three million years old. The Pleistocene-Tertiary boundary is of particular importance to Pleistocene geologists and paleontologists in the Central Plains states of Kansas and Nebraska because of the occurrence of abundant fossils in sediments of both ages. It is axiomatic that the lower boundary of the Pleistocene will be the last to receive general agreement.

Continental ice sheets were agents of both erosion and deposition. Erosion by ice was most active in the areas of thicker ice to the north and east of the Central Plains states. In the absence of evidence to the contrary, it is assumed that continental ice sheets moved into a region of moderate climate in the Central Plains states. Deposition of relatively thick sediments took place along the margins of the ice sheets as glacial till, as outwash from streams issuing from the ice, and as lake deposits marginal to the ice.

Stream erosion was extremely active and probably concurrent with continental glaciation at considerable distances beyond the ice sheets. Although due partly to climatic and sea-level change, the increase in this type of erosion probably was due mostly to rejuvenation of streams by crustal adjustment, including the tilting of surfaces by downwarping under the load of ice and epeirogenic upwarping in the Rocky Mountain area. Relatively thick alluvial sediments were deposited in the periglacial region in Kansas and Nebraska and, to a lesser extent, in the adjacent states. Such deposition took place under basin-like conditions.

Eolian deposits constitute an important part of the Pleistocene deposition in the Central Plains. Loess of Wisconsinan age (Thorp, Smith, et al., 1952) is the surface deposit in much of both the glacial and periglacial regions of the Central Plains states, and ranges in thickness from a few inches to many tens of feet. Also, dune sand of Wisconsinan age is the surface deposit in about one-third of Nebraska and smaller parts of Kansas and South Dakota. The typically reddish-brown loess of the Loveland Formation of Illinoian age has been described in the literature of each of the Central Plains states. Reed and Dreeszen (1965, fig. 3, p. 23) have suggested that eolian deposition also was contemporaneous with each of the periods of older alluviation in the Pleistocene. The cyclic nature of the alluvial and eolian deposits, along with profiles of weathering and soil formation on their surfaces, reflect not only the changing conditions of glacial advance and retreat but also crustal adjustment. Detailed chronology and stratigraphic classification of the Pleistocene deposits have been attempted by field and laboratory study of their nature and position in both the surface and the subsurface. Evidence from the interpretation of vertebrate and invertebrate faunas has contributed significantly to this study. Radiocarbon dating has been a valuable tool in the study of deposits of Wisconsinan age. The importance of the Pearlette Volcanic Ash as a marker horizon and the significance of the presumption that there is but one recognizable ash bed in the Pleistocene deposits of the Central Plains states cannot be overemphasized (Reed, 1948, p. 161; Frye et al., 1948, pp. 504-506; Jewett, 1963, pp. 353-356; Swineford, 1963, pp. 359-362; Reed and Dreeszen, 1965, pp. 6-8; Wilcox, 1965, pp. 811-812).

Although the deposits in the periglacial region merge with those in the glaciated area, correlation problems between the two areas still exist. The chronology in both areas has presumed the accuracy of the Pleistocene framework concept of four major glacial periods and three interglacials.

MULTIPLE GLACIATION

The concept of multiple glaciation within each of the four major glacial periods has been suggested by Frye and Willman (1963, p. 504), Reed et al. (1966, pp. 1-2), and others. The evidence for multiple glaciations within the last major glaciation, the Wisconsinan Stage, has been documented by many investigators. The more complete preservation of deposits and morphologic features of the Wisconsinan have led to a far greater knowledge of this glacial episode than of any of the older episodes. Paleontological evidence has contributed greatly to this knowledge, yet the present classification of the Wisconsinan is based to a considerable extent on radiocarbon dating. Documentation of multiple glaciations in the older episodes is far more difficult, not only because of the lack of dating tools and the destruction or disturbance of morphological features (and of the deposits themselves), but by the fact that the sediments are, to a large extent, buried by younger deposits. Since much of the record is obscured by burial, subsurface methods of investigation are essential tools for the study of the Pleistocene.

A classification chart of the Pleistocene deposits in Nebraska (Fig. 1) was prepared by Reed and Dreeszen (1965) in an attempt to relate evidence obtained from extensive subsurface investigation to surface features, at the same time retaining the framework concept of four major continental glaciations and three interglacial periods. Rock-stratigraphic units were named to replace time-stratigraphic names, the complexity of the pre-Wisconsinan glaciations was suggested, loessic equivalents of fluviatile deposits were recognized, and the relation of these deposits to glacial deposits was inferred. They warned fellow workers, however (1965, p. 3): "It is not proposed that names herewith suggested for use in Nebraska be utilized in neighboring states unless there is a need for a more detailed subdivision of units now recognized in those states."

In an effort to facilitate the understanding of the regional geomorphology of the glaciated region of eastern Nebraska and adjacent areas, a relief map (Fig. 2) was prepared by shading valley and valley-slope features on 1:250,000 topographic maps. Reed and Dreeszen (1965, pp. 8-9) interpreted the prominent drainage divides in eastern Nebraska (Fig. 3) to be morainic ridges, and they related the ridges to

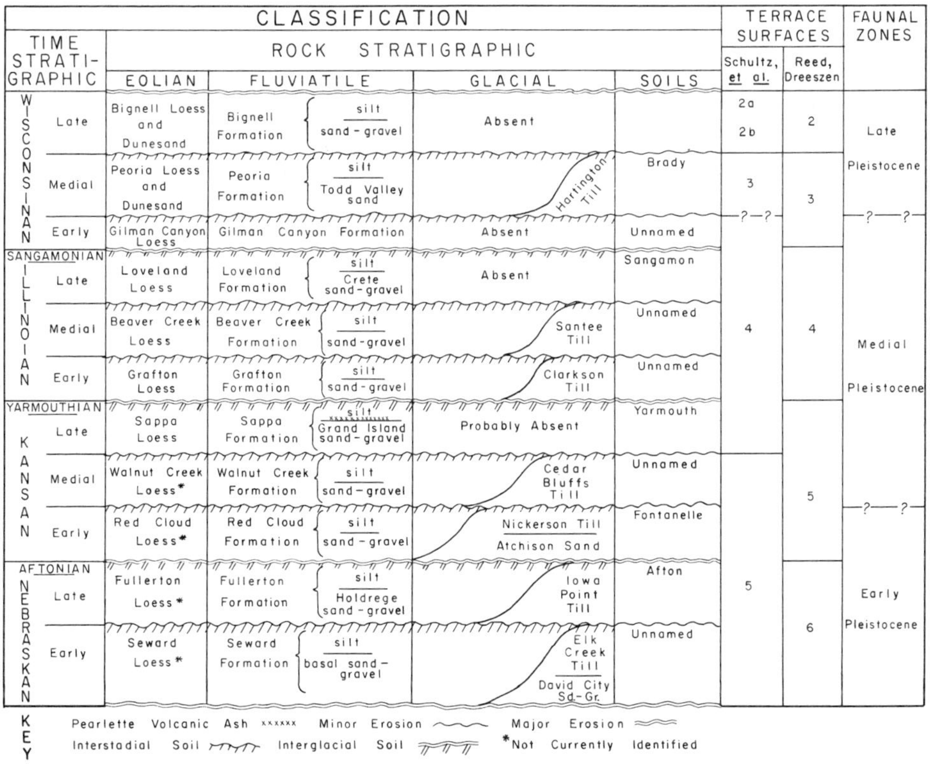

TIME STRATI-GRAPHIC	CLASSIFICATION					TERRACE SURFACES		FAUNAL ZONES
	ROCK STRATIGRAPHIC					Schultz, et al.	Reed, Dreeszen	
	EOLIAN	FLUVIATILE		GLACIAL	SOILS			
WISCONSINAN Late	Bignell Loess and Dunesand	Bignell Formation	silt / sand-gravel	Absent	Brady	2a / 2b	2	Late Pleistocene
WISCONSINAN Medial	Peoria Loess and Dunesand	Peoria Formation	silt / Todd Valley sand	Hartington Till		3	3	
WISCONSINAN Early	Gilman Canyon Loess	Gilman Canyon Formation		Absent	Unnamed	?—?—		?—?—
SANGAMONIAN					Sangamon			
ILLINOIAN Late	Loveland Loess	Loveland Formation	silt / Crete sand-gravel	Absent	Unnamed	4	4	Medial Pleistocene
ILLINOIAN Medial	Beaver Creek Loess	Beaver Creek Formation	silt / sand-gravel	Santee Till	Unnamed			
ILLINOIAN Early	Grafton Loess	Grafton Formation	silt / sand-gravel	Clarkson Till	Yarmouth			
YARMOUTHIAN								
KANSAN Late	Sappa Loess	Sappa Formation	silt / Grand Island sand-gravel	Probably Absent	Unnamed	5		?—?—
KANSAN Medial	Walnut Creek Loess*	Walnut Creek Formation	silt / sand-gravel	Cedar Bluffs Till	Fontanelle			
KANSAN Early	Red Cloud Loess*	Red Cloud Formation	silt / sand-gravel	Nickerson Till Atchison Sand	Afton			Early Pleistocene
AFTONIAN					Unnamed			
NEBRASKAN Late	Fullerton Loess*	Fullerton Formation	silt / Holdrege sand-gravel	Iowa Point Till		5	6	
NEBRASKAN Early	Seward Loess*	Seward Formation	silt / basal sand-gravel	Elk Creek Till / David City Sd-Gr.				

KEY: Pearlette Volcanic Ash ××××× Minor Erosion ~~~ Major Erosion ≈≈≈
Interstadial Soil ⌒⌒⌒ Interglacial Soil ⫝⫝⫝ *Not Currently Identified

Fig. 1. Classification chart of the Pleistocene deposits of Nebraska (from Nebraska Geological Survey Bulletin 23).

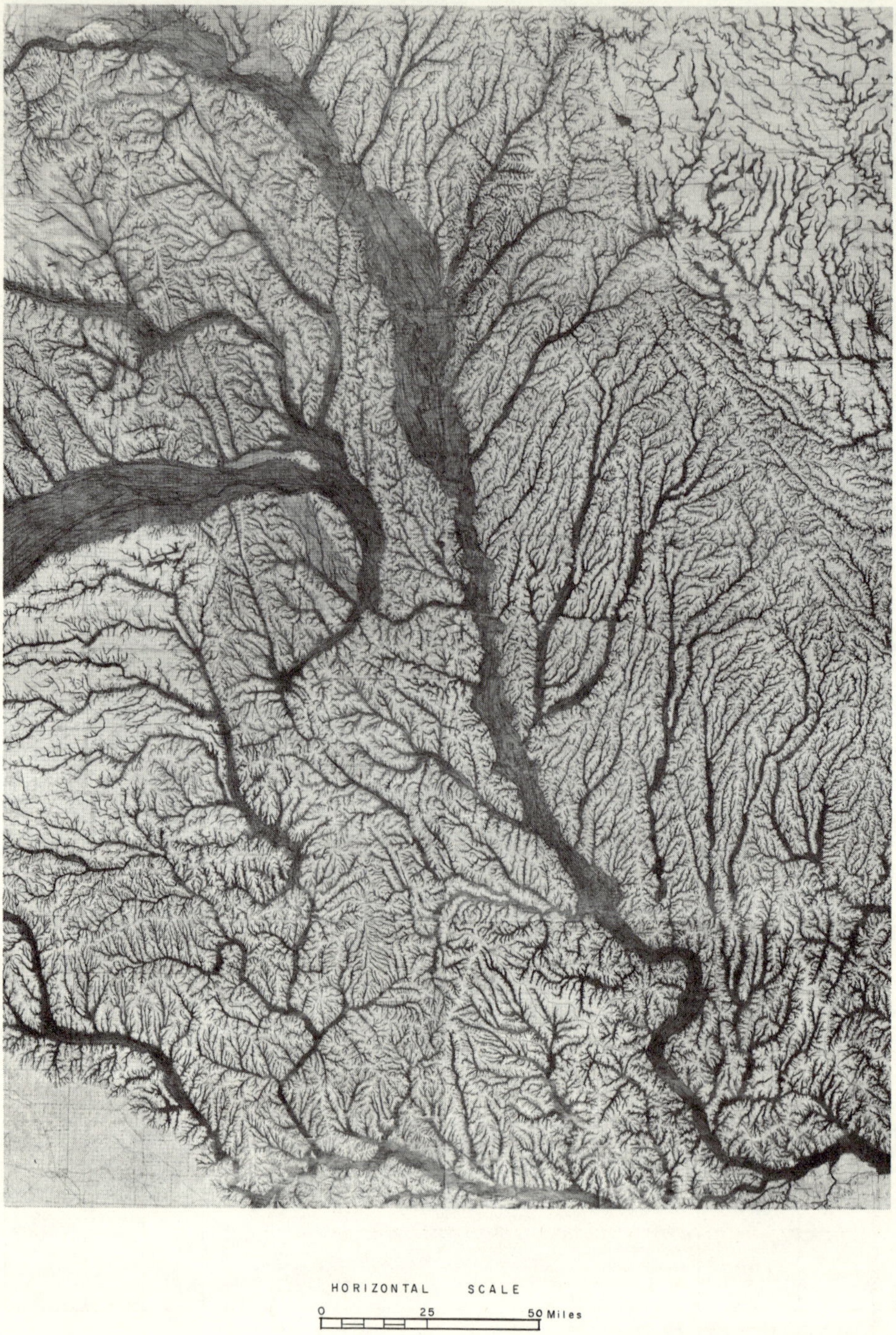

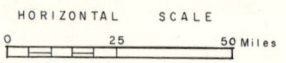

FIG. 2. Relief map of eastern Nebraska and adjoining area based on U.S. Geological Survey 1:250,000 scale topographic maps (from Nebraska Geological Survey Bulletin 23).

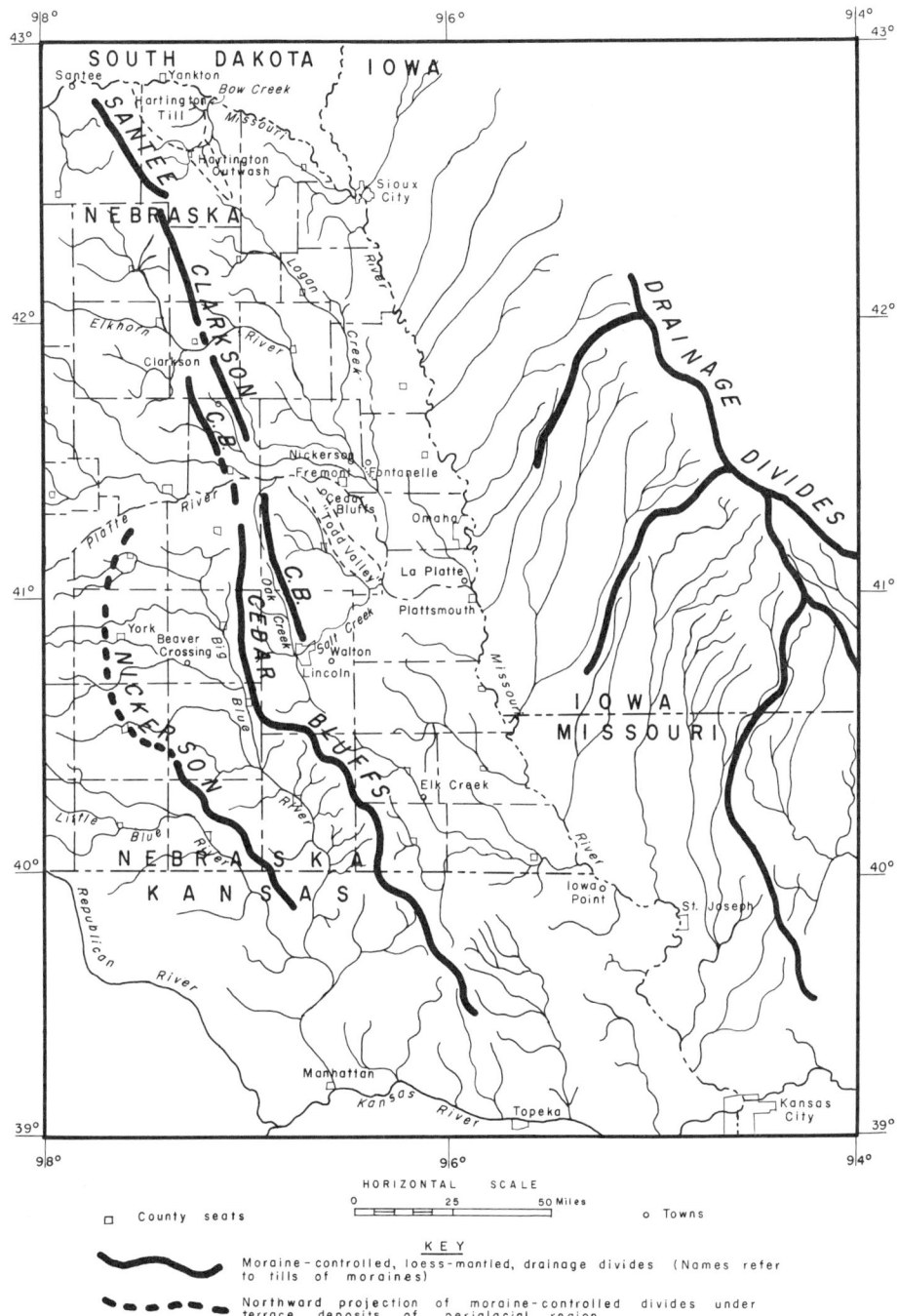

Fig. 3. Drainage pattern and moraine-controlled drainage divides in eastern Nebraska and adjoining area (from Nebraska Geological Survey Bulletin 23).

major advances and stands of ice sheets. Subsurface information from the logs and samples of several hundred test holes drilled on a systematic pattern in the eastern part of the state supports this view. A map showing the areal distribution of glacial till and the areas of occurrence of combined thickness of glacial tills greater than 150 feet also has been prepared (Fig. 4). The distribution and thickness of the tills correspond closely to drainages and drainage divides as suggested in figures 2 and 3. The maximum thickness of combined tills is slightly more than 300 feet and is attained only in northeastern Nebraska. The thickness of combined tills in those areas of till occurrence on the rest of the map is generally less than 50 feet. However, the same number of tills (often two to three as distinguished by soils, weathered horizons, pebble counts, or intermediate beds) is found in each of the thicker and thinner areas of till occurrence.

The present distribution and thickness of the tills are an expression of extent of glacial advance, configuration of the surface overridden by the ice, original till thickness, and glacial, nonglacial, and post-glacial erosion. In parts of the area of thick tills, a single till with a thickness of more than 100 feet fills preexisting valleys. The Nickerson Till of Kansan age is thin and occurs as outliers along the western margin of the glaciated area as the result of erosion by streams diverted southward by ice of later advances. Surely this route must have been followed by the Missouri River in Kansan time and much of Illinoian time. The Cedar Bluffs moraine-divide has been breached and till removed by headward erosion and capture of the Platte River in the central part of the area. The history of the development of the drainage system in the glaciated region of eastern Nebraska and adjacent areas is not known in detail. Preliminary analysis suggests that each of the major modern streams follows paths conecting segments of older drainages. Regional and local structure influenced the direction and extent of the earliest ice flows. Later stream patterns were reestablished and controlled by divide ridges believed to be an expression of lateral and marginal moraines. Continental ice flow of the later episodes was controlled to some extent by the then existing drainage pattern. Although segments of the modern streams and some of the features of their valleys are inherited from pre-Pleistocene and later Pleistocene morphological features, the streams and their valleys are essentially Wisconsinan in age.

TYPE LOCALITIES

Type localities for the Nebraskan, Aftonian, and Kansan occur in the respective states of Nebraska-Iowa, Iowa, and Kansas and were named from deposits associated with the Keewatin or western glacial advance. The Yarmouth(ian) Stage was named by Leverett (1898) in reference to sediments from the spoil of a well dug near Yarmouth, Iowa. Type localities for the Illinoian, Sangamon(ian), and Wisconsin(an) stages are in the states of Wisconsin and Illinois and were named from deposits associated with the eastern glacial advance, the Lake Michigan Lobe (Frye et al., 1965, p. 43).

Stratigraphic correlation of units having widely spaced geographical centers and deposited in different environments has been attempted by many workers, largely through extrapolation and reference to the standard time-stratigraphic classification. The success of this and previous generations of Pleistocene stratigraphers in developing a classification framework and relating sediments to it appears to need reevaluation. A perusal of the literature suggests that present workers are divided between those having supreme confidence in, and those who are somewhat dubious of, the success.

NEBRASKAN AND KANSAN STAGES

Continental ice of Nebraskan age moved into a lowland area of the Central Plains states across regionally low divides. Ice as prominent and partly coalescing lobes probably moved westward into Nebraska and Kansas up the valleys of drainages that trended generally west to east across the valley of the modern Missouri River (Reed *et al.*, 1965, fig. 2). The valleys had relatively low gradients of a few feet per mile, but locally the relief between the valleys and uplands was as much as 300 feet or more. Westward movement of ice was controlled by the gradient of the valleys and the regional slope and by the thickness and load of ice. Till of Nebraskan age is recognized mostly in the subsurface, and then most often as a remnant fill of the early Pleistocene valleys. Remnants also are found where some protection from erosion was afforded by nearby bedrock knobs or ridges.

Generally, only one till each of Nebraskan and Kansan ages has been recognized in Iowa, Kansas, and Missouri. Dort (1965, 1966a, 1966b) and Bayne (1969) have suggested the occurrence of more than one till of Kansan age in Kansas. In Nebraska at least two tills of Kansan age and multiple tills of Nebraskan age are recognized. According to Reed *et al.* (1966, figs. 19 and 20 and table 4), the Cedar Bluffs and Nickerson tills of Kansan age can be distinguished from each other and from till of Nebraskan age by the ratio of sedimentary to nonsedimentary pebbles and by heavy mineral counts. These ratios have been determined at a number of outcrop localities and from samples of several test wells in southeastern Nebraska, but these studies have not yet been extended into northeastern Nebraska. Several problems of correlation still have not been resolved in regard to the tills and their relation to the deposits of the periglacial region.

The extensive use that has been made of the Pearlette Volcanic Ash bed and the considerable reliance that has been placed on it as a stratigraphic marker for correlating Pleistocene deposits in the Central Plains states is well known to most Pleistocene stratigraphers. Most workers have considered the Pearlette to be Late Kansan or Yarmouthian in age and believe that it represents the one and only recognizable Pleistocene accumulation of volcanic ash in the Central Plains states. Frye *et al.* (1948) quite convincingly used petrographic characteristics of volcanic ash samples from a large area of the Central Plains states, plus the associated molluscan fauna, to validate the use of the Pearlette as a tool for correlation. However, Smith (1940, p. 119) expressed some doubt that the volcanic ash deposits in Meade County, Kansas, were all of the same age. Moreover, Reed and Dreeszen (1965, p. 8) noted the occurrence of a volcanic ash that they believed to be of Nebraskan age. In 1966 a core from a three-foot volcanic ash bed was recovered from a test hole in SE cor., sec. 32, T. 15 N., R. 3 E., Butler County, Nebraska (David City Locality). The bed underlies the Cedar Bluffs and Nickerson tills and occurs at a depth of 184.5 to 187.5 feet below land surface in a thick sequence of clay and silt believed to be the Fullerton Formation of Nebraskan age. Preliminary laboratory and petrographic analysis of the volcanic ash in the core suggest that it is indistinguishable from the Pearlette although Swineford (personal communication, 1948) examined a weathered sample of the ash collected in cuttings from a rotary test drilled a few feet away from the 1966 core and judged the ash to be unlike the Pearlette in index of refraction and shape of the shards. A part of the core has been sent to the U.S. Geological Survey in Denver for further study.

Three samples of volcanic ash from Nebraska, including the ash from the David City test hole, were submitted in 1967 to Isotopes, Inc., for K/Ar age determination. The following re-

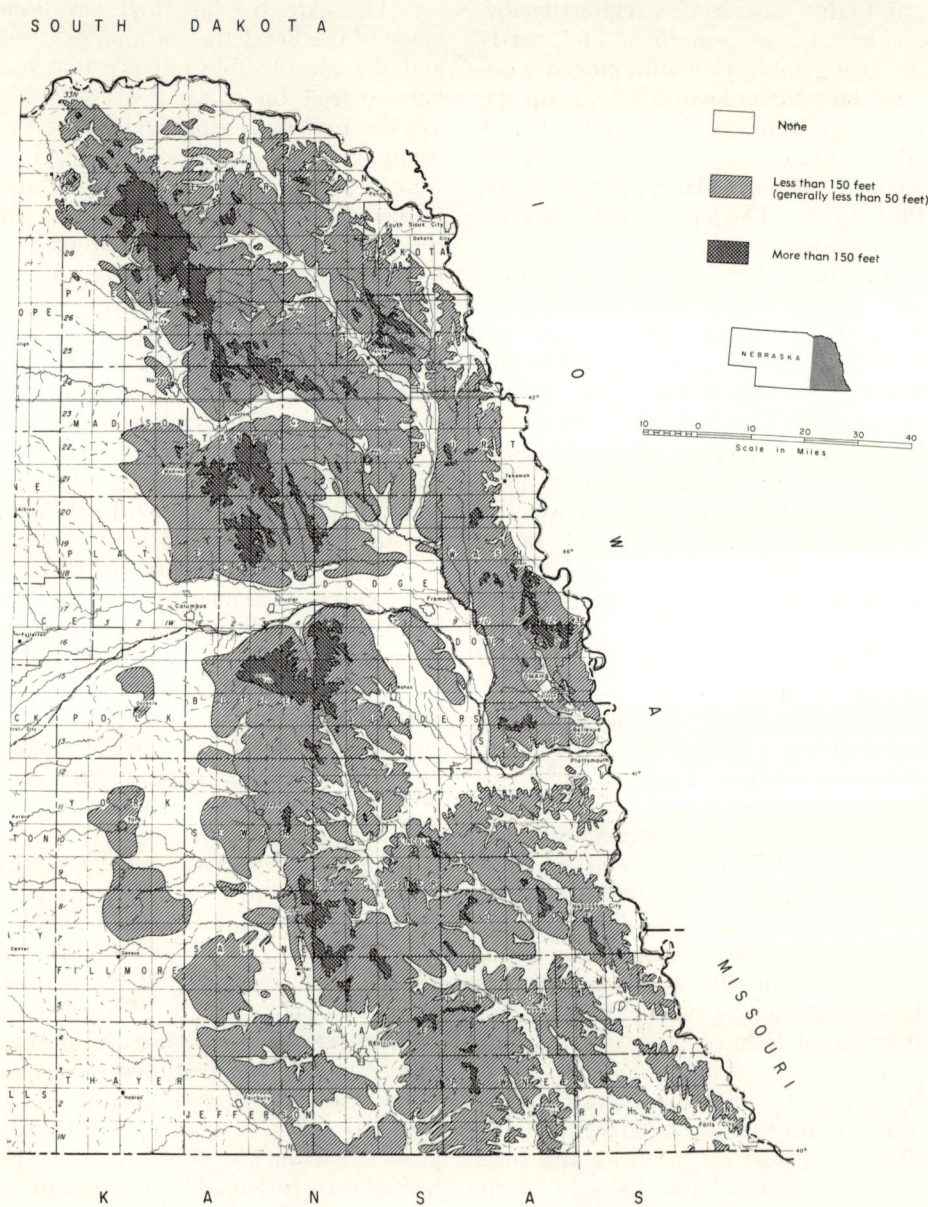

Fig. 4. Distribution and thickness of glacial till in Nebraska.

sults were reported and are presented for reader interpretation, with no intent to imply that the dates are absolute, but as evidence for the probable existence of more than one volcanic ash bed.

Sample	Locality Name	Remarks	Isotopic Age (m.y.)	Ratio, Radiogenic Argon to Total Argon	Percent K
1	Dam 7 or *Mammut moodiei* Site, center W. ½ sec. 26, T. 9 N., R. 2 E., Seward County, Nebraska	Outcrop sample, volcanic ash in Sappa Formation (Schultz and Smith, 1965b, p. 111 and figs. 14-62)	2.0±1.0	0.06	4.32
2	David City Locality, Test Hole SE corner sec. 32, T. 15 N., R. 3 E., Butler County, Nebraska	Core of volcanic ash from Fullerton Formation	7.8±1.2	0.48	4.67
3	Coleridge Ash Site, NE ¼ sec. 11, T. 29 N., R. 1 E., Cedar County, Nebraska	Outcrop sample of volcanic ash in silts below till of Illinoian Age? (Schultz and Smith, 1965b, pp. 37-38)	4.9±1.5	0.07	4.59

ILLINOIAN STAGE

The Illinoian Stage, too, is represented by sediments deposited during several advances of Illinoian ice. Frye *et al.* (1965, pp. 50-52), in Illinois, and Wayne (1963, p. 53) and Wayne and Zumberge (1965, pp. 67-68), in Indiana, recognized three substages of the Illinoian. The Illinoian Stage is defined in Illinois as consisting of those deposits lying above the Yarmouth Soil and below the Sangamon Soil (Frye *et al.*, 1965, p. 50). According to Wayne (1966, p. 32), ice sheets of the Illinoian reached farther south—south of the 38th parallel—than did any of the other Pleistocene glaciers.

The principal area of occurrence of till that is now recognized as Illinoian is in the states of Illinois and Indiana. The only Illinoian till recognized in Iowa is continuous with the earliest of the Illinoian tills in Illinois (Wright and Ruhe, 1965, p. 30) and is restricted to a narrow belt close to the Mississippi River in southeastern Iowa (Kay and Graham, 1943, pp. 15-44). Although Wright and Ruhe (1965, pp. 30-31) questioned the occurrence of any till of Illinoian age in Minnesota, its presence in southeastern South Dakota has been suggested by Flint, Steece, and Tipton (in Lemke *et al.*, 1965, p. 19), and in northeastern Nebraska by Reed and Dreeszen (1965, p. 38). The principal basis for these identifications is the occurrence of till over volcanic ash presumed to be the Pearlette at localities near Hartford, South Dakota (Schultz and Smith, 1965a, pp. 21-23), and at Coleridge, Nebraska (Schultz and Smith, 1965b, pp. 37-38). Perhaps the most convincing evidence for Illinoian deposition and presumed glaciation in the northern Central Plains states is the widespread occurrence of the Loveland Loess in Nebraska, Kansas, southwestern Iowa, and northern Missouri. The type locality for the Loveland Formation (Shimek, 1909) is near Loveland, Iowa. Because the type section was destroyed, Daniels and Handy (1959) described and designated a new type section for the Loveland Loess in western Iowa (see Schultz and Smith, 1965b, pp. 28-29, for discussion). The basis for assigning an Illinoian age to the Loveland Formation (or Loess) is that many workers have traced or extrapolated deposits of similar lithology from the Loveland type area into deposits of the glaciated area of western Illinois. The key horizon that has been used in this correlation is the Sangamon Soil, which was named from Sangamon County, Illinois. Frye *et al.* (1965, p. 52) stated that "the name Sangamon Soil, as a soil-stratigraphic unit, is applied to the soil where developed on Liman, Kansan, or older deposits. The stage is named from the Sangamon Soil as developed in Buffalo Hart till, and therefore it encompasses only the span of time from the retreat of the youngest Illinoian glacier to the deposition of the earliest Roxana (Wisconsinan) Loess."

Reed and Dreeszen (1965, pp. 35-40) proposed an Illinoian age for a complex sequence of three separate alluvial deposits that form a terrace plain in south-central Nebraska. They also suggested that three relatively thick loesses occurring on uplands in central Nebraska are equivalent in age to the proposed Illinoian alluvial sequences and in Nebraska restrict the Loveland Formation and Loess to the Late Illinoian. These deposits were demonstrated to lie at a position above an ash believed to be the Pearlette or in valleys incised through that ash. Soils are developed on each of three alluvial and eolian deposits (Schultz and Smith, 1965b, figs. 10-48 and 10-49). The name Sangamon Soil is restricted in Nebraska generally to that soil developed in the Loveland Loess or sediments of the Loveland Formation. Frye and Leonard (1954, 1965) noted the multiple character of the Sangamon Soil in northern Kansas and suggested that (1965, p. 212) "there were relatively brief intervals of stability and soil formation, followed by alluviation in the Central Great Plains, during middle to late Illinoian time." Frye *et al.* (1968, p. 9) noted the possibility in the Central Great Plains of confusing the one or more soils in the Loveland Loess (unrestricted) with Sangamon Soil. Frye *et al.* (1965, p. 50) recognized that in Illinois at least one minor soil occurs within Loveland Silt in western Illinois.

Frye *et al.* (1965, p. 43) suggested that the "stratigraphic relation of the younger Pleistocene deposits to the Kansan is based on their overlapping relations in the Mississippi Valley, particularly in western Illinois." Frye *et al.* (1964, p. 15) assigned a Kansan age to a till in western Illinois that previously had been called Illinoian. They judged the till to be Kansan on the basis of stratigraphic sequence, topographic expression, and mineral composition. No evidence of a Yarmouth Soil was found on the Kansan till (or even a leached zone) where the till was overlain by a younger till.

WISCONSINAN STAGE

There has been no intent to imply that problems of Wisconsinan correlation do not exist either in the extensive areas of Wisconsinan glaciation or between that area and the Central Plains states. The assignment of Wisconsinan or Sangamonian and Illinoian ages to some soils and deposits in the Central Plains is not without some question when regarded in light of chronologies being used in the Mississippi Valley and the Rocky Mountains. A number of Pleistocene geologists writing in Wright and Frey (1965) followed Frye and Willman (1963) in placing the lower boundary of the Wisconsinan at approximately 70,000 years BP. However, Wright (1964) placed the lower boundary of the main Wisconsinan at the base of the Tazewell, at about 23,000 years ago (Wright, this volume).

The base of the Peoria Loess in Illinois is placed at about 22,000 BP, the Farmdale silt and peat range from about 22,000 BP to 28,000 BP, and the Roxana Silt (formerly called Late Sangamon Loess and Farmdale Loess) is dated from about 28,000 BP to 70,000

BP (Frye et al., 1965, pp. 52-58). Only a few radiocarbon dates older than 28,000 BP have been reported from the Central Plains states. Radiocarbon dates have been obtained in Nebraska for organic carbon in the Gilman Canyon Formation—a dark-colored humus-enriched horizon one to several feet thick and formerly regarded by most workers as the A-horizon of the Sangamon Soil (Reed and Dreeszen, 1965, pp. 40-41). A previously unpublished radiocarbon date of 27,900 (+1100, −1000) years BP (I-2188) was obtained from the upper 18 inches of the 44-inch basal layer of the formation at the type locality (Reed and Dreeszen, 1965, p. 62), and a date of 32,000 (+2000, −1600) (I-1851) from the lower 18 inches (the latter date was erroneously reported in Reed et al., 1966, p. 3, as having been obtained from the upper 18 inches of the humic silt). Other dates not previously published were obtained from the Gilman Canyon at the Yankee Hill Brick Plant near Lincoln (Reed et al., 1966, p. 7, figs. 5 and 6), where the radiocarbon age of the upper few inches of a 30-inch layer was 26,900 (+1000, −900) (I-2189), and of the lower few inches was 34,900 (+2100, −1700) (I-2190). At another site in east-central Nebraska near Winslow, the organic carbon in the upper 17 inches of a 39-inch layer of the Gilman Canyon was 23,000±600 years BP (I-2191) and in the lower 15 inches was 31,400 (+1800, −1500) years BP (I-2192). The Gilman Canyon at the Winslow site (highway roadcut, NE ¼ sec. 10, T. 19 N., R. 8 E., Dodge County) is 26 feet 8 inches below the ground surface, and the formation at the Yankee Hill Brick Plant is about 15 feet below the surface. At the type locality it is under 83 feet or more of Peoria Loess. In northern Iowa a date of 29,000±3500 years is reported from organic carbon in a basal buried A-horizon under 44 feet of younger loess and above a Yarmouth-Sangamon paleosol (Ruhe et al., 1968, p. 19). Ruhe (this volume) reports similar dates for basal Wisconsinan deposits at other places in Iowa.

The Gilman Canyon lies with apparent conformity on a soil that has been traced over much of southwestern Iowa and much of Nebraska and Kansas by many workers who have regarded the Gilman Canyon and the underlying reddish-brown clay-enriched soil on the Loveland Formation as the Sangamon Soil. If the start of the Wisconsinan is accurately placed at about 70,000 BP, what was happening in the Central Plains during approximately the first half of the Wisconsinan? Even if the dates obtained in Nebraska were proved to be in error by a few thousand years (because of the contamination of the roots of plants which grew at higher and younger levels in the loess), the question would be only partly answered.

A similar problem exists in regard to radiocarbon dates obtained from organic carbon in the Brady Soil at its type locality in western Nebraska. A radiocarbon date of 9160±250 (W-234) was obtained in 1954. The soil was resampled in 1965, and a date of 9750±300 (W-1676) was determined. Contamination of the soil by roots of more recent plants is certain. Radiocarbon dates varying from 12,550±400 (W-231) to 12,700±300 (W-233) have been reported from snail shells in the basal part of the Bignell Loess in Doniphan County, Kansas. Frye et al. (1968, p. 18) suggested a compromise date of 11,000 BP, because the fossil shells contained an unknown amount of dead carbonate contamination and accordingly gave dates that are too old.

The Bignell Loess, which overlies the Brady in Nebraska, is relatively thin and at the present time is recognized only as a "lip" on bluffs along some stretches of the major streams. Elsewhere the Brady Soil is a part of the modern soil, and the thin Bignell increment forms a part of the thick A-horizon of the modern soil.

Correlation of Wisconsinan glaciation in the Rocky Mountains with the plains sequence presents interesting problems. Richmond (1965, table 2) suggested a possible correlation of the Rocky Mountain glaciations with the midcontinent region (after Frye and Willman, 1960). He suggested a three-stade Pinedale Glaciation dating from about 6500 BP to 25,000 BP and noted that a mature zonal soil separates the Pinedale deposits from the underlying Bull Lake deposits. He assigned a period of time for that interglacial period (or interglaciation) ranging from about 25,000 BP to 32,000 BP—nearly the same range of time suggested for the Gilman Canyon in Nebraska—and suggested correlation of the Bull Lake Glaciation with the Altonian Substage of the Wisconsinan of Illinois. Richmond also suggested that the glaciation in the Rocky Mountains may have extended into the Sangamonian and noted that at least two rather strongly developed soils were formed during deglacial or "nonglacial" intervals that separated advances of two or perhaps three Bull Lake glacial advances. Again, one wonders about the record of this glacial period in the plains states to the east. The similarity between the Bull Lake sequence, with the overlying soil, and the Gilman Canyon–Illinoian sequence proposed by Reed and Dreeszen (1965) is impressive.

SUMMARY

The classification of glacial and periglacial deposits and the understanding of the complex glacial history of the midcontinent area have been and will continue to be subjects of research, speculation, and analysis by students of the Pleistocene. Three generations of workers have been concerned with the details of Pleistocene stratigraphy. The framework within which we are now working was established largely through the efforts of the early workers. Their work was monumental. Nevertheless, the early workers, like those of each of the succeeding generations, attempted to write the first chapter last. Possibly the framework they established is completely satisfactory, but, on the other hand, a reexamination of all the data now available might show that it is not. The early workers were few, outcrops were scarce, and modern topographic maps and aerial photographs were not available. Furthermore, the fossil record was only partially known, and relatively little subsurface information was available. The tool of carbon-14 dating and several laboratory techniques now in use had not yet been developed.

During the several decades that have passed since establishment of the Pleistocene framework, an astronomical amount of detailed information on the Pleistocene has accumulated. Generally the information, much of it obtained by exploratory drilling, has been interpreted in accordance with the established framework, even though new discoveries were demonstrating that glacial history is far more complex than was conceived by early workers. The time is ripe, therefore, not only for a reevaluation of the framework but also for a reexamination of the conclusions reached with that framework in mind.

The problems of geography, paucity of exposures, difficulty in recognizing glacial land forms, erosion or complete removal of older sediment, and scarcity of vertebrate fossils will have to be met. Interdisciplinary action among Pleistocene scientists will be mandatory. Additional exploratory drilling almost surely will need to be done, and absolute dating methods will need to be applied before other-than-tacit agreements can be reached. Unfortunately, the precision of correlation between the glacial sequence of the Mississippi River Valley region and the glacial and periglacial sequence of the

Missouri River Valley region appears to diminish beyond the reach of radiocarbon dating, making it necessary to rely on other criteria for correlation of the older deposits. Iowa, because of its geographical location, may hold the key to the solution of several important problems of correlation between the glacial sequences in the two principal areas of past studies.

LITERATURE CITED

Bayne, C. K.
  1969. Evidence of multiple stades in the Lower Pleistocene of northeastern Kansas. Trans. Kansas Acad. Sci., 71:340-349.
Curray, J. R.
  1965. Late Quaternary history, continental shelves of the United States. Pp. 723-735, in The Quaternary of the United States (H. E. Wright, Jr., and D. G. Frey, eds.), Princeton Univ. Press, Princeton, New Jersey, x+922 p.
Daniels, R. B., and R. L. Handy
  1959. Suggested new type section for the Loveland loess in western Iowa. Jour. Geol., 1:114-119.
Dort, W., Jr.
  1965. Nearby and distant origins of glacier ice entering Kansas. Amer. Jour. Sci., 263:598-605.
  1966a. Multiple Early Pleistocene glacial stades, northeastern Kansas. Geol. Soc. Amer. Spec. Paper, 87:47.
  1966b. Nebraskan and Kansan stades: complexity and importance. Science, 154:771-772.
Evernden, J.F., D. E. Savage, G. H. Curtis, and G. T. James
  1964. Potassium-argon dates and the Cenozoic mammalian chronology of North America. Amer. Jour. Sci., 262:145-198.
Frye, J. C., and A. B. Leonard
  1954. Significant new exposures of Pleistocene deposits at Kirwin, Phillips County, Kansas. Bull. Kansas Geol. Surv., 109:33-48.
  1965. Quaternary of the Southern Great Plains. Pp. 203-216, in The Quaternary of the United States (H. E. Wright, Jr., and D. G. Frey, eds.), Princeton Univ. Press, Princeton, New Jersey, x+922 pp.
Frye, J. C., A. Swineford, and A. B. Leonard
  1948. Correlation of Pleistocene deposits of the Central Great Plains with the glacial section. Jour. Geol., 56:501-525.
Frye, J. C., and H. B. Willman
  1960. Classification of the Wisconsinan stage in the Lake Michigan glacial lobe. Illinois State Geol. Surv. Circ., 285:1-16.
  1963. Development of Wisconsinan classification in Illinois related to radiocarbon chronology. Bull. Geol. Soc. Amer., 74:501-506.
Frye, J. C., H. B. Willman, and R. F. Black
  1965. Outline of glacial geology of Illinois and Wisconsin. Pp. 43-61, in The Quaternary of the United States (H. E. Wright, Jr., and D. G. Frey, eds.), Princeton Univ. Press, Princeton, New Jersey, x+922 pp.
Frye, J. C., H. B. Willman, and H. D. Glass
  1964. Cretaceous deposits and the Illinoian glacial boundary in western Illinois. Illinois State Geol. Surv. Circ., 364:1-28.
  1968. Correlation of midwestern loesses with the glacial succession. Pp. 3-21, in Loess and related eolian deposits of the world (C. B. Schultz and J. C. Frye, eds.), Univ. Nebraska Press, Lincoln, 369 pp.
Heim, G. E., Jr., and W. B. Howe
  1963. Pleistocene drainage and depositional history in northwestern Missouri. Trans. Kansas Acad. Sci., 66:378-392.
Hibbard, C. W., D. E. Ray, D. E. Savage, D. W. Taylor, and J. E. Guilday
  1965. Quaternary mammals of North America. Pp. 509-525, in The Quaternary of the United States (H. E. Wright, Jr., and D. G. Frey, eds.), Princeton Univ. Press, Princeton, New Jersey, x+922 pp.
Howe, W. B., and G. E. Heim, Jr.
  1968. The Ferrelview formation (Pleistocene) of Missouri. Missouri Geol. Surv. and Water Resources, Rep. Inv., 42:1-32.
Jewett, J. M.
  1963. Pleistocene geology in Kansas. Trans. Kansas Acad. Sci., 66:347-358.
Kay, G. F., and J. B. Graham
  1943. The Illinoian and post-Illinoian Pleistocene geology of Iowa. Bull. Iowa Geol. Surv., 38:1-262.
Lemke, R. W., W. M. Laird, M. J. Tipton, and R. M. Lindvall
  1965. Quaternary geology of Northern Great Plains. Pp. 15-27, in The Quaternary of the United States (H. E. Wright, Jr., and D. G. Frey, eds.), Princeton Univ. Press, Princeton, New Jersey, x+922 pp.
Leverett, F.
  1898. The weathered zone (Yarmouth) between the Illinoian and Kansan till sheets. Jour. Geol., 6:238-243.
Reed, E. C.
  1948. Stratigraphy and geomorphology of the Pleistocene of Nebraska. Bull. Geol. Soc. Amer., 59:613-616.
Reed, E. C., and V. H. Dreeszen
  1965. Revision of the classification of the

Pleistocene deposits of Nebraska. Bull. Nebraska Geol. Surv., 23:1-65.

Reed, E. C., V. H. Dreeszen, C. K. Bayne, and C. B. Schultz
1965. The Pleistocene in Nebraska and northern Kansas. Pp. 187-202, *in* The Quaternary of the United States (H. E. Wright, Jr., and D. G. Frey, eds.), Princeton Univ. Press, Princeton, New Jersey, x+922 pp.

Reed, E. C., V. H. Dreeszen, J. V. Drew, V. L. Souders, J. A. Elder, and J. D. Boellstorff
1966. Evidence of multiple glaciation in the glacial-periglacial area of eastern Nebraska. Pp. 1-25, Guidebook, 17th Ann. Meeting, Midwestern Sect. Friends of the Pleistocene, Univ. Nebraska, Lincoln.

Richmond, G. M.
1965. Glaciation of the Rocky Mountains. Pp. 217-230, *in* The Quaternary of the United States (H. E. Wright, Jr., and D. G. Frey, eds.), Princeton Univ. Press, Princeton, New Jersey, x+922 pp.

Ruhe, R. V., W. P. Dietz, T. E. Fenton, and G. F. Hall
1968. Iowan drift problem, northeastern Iowa. Iowa Geol. Surv., Rep. Inv., 7:iv+1-40.

Schultz, C. B., and H. T. U. Smith (eds.)
1965a. Upper Mississippi Valley. Guidebook [INQUA] Field Conf. C, 7th Cong., Internat. Assoc. Quaternary Res., 126 pp.
1965b. Central Great Plains. Guidebook [INQUA] Field Conf. D, 7th Cong., Internat. Assoc. Quaternary Res., 128 pp.

Shimek, B.
1909. Aftonian sands and gravels in western Iowa. Bull. Geol. Soc. Amer., 20:399-498.

Smith, H. T. U.
1940. Geologic studies in southwestern Kansas. Bull. Kansas Geol. Surv., 34:1-244.

Swineford, A.
1963. The Pearlette Ash as a stratigraphic marker. Trans. Kansas Acad. Sci., 66:359-362.

Thorp, J., H. T. U. Smith, and others
1952. Pleistocene eolian deposits of the United States, Alaska, and parts of Canada. Geol. Soc. Amer., map, scale 1:2,500,000.

Wayne, W. J.
1963. Pleistocene formations in Indiana. Bull. Indiana Geol. Surv., 25:1-85.
1966. Ice and land: a review of the Tertiary and Pleistocene history of Indiana. Pp. 21-39, *in* The Indiana sesquicentennial volume, natural features of Indiana, Indiana Acad. Sci., Indianapolis.

Wayne, W. J., and J. H. Zumberge
1965. Pleistocene geology of Indiana and Michigan. Pp. 63-83, *in* The Quaternary of the United States (H. E. Wright, Jr., and D. G. Frey, eds.), Princeton Univ. Press, Princeton, New Jersey, x+922 pp.

Wilcox, R. E.
1965. Volcanic-ash chronology. Pp. 807-816, *in* The Quaternary of the United States (H. E. Wright, Jr., and D. G. Frey, eds.), Princeton Univ. Press, Princeton, New Jersey, x+922 pp.

Wright, H. E., Jr.
1964. Classification of the Wisconsin glacial stage. Jour. Geol., 72:628-637.

Wright, H. E., Jr., and D. G. Frey (eds.)
1965. The Quaternary of the United States. Princeton Univ. Press, Princeton, New Jersey, x+922 pp.

Wright, H. E., Jr., and R. V. Ruhe
1965. Glaciation of Minnesota and Iowa. Pp. 29-41, *in* The Quaternary of the United States (H. E. Wright, Jr., and D. G. Frey, eds.), Princeton Univ. Press, Princeton, New Jersey, x+922 pp.

The Lubbock Subpluvial

Fred Wendorf

ABSTRACT

The Lubbock subpluvial was an episode of markedly cooler summers and more effective precipitation, recorded at two localities on the Llano Estacado and dated by radiocarbon between 8300 and 8600 B.C. The Folsom occupation of this area overlaps this climatic event. Pollen spectra indicate that a boreal forest of pine and occasional spruce developed on the Llano at this time. Several invertebrate and diatom assemblages provide additional data on the changes in the environment just prior to, during, and immediately after, this event. Several hypotheses concerning the cultural responses of the Folsom hunters to these environmental changes are proposed but not tested.

INTRODUCTION

This paper is concerned with the definition of a terminal Late Pleistocene climatic event on the Llano Estacado, the southernmost of the High Plains, in eastern New Mexico and western Texas. This event, named the Lubbock subpluvial, was a brief period of probably less than 400 years duration when there appear to have been environmental changes of considerable magnitude. The Lubbock subpluvial also coincides with at least part of the period when Folsom points were the predominant style of projectile used throughout the High Plains; however, Folsom points have a longer time range than the Lubbock subpluvial. They were made both earlier and later than the indicated duration of this event.

The Llano Estacado is an area of approximately 20,000 square miles, bounded on all sides by steep vertical escarpments (Fig. 1). The surface of the Llano is one of the flattest landscapes on earth, with the only topographic variety occurring in the few shallow, dry, and dune-choked drainage channels, in the few low hummocks of dunes that are found most commonly along the western edge, and in the relief provided by literally thousands of playas and lakes, some of which reached considerable size (one, Arch Lake, covered more than 100 square miles). Most of the playas are now dry except after heavy rains, but the larger lakes frequently contain highly saline water.

The surface of the Llano is slightly tilted toward the south and east, with a slope averaging around 10 feet per mile. Elevation ranges from 4800 feet at the northwest corner, near Tucumcari, New Mexico, to around 2700 feet along the southeastern edge beyond Midland, Texas. The climate is semi-arid, with hot summers and moderate winters, although occasional brief, sharp cold fronts penetrate the region each winter. Average temperatures range in summer from 79° to 81°F, and in winter from 37° to 44°F. Average annual precipitation ranges from 14 to 19 inches, but there is wide variation from year to year, and drouths are common.

During the Wisconsin maximum, at a time probably equivalent to the Tazwell and Cary glacial maxima, the literally thousands of playas and lakes in this area were brimful with water and the vegetation was dominated by a pine and spruce forest. Pollen profiles of this interval, known as the Tahoka pluvial, have been recovered from 10 localities scattered throughout the Llano Estacado, and the Tahoka now may be regarded as reasonably well documented. There is no evidence that man was present at this time.

Climatic reconstructions for the Tahoka pluvial, based on the calculated frequency of spruce in the pollen spectra, together with the associated invertebrate fauna, indicate that summer temperatures then were between 60° and 65°F, or from 15° to 20°F cooler than today (Wendorf, 1961, p. 129). Reeves (1965), on the basis of snow-line depression in the adjacent mountains, has refined this temperature range and suggested that the average July temperatures at this time were reduced 18°F. He also calculated the

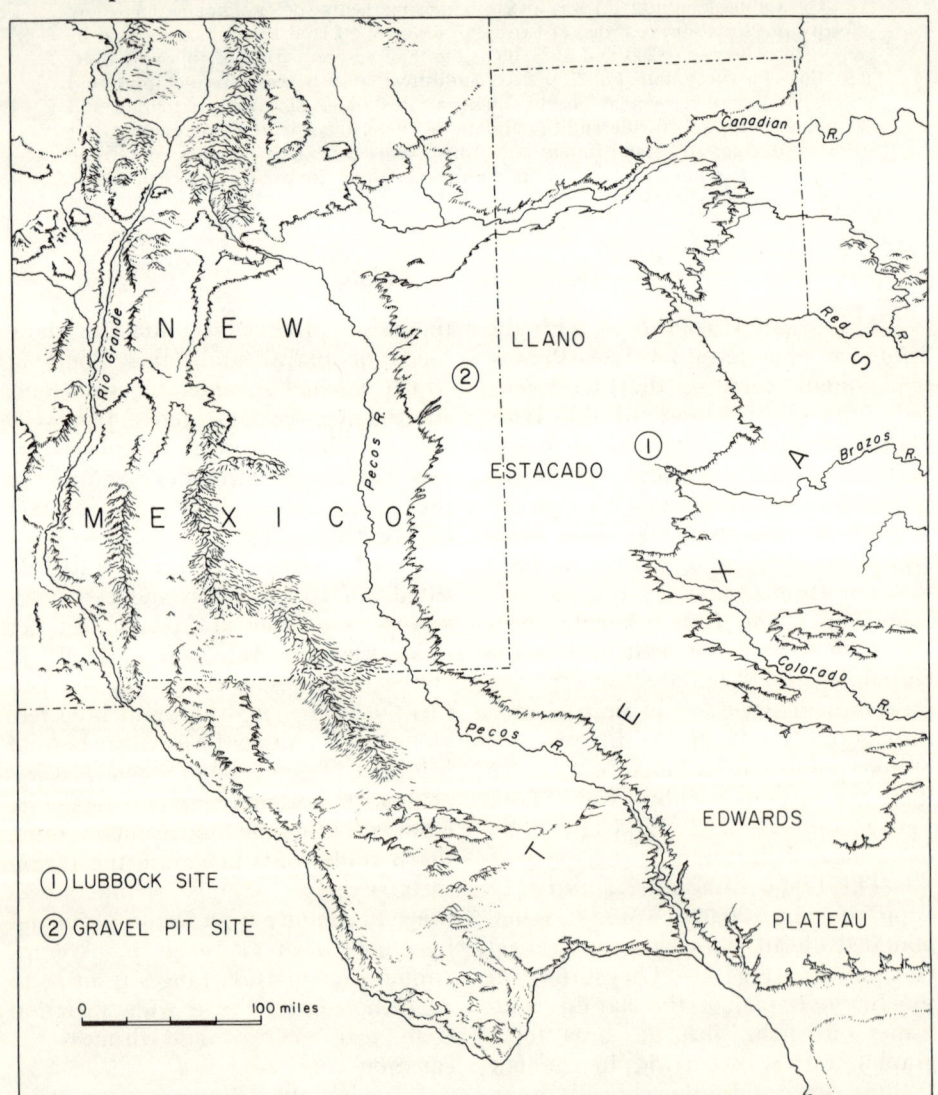

FIG. 1. Map of the Llano Estacado and adjacent areas, showing location of the Lubbock Site (1) and the Gravel Pit Site on Blackwater Draw (2). Map redrawn from "A Physiographic Diagram of the United States," by A. K. Lobeck, Columbia University.

evaporation parameters using this temperature, and he concluded that in spite of the abundant evidence of pluvial conditions, no significant increase in precipitation occurred. If his calculations are correct, the required rise in the water table, the saturation of the subsoil, and the filling of the pluvial basins could have occurred solely by a reduction of this magnitude in the summer temperatures.

The end of the Tahoka is signaled in a number of diagrams by the abrupt decline in pine and spruce frequencies. The precise date when this occurred, however, cannot yet be determined. Applicable radiocarbon dates are not available, and thus the end date for the Tahoka pluvial must be estimated on other grounds. There are at least two conflicting models proposed. In one, the pine-spruce forest is assumed to have survived on the Llano until the warming conditions of the Two Creek interstade had reached their maximum, now dated around 9900 B.C. (Frye et al., 1965, pp. 55-57). In the other model, the Llano Estacado is seen as an area in delicate ecological tension, which responded in a highly sensitive manner to the climatic changes of the Pleistocene. Thus, those climatic phenomena that led to the expansion and retraction of the main glacial ice sheets in central North America were reflected on the Llano Estacado perhaps even more abruptly than in the areas immediately adjacent to the ice. This is because of the neutralizing effects of the ice mass itself, which probably would reduce any general tendency toward an increase in temperature. Thus, a general rise in temperature, which could result in a minor retreat in the ice front and might have little or no effect on the environment of the area adjacent to the ice, could produce dramatic results in an area such as the Llano Estacado. The increase in temperature would cause an increase in evaporation, and lakes would then become smaller. The vegetation might also change. Pine and spruce, which could colonize the area only under conditions of maximum depression of temperature and consequently increased availability of moisture, would then be faced with an unfavorable environment and would either retreat or produce less pollen. According to this model, then, the abrupt decline in pine and spruce values which marks the end of the Tahoka pluvial occurred shortly after the onset of the retreat that followed the Cary maximum, and not several thousand years later when this retreat was geographically most extensive. This would place the end of the Tahoka pluvial slightly after 12,000 B.C.

After the end of the Tahoka pluvial there followed a period of perhaps 900 years according to the first model, or around 3000 years in the second model, for which there is limited environmental data for the Llano Estacado, except near the end, when the first appearance of man is recorded by the presence of Clovis points associated with mammoth and bison. A series of radiocarbon dates places these finds between 9300 and 9200 B.C. (Haynes, 1964). The invertebrate and vertebrate fauna associated with this period suggest that some wooded areas were still present and that summer temperatures remained cool, whereas winter temperatures may have been warmer than at present (Slaughter, 1970).

After the appearance of man, the continued decline in available moisture is recorded stratigraphically by a reduction in stream flow, by shifts in the diatom flora from fresh to saline species, by an increase in those species of invertebrates that prefer ponds as against those that live in running water, and in the pollen by extremely high frequencies of chenopods and extremely low values of pine. This period of relative aridity has been previously referred to as the Scharbauer interval (Wendorf, 1961, pp. 121-122). The climax of this period also marks the disappearance both of the mammoth and of Clovis artifacts from the Llano.

THE LUBBOCK SUBPLUVIAL

The Lubbock subpluvial was proposed initially (Wendorf, 1961, pp. 115-133) as the name for an interval of cool, moist conditions indicated by invertebrate and diatom assemblages from the Lubbock Site. There was no evidence available on the pollen rain during this period, other than that obtained at the San Jon Site, which is now regarded as of questionable age. Recently, however, several fossil pollen spectra were recovered from both the Lubbock Site and the Clovis Gravel Pit at Blackwater Draw from levels that may be referred to the Lubbock subpluvial. These analyses by F. C. Oldfield and James Schoenwetter provide the first secure botanical data on this interval. Since the pollen diagrams are to be published in detail elsewhere (Wendorf and Hester, 1970), only the broad outlines will be considered here.

STRATIGRAPHY

The Lubbock subpluvial occurs in a similar stratigraphic position at the Lubbock and Clovis Gravel Pit localities, both of which have been previously discussed in several publications (Sellards, 1952; Wendorf, 1961, pp. 115-117; Haynes, 1965; Green, 1962). Both localities are located well back from the edge of the Llano: the Gravel Pit is some 25 miles east of the western escarpment, and the Lubbock Site is 15 miles west of the eastern escarpment. Neither is situated where stream flow from areas beyond the Llano could conceivably contaminate their deposits. Seven units are recorded at the Lubbock Site (Fig. 2). At the base, resting on coarse gravel of earlier Pleistocene age, is a fine gravel grading upwards into sand and clay (Unit 1). It contains elephant and other members of the Late Pleistocene fauna. One nondiagnostic artifact was obtained from the upper part of this unit, and there is a single radiocarbon date of 10,700 B.C.±250 years from below the zone where the artifact was recovered. Above this is a diatomite (Unit 2) near the base of which several Folsom artifacts were recovered. A single Agate Basin point was recovered from near the top of this unit. Two radiocarbon

FIG. 2. Schematic section of the Lubbock Site, with superimposed frequencies of pine and spruce pollen (as percentages of total pollen recovered from the sample) in three pollen profiles.

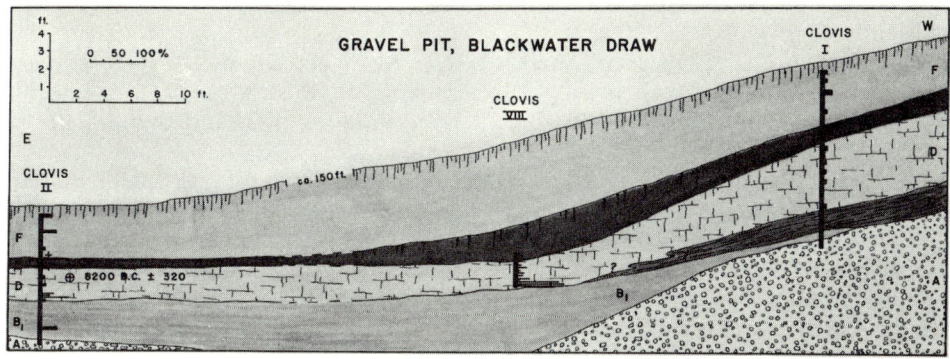

FIG. 3. Schematic section of the west wall of the Gravel Pit Site in Blackwater Draw, with superimposed frequencies of pine and spruce pollen (as percentages of total pollen recovered from the sample) in three pollen profiles.

dates have been obtained from Unit 2: 7933 B.C.±350 years (C-558) and 7750 B.C.±450 years (L-283G). These dates, one on burned bone, the other on shell, may be slightly too young. Above the diatomite are two thick deposits (Units 3 and 4) of friable and carbonaceous silts. *Bison antiquus* was recovered from the top of Unit 4, but diagnostic artifacts have not been found in either deposit. Unit 5 is a humus-stained sand and silt, probably a dune that accumulated in a mat of thick vegetation. Units 6 and 7 are recent, and probably date in the last thousand years or so.

The Clovis Gravel Pit at Blackwater Draw has at the base a gray-colored spring and pond sand (Unit B1), containing elephant, horse, and camel (Fig. 3). Clovis artifacts have been recovered from this deposit; however, this association has been questioned by Haynes (1964, p. 1408), who feels that the Clovis material was slightly later and was accidentally introduced into the gray sand when that unit was fluid. According to Haynes, the Clovis occupation was contemporary with the deposition of the basal part of the overlying spring deposits of the massive brown sand wedge (Unit C). The upper part of the brown sand wedge is seen as partially contemporary with the deposition in the center of the pond of the lower part of the third unit at this site, the diatomite (Unit D). A slightly different version of this interpretation is offered by Hester, who identified a laminated brown sand wedge, which was distinct from the lower massive brown sand wedge and which interfingered with the diatomite and contained Folsom artifacts. The diatomite at this site has yielded four radiocarbon dates ranging from 8450 B.C.±200 years to 8220 B.C.±250 years (A-386, A-379, A-380, A-488). Folsom artifacts occur throughout the diatomite at this site, and they are associated with abundant remains of the large Pleistocene bison *(B. antiquus)*. Agate Basin points were recovered from near the top of this unit, as at Lubbock. Within the range of statistical error, the Gravel Pit dates indicate that the deposition of the diatomite occurred over a comparatively short period, probably no more than 700 years, between 8000 and 8700 B.C. Because of the unique lithologic and cultural similarities shared by the diatomite at these two localities, it seems reasonable to suggest that they are of about the same age, in spite of the two slightly younger radiocarbon dates from the Lubbock Site.

Over the diatomite at the Clovis Gravel Pit is a carbonaceous silt (Unit E) containing a large bison (possibly *B. occidentalis*) and artifacts of what has been termed the "Plano" tradition.

There are two radiocarbon dates from this unit of 7940 B.C.±290 years (A-489) and 6520 B.C.±350 years (A-512) on charcoal, and two others of 4350 B.C.±150 years (O-169) and 4280 B.C. ±150 years (O-170). The last two dates are on burned bone and are rejected as being too recent in view of the associated industry and numerous considerably older radiocarbon dates from other sites associated with similar kinds of artifacts. An age from 8000 to 6500 B.C. is indicated both by the associated archaeology and the two dates on more acceptable materials.

Above the carbonaceous silt is a jointed sand, containing Archaic artifacts. There is a single radiocarbon date of 3000 B.C.±130 years (O-157) on this unit.

ENVIRONMENTAL DATA

The basic data on the environment during the Lubbock subpluvial is provided by a series of pollen diagrams, diatom spectra, and invertebrate assemblages from the diatomite at the Lubbock and Clovis Gravel Pit localities. The precise physical or environmental factors that caused the formation of the diatomite at these two localities are not known; however, on the basis of their similar archaeological and faunal contents, it is believed that the deposition occurred synchronously at both localities and probably at the same time at numerous other places on the Llano. Diatomite has been observed in several exposures on Blackwater Draw, at the Plainview Site, and at several other localities in various parts of the Llano in stratigraphic positions closely similar to those at the Gravel Pit and Lubbock sites.

In the base of the diatomite at Lubbock, at about the same level as the Folsom occupation at this site, pine and spruce are present, but only in low frequencies. Pine values are even less than they are today. Presumably this is the terminal phase of the Scharbauer interval. It is followed by a sudden change in the pollen spectra. Pine (with some spruce) abruptly increases to over 75 percent of the total pollen, remains at this frequency in several succeeding samples, and then abruptly declines.

At the Clovis Gravel Pit the pollen diagrams (Fig. 3) disclose a shift from low values for pine and spruce to high frequencies (more than 70 percent) followed by a decline—a sequence closely similar to that noted at Lubbock. At both localities the high pine and spruce percentages occurred approximately midway during the period of diatomite deposition, as this may be reconstructed from the various stratigraphic sections. In both instances, in the center of the pond, where the initial deposition of diatomite presumably occurred, the lower part of the diatomite has low pine-spruce frequencies and is assigned to the Scharbauer interval. In some profiles, where subsequent erosion has removed the upper part of the diatomite, only the zone of low pine values is preserved. Toward the edge of the pond, however, where the upper part of the diatomite remains, the high pine-spruce zone occurs above a thin zone of low pine values. In the same sections and in the uppermost part of the diatomite, the zone of high pine is followed by an abrupt decline in pine, coupled with a disappearance of spruce.

The consistency of these changes in pollen frequencies lends considerable confidence in their validity as recording a short-lived invasion of pine and spruce on the Llano Estacado. They occur at both localities in several different profiles in similar stratigraphic positions. They were obtained by different analysts using different recovery techniques and from samples collected at different times. This consistency leads us to infer that the high-pine zones are not due to accidents of sampling or analysis, or to differential preservation, but truly reflect the flora existing on the Llano at this time. A comparison with modern pollen rain from numerous localities throughout the Southwest and elsewhere (Faegri and Iversen, 1950; Potter and Rowley, 1960; and Hevly et al., 1965), indicates that only a dense boreal pine forest with occasional spruce can produce the high values of pine-spruce pollen recovered from these two localities. The existence of such a forest on the Llano during this period is strongly indicated.

The invertebrate fossils from these two sites also yield additional information on the environment during the Lubbock subpluvial. Two collections, one slightly above the other, were obtained from approximately the lowest third of the diatomite at the Lubbock Site (Collections IV and V, Fig. 4), and three collections were recovered from the lower, middle, and upper parts of the diatomite at the Gravel Pit Site in Blackwater Draw. The indicated environmental preferences for the species identified in each collection are shown in Figure 5. Both sites contain a number of aquatic forms; however, those species that require permanent water were present only at Lubbock, suggesting that this locality had somewhat better water sources. The most significant difference between the two sites is the absence at Lubbock of those terrestrial forms that live in wooded habitats. This absence does not, however, conflict with the pollen data, since both of the Lubbock collections were taken from the zone below the levels of high pine-pollen values at that site.

At the Gravel Pit the two species that require wooded areas, *Euconvlus fulvus* and *Zonitoides arboreus*, have an interesting shift in frequency in the three collections studied. They are most common in the lower part of the diatomite, as may be seen in these restricted percentages calculated on the basis of the total collection, less the immature pupillids and lymaeids:

Level	Total Number Identified	Number of Woodland Species	Restricted Percentage of Woodland Species
Upper Diatomite	1472	11	.07
Middle Diatomite	2939	165	5.6
Lower Diatomite	721	86	11.9

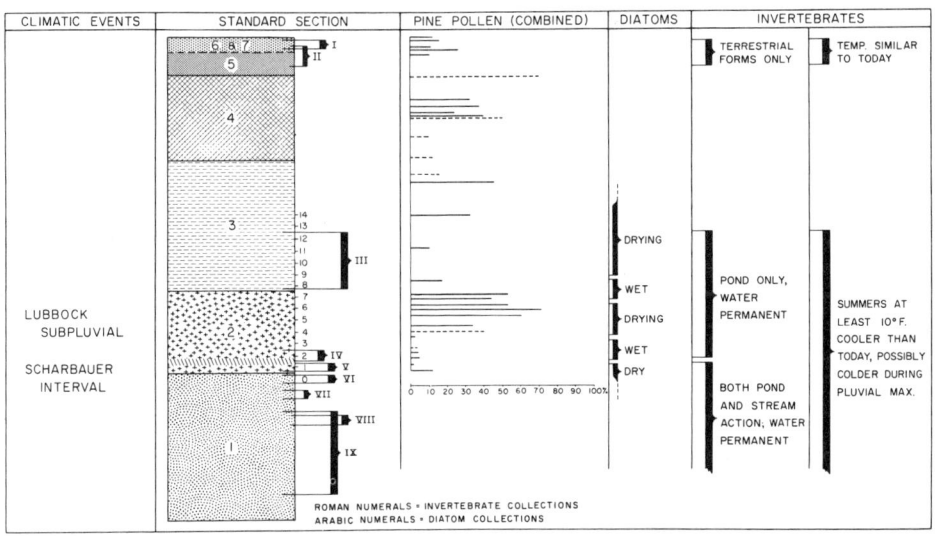

Fig. 4. The standard section at the Lubbock Site, showing the relationship between the combined pollen, diatom, and invertebrate collections and the environmental interpretation derived from them. Broken line in the combined pollen diagram indicates that the total pollen recovered was insufficient for reliable results.

HABITAT	SPECIES		LUBBOCK IV	LUBBOCK V	GRAVEL PIT L	GRAVEL PIT M	GRAVEL PIT U
Permanent Water Pond or Stream	*Ferrissia rivularis*	(C)	–	X	–	–	–
	Sphaerium solidulum	(C)	–	X	–	–	–
Pond	*Pisidium abditum*		X	X	–	–	–
Permanent to Subpermanent Water	*Helisoma trivolvis*		X	–	–	–	–
	Physa anatina		X	X	–	–	–
Permanent or Temporary Water	*Gyraulus crista*	(C)	X	–	–	–	–
	G. circumstriatus	(C)	–	X	X	X	X
	G. parvus		–	X	–	–	–
	Lymnaea palistris	(C)	X	–	–	–	–
	Menetus exacuous	(C)	X	X	–	–	–
	Physa gyrina		–	–	–	X	–
	Pisidium castertanum		–	–	X	X	X
Temporary Ponds	*Lymnaea caperata*	(C)	X	X	–	–	–
	L. parva		–	X	–	–	–
Semiaquatic Riparian	*Carychium exiguum*		–	–	X	X	X
	Deroceras laeve		–	–	–	X	–
	Oxyloma retusa		X	–	–	–	–
	Vertigo milium		–	X	–	–	X
	V. ovata		–	X	X	–	–
Damp Humus or Grass	*Gastrocopta tappaniana*		–	–	X	X	X
Damp Humus of Wooded Area	*Euconulus fulvus*	(C)	–	–	X	X	X
	Zonitoides arboreus		–	–	X	X	–
Uncertain or Varied Habitat	*Gyraulus labiatus*		X	X	–	–	–
	G. similaris		X	X	–	–	–
	Lymnaea bulimoides		–	X	–	–	–
	Menetus pearlettei		X	X	–	–	–
	Succinea grosvenori		–	–	X	X	X
	Vallonia cyclophorella	(C)	–	–	X	X	X

Fig. 5. Habitat preferences for invertebrates identified from the Gravel Pit Site and from collections IV and V at the Lubbock Site. Identification of Lubbock invertebrates by A. Byron Leonard, from Wendorf, 1961, p. 110; those from the Gravel Pit were identified by Robert J. Drake.

The pollen diagrams from the Gravel Pit indicate that the highest pine-pollen frequencies occur near the middle of the total diatomite accumulation at this site; however, away from the center of the pond, where the earliest diatomite accumulation occurred, the high-pine zone might be in the lower part of that unit (i.e., Clovis VIII profile, Fig. 3). This appears to be the situation here, although it cannot be checked in this instance because no pollen could be recovered from the sediments at the exposure where the snails were collected.

The invertebrates also yield significant data on the prevailing temperatures during the Lubbock subpluvial.

Of the 28 species identified, nine have distributions today either much farther north or at much higher elevations than the Llano Estacado. Both northeastern and northwestern species (i.e., *Ferrissia rivularis* and *Vallonia cyclophonella*) are included in the fossil populations. A cool summer temperature probably is the critical factor in the distribution of these northern snails (Taylor, 1960, pp. 6-7). If so, then it is significant that on the plains, *V. cyclophonella* occurs today no farther south than central South Dakota, an area where the average July temperature is 10°F cooler than at Clovis, New Mexico. A depression of summer temperatures of at least this magnitude is indicated for the entire period when the diatomite was being deposited. It may be significant, however, that the highest frequency of this the most "arctic" member of the fauna occurs in the lower part of the diatomite, in the same zone where the highest percentages of woodland forms were present:

Level	Total Number Identified	Number of *V. cyclophonella*	Restricted Percentage
Upper Diatomite	1472	2	.001
Middle Diatomite	2939	5	.002
Lower Diatomite	721	25	.035

In the same levels where the decline in *V. cyclophonella* occurs, there is also an increase in *Gyraulus circumstriatus:*

Level	Total Number Identified	Number of *G. circumstriatus*	Restricted Percentage
Upper Diatomite	1472	1200	81.5
Middle Diatomite	2939	2100	71.3
Lower Diatomite	721	159	22.0

G. circumstriatus is also a northern species, with a distribution across the northern United States from coast to coast and southward in the Rocky Mountains. It appears, however, to tolerate slightly higher summer temperatures than *Vallonia cyclophonella*, since the southern limit of *G. circumstriatus* on the plains is a little farther south, in southern Nebraska (Hibbard and Taylor, 1960, p. 98). This may indicate that a slight increase in summer temperatures occurred between the periods when the lower and middle levels of the diatomite were deposited. While a change in temperature is suggested, it is important to keep in mind that other factors not related to temperature, such as a minor change in the microenvironmental situation, might have caused the frequency shifts in these two snails. This can only be tested by similar comparisons at other localities, and at the moment the needed collections are not available.

The diatom assemblages also provide useful data on this interval (Fig. 4). At Lubbock a high frequency of saline diatoms occurs at the base of the diatomite, but slightly higher there is a thin zone where fresh-water species predominate. A brief interval of increased moisture is indicated. The increase, however, was not sufficient to support a permanent flowing stream of the sort that was present when the clay and sand of Unit 1 below the diatomite was formed. Above this brief moist episode the diatoms are again saline until near the top of the diatomite where fresh-water forms again occur.

Ideally, the moist intervals should coincide with an increase in pine and spruce pollen. However, the correlation is not that close; instead, the pine-spruce peak occurs between the two moist phases. In part this situation may be due to the fact that slightly different profiles were sampled for pollen and invertebrates. There may be another explanation, however. The pollen spectra from the base of the diatomite (Lubbock III profile) is dominated by chenopods. In the sample immediately above, however, the dominance shifts abruptly to tubuliflora.

This change undoubtedly has environmental significance. High values for chenopods in this area are generally regarded as an indication of relative aridity (Martin, 1963; Hafsten, in Wendorf, 1961, p. 80). The tubuliflora, on the other hand, are aquatic, and an increase in these plants may suggest that water had become more abundant. This change in pollen frequencies, therefore, seems in remarkably close agreement with the dry-to-moist shift indicated by the diatom data. It is likely that the zone of tubuliflora dominance records the interval immediately after the beginning of more pluvial conditions, when the aquatic plants would increase rapidly, but the slower responding pine and spruce were not yet evident. After this brief maximum there must have been a longer period when available moisture was seemingly reduced, but not sufficiently to inhibit the increase in pine and spruce.

The pollen diagrams from the Lubbock Site indicate that there may have been one, or possibly two, intervals of high pine frequencies after the Lubbock subpluvial. These suggested high-pine zones occur about midway in Units 3 and 4. The precise age of these two units is not known, but the top of Unit 4 may date as recent as 4000 B.C. In both instances, however, the amount of pollen recovered from these zones was either marginal or insufficient for statistical reliability, and furthermore, there is no other evidence at any other site on the Llano (with the possible exception of the San Jon Site) for an event of this magnitude after the Lubbock subpluvial. Additional work will be necessary before the validity of these high-pine zones can be evaluated.

DISCUSSION

The radiocarbon dates indicate that the deposition of the diatomite occurred over a comparatively short period, of probably no more than 700 years, between 8000 and 8700 B.C. Within this period of diatomite deposition, in ponds in several widely separated localities on the Llano Estacado, the Lubbock subpluvial is defined informally as a climatic event that began with the onset of pluvial conditions recorded initially by changes in the diatom flora from saline to fresh-water species, and subsequently recorded by an abrupt increase in pine and spruce pollen frequencies. The event terminates with an abrupt decline in pine and spruce values in the upper part of the diatomite. As defined, the Lubbock subpluvial has a duration somewhat shorter than that indicated for the deposition of the diatomite. It may have lasted only 300 years, and may be approximately dated between 8600 and 8300 B.C.

The indicated duration of the Lubbock subpluvial may be sufficient for pine and spruce to invade and recolonize the Llano from the refuge areas in the mountains to the west, where these boreal trees presumably retreated at the end of the Tahoka pluvial; however, the time involved seems too short for a movement of this extent. It is possible that the Llano was not completely deforested at the end of the Tahoka pluvial and that considerable numbers of pine and a few spruce managed to survive in protected areas. From these positions they could have quickly spread back onto the Llano when conditions became more favorable. That such was indeed the case is indicated by the fauna from the "brown sand wedge" recovered by Slaughter (1970) at the Clovis I profile. Several "woodland" species were identified in this fauna, the most notable of which was the Arizona gray squirrel *(Sciurus arizonensis)*, whose principal food is the nut from the western yellow pine and whose habitat today is restricted to the upland pine forests of southern New Mexico and Arizona. A few pine trees must have been growing in the near vicinity of the Gravel Pit locality for these tree squirrels to have been present, and yet the single pollen spectrum available from this unit has very low values of both pine and spruce. Blackwater Draw, with its fringe of deep dune sand along the edge of the stream channel, might have been a "refuge" area where a few pine and spruce survived throughout the Scharbauer interval.

Vegetation changes of the magnitude proposed for the Lubbock subpluvial on the Llano Estacado have not been recorded elsewhere on the plains or in the Southwest. In southern Arizona, at the Lehner Site (Mehringer and Haynes, 1965), and in southern Nevada, at Tule Springs (Mehringer, 1967), only minor vegetation shifts at about this time are indicated. The pollen profile from Domebo Site in central Oklahoma does not indicate even a minor change during this period (Leonhardy, 1966). It is difficult to account for this discrepancy. It is possible in at least one instance (Domebo) that the samples were not collected at close enough intervals and the event was accidentally missed; however, this does not seem to be the case at either Lehner or Tule Springs. Nevertheless, the consistency of the data from the two localities on the Llano strongly argues for the reality of the environmental change postulated for this particular area. Unusually thick deposits of terminal Pleistocene age have been preserved at these two localities, and they undoubtedly contain nearly unique records for this interval.

SPECULATIONS

The cultural reaction of the human population that was present on the Llano Estacado during the Lubbock subpluvial cannot be determined with the data now at hand. For the most part, American archaeologists working with materials of Folsom age have emphasized problems of chronology and the typology of the projectile points. This is in part because most of the excavated localities are "kill sites" rather than camps, and also because of the paucity of the materials available. Few archaeologists working with the rich and varied tool inventories found at contemporary sites in the Old World can appreciate the magnitude of effort required to recover even the meager artifacts available from most Folsom sites. This paucity, moreover, has limited interest in the nonprojectile point portion of the tool complex; and as a consequence, no systematic scheme for their description has been proposed. Frequently these tools are described in a most cursory manner, if at all. It is the purpose of this section to indicate the potential value of these nonprojectile tools, and to suggest that they be reevaluated within the context of the environmental framework now available.

As we have seen, Folsom points occur throughout the deposition of the diatomite at the Gravel Pit locality; thus at this site, and probably elsewhere in this area, they occur before, during, and after the rather significant climatic changes of the Lubbock subpluvial. The evidence suggests that these points were used primarily to kill the extinct bison (*B. antiquus*) and that these animals were hunted throughout the interval when the vegetation changed from open sage and grass savanna to boreal forest and then back to grasslands.

It seems reasonable to assume that grassy parklands persisted in many areas, even during the maximum extension of the boreal forest in the Lubbock subpluvial. The bison, and their human predators, may have continued to cluster in these grassy parklands. In this event, the Folsom hunters may have been so adapted to this microenvironmental niche that their tool complex may have changed little during this period of macroenvironmental shifts.

On the other hand, it is also possible that there was at least a partial adaptation to the forested environment during the Lubbock subpluvial. The expansion of pine and spruce presumably brought more forest animals into the area, and in time the Folsom hunters may have developed effective techniques for their exploitation. This possibility cannot be tested by comparison of the tool complexes from bison kill sites. The hunting and butchering tools used on bison may have remained virtually unchanged, and indeed the limited evidence now available indicates that this was the situation.

The degree of cultural change that may have occurred during the Lubbock subpluvial perhaps can best be observed by comparison of the total tool inventories from a series of camp sites occupied before, during, and after this event. A comparison between these assemblages may show significant differences, for it is unlikely that the Folsom tool kit was so simple and generalized that it could be utilized in any environmental situation. However, even if the tool kit does prove to be this generalized, we will have gained useful information.

At the moment, the data to test these hypotheses are not available, inasmuch as total tool complexes have been reported from few Folsom campsites. It is not that such sites are unknown. For example, of the 29 Folsom Period sites known from surveys on the Llano Estacado, 19 are believed to be campsites (Wendorf and Hester, 1962, table 3). The tools have been reported in detail from only one of these sites (Hester, 1962). When more of these Folsom sites have been studied and dated, we will then be in a position to evaluate what effects, if any, the climatic changes of the terminal Pleistocene had on the early inhabitants of North America.

ACKNOWLEDGMENTS

This paper has drawn heavily on the results of unpublished research obtained by several of my colleagues on an interdisciplinary project conducted with the aid of NSF grant G-22086 on which I was Principal Investigator. In particular, I wish to acknowledge use of data obtained by Robert J. Drake, James Hester, C. Vance Haynes, Frank Oldfield, James Schoenwetter, and Bob Slaughter. While I have used extensively the joint data from this project, the conclusions offered here are my own. Publication of the final report on this project has been long delayed, but it should appear either before, or at about the same time as, this paper. This final report contains full documentation of the information summarized here.

LITERATURE CITED

Faegri, K., and J. Iversen
  1950. Text-book of modern pollen analysis. Copenhagen, Denmark, 168 pp.
Frye, J. C., H. B. Willman, and R. F. Black
  1965. Outline of glacial geology of Illinois and Wisconsin. Pp. 43-61, in The Quaternary of the United States (H. E. Wright, Jr., and D. G. Frey, eds.), Princeton Univ. Press, Princeton, New Jersey, x+922 pp.
Green, F. E.
  1962. The Lubbock Reservoir site. Mus. Jour., West Texas Mus. Assoc., 6:85-123.
Haynes, C. V.
  1964. Fluted projectile points: their age and dispersion. Science, 145:1408-1413.
  1965. Carbon-14 dates and early man in the New World. Interim Res. Rep., Geochronology Lab., Univ. Arizona, 9:1-24.
Helvy, R. H., P. J. Mehringer, Jr., and H. G. Yocum
  1965. Modern pollen rain in the Sonoran desert. Jour. Arizona Acad. Sci., 3:123-135.
Hester, J. J.
  1962. A Folsom lithic complex from the Elida site, Roosevelt County, New Mexico. El Palacio, 69:92-113.
Hibbard, C. W., and D. W. Taylor
  1960. Two Late Pleistocene faunas from southwestern Kansas. Contrib. Mus. Paleontol., Univ. Michigan, 16:1-223.
Leonhardy, F. C. (ed.)
  1966. Domebo: a Paleo-Indian mammoth kill in the prairie-plains. Contrib. Mus.

Great Plains, 1:1-53.

Martin, P. S.
1963. The last 10,000 years. . . . Univ. Arizona Press, Tucson, 87 pp.

Mehringer, P. J., Jr.
1967. Pollen analysis of the Tule Springs area, Nevada. Pp. 129-200, *in* Pleistocene studies in southern Nevada (H. M. Wormington and D. Ellis, eds.), Nevada State Mus., Anthro. Papers, 13:1-411.

Mehringer, P. J., Jr., and C. V. Haynes
1965. The pollen evidence for the environment of early man and extinct mammals at the Lehner mammoth site, southeastern Arizona. Amer. Antiquity, 31:17-23.

Potter, L. D., and J. Rowley
1960. Pollen rain and vegetation, San Augustin Plains, New Mexico. Bot. Gaz., 122:1-25.

Reeves, C. C., Jr.
1965. Pleistocene climate of the Llano Estacado. Jour. Geol., 73:181-189.

Sellards, E. H.
1952. Early man in America: a study in prehistory. Univ. Texas Press, Austin, 211 pp.

Slaughter, B. H.
1970. An ecological interpretation of the brown sand wedge local fauna, Blackwater Draw, New Mexico; and a hypothesis concerning extinction. *In* Paleoecology of the Llano Estacado (F. Wendorf and J. J. Hester, eds.), vol. 2, Publ. Fort Burgwin Res. Center, in press.

Taylor, D. W.
1960. Late Cenozoic molluscan faunas from the High Plains. U.S. Geol. Surv. Prof. Paper, 337:iv+1-94.

Wendorf, F. (ed.)
1961. Paleoecology of the Llano Estacado. Publ. Fort Burgwin Res. Center, 1:1-144.

Wendorf, F., and J. J. Hester
1962. Early man's utilization of the Great Plains environment. Amer. Antiquity, 28: 159-171.
1970. Paleoecology of the Llano Estacado. Publ. Fort Burgwin Res. Center, vol. 2, in press.

Soils, Paleosols, and Environment

ROBERT V. RUHE

ABSTRACT

Certain features of environment may be read from the properties of paleosols if comparison is made to land-surface soils that are somewhat better understood in relation to environment. Using this principle, a reconstruction of the past is possible. The present rigorous semiarid regime of the Central Plains dates from Brady time 11,400 to 9100 years ago. The prairie in Iowa began 7000 to 8000 years ago. In pre-Brady time to the beginning of Wisconsin time, the climate in general was cooler and relatively more moist. Coniferous forest dominated Iowa to the Missouri River Valley in the southwestern corner of the state 24,500 years ago. The limits of extension of such vegetation westward on the plains is unknown. The forested Sangamon landscape was 175 to 350 miles northwest and southwest of present humid southern and northern forest and woodland zones. Sangamon Soil zones were displaced as much as 100 miles west of land-surface soil-analogue zones. Sangamon climates were more moist and may have been warmer than those of today.

INTRODUCTION

Environment in the Quaternary of the plains may be examined from several points of view in regard to soils and paleosols. If paleosols are studied in comparison to soils on the land surface, the environments of which are better understood, certain inferences may be made about the environments in which the paleosols formed.

Buried soils also should be studied from the viewpoint of soil-stratigraphic units (Richmond and Frye, 1957; Amer. Comm. Strat. Nomen., 1961). These soils separate deposits that may contain both faunal and floral assemblages, which may be used for ecological interpretations. The latter includes both macrofossils of wood and microfossils of pollen. Wood contained in the deposits below and above the buried soil, and organic carbon of the A-horizon of the buried soil may be radiocarbon dated. A time framework of the Late Pleistocene and Recent environmental regimes may be constructed.

The term "plains" is used in the broadest sense in this paper. It includes the younger and older glacial drift plains, some loess-mantled, of the Central Lowlands and, of course, the Great Plains of the mid-continent. The reason for such latitude of definition is basic—a proper evaluation of the plains must be related to the known standard areas of the Pleistocene in the Upper Mississippi River Valley region. Such alignment also permits the layout of a transect model for illustrative purposes that extends from the younger drift plains across the Great Plains to the southwest (Fig. 1). Considerable detail is known at the ends of the transect. The "in-between" perhaps may be evaluated from the end members.

Specific models along the transect will be developed, and a plea is made that much more work has to be done before a reasonably definitive system can be formulated across the plains. The discussion will be restricted to the Sangamon and younger intervals. This contribution is Journal Paper no. J-6094, project no. 1250, Iowa Agriculture Experiment Station, Ames.

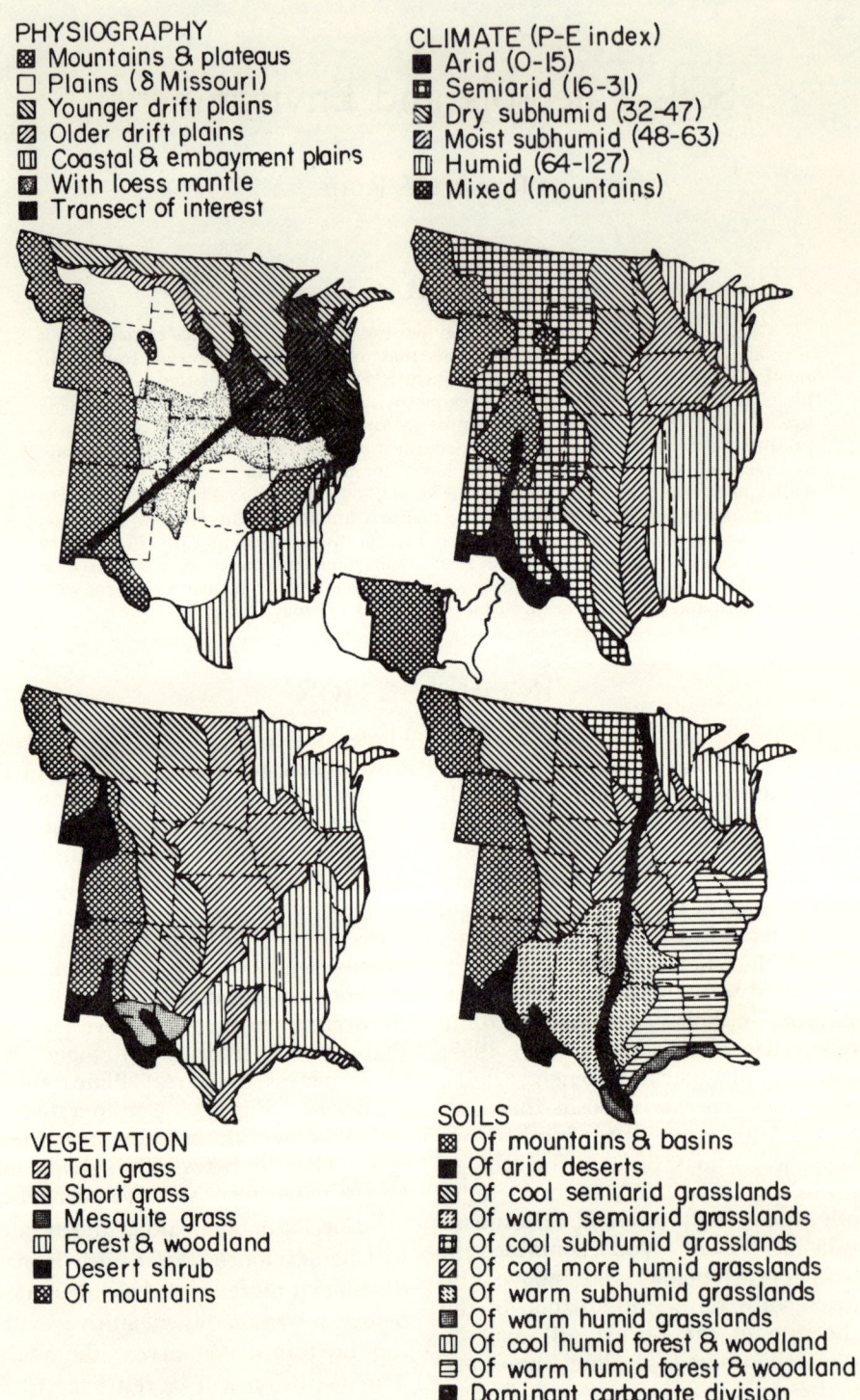

Fig. 1. Maps: Generalized physiography of the plains region. Climate based on P-E index after Thornthwaite (1941). Vegetation zones after Barnes (1948). Soils after Kellogg (1941) and adjusted N.C.R.-3 Tech. Comm. Soil Survey (1960) and U.S.D.A. Soil Conservation Service Map (1967).

Fig. 2. Soils and paleosol in Iowa. A. Tama Soil formed under grassland. B. Late Sangamon paleosol. C. Fayette Soil formed under forest. Compare gross features of B with A and C. Photos A and C by R. W. Simonson.

LAND-SURFACE SOILS

SOIL PROPERTIES AND ENVIRONMENT

Soils that form under grass or under trees generally have distinctive properties that are associated with each of these covers. A model that illustrates this generalization is comprised of the Tama and Fayette soils in Iowa, both of which formed in similar Wisconsin loess on the same land surface, are 14,000 years old, have similar local topographic positions, and presently have a similar climate. Mean annual rainfall is 32 to 34 inches, with average winter temperatures of 16 to 20°F and average summer temperatures of 72 to 74°F. The Tama Soil formed under grass (Fig. 2), and the Fayette Soil formed under forest. The differences in morphologies of the profiles of these two soils are pictorially self-evident. The Tama Soil has a thick, dark-colored surface horizon that grades downward into a lighter-colored B-horizon that has pedogenic structure and accumulation of clay. The Fayette Soil has a thin, dark-colored A1-horizon overlying a lighter-colored, gray, platy A2-horizon, in turn over a darker-colored B-horizon, which has pedogenic structure and accumulation of clay.

Other generalizations can be made about other properties of these soils (Table 1). Eluviation of clay has been more intensive under forest than under grass. Organic carbon has accumulated to greater depths under grass than under forest. In the Tama Soil the value is 0.68 percent at 20 inches, but is 0.66 percent at only two inches in the Fay-

TABLE 1. Comparison of Properties of Tama and Fayette Soils of Iowa

Horizon	Depth (in.)	Clay (%)	pH (1:1)	Organic carbon (%)	Fe_2O_3 (%)	Base saturation (%)
Tama Soil under grass						
Ap	0-6	28.6	5.7	2.35	0.9	66
A1	6-11	32.2	5.8	1.95	1.5	62
A3	11-16	34.2	5.7	1.42	1.6	68
B1	16-20	35.6	5.8	0.97	1.6	72
B21	20-25	35.4	5.7	0.68	1.7	73
B22	25-29	33.2	5.7	0.45	1.7	74
B23	29-35	30.5	5.7	0.34	1.7	76
B3	35-45	28.2	5.8	0.21	1.7	77
C	45-51	28.5	6.1	0.15	1.6	81
C	51-61	27.6	6.5	0.12	1.5	84
Fayette Soil under forest						
A1	0-2	13.0	5.6	5.63	0.9	56
A2	2-9	13.5	5.0	0.66	1.0	20
B1	9-17	16.9	5.0	0.30	1.2	34
B21	17-24	23.8	5.0	0.20	1.6	47
B22	24-35	27.7	4.9	0.16	1.7	51
B23	35-42	30.7	5.0	0.17	1.8	64
B3	42-48	29.6	5.2	0.14	1.8	65
C	48+	26.3	5.1	0.12	1.8	66

ette Soil. Leaching of bases has been more intensive under forest as shown by the pH and base saturation values, and removal of iron oxide in upper parts of the soil has been greater under forest. Obviously, vegetation, whether grassland or forest, can be read from the soils.

Wetness or dryness of soils usually can be determined, with limitations, from the color or patterns of color in soils (U.S.D.A. Soil Survey Staff, 1951). Red, brown, and yellow colors generally are related to the nature of iron oxides, and in the Midwest they usually indicate good internal drainage and aeration within the soil. Gray, blue, or green colors usually indicate poor internal drainage and aeration. Soil color may be inherited from the color of the material in which the soil formed. Hence, this factor must always be weighted in any analysis. Patterns of color are known as mottling, which means marked with spots of color. Mottles are mainly gray, yellow, brown, and red, and usually indicate some condition of impeded drainage in the soil.

These colors or patterns of color may be within the soil or beneath it. Toward the plains in southwestern Iowa, a common feature in the loess on summits of flat hills is a zoning of color and color patterns to depth. The lower part of the soil solum is brown or yellowish brown. The next lower zone is yellowish brown mottled with gray, and a downward reversal of gray mottled with brown. A lower zone is gray with sparse reddish brown nodules and tubules. Such zoning has been observed in the loess-mantled older drift plains of Nebraska and Kansas (Fig. 1).

These zones comprise the weathering profile beneath the soils. The gray zone with nodules and tubules is the deoxidized zone, which is a relict gleyed zone that formed when subsurface water tables and zones of saturation perched on more impermeable paleosols and were much higher than they are today (Ruhe and Scholtes, 1956). By relating the deoxidized zones to faunal zones in the loess, it is evident that the weathering zones are younger than the faunal zones and the loess itself. They were correlated with cool, relatively moist conditions of the Late Pleistocene and the earlier post-glacial time. Similar features have been noted in Kansas, in Brown County near the Missouri River Valley but also as far west on the plains as Riley and Republic counties (Bidwell *et al.*, 1968).

From all of these indicators, possible vegetative cover and conditions of wetness or dryness of the soils and in the subsurface may be estimated. However, the wetness or dryness refers to the soil climate and not to the atmospheric climate. Well-drained and poorly drained soils can be near one another on the landscape under the same atmospheric climate today.

PRESENT REGIONAL PATTERNS OF SOILS AND ENVIRONMENT

Certain broad generalizations can be drawn on a regional basis between climate, vegetation, and soils across the plains. Use of the P-E index (Thornthwaite, 1941; Blumenstock and Thornthwaite, 1941), shows that climatic zones approximately parallel the trend of the Mississippi River Valley and the Rocky Mountain front, and extend north to south across the plains. The zones in order from east to west are humid, moist subhumid, dry subhumid, semiarid, and arid (Fig. 1). A broad salient of the moist subhumid zone splits the humid zone in two across Iowa and Illinois. The arid zone is just west of the plains in New Mexico, mainly along the Rio Grande, but a part of this zone is on the plains along the Pecos River Valley.

Broad zones of vegetation (Fig. 1) conform in general to the climatic zones but are not exactly coincident (Barnes, 1948). Forest and woodland relate to the humid zone. Tall-grass prairie mostly fits the subhumid zones. Short-grass prairie is in the semiarid zone, and mesquite grassland and desert shrub relate to the arid zone.

According to Kellogg (1941), groups of soils can be broadly associated throughout the region (Fig. 1). Some of his boundaries have been modified and adjusted with reference to later regional groupings (N.C.R.-3 Tech. Comm. Soil Survey, 1960; U.S.D.A. map, 1967). The grouping of the soils relates in general to climatic and vegetation zones but with an additional stratification by temperature from north to south (Fig. 1).

If soils at better-drained sites under grassland are examined, a definite sequence in soil development is related to environment from the more humid to the more arid country (Fig. 3). The sequence is also restricted to the part of the plains that is mantled with loess and other eolian deposits of Wisconsin age (Fig. 1). From more humid to more arid zones the A-horizons of soils become thinner; the B-horizons have lesser development as indicated by illuvial clay; the profiles have less leaching of bases; and carbonates are present in, and immediately beneath, the sola. In

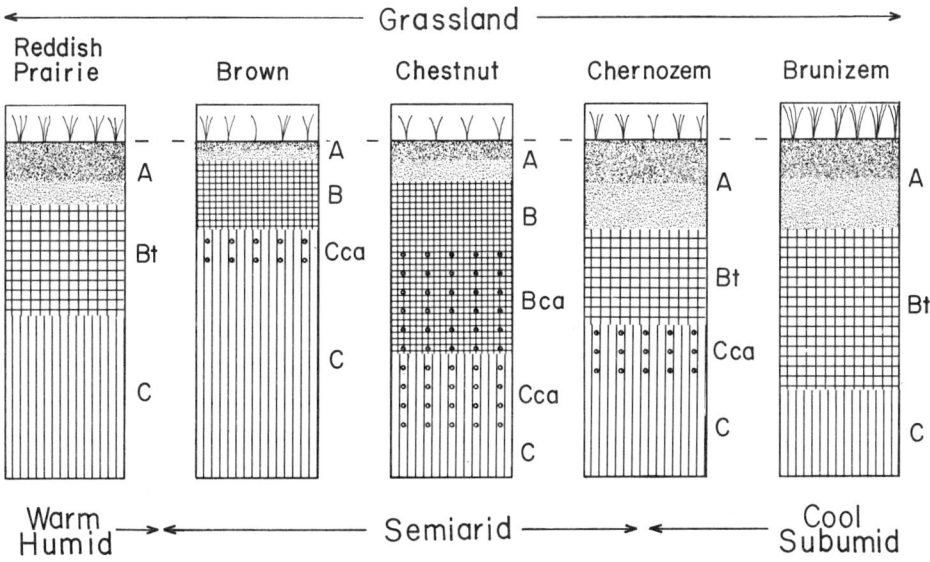

FIG. 3. Sequence of soils formed under grassland, from more humid to more arid region. From N.C.R.-3 Tech. Comm. Soil Survey (1960).

regard to carbonates, the old line of Marbut (1935) separating Pedocals, with carbonates, from Pedalfers, without carbonates, remains valid (Fig. 1). While one is driving along any east-west highway throughout the plains region and scanning road cuts while en route, the absence of carbonate in the soils east of the line and the presence of carbonate west of the line is striking. This line approximately coincides with the boundary between Thornthwaite's (1941) dry-subhumid and wet-subhumid climatic zones.

One further complication enters the regional soil picture on the present land surface. From north to south, the colors of the soil B-horizons have stronger chromas and redder hues. Hue refers to a specific color, such as red, whereas chroma refers to the strength or departure from a neutral of the same lightness or value of the color. Previously, a north-south stratification of soils on the plains was related to temperature, and, in fact, the temperature stratification was read from the red color. But is red color only related to temperature?

All of the foregoing are extremely broad generalizations about the soil-environment relationships on the land surface of the plains. Many kinds of soils are present (Bidwell, 1956; Elder, 1964; N.C.R.-3 Tech. Comm. Soil Survey, 1960; Oschwald et al., 1965; Baldwin et al., 1938; U.S.D.A. Soil Survey Staff, 1960; U.S.D.A. map, 1967), so, in detail, soils and landscape and environment are more complex. For proper understanding, however, one must proceed from the generalized to the detailed.

ARE SOIL PROPERTIES TRUE INDICATORS OF ENVIRONMENT?

A question may be raised concerning the broad generalizations of soil properties and environment, so an examination is necessary. Does accumulation of carbonate in the solum or beneath the solum indicate more arid environment? This problem can be examined just west of the plains in southern New Mexico. There a soil formed on the Picacho surface of Late Pleistocene age (Ruhe, 1965, 1967). This soil formed in rhyolitic gravel and has a thin A-horizon overlying a textural B-horizon, which, in turn, rests abruptly on a K-horizon. The K-horizon is a soil horizon of carbonate accumulation (Gile et al., 1965).

Clay in the soil profile increases systematically downward from 12.7 to 20.3 percent, and abruptly abuts the upper surface of the K-horizon at a depth of 11 inches. Organic carbon and free iron oxides have similar vertical distributions. When carbonate is removed from the system in the laboratory, the clay, organic carbon, and free iron oxides continue downward at relatively large amounts in the subhorizons of the profile designated as K1 and K2. These constituents then decrease into the C-horizon. These related subparallel vertical distributions of clay, organic carbon, and free iron oxides are characteristic of B-horizons of many soils. On a noncarbonate basis, the B-horizon extends downward to a depth of 25 inches.

But abundant carbonate (44 percent carbonate equivalent) forms an indurated laminated layer at the 11- to 12-inch depth and fills interparticle space (15 percent carbonate equivalent) to the 25-inch depth. Sealed in the laminated layer is organic carbon whose radiocarbon age is 9550 ± 300 years. The sealant, carbonate of the K-horizon, must be younger. The clay, organic-carbon, and free-iron-oxide distribution of the carbonate-engulfed B-horizon must be as old and older. The radiocarbon age dates back toward the end of the last Wisconsin glacial episode, the pluvial of the plains, when relatively more moist climatic conditions existed in that part of New Mexico (Clisby and Sears, 1956; Harbour, 1958; Leopold, 1951). The carbonate in the soil, being younger, must record the more arid environment that followed the last pluvial.

The organic carbon in the A-horizons of surface soils probably reflects the very latest Recent time and environment. Organic matter is continually added to the A-horizon through the decomposition of vegetation that is provided in its cyclic growth. Within the organic-carbon system in the soil, some parts of it are "stable" and some are "relatively mobile and unstable" (Campbell et al., 1967a, 1967b). Thus, various fractions of the organic carbon can be separated by chemical extraction with alkaline or acid solutions, and the various fractions can be radiocarbon dated. The age of an unfractionated sample is a combination of the ages of all of the fractions and has been termed the "mean residence time," or MRT.

In some Iowa soils under grassland the MRT of A-horizon organic carbon ranges from less than 100 to 440 ± 120 years (Ruhe, 1967). Data are not available for organic-carbon fractions. However, in grassland soils on the younger drift plains in Saskatchewan, Canada, where such study has been made, the radiocarbon ages of fractions range from 25 ± 50 years to 1410 ± 95 years, where MRT of the unfractionated carbon is 870 ± 50 years (Campbell et al., 1967a, 1967b). Therefore, the organic-carbon system, which is the basis for recognition of A1-horizons, should be related to, and, in turn, be an indicator of, the very latest environment.

Grassland soils on the younger drift plains in Iowa may be placed in their proper environmental framework by relating them geomorphically and stratigraphically to their associated soils that occur in poorly drained depressions. Most peat and muck bogs on the Des Moines drift lobe contain two organic soil profiles (Farnham, 1960) that are stratigraphically superposed (Walker, 1966). The lower profile is a peat or muck formed on silts, and a similar upper profile buries the lower profile. The silts record a period of instability on hillslopes bounding the bogs, so that mineral sediments were eroded, transported, and deposited in the depressions. Deposition of mineral sediment dominated accumulation of organic matter *in situ*. Peats and mucks record periods of greater stability of the bounding hillslopes, so that lesser amounts of mineral sediment were eroded, transported, and deposited in the depressions. Organic matter accumulating *in situ* dominated deposition of mineral sediment. The lower sedimentologic-pedogenic cycle culminated about 7000 to 8000 years ago. The upper mineral sediment was deposited to about 3000 years ago, and the uppermost peat and muck formed, with associated hillslope stability, during the past 3000 years (Fig. 4).

Associated with the two organic soils are pollen profiles (Walker, 1966). The pollen in the lower profile is dominantly arboreal, including conifers, of which spruce is prominent (Brush, 1967). Nonarboreal pollen, including grasses, dominates the upper profile (Fig. 4). Recorded herein is the postglacial shift from forest to prairie at about 7000 to 8000 years ago.

As demonstrated by Walker (1966), a layer as thick as five feet had to be stripped from the adjacent hillslopes to provide the amount of sediment of the upper silt. This happened in post-forest time, so any evidence of forest influence in the soils was removed by erosion. Consequently, the soils on the adjoining hillslopes are grassland soils that formed during the more stable period of the last 3000 years.

Spruce is not a native tree in Iowa today, but it grows to the north in northern Minnesota and the upper peninsula of Michigan, where mean annual temperatures are about 11°F cooler. Annual rainfall amounts and distributions are similar. Cooler temperatures, however, would have con-

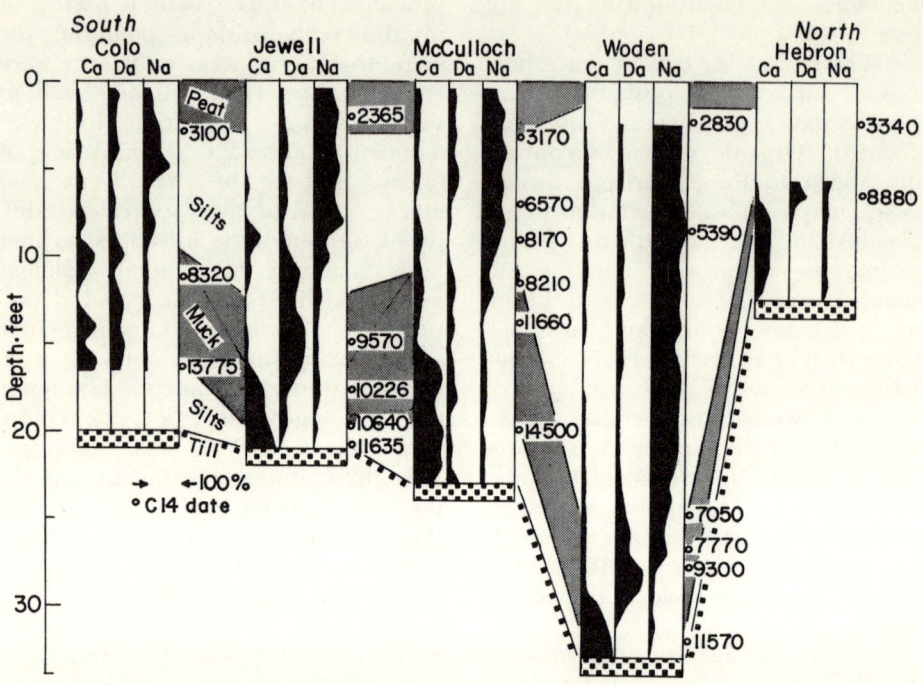

FIG. 4. Pollen profiles related to surface and buried organic soils in peat bogs on Des Moines drift lobe, from central to northern Iowa. From Ruhe, © 1969, Iowa State University Press, Ames.

served moisture, so that the climate would have been relatively more moist. The environment zone at that time in central Iowa was about 350 miles south of the zone of the same environment today. A similar distance to the south and west would be the plains, and the question may be asked, Does a record exist in the organic soils there?

In the northern Sand Hills of Nebraska, pollen and seed analyses indicate the occurrence of a boreal forest with spruce at about 12,600 years ago (radiocarbon dating), which was abruptly succeeded by grassland with somewhat more pine than today (Sears, 1961; Watts and Wright, 1966). In central Texas, just to the east of the grassland belt (Fig. 1), the pollen profile in a bog shows a change from spruce-fir to chestnut-alder and grassland (Potzger and Tharp, 1954). Although these are only a few isolated sites, post-glacial environmental changes are apparent. But a great deal more work is needed on organic soils of the plains for a more complete story.

Also associated with soils and Pleistocene deposits of the younger drift plains and loess-mantled older drift plains in Iowa are macrofossils of wood that universally represent coniferous species such as larch, hemlock, yew, and spruce. Their radiocarbon ages are as young as 8320±275 years (Walker, 1966) and as old as 37,600 ±1500 years (Daniels and Jordan, 1966). The area covered by 30 dates extends from central Iowa to the Missouri River Valley in the southwestern part of the state (Ruhe, 1969). Presumably, similar macroflora of similar age extended westward on the plains. A cool, relatively moist environment is indicated during these times.

In summary, then, certain soil properties are reasonable indicators of environment. Carbonate accumulation in better-drained soils shows more arid conditions. The nature and distribution of organic carbon in soil A-horizons relate to the latest environments. Pollen and macrofossils of plants associated with organic soils not only permit estimates of vegetation but inferences regarding climate as well.

PALEOSOLS

Applying the principle of uniformitarianism, paleosols may be used to interpret past environments if they are recognized as analogues of land-surface soils, the environments of which are better known. However, there is a major problem concerning paleosols. They have not been mapped on a local or regional basis as have land-surface soils, except in isolated cases (Ruhe, 1956). Consequently, geographical changes in the morphology of the paleosols are known only from spot sites. In addition, little systematic laboratory study has been coupled with systematic geographic study of paleosols. When such study is made, some striking property distributions become evident (Ruhe, 1965). As a result, the deduction of environment from paleosols can only be done at a broad, generalized level. There is great need for systematic mapping and study of paleosols.

MAJOR BURIED SOILS OF THE PLAINS

Three buried soils have widespread distribution on the plains. From the land surface downward, they are the Brady Soil, the basal soil of the Wisconsin Loess, and the Sangamon Soils. Other paleosols are present; but because of limited knowledge about them, they will be excluded from the following discussion. Note that the previously discussed lower organic soil in the bogs on the younger drift plain in Iowa is a buried soil, and that specific environmental interpretations could be made from associated pollen and plant macrofossils.

Brady Soil.—This soil was named by Schultz and Stout (1945). It is restricted to the Wisconsin (Peoria-Bignell) Loess region west of the Missouri River and has widespread distribution in Nebraska and Kansas (Schultz and Stout, 1948; Thorp *et al.*, 1951; Frye and Leonard, 1952; Reed and Dreeszen, 1965). The Brady Soil separates the Peorian and Bignell members of the Wisconsin Loess. East of the Missouri River Valley in Iowa, the Brady Soil has not been formally recognized, although numerous dark-colored bands have been observed in the Wisconsin Loess (Daniels *et al.*, 1960). No one of them, however, has been correlated with the Brady Soil, even though the bands resemble weak A-horizons of buried soils. Similar bands were described in the Bignell Loess in Kansas and Nebraska and were considered to be weak soils that represent short pauses in loess deposition (Thorp *et al.*, 1951).

Properties of the Brady Soil suggest that it is a "Chernozem-like silt loam without textural variation in the profile, and some of it has a mottled subsoil, suggesting that it may be a Humic-Gley soil" (Thorp *et al.*, 1951). In the northwestern quadrant of Kansas the soil is moderately to poorly drained, and the depth of leaching of carbonates ranges from one to three feet. The A-horizon is generally dark, and the carbonate nodules one-half to one inch in diameter are in the lower part of the B-horizon. The dark gray to grayish brown B-horizon has accumulated clay and good structure at some places, but elsewhere little textural contrast occurs between the A- and B-horizons (Frye and Leonard, 1952).

A comparison of this buried soil with the land-surface soils shows much similarity. The better-drained Brady Soil is Chernozem-like (Thorp *et al.*, 1951), and a broad belt of Chernozems is on the land surface in the same areas today (N.C.R.-3 Tech. Comm. Soil Survey, 1960). In fact, the Brady Soil shows no westward offset from the present similar surface soils, indicating that the present semiarid climate of the plains was initiated during Brady time (Frye and Leonard, 1957a).

This climatic inference is substantiated by the ecology of molluscan fauna below and above the Brady soil-stratigraphic unit. In pre-Brady Wisconsin deposits, aquatic pulmonate snails are widespread, but in post-Brady deposits only hardy species that are capable of enduring extremes of summer heat and winter cold survived (Frye and Leonard, 1957a). As far south as western Texas, during and following Brady time, ecological conditions changing toward greater aridity resulted in extinction of the greater part of the earlier Wisconsin molluscan faunas (Frye and Leonard, 1964). Throughout the plains, post-Brady molluscan faunal assemblages are generally depauperate and characterized by species that are now living in the region (Frye and Leonard, 1963).

When was Brady time? It is considered equivalent to the Two Creeks interval of the Wisconsin (Frye and Leonard, 1964; Reed and Dreeszen, 1965), which is radiocarbon dated at about 11,400 years (Rubin, 1960). However, a radiocarbon date of 9160 ±250 years (W-234) was determined on organic carbon from the A-horizon of the Brady Soil at the type locality. The possibility of contamination of the sample by modern rootlets and otherwise was noted previously (Rubin and Suess, 1956; Ruhe, 1968). A second organic carbon sample was recently analyzed and was dated at 9750±300 years (W-1767). But again the possibility of contamination by a few modern rootlets is possible (personal communication, J. A. Elder, 1965, and E. C. Reed, 1965). If a Two Creeks correlation of the Brady Soil is valid, contamination of the buried A-horizon must amount to about 9 to 12 percent modern organic carbon in the total organic carbon in order to reduce the age 1650 to 2240 years. The possible contamination can be determined by manipulating radioactive decay equations (Ruhe, 1969).

Another dating of the Brady Soil is also unsatisfactory. Mollusk shells from Bignell Loess above the Brady Soil in the Iowa Point section, Doniphan County, Kansas, are 12,550±400 years (W-231) and 12,700±300 years (W-233). These dates are older than Two Creeks, but because of the uncertainty in radiocarbon assay of land snails, the dates could differ from actual values by 1000 years or more (Rubin and Suess, 1956; Rubin *et al.*, 1963). Both dating systems are unsatisfactory because of possible contamination in each. A special effort should be made to finitely date the Brady Soil so that its proper placement in the Late Pleistocene or Recent can be made. Such dating is of particular importance in determining the time sequence of environment changes.

Basal soil of Wisconsin Loess.—A common feature at the base of the Wisconsin Loess in Iowa (Ruhe, 1968) and at the base of the Peorian Loess in Nebraska and Kansas (Thorp *et al.*, 1951; Frye and Leonard, 1952) is a buried A-C soil that overlies the pre-Wisconsin paleosols. At some places a dark grayish brown A-horizon has an underlying leached silty C-horizon. At other places, only a series of bands with organic carbon are intercalated with the leached silts. The whole zone has been interpreted as the initial slow accumulation of Wisconsin Loess so that carbonates were leached and so that organic matter decayed and darkened the first increments of Wisconsin Loess (Thorp *et al.*, 1951). Where banding occurs, minor pauses in loess accumulation are represented (Frye and Leonard, 1952). This zone is overlain by many feet of calcareous loess.

This zone is also the *Citellus* zone of the Nebraska literature (Frye and Leonard, 1952), or the soil horizons in which the ground squirrel of that genus dug its burrows (Thorp et al., 1951; Reed and Dreeszen, 1965; Ruhe, 1968). This zone through Kansas, Nebraska, and Iowa was stratigraphically correlated with the Farmdale of Illinois; but on the basis of radiocarbon dating, it is not (Frye and Leonard, 1965; Ruhe, 1968). The basal soil correlates only partially in time. The Farmdale in Illinois dates from 22,000 to 28,000 years ago (Frye and Willman, 1960). Organic carbon from the A-horizon of the basal soil of Wisconsin Loess in Iowa dates from 16,500 to 29,000 years (Table 2). If a sequence of samples is aligned from the Missouri River to the south-central part of the state, the basal age (Y) decreases systematically with distance in miles (X) as expressed by $Y=26,500-55X$ (Ruhe, 1969). Farther to the west in Nebraska, at the Buzzards Roost section in Lincoln County, the date (I-1851) of the correlative basal soil A-horizon is 32,300 (+2000, −1600) years (personal communication, E. C. Reed, 1965).

The morphologic properties of the basal soil are not diagnostic of either forest or grassland as they are generally weakly developed A-C or O-C profiles at best. However, pollen from the peat of sample I-1403 (Table 2) in Iowa is dominantly coniferous and wood fragments were larch (Graham, 1962). In Pottawattamie County, wood from the basal soil A-horizon is larch (W-141) and spruce (I-1023, cf. I-1420). Higher in the loess section above the basal soil, spruce wood in Harrison County is $19,050 \pm 300$ years old (W-879). As pointed out previously, conifers were dominant to 10,000 years ago and were still present 8000 years ago in Iowa. The flora associated with the basal soil indicates a cool, relatively moist forest environment.

Westward on the plains, the fauna associated with the deposits of pre-

TABLE 2. Radiocarbon Dates of A-horizon of Basal Soil of Wisconsin Loess in Iowa

Sample*	Date in years before present	Location	Notes
I-1419A	16,500±500	Humeston, Wayne County	Soil organic carbon residue (OC)
W-1687	18,300±500	Salt Creek, Tama County	OC
I-1411	18,700±700	Greenfield, Adair County	OC
I-1419B	19,000+6000 −3000	Humeston, Wayne County	Humic acid fraction of I-1419A
W-879	19,050±300	Logan, Harrison County	Spruce wood
I-1408	19,200±900	Harvard, Wayne County	OC
I-1409	20,300±400	Hayward paha, Tama County	OC
I-1864A	20,500±400	Sheldon, O'Brien County	OC
I-2332	20,700±500	Alburnett paha, Linn County	OC
I-1410	20,900±1000	Murray, Clarke County	OC
I-3655	21,000±500	Muscatine, Muscatine County	OC
I-3702	21,150±450	Adair, Adair County	OC
I-3653	21,350±750	Hills, Johnson County	OC
I-1023	21,360±850	Bentley, Pottawattamie County	Spruce wood
W-1681	21,600±600	Palermo area, Grundy County	OC
I-1404	22,600±600	Palermo area, Grundy County	OC
I-1420	23,900±1100	Bentley, Pottawattamie County	OC
I-1403	23,900±1100	Grinnell, Poweshiek County	Peat, conifer zone
W-141	24,500±800	Hancock, Pottawattamie County	Larch wood from peat
I-1406	24,600±1100	Kinross, Keokuk County	OC
I-1267	25,000±2500	Hayward paha, Tama County	OC
I-1269	29,000±3500	Salt Creek, Tama County	OC

* Sample numbers are I for Isotopes, Inc., and W for U.S. Geological Survey, Washington, D.C.

Brady time contains a widespread distribution of aquatic pulmonate gastropods (Frye and Leonard, 1957a). As far south as Texas, the shallow undrained depressions were occupied by lakes of semipermanent bodies of water (Frye and Leonard, 1957b). Both faunal and sediment records indicate that early Wisconsin time on the plains was characterized by a cooler and more moist climate than exists at present (Frye and Leonard, 1957b). Currently unknown is the extent of forest cover of that time, but in Iowa northern conifers were present throughout the state, including the southwest part along the Missouri River Valley (Ruhe, 1969).

Sangamon Soils.—The Sangamon Soils are by far the most extensive of all of the strongly developed buried soils of the Central Plains (Thorp *et al.*, 1951; Frye and Leonard, 1952; Reed and Dreeszen, 1965; Ruhe, 1965, 1968). They have wide distribution east of the Mississippi River Valley in Illinois and Indiana, and westward across Iowa and beyond the Missouri River Valley on the plains in Kansas and Nebraska. Two Sangamon Soils are recognized in Iowa. One is the well-known Sangamon Soil that is formed in Loveland Loess (Ruhe and Scholtes, 1956). The other is the Late-Sangamon Soil that is formed on pediments below the level of the Sangamon and Yarmouth-Sangamon surfaces (Ruhe, 1956). In general, in eastern Nebraska and to the eastward these soils are deeply leached and strongly developed. But to the westward the soils are leached to lesser depths (Thorp *et al.*, 1951).

In western Iowa, where these buried soils have been studied in great detail, variations in the properties of the soils geographically and geomorphically are known (Ruhe, 1965). The B-horizons of the better-drained Sangamon Soil have redder hues and stronger chromas than surface soils. The paleosol B-horizons also have heavier texture. For example, as the Loveland Loess thins with distance eastward from the Missouri as $T = 1/(0.043 + 0.00104D)$, where T is thickness in feet and D is distance in miles, the clay content (C) of the B-horizons of the Sangamon Soil increases with decrease in loess thickness as $C = 1/(0.0016 + 0.0011T)$. The Late-Sangamon Soil has color and development of horizons similar to the Sangamon Soil (Ruhe, 1956).

Where buried in western Iowa, these soils have properties that are analogous to land-surface soils formed under forest, with the exception of the redder hues and stronger chromas. Consequently, they are considered as analogues of Gray-Brown Podzolic soils (Simonson, 1941, 1954; Ruhe and Scholtes, 1956; Prill and Riecken, 1958; Ruhe 1965). Even visual inspection of the Late Sangamon paleosol shows it similar in gross morphology to surface soils formed under forest instead of those formed under grass (Fig. 2, cf. B to C and to A). Where buried, the Late-Sangamon Soil has been resaturated with bases brought down from the overlying loess, but upon exhuming by erosion of the loess, releaching has occurred (Prill and Riecken, 1958). Where currently under grassland, the surface horizon has darkened and thickened due to the addition of organic matter (Ruhe and Daniels, 1958).

How far west of the Missouri River can the forested Sangamon Soils be traced? The Late Sangamon paleosol has been exhumed in large areas of southwest Iowa, so that in standard, operational soil survey that soil is recognized in part as the Adair soil series (Ruhe *et al.*, 1967). The series was established during the studies of the Greenfield Quadrangle, Adair County, Iowa, during the period 1955-57. The same series is identified in Pawnee County, Nebraska, about 60 to 70 miles west of the Missouri River Valley and along the Kansas state line (Millet and Drew, 1963). Farther west, the Sangamon Soils, where buried, have dark-colored A-horizons and thick, reddish brown B-horizons, with maximal textural contrasts between the A- and B-horizons, suggesting that the well-drained soils were Reddish Prairie and reddish Chernozem soils formed under grass (Thorp *et al.,* 1951).

In Kansas the Sangamon Soil changes from a forest border-grassland soil in the northeastern part of the state through a Chernozem belt to a soil of a semiarid type in the northwest (Frye and Leonard, 1952). When the paleosol is compared to surface soils of similar drainage and parent material, the belts of the paleosolic "great soil groups" are offset to the west by as much as 100 miles (Frye and Leonard, 1957a).

In any event, the occurrence in southeastern Nebraska and northeastern Kansas of Sangamon paleosols that are analogous to forest soils indicates an environment in that region that was drastically different from the present environment. The area today is grassland under a dry subhumid climate (Fig. 1). Currently the nearest northern forest and woodland is about 350 miles to the northeast, and the nearest southern forest and woodland is about 175 miles to the southeast. The conclusion must be reached that the Sangamon climate was relatively more moist than the present climate in that part of the plains.

The redder hues and stronger chromas of the B-horizons of the Sangamon Soils, in contrast to surface soils, have been discussed at some length in the literature (Scholtes *et al.,* 1951; Thorp *et al.,* 1951; Frye and Leonard, 1952; Simonson, 1954; Ruhe and Scholtes, 1956; Ruhe, 1965, 1967; Frye and Leonard, 1967; Ruhe, 1968), and there is no need to repeat the discussion. In general, red colors in soils have been interpreted as reflecting warmer temperatures during soil formation or longer time of weathering. Where parent material colors are excluded, the red colors are believed to be caused by the nature of iron oxide in the soil. Generally, hematite is believed to cause red color, and high temperatures are necessary for its formation (Schwertmann, 1959a, 1959b; Mackenzie, 1959; Oades, 1963). However, some soils of the Coastal Plains in the United States have red B-horizons and the dominant iron oxide is goethite (Soileau and McCracken, 1967). Goethite is usually considered to yield yellowish and brownish colors (Schwertmann, 1959a, 1959b; Mackenzie, 1959; Oades, 1963). So, the dilemma continues—warmer temperature or age.

RELICT SOILS OF THE PLAINS

The problem of the red color brings us to relict soils of the plains. A relict soil is a paleosol that formed on a preexisting landscape but which has not been buried by younger sediments (Ruhe, 1965). These soils emerge on the land surface from beneath the Wisconsin loess and other eolian deposits in southern Kansas, western Oklahoma, and Texas.

The Reddish Prairie Soils (Baldwin *et al.,* 1938) relate to a pre-Wisconsin land surface without a blanket of Wisconsin Loess, and appear to be the emergent Sangamon Soil (Thorp *et al.,* 1951). Consequently, the Reddish Prairie Soils, which were related to the north-south temperature stratification of the plains (Baldwin *et al.,* 1938; Kellogg, 1941), may owe their color to age rather than to present climatic effects (Thorp *et al.,* 1951).

The same reasoning can be applied to the Reddish Brown and Reddish Chestnut Soils (Baldwin *et al.,* 1938) of the plains (Ruhe, 1965). In fact, some of these soils have formed in Illinoian "Cover sands" and have been weathering since Sangamon time (Frye and Leonard, 1957b).

The dilemma of whether red color is due to warmer temperature or to age remains, even though the temperature factor may be preferred (Frye and Leonard, 1967).

SUMMARY

Soils formed under grassland on the land surface of Wisconsin deposits have thicker A-horizons and are leached deeper in the subhumid areas than in the semiarid areas of the plains. Accumulation of calcium carbonate in the lower part of the solum or in the C-horizon characterizes the more arid environment. Radiocarbon dating of a specific soil in New Mexico shows that carbonate did accumulate in the profile during the arid time following the last pluvial.

Based on a comparison of the distribution of properties of the Brady Soil and the surface soils and a comparison of fauna above and below the Brady soil-stratigraphic horizon, the zonality of current aridity and grassland began in Brady time in Kansas and Nebraska. Brady time was 11,400 to 9100 radiocarbon years ago. In Iowa the prairie environment began 7000 to 8000 years ago, as shown by pollen profiles that are related to buried organic soils in bogs on the younger drift plains.

Fauna and flora below the buried soils indicate cooler and relatively more moist conditions at those places and prior to those times to the beginning of Wisconsin time. Iowa, to its western and southern boundaries, was under coniferous forest. How far westward on to the Central Plains such forest existed is a question, but apparently boreal forest may have been in the Sand Hills of Nebraska 12,600 years ago. Perhaps pollen studies of the basal soil of Wisconsin Loess (Gilman Canyon Formation in Nebraska) could resolve this problem.

Late Sangamon and Sangamon paleosols in Iowa formed under forest. The Late-Sangamon Soil can be traced westward 60 to 70 miles beyond the Missouri River Valley. The Sangamon paleosols in southeastern Nebraska and northeastern Kansas also have forest influence, although they are 175 to 350 miles northwest and southwest of present humid forest and woodland zones. Westward on the plains the Sangamon Soils are more intensively developed than surface soils but have gross features similar to land-surface soils. The paleosol belts are displaced as much as 100 miles westward of the surface-soil belts, indicating a possible similar displacement of moisture belts.

The red-colored B-horizons of the pre-Wisconsin paleosols may indicate warmer climate during the Sangamon, but the factor of time cannot be excluded. Some of the reddish-colored surface soils on the plains in Kansas, Oklahoma, and Texas probably are relict from Sangamon time.

LITERATURE CITED

American Commission Stratigraphic Nomenclature
  1961. Code of stratigraphic nomenclature. Bull. Amer. Assoc. Petrol. Geol., 45:645-665.
Baldwin, M., C. E. Kellogg, and J. Thorp
  1938. Soil classification. Pp. 979-1001, *in* Soils and men, U.S.D.A. Yearbook Agric., 1232 pp.
Barnes, C. P.
  1948. Environment of natural grassland. Pp. 45-49, *in* Grass, U.S.D.A. Yearbook Agric., xiv+892 pp.
Bidwell, O. W.
  1956. Major soils of Kansas. Kansas Agric. Exp. Sta. Circ., 336:1-16.
Bidwell, O. W., D. A. Gier, and J. E. Cipra
  1968. Ferromanganese pedotubules on roots of *Bromus inermis* and *Andropogon gerardii*. Trans. IX Internat. Cong. Soil Sci., Adelaide, 4:683-692.
Blumenstock, D. I., and C. W. Thornthwaite
  1941. Climate and the world pattern. Pp. 98-127, *in* Climate and man, U.S.D.A. Yearbook Agric., xii+1248 pp.
Brush, G. S.
  1967. Pollen analyses of late-glacial and postglacial sediments in Iowa. Pp. 99-115, *in* Quaternary paleoecology (E. J. Cushing and H. E. Wright, Jr., eds.), Yale Univ. Press, New Haven, Connecticut, vii+433 pp.
Campbell, C. A., E. A. Paul, D. A. Rennie, and K. J. McCallum

1967a. Factors affecting the accuracy of the carbon-dating method in soil humus studies. Soil Sci., 104:81-85.

1967b. Applicability of the carbon-dating method of analysis to soil humus studies. Soil Sci., 104:217-224.

Clisby, K. H., and P. B. Sears
1956. San Augustin Plains—Pleistocene climatic changes. Science, 124:537-539.

Daniels, R. B., and R. H. Jordan
1966. Physiographic history and the soils, entrenched stream systems, and gullies, Harrison County, Iowa. U.S.D.A. Tech. Bull., 1348:1-116.

Daniels, R. B., R. L. Handy, and G. H. Simonson
1960. Dark colored bands in the thick loess of western Iowa. Jour. Geol., 68:450-458.

Elder, J. A.
1964. Nebraska soils. Nebraska Conserv. Bull., 36:1-34.

Farnham, R. S.
1960. Organic soils. Pp. 34-36, in Soils of the north central region of the United States. Wisconsin Agric. Exp. Sta. Bull. (North Central Reg. Publ. 76), 544:1-192.

Frye, J. C., and A. B. Leonard
1952. Pleistocene geology of Kansas. Bull. Kansas Geol. Surv., 99:1-230.

1957a. Ecological interpretations of Pliocene and Pleistocene stratigraphy in the Great Plains region. Amer. Jour. Sci., 255:1-11.

1957b. Studies of Cenozoic geology along eastern margin of Texas High Plains, Armstrong to Howard counties. Texas Bur. Econ. Geol. Rep. Inv., 32:1-62.

1963. Pleistocene geology of Red River Basin in Texas. Texas Bur. Econ. Geol. Rep. Inv., 49:1-48.

1964. Relation of Ogallala Formation to the Southern High Plains in Texas. Texas Bur. Econ. Geol. Rep. Inv., 51:1-25.

1965. Quaternary of the Southern Great Plains. Pp. 203-216, in The Quaternary of the United States (H. E. Wright, Jr., and D. G. Frey, eds.), Princeton Univ. Press, Princeton, New Jersey, x+922 pp.

1967. Buried soils, fossil mollusks, and Late Cenozoic paleoenvironments. Pp. 429-444, in Essays in paleontology and stratigraphy (C. Teichert and E. L. Yochelson, eds.), Spec. Publ. Univ. Kansas Dept. Geol., 2:1-626.

Frye, J. C., and H. B. Willman
1960. Classification of the Wisconsinan stage in the Lake Michigan glacial lobe. Illinois Geol. Surv. Circ., 285:1-16.

Gile, L. H., F. F. Peterson, and R. B. Grossman
1965. The K horizon: a master soil horizon of carbonate accumulation. Soil Sci., 99: 74-82.

Graham, B. F.
1962. A post-Kansan peat at Grinnell, Iowa: a preliminary report. Proc. Iowa Acad. Sci., 69:39-44.

Harbour, J.
1958. Microstratigraphic and sedimentological studies of early man site near Lucy, New Mexico. Unpubl. M.S. thesis, Univ. New Mexico, Albuquerque.

Kellogg, C. E.
1941. Climate and soil. Pp. 265-291, in Climate and man, U.S.D.A. Yearbook Agric., xii+1248 pp.

Leopold, L. B.
1951. Pleistocene climate in New Mexico. Amer. Jour. Sci., 249:152-167.

Mackenzie, R. C.
1959. The ageing of sesquioxide gels. I. Iron oxide gels. Mineral. Mag., 32:153-165.

Marbut, C. F.
1935. Soils of the United States. In Atlas of American agriculture, pt. 3, U.S.D.A. Bur. Chem. and Soils, 98 pp.

Millet, J. L., and J. V. Drew
1963. Characterization and genesis of Pawnee and Adair soils in southeastern Nebraska. Proc. Soil. Sci. Soc. Amer., 27:683-688.

N.C.R.-3 Tech. Comm. Soil Survey
1960. Soils of the north central region of the United States. Wisconsin Agric. Exp. Sta. Bull. (North Central Reg. Publ. 76), 544: 1-192.

Oades, J. M.
1963. The nature and distribution of iron compounds in soils. Soils and Fertilizers, 26:69-80.

Oschwald, W. R., F. F. Riecken, R. I. Dideriksen, W. H. Scholtes, and F. W. Schaller
1965. Principal soils of Iowa. Iowa State Univ. Ext. Serv. Spec. Rep., 42:1-76.

Potzger, J. E., and B. C. Tharp
1954. Pollen study of two bogs in Texas. Ecology, 35:462-466.

Prill, R. C., and F. F. Riecken
1958. Variations in forest-derived soils formed from Kansan till in southern and southeastern Iowa. Proc. Soil Sci. Soc. Amer., 22:70-75.

Reed, E. C., and V. H. Dreeszen
1965. Revision of the classification of the Pleistocene deposits of Nebraska. Bull. Nebraska Geol. Surv., 23:1-65.

Richmond, G. M., and J. C. Frye
1957. Note 19: status of soils in stratigraphic nomenclature. Bull. Amer. Assoc. Petrol. Geol., 41:758-763.

Rubin, M.
1960. Changes in Wisconsin glacial stage chronology by C14 dating. Trans. Amer. Geophy. Union, 41:288-289.

Rubin, M., R. C. Likens, and E. G. Berry
1963. On the validity of radiocarbon dates from snail shells. Jour. Geol., 71:84-89.

Rubin, M., and H. E. Suess
1956. U.S. Geological Survey radiocarbon dates III. Science, 123:442-448.

Ruhe, R. V.
  1956. Geomorphic surfaces and the nature of soils. Soil Sci., 82:441-455.
  1965. Quaternary paleopedology. Pp. 755-764, in The Quaternary of the United States (H. E. Wright, Jr., and D. G. Frey, eds.), Princeton Univ. Press, Princeton, New Jersey, x+922 pp.
  1967. Geomorphic surfaces and surficial deposits in southern New Mexico. Mem. New Mexico Bur. Mines, 18:1-66.
  1968. Identification of paleosols in loess deposits in the United States. Pp. 49-65, in Loess and related eolian deposits of the world (C. B. Schultz and J. C. Frye, eds.), Univ. Nebraska Press, Lincoln, 369 pp.
  1969. Quaternary landscapes in Iowa. Iowa State Univ. Press, Ames, 255 pp.

Ruhe, R. V., and R. B. Daniels
  1958. Soils, paleosols, and soil-horizon nomenclature. Proc. Soil Sci. Soc. Amer., 22:66-69.

Ruhe, R. V., R. B. Daniels, and J. G. Cady
  1967. Landscape evolution and soil formation in southeastern Iowa. U.S.D.A. Tech. Bull., 1349:1-242.

Ruhe, R. V., and W. H. Scholtes
  1956. Ages and development of soil landscapes in relation to climatic and vegetational changes in Iowa. Proc. Soil Sci. Soc. Amer., 20:264-273.

Scholtes, W. H., R. V. Ruhe, and F. F. Riecken
  1951. Use of the morphology of buried soil profiles in the Pleistocene of Iowa. Proc. Iowa Acad. Sci., 58:295-306.

Schultz, C. B., and T. M. Stout
  1945. Pleistocene loess deposits of Nebraska. Amer. Jour. Sci., 243:231-244.
  1948. Pleistocene mammals and terraces in the Great Plains. Bull. Geol. Soc. Amer., 59:533-588.

Schwertmann, U.
  1959a. Mineralogische und chemische Untersuchungen an Eisenoxyden in Böden und Sedimenten. Neues Jb. Miner., Abh., 93:67-86.
  1959b. Die frakionierte Extrakion der Freien Eisenoxyde in Böden, ihre mineralogischen Formen und ihre Entstehungsweisen: Zeits. Pflanzenernährung, Düngung, Bodenkünde, 84:194-204.

Sears, P. B.
  1961. A pollen profile from the grassland province. Science, 134:2038-2040.

Simonson, R. W.
  1941. Studies of buried soils formed from till in Iowa. Proc. Soil Sci. Soc. Amer., 6:373-381.
  1954. Indentification and interpretation of buried soils. Amer. Jour. Sci., 252:705-732.

Soileau, J. M., and R. J. McCracken
  1967. Free iron and coloration in certain well-drained coastal plain soils in relation to their other properties and classification. Proc. Soil Sci. Soc. Amer., 31:248-255.

Thornthwaite, C. W.
  1941. Atlas of climatic types in the United States. U.S.D.A. Misc. Publ., 421:1-7.

Thorp, J., W. M. Johnson, and E. C. Reed
  1951. Some post-Pliocene buried soils of central United States. Jour. Soil Sci., 2:1-19.

U.S. Department of Agriculture, Soil Conservation Service
  1967. Patterns of soil orders and suborders of the United States (map).

U.S. Department of Agriculture, Soil Survey Staff
  1951. Soil survey manual. U.S.D.A. Handbook 18, 503 pp. (suppl. 1962, pp. 173-188).
  1960. Soil classification, a comprehensive system. U.S.D.A. Soil Conserv. Serv., 265 pp. (suppl. 1967, 207 pp.).

Walker, P. H.
  1966. Postglacial environments in relation to landscape and soils on the Cary drift, Iowa. Iowa Agric. Exp. Sta., Res. Bull., 549:838-875.

Watts, W. A., and H. E. Wright, Jr.
  1966. Late-Wisconsin pollen and seed analysis from the Nebraska Sandhills. Ecology, 47:202-210.

The Character of Late-Glacial and Post-Glacial Climatic Changes

REID A. BRYSON, DAVID A. BAERREIS, AND WAYNE M. WENDLAND

ABSTRACT

Computer analysis of world-wide radiocarbon dates for significant climatically related events of the last 10,000 years yields an objective consensus that these events occurred at preferred times. These dates match recent datings of the breaks in the Blytt-Sernander sequence, and are also applicable to North America.

Estimates of the time duration of the transition between the Blytt-Sernander episodes, which can be equated to climatic episodes, suggest that significant ecological effects can occur in just a few decades. This is perhaps related to the nonlinear behavior of the atmosphere, and to the nonlinear response of biota to climate.

These results confirm the suggestion that the concept of a gradual anathermal-hypsithermal-medithermal trend is oversimplified and should be replaced.

The response of biota to climatic change is much faster than the response of glaciers. Care must be exercised in comparing evidence of such diverse character. It is suggested that some ecological changes might be directly caused by atmospheric changes of global scope, but that others of a regional nature may be due to "feedback" mechanism, or delayed effects of earlier causes. An example is the change of climate on the Great Plains that was associated with the opening of a low corridor to the Arctic as the Laurentide ice sheet retreated in response to the change from glacial to post-glacial climate. This probably occurred on the order of 2000 years after the basic global climatic change.

The evidence points to a steplike climatic variation and ecological response. There are modern analogues to many of the post-glacial climatic patterns. It is suggested (1) that the Blytt-Sernander terminology be adopted for North America, because the sequence in time is comparable, and (2) that interdisciplinary research efforts might be fruitfully concentrated on the characterization of the climatic episodes, using more sophisticated descriptions than "warm," "dry," "cool," and so forth.

INTRODUCTION

Ten volumes of *Radiocarbon* (1959-1968) now provide nearly 10,000 dates from the Late Pleistocene and Holocene. This is a sufficiently large body of data that one may begin treating certain chronological problems statistically. Two examples of analyses that may be performed on bulk radiocarbon data after stratification will be used in this paper.

Many of the descriptions of dated material found in the pages of *Radiocarbon* include such terms as "significant break in rate of peat accumulation," or "highest sea level," or variants of these. Many scholars from many countries are represented in the list of those who have carefully selected the samples that bear such descriptions. From the body of dates selected by the original investigators as environmentally significant, it is possible to obtain an objective consensus.

To do this, we perused the entire list of dates, rejecting all but those thought to be significant by the person who wrote the sample description. We rejected dates on archaeological materials, for they are mostly selected for the cultural rather than directly environmental context. We rejected dates aimed at establishing correspondence of radiocarbon and calendar dates, and those simply establishing time points along stratigraphic sequences unless identified as suggesting a significant environmental change. We rejected those with cryptic descriptions (unfortunately, a large number). We collected together those marking recurrence surfaces, stratigraphic breaks, sea level maxima and minima, glacial maxima or minima, and taxon or species maxima—in short, those dates indicating discontinuities in the basic variable or its derivative that were thought to be sufficiently significant to justify the considerable expense of a radiocarbon assay. This reduced the number of dates used to 620, representing the work of scores of scientists.

The frequency with which these dates fell within each two centuries of the last 10,000 years was counted and subjected to a filter of the form $N_i = 0.25 N_{i-1} + 0.50 N_i + 0.25 N_{i+1}$, where N_i is the number of significant dates that fell in the ith interval. "Normal distributions" were fitted simultaneously to the major peaks of frequency (Fig. 1) by the least-squares method (Johnson, 1966). The time series of significant date frequencies also was combined into 100-, 300-, and 400-year class intervals. The coarser intervals expectedly show a smaller number of peaks, but apparently all peaks that we regard as identifying significant dates tend to converge on about seven major times of discontinuity (Table 1). It is believed significant that the absolute numbers of dates collected as representing many of these individual times are about equal. Allowing for interlaboratory differences, sampling errors, inherent statistical scatter, and the difficulty of collecting samples from exactly on a discontinuity, one should only expect the dates to approximate, or converge on, the most likely date.

TABLE 1. Modal and Median Dates of Significant Environmental Changes

Years BP selected "by eye" Class interval used (years)				Median date BP from least-square computer fit of normal distributions to actual radiocarbon ages (only those distributions representing more than 5 percent of total)
100	200	300	400	
250				
500				
750	800			760
1250	1200	1150		
1650	1600	1750		1690
2050			1900	
2450				
2750	2800	2650	2700	2890
3150	3200	3250		
3550	3600			
3750				
4100	4200			
4550				
4750	4800	4750	4700	4680
4950				
5150				
5450	5300	5350		
5950	6000	5950		5980
6300				
6550				
6950				
7150	7200	7150	7100	
7750	7800	7750	7900	
8150	8200			
8550	8600	8650		8450
8850				
9100	9200	9250	9100	9140
9450				
9850				

Table 1 thus represents an objective consensus of the times at which major environmental changes occurred. These dates have been recognized as those dividing the episodes of the well-known Blytt-Sernander sequence in Europe (see Nilsson, 1964). They and some of the minor peak dates are also the dates that appear most frequently as marking significant environmental changes in North American studies (Nichols, 1967a; Bryson and Wendland, 1967b).

THE POST-GLACIAL CLIMATIC SEQUENCE

The model of the post-glacial climatic sequence postulated by Antevs (1948, 1952, 1955) and often used by North American scientists consists pri-

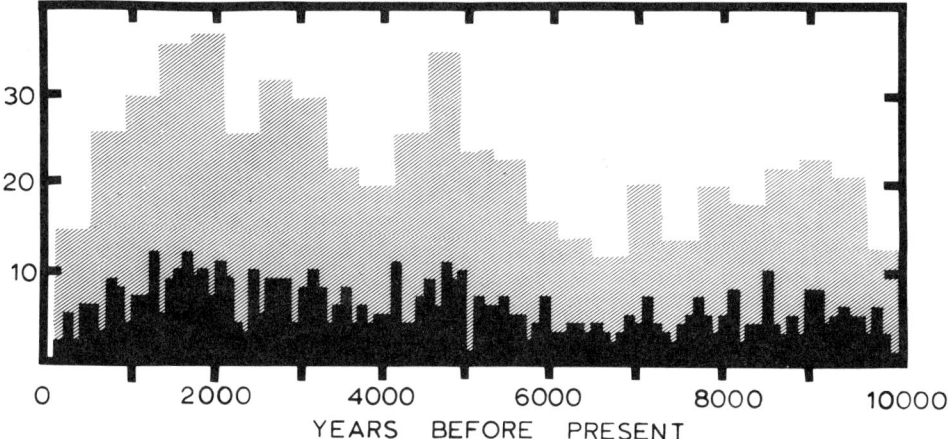

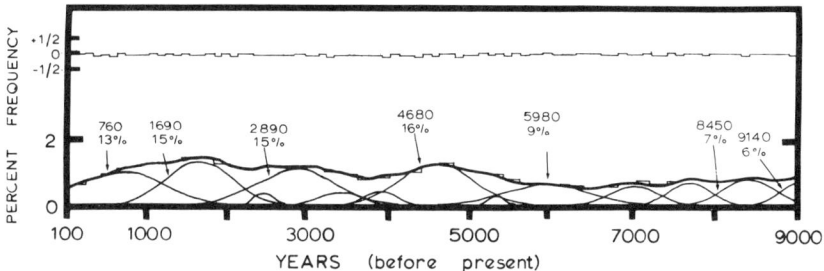

FIG. 1. Upper: Frequency histogram of carbon-14 dates by 100-year and 400-year class intervals. Lower: Least-squares estimates of the partial collective parameters for histogram of carbon-14 dates.

marily of temperatures slowly rising to a broad, flattened peak about 7000-5000 BP and then slowly declining to the present. It was further suggested that the "altithermal" peak was accompanied by a long drought or at least maximum aridity (Antevs, 1955, pp. 328-329; see also Malde, 1964). At a time when absolute dating was not possible, the concept of a post-glacial "climatic optimum" with later periods tending back towards glacial conditions was useful. Indeed, this may be all that is indicated if the last 10 millennia are represented by half a dozen samples, as they often were on early pollen diagrams.

There are major difficulties in the application of the concept on a more sophisticated level, however. The atmosphere simply does not behave in such a way that the world, or even a continent, gets everywhere warmer or drier or colder or wetter as the climate changes (Sawyer, 1966), even though the world mean may change that way. One does not find temperature and precipitation anomalies everywhere of the same sign, nor is the temperature anomaly highly correlated with the precipitation anomaly. Even during the smoothly varying change of solar radiation intensity from winter through summer and into autumn, the climatic response to this regular variation of the "forcing function" is irregular in space and time and quite nonlinear. The march of the seasons in Arizona is not like that in New Mexico, and certainly not like that in Nebraska. When the jet stream shifts southward in the autumn, the rains increase in California but decrease in eastern Colorado; and when a strong ridge forms over the Canadian Rockies, British Columbia may get warmer and Saskatchewan colder. There is abundant evidence that the nonlinearities of the atmosphere are such that even in response to a slowly and steadily varying external cause, the response of the atmosphere will be to shift through a series of quasi-stable states with rapid transitions between states (Fultz, 1959; Bryson and Lahey, 1958; Wahl, 1953). Even erratic or discontinuous behavior of the forcing function cannot change the character of the response to smooth, simple, and universal.

The problem of the paleoclimatologist is thus complicated by the complexity of the pattern at each given time, but simplified by the existence of a finite number of quasi-steady states. If each of these periods can be characterized by a dominant pattern and a time interval for which the pattern is valid, we shall have progressed far in the interpretation of past climates. This paper is concerned primarily with the time of transition between quasi-steady patterns and the rapidity of the transitions. Secondary attention will be paid to the use of physical principles and modern analogues to aid in the characterization of the patterns of past climates. Even though temperature and precipitation anomaly patterns are complex, as are the biological and geomorphic patterns from which they are most often inferred, the physical continuity of the atmosphere is a powerful aid, as is the immobility of topography on a multimillenia scale.

Putting aside the Antevs model as having served its purpose but being no longer sufficiently sophisticated, we may ask when the step-transitions between climatic episodes occurred. The most attractive explanation of world-wide correlative environmental changes is climatic change, unless one wishes to postulate major meteor impacts, instability of the earth's axis of rotation, or some other cosmic factor. The evidence, in the opinion of the authors, strongly points to the key dates of Table 1 as the times of climatic change from one quasi-stable pattern to another. We suggest that the time is appropriate to name the climatic episodes between the key dates, or at least to identify the periods between these dates in the same way that geological times are named. Since climate is a global phenomenon and growing evidence suggests that the major episodes of quasi-stable climate have been identified, it appears appropriate to adopt a common world-wide terminology for the subdivisions of post-glacial time. Rather than confuse the issue, we suggest adoption of the Blytt-Sernander nomenclature as the current climatic or ecologic characterization of the episodes (Table 2).

The adoption of the suggested division of post-glacial time does not require one to force biologic and geomorphic evidence to fit these dates. The encroachment of vegetation into the formerly glaciated area of Canada took all of post-glacial time, so that dates of successional changes are dependent on when the terrain became available (Fig. 2). However, changes in biotic community structure or geomorphology initiated by climatic change most frequently started around

TABLE 2. Subdivision of Post-Glacial Time and Tentative Division Dates

Episode	Tentative date BP	Subepisode division dates	
(Late Glacial)*			
	ca. 10,500		
Pre-Boreal			
	9650		
Boreal		9140	
	8450		
Atlantic		7730	5980
	4680		
Sub-Boreal		3970	3480
	2890		
Sub-Atlantic			
	1690		
(later episodes or subepisodes)**		760	

*Glacial chronology and terminology is not considered here.
** A nomenclature of post-Sub-Atlantic time has been suggested elsewhere by Baerreis and Bryson (1965).

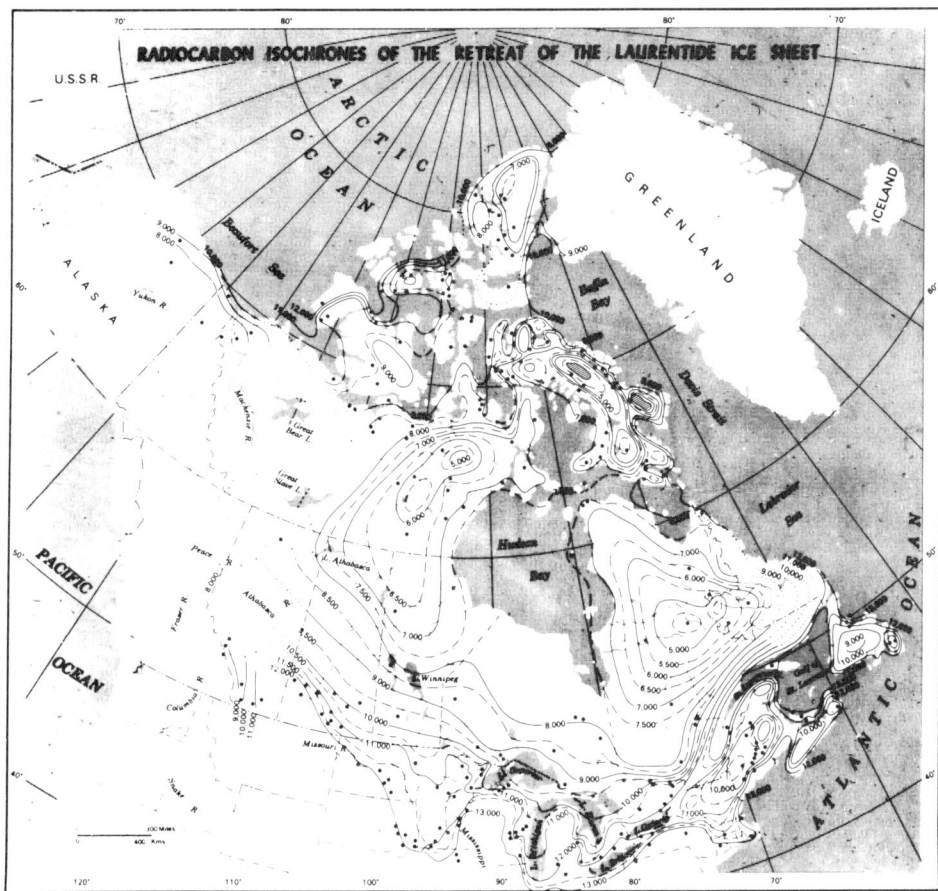

Fig. 2. Radiocarbon isochrones of the retreat of the Laurentide ice sheet (Bryson and Wendland, 1967a).

the times given in Tables 1 and 2, if our objective consensus is correct.

Can we now characterize the postulated succession of post-glacial climatic episodes? Unfortunately, we cannot describe them in detail, for as pointed out above, no climatic anomaly can be described (other than locally) with such simple terms as warmer, wetter, and so forth. It is the hope of paleoecologists and paleoclimatologists that some day a sufficiently complete description of past climates may be possible. It is clear that verbal descriptions will be inadequate and that maps will be necessary to condense the complex patterns of climate into comprehensible form. Some tentative sketch maps are given in Figures 3, 4, and 5, which hopefully can be elaborated some day with quantitative data and corrections. These were obtained by using the relationship of modern mean patterns of airstreams and frontal boundaries to the modern distribution of biota, as shown in Figures 6 and 7 (Bryson, 1966), and by reconstructing the patterns that would fit both fossil data and climatic continuity.

Comparison of these sketch maps with the modern mean pattern of air masses and fronts suggests that a grassland peninsula (see Fig. 3) was almost restricted to western Kansas and eastern Colorado in late-glacial times. A modern analogue for the climate is the January of 1963. During Atlantic time the wedge of Pacific air that characterizes the grassland climate was ex-

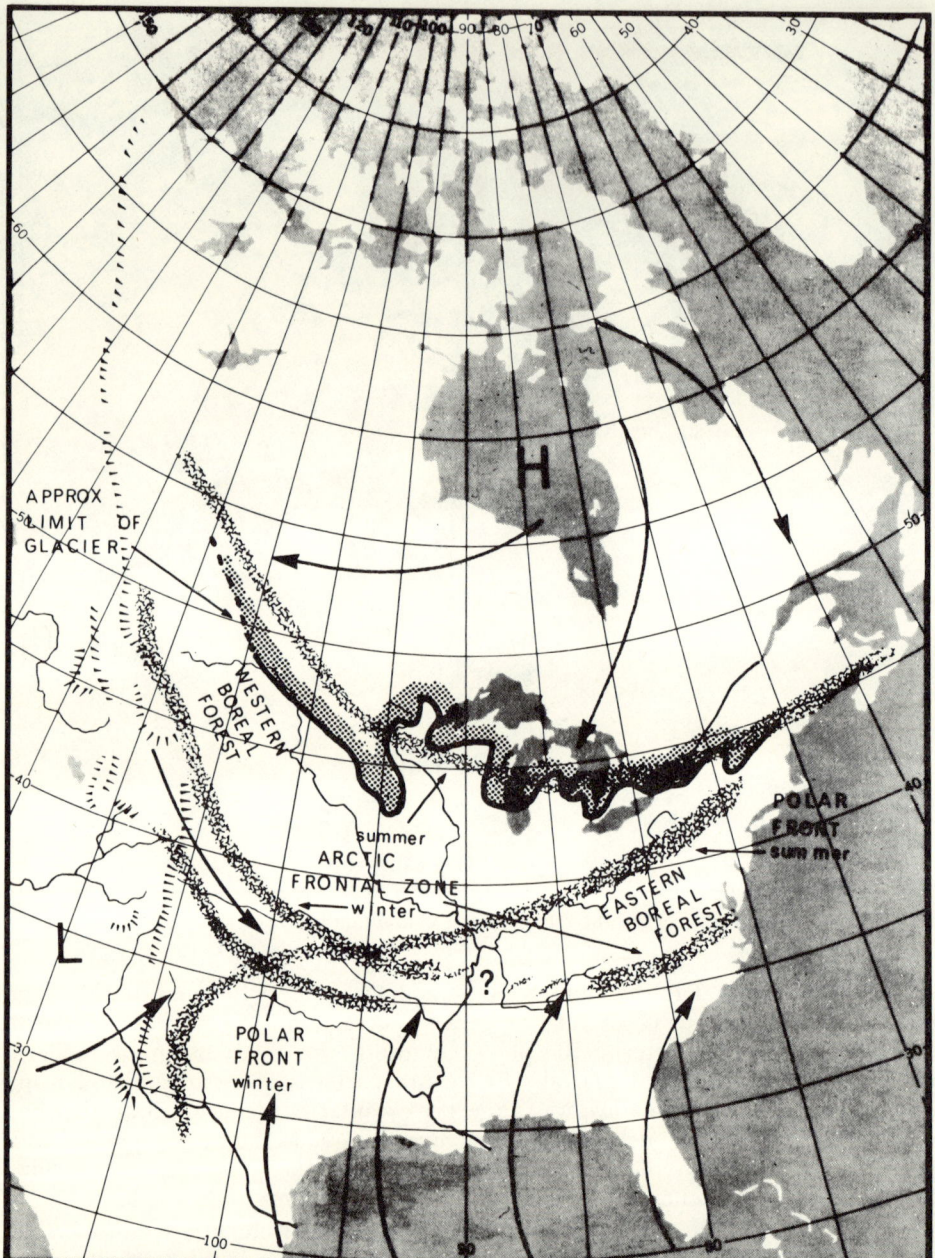

Fig. 3. Partial reconstruction of mean frontal zones during late-glacial time (13,000 to 10,000 years ago (Bryson and Wendland, 1967b).

panded northeastward into central Minnesota and eastward towards the Atlantic seaboard, but modern correlation of biota and climate would suggest that grassland probably did not extend beyond western Pennsylvania (see also Wright, 1968). In early Sub-Boreal time it appears that the grassland had contracted westward and the prairie peninsula had become a stub in Iowa.

In late Sub-Boreal time the evidence suggests that a strong ridge of high pressure developed in summer over the Canadian Rockies. Such an at-

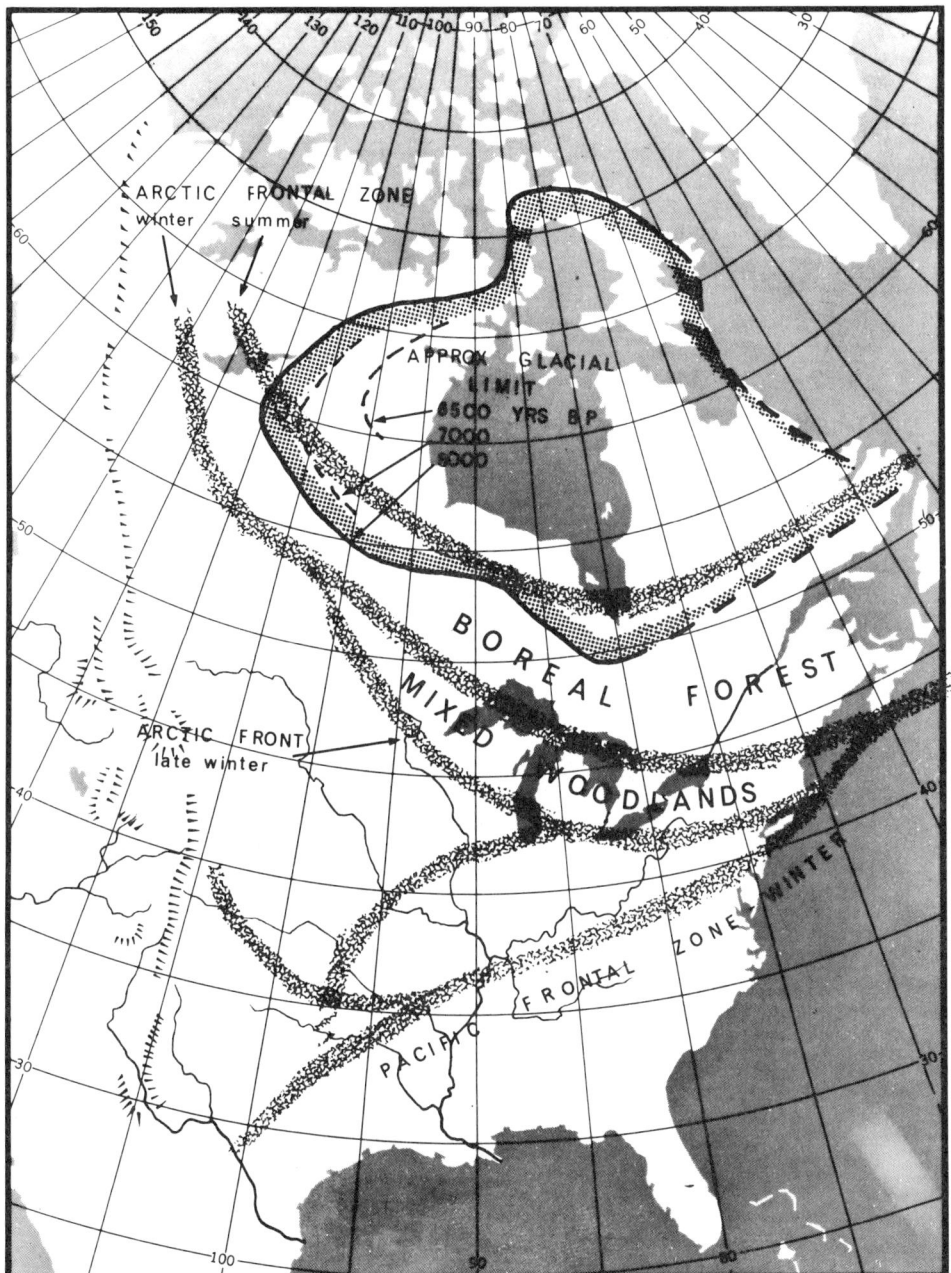

Fig. 4. Partial reconstruction of mean frontal zones during Cockburn-Cochrane time (*ca.* 8000 years ago) (Bryson and Wendland, 1967b).

mospheric pattern should have shifted West Coast climates northward, and this is indicated by the use of acorns by the Indians of British Columbia starting about 3500 BP (C. Borden, personal communication). By contrast there should have been a stronger flow from the Arctic in central Canada and displacement of the climata and biota southward. This is verified by paleosol and macrofossil evidence from Keewatin (Bryson *et al.*, 1965; Nichols, 1967b) and pollen evidence for change from grassland towards forest at Peace

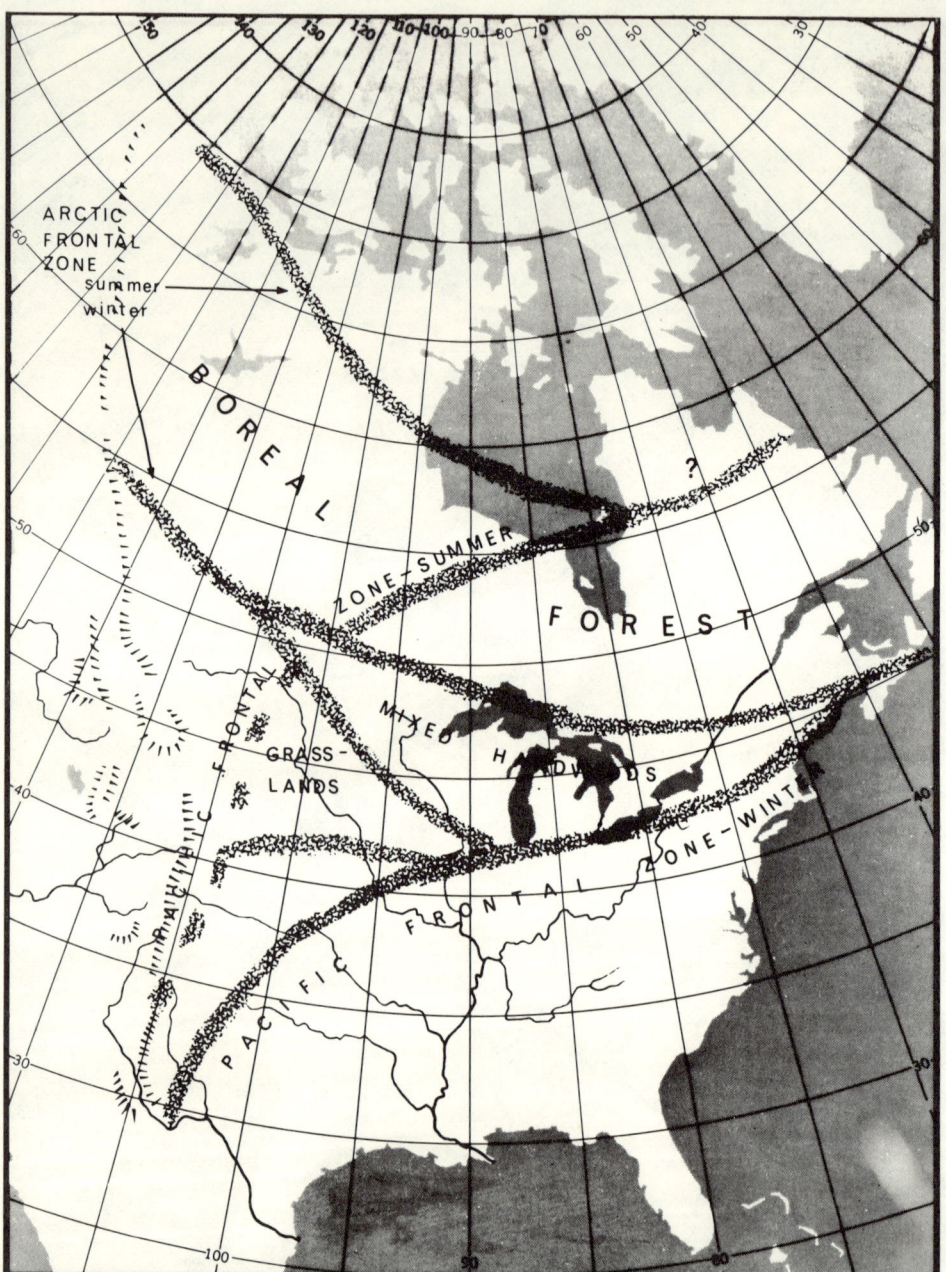

Fig. 5. Partial reconstruction of mean frontal zones during early Sub-Boreal time (*ca.* 5000 to 3500 years ago). Summer position of the Arctic front is the most certain feature of this time (Bryson and Wendland, 1967b).

River, Alberta (H. Nichols). If the more meridional motion over the Rockies persisted throughout the year, chinooks east of the Rockies should have been less frequent and temperatures on the High Plains lower, thus lowering the lower tree line on the east slopes of the Cordillera. The summer of 1967 might be a modern analogue.

Since past climates differ from present climates more in quantity than in kind, modern analogues and analy-

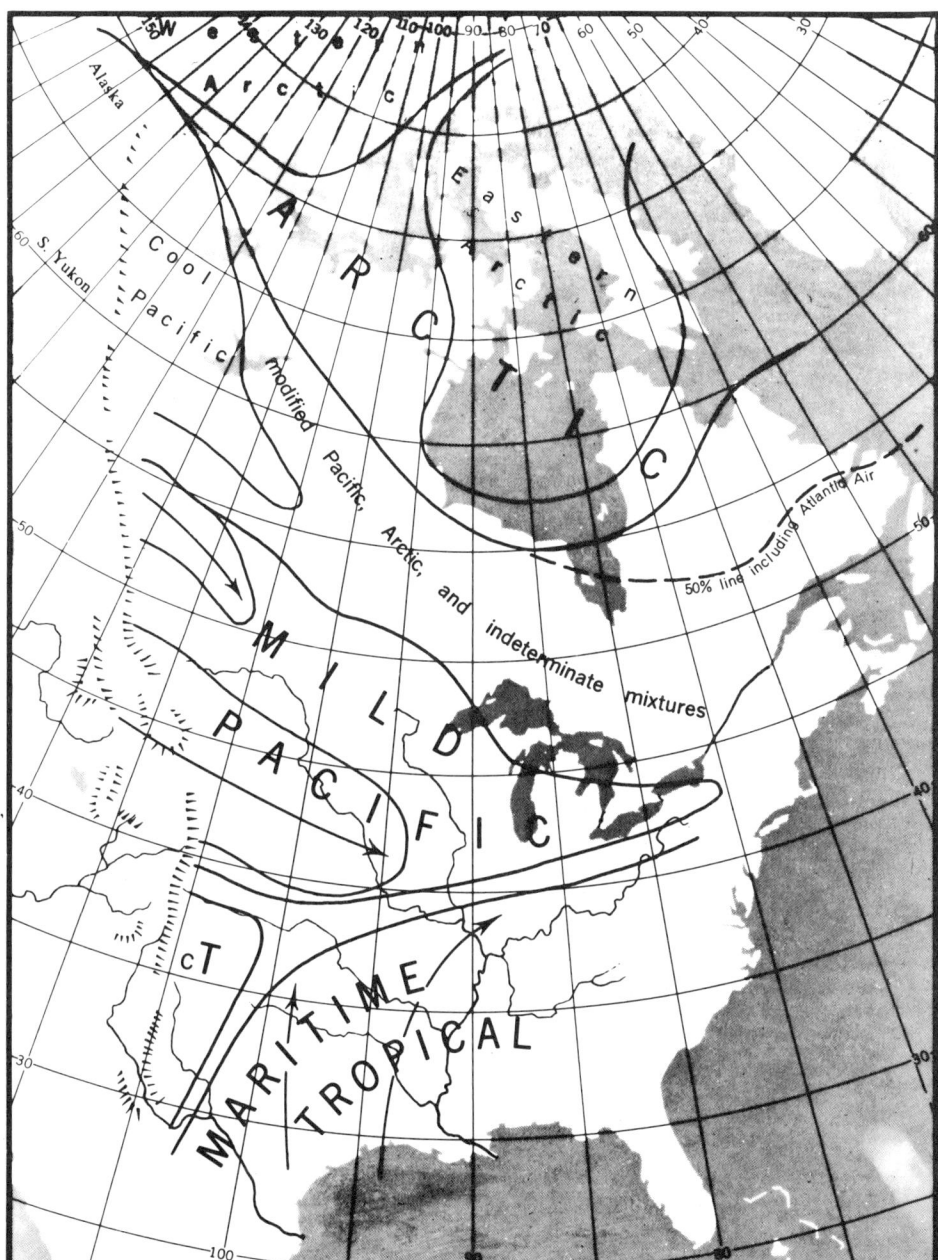

Fig. 6. Composite chart of regions dominated by the various air-mass types at the present time. The area within the lines of demarcation are occupied more than 50 percent of the time by the indicated air mass (Bryson, 1966).

sis provide a powerful framework for typing together isolated bits of field data (see Blasing, 1968). We believe that a combination of modern climatic analysis with a collation of field data from the past will "flesh out" the "bare bones" of climatic frameworks such as those suggested above. (For an example of the use of modern climatic data to estimate patterns of climatic change, see the summary chapter of Baerreis and Bryson, 1967.)

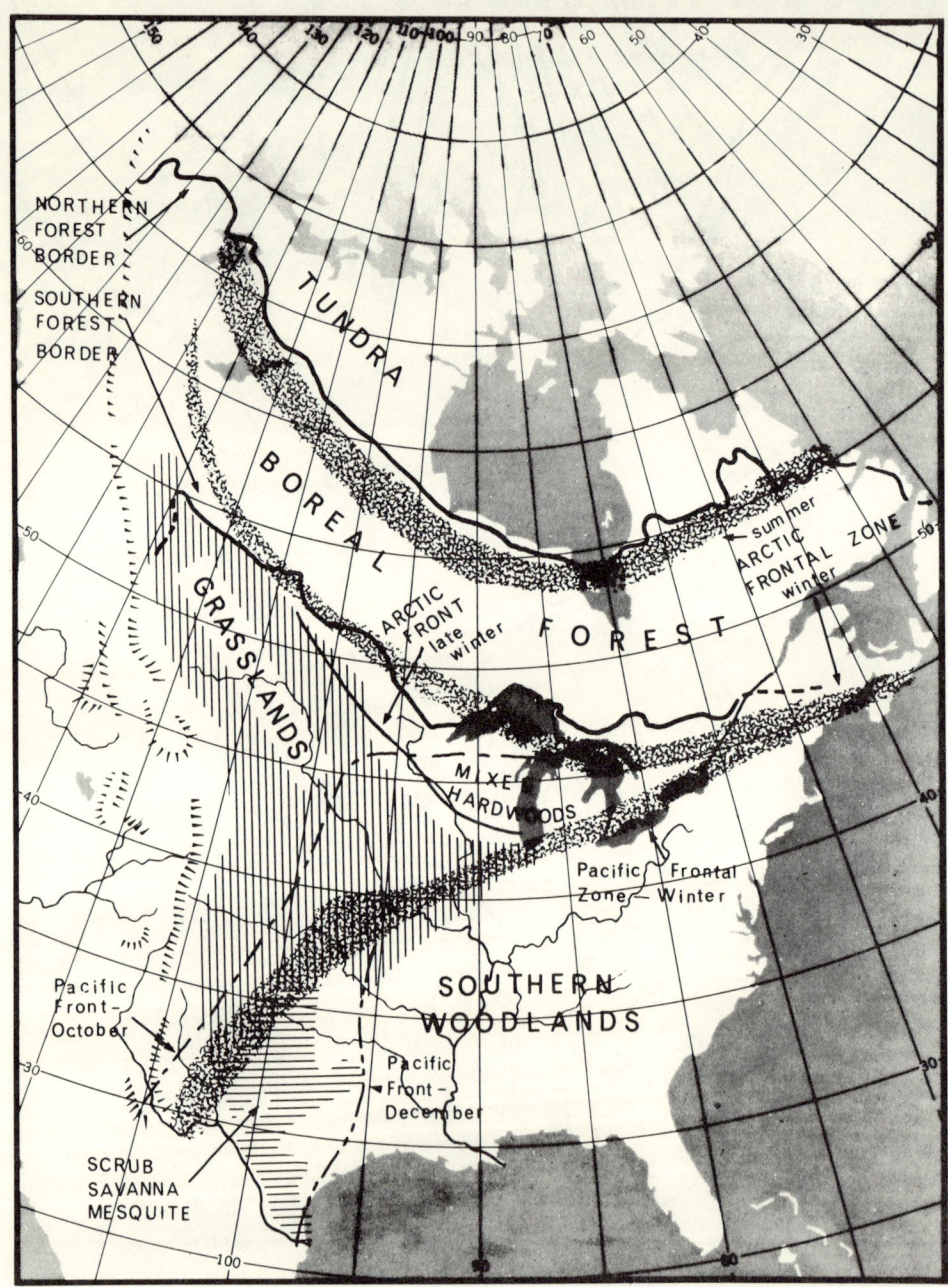

Fig. 7. Present day coincidence of biotic regions with meteorologically defined climatic regions. Climatic regions taken from Bryson (1966); "grasslands" and "scrub savanna-mesquite" from Küchler (1964).

THE MAGNITUDE AND RAPIDITY OF CLIMATIC CHANGES

If climate does indeed change in a steplike fashion, then the duration of the transition from one quasi-stable climatic state to another should be appreciably shorter than the length of a climatic episode. Furthermore, unless the mean state of the climate changes by a significant amount during these transitions, research along the lines discussed above is an exercise in futility. (It is clearly recognized here that "mean state" and "significant" ultimately require careful definition and qualification.) For our present purposes it is perhaps sufficient to say that the duration of the transitions should, in general, be less than one-third the length of the episodes they separate. Inasmuch as the minor intervals in Table 1 average somewhat over 700 years, and the major intervals about 1500 years, transitions should take less than 230 years between sub-episodes and less than 500 years between main post-glacial episodes. This would mean that stable climates would last twice as long as transitions between climates. The definition of a significant climatic change is more complex, but to the paleoecologist it should at least result in an observable ecological change, and perhaps also in a cultural change among the human inhabitants of the area. The problem is complicated by the fact that even global climatic pattern changes, as long as they produce both positive and negative anomalies, must also leave some areas with no change.

Referring back to Table 1, which is a list of dates marking significant changes in the environment by the consensus of scholars, we might ask whether they show any evidence of rapidity of change. If one realizes that there is no standardization in this stratigraphic sampling, then radiocarbon dates should scatter around the true date of the causal climatic change. In fact, some dates will be unrelated to climate, but instead will depend on other local, biological, or cultural causes. Some palynologists appear to attach significance to maxima of the pollen-vector function as times of significant change, while others use the times of most rapid change, that is, the maxima of the time derivative of the function. Since recurrence surfaces may be time-transgressive and samples may be collected just below, crossing, or just above, the "true" horizon, a considerable scatter around the time at which the initiating cause of the renewed peat growth occurred should be found. This source of scatter should be unrelated to the actual duration of the climatic transition, but should simply make the climatic transition look less abrupt than it actually was. Yet the standard deviation of the dates around the median values given in Table 1 ranges from about 240 years to 530 years!

Table 3, in addition to listing the standard deviations around each of the median significant dates, also lists the percentage of all dates used that appear to refer to that central time. This should be a measure in part of the obviousness or importance *environmentally* of each central date. From these percentages it appears that 9140, 8450, 5980, 4680, 2890, 1690, and 760 BP represent major environmental events.

TABLE 3. Environmentally Significant Dates of Change, Standard Deviation, and Frequency of Occurrence in Radiocarbon (in percentage of dates identified as significant environmentally)

Date BP	Episode	δ (Years)	Percentage
	Pre-Boreal		
9650	— — — —	240	3
	Boreal I		
9140		280	6
	Boreal II		
8450	— — — —	320	7
	Atlantic I		
7730		260	5
	Atlantic II		
7050		300	5
	Atlantic III		
5980		530	9
	Atlantic IV		
4680	— — — —	490	16
	Sub-Boreal		
2890	— — — —	510	15
	Sub-Atlantic		
1690	— — — —	410	15
760		470	13

THE CLIMATIC CHANGE OF THE 12TH CENTURY IN THE PLAINS AREA

An event late enough in history to be documented with climatic data, and for which we have ecological and cultural evidence, is the change that occurred in the 12th century A.D. Lamb (1966) has ably documented the climatic change of that time for Europe. In essence, the climate of Western Europe became milder in winter, with good summers and uniformity for some distance inland, starting about A.D. 1160 (calendar date). Lamb (1966, p. 99) attributed this to the inland sweep of the westerlies, and it is easily recognizable as the "Westwetter" of the German meteorologists. Operating on the assumption that the circulation of the atmosphere must be globally consistent, we postulated a period of increased westerlies across North America starting at the same time. Strong westerlies in North America do not have the same effect as in Europe, for the western cordillera changes the character of the air that penetrates into the interior. As Borchert (1950) has shown, the prairie peninsula is occupied by a wedge of air, dried by subsidence on crossing the Rockies, which is driven far eastward by the westerlies. The stronger the westerlies, the farther east the dry wedge should push, and with it the associated biota.

Assuming further that modern data provide analogues for the interpretation of the past, we compiled Figure 8, which shows the change in July rainfall that one might expect with an increase in the strength of the hemispheric westerlies equal to the difference between the below-normal westerlies and the above-normal westerlies of recent years.

We selected the prehistoric Mill Creek culture of western Iowa as a research station because the economy involved agriculture as well as hunting and because the settlement pattern resulted in the development of large, thick middens in which changes in the culture and associated faunal and floral resources could be anticipated. Fur-

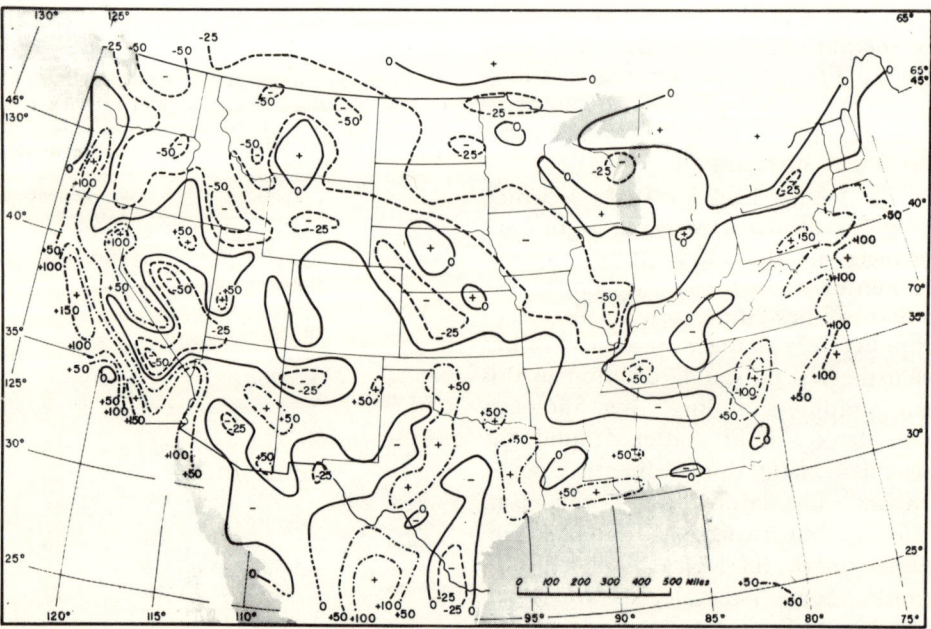

FIG. 8. July precipitation with high zonal index minus precipitation with low zonal index (after Baerreis and Bryson, 1967).

ther, available dates indicated that the occupation bracketed A.D. 1200. The map (Fig. 8) shows that an increase in the strength of the westerlies should have produced a 30 to 50 percent decrease in summer precipitation, enough to markedly affect the ecology and economy of a culture in such a marginal region. If our hypothesis was correct, we reasoned that not only should we find evidence of drought putting a stress upon the agricultural economy, but since it would modify the local vegetation, it should also have an impact upon the game hunted. Excavations in several sites of the Mill Creek culture carried out in the summer of 1963 confirmed these predictions and provided data for estimating rates of ecological response to climatic change and the impact of such a change upon man (Baerreis and Bryson, 1967).

Of the three major sites in which strata pits were excavated, the clearest stratigraphy, chronological placement, and ecological record came from the Phipps site (13CK21). The village was first occupied about A.D. 900 and abandoned about A.D. 1400. It was located on the flood plain of Mill Creek, in the midst of the village cornfields. The pollen spectra from the 10th and 11th centuries suggest prairie openings in the bottoms, with willows and elms in the wetter places, oaks along the terraces and better drained areas, and grassland on the uplands. This would compare rather favorably with the pattern at the present time if one substitutes "cornfield" for "prairie" in the description. Charcoal samples indicate ash, walnut, ironwood, hickory, basswood, maple, and probably cottonwood as also present. In the woods of the bottoms (and probably in the Indian cornfields) deer and elk were hunted, and provided 88 percent of the large-game meat for the Indians who lived in the villages. Bison provided only 12 percent of the meat, despite its larger size.

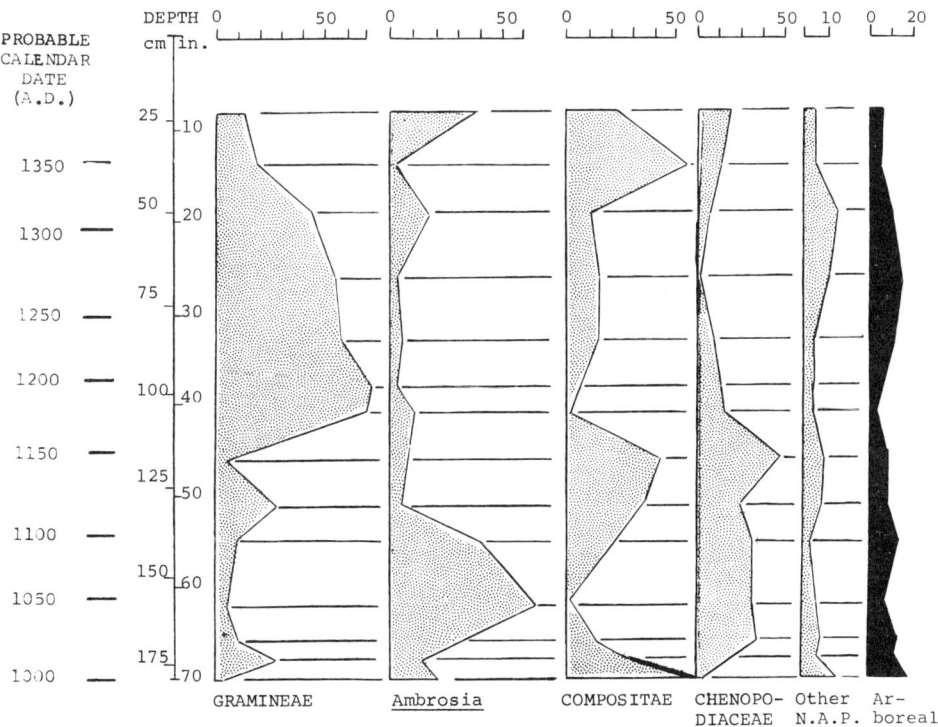

Fig. 9. Nonarboreal pollen diagram from Phipps site (13CK21) in northwestern Iowa.

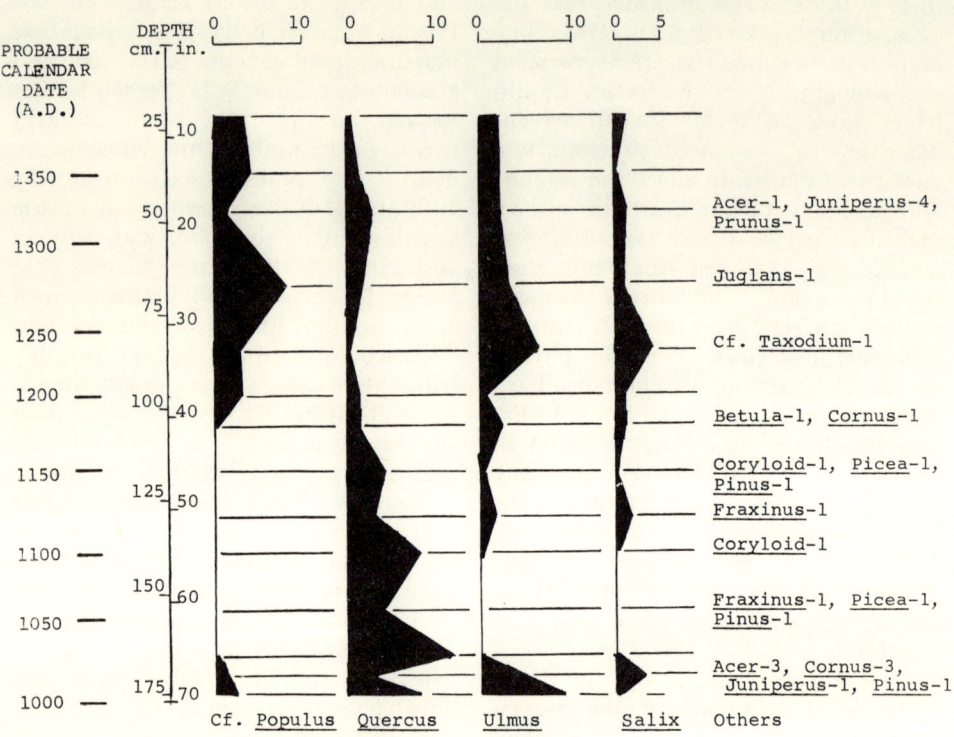

FIG. 10. Arboreal pollen diagram from Phipps site (13CK21) in northwestern Iowa.

The pollen spectra from the village midden show a rapid decline of oak, a more rapid rise of grass, and a sudden rise to dominance of *Populus*, probably cottonwood, in the 12th century (Figs. 9 and 10). Though the proportion of bison meat eaten had been slowly rising, it rose abruptly at this time to 64 percent of the total. It appears quite clear that the decline of browse and cover associated with the onset of drought conditions greatly reduced the availability of deer. One cannot argue that bison simply became more abundant and were preferred, because with the onset of the 12th century there is a sharp decline in the absolute number of bone elements recovered. Since this decline may also be measured as a proportionate decline in the percentage of bone in relation to artifact categories, such as potsherds, it would seem to reflect a decline in the yield of hunting, rather than, perhaps, a population shift. Despite the stress conditions, even more reliance may have been placed on vegetable foods after the 12th century than before.

Apparently the climatic change of the late 1100's was sufficiently disruptive to be reflected in cultural change as well. The division between the lower and upper Little Sioux phases of the Mill Creek culture coincides with the climatic change.

Some statement may also be made as to the rapidity of this change, which was clearly so significant in northwestern Iowa. At this time scale, one must also consider the relation of radiocarbon dates to calendar dates, for the climatic data are by calendar date, the ecological, by radiocarbon assay.

Lamb (1966) collected the European climatic data by decades, then into overlapping half-century means. It appears that the change of climate started rather abruptly in the 1160's. Adjusting the radiocarbon ages for the pollen diagram of Figure 9 by the

Stuiver-Suess radiocarbon-calendar age relationship, presented in Figure 11, shows that the precipitous rise of grass pollen started at the same time.

The accumulation rate on the midden was about 2.9 years per centimeter. Examination of the pollen profiles shows that the decline of the oak-pollen rain from its predrought level to the low level of the post-1200 era took less than a century. The rapid rise of grass pollen from less than 5 percent of the nonarboreal pollen to about 70 percent took 45 years at the most, while the rise of the phreatophytes such as cottonwood and willow required only 15 years or less. The times may have been shorter, but the sampling interval covered about 15 years. There can be little doubt that these were catastrophic changes.

Farther west on the plains, farming should have been more marginal than in Iowa, and the onset of drought associated with increased strength of the zonal westerlies should have terminated successful farming. That the occupation of what W. R. Wedel (1956, p. 87) called the Small Village Complexes in the western portion of the Southern Plains was indeed terminated at the time of the 13th-century "dust-bowl" has been amply demonstrated (Wedel, 1937, 1941, 1953, 1961).

By contrast with the predicted and verified drought of the 13th and 14th centuries in Iowa and Nebraska, Figure 8 suggests that the July rainfall in western Oklahoma and northwestern Texas should have increased substantially at the same time. Similar results for August as well suggest that agricul-

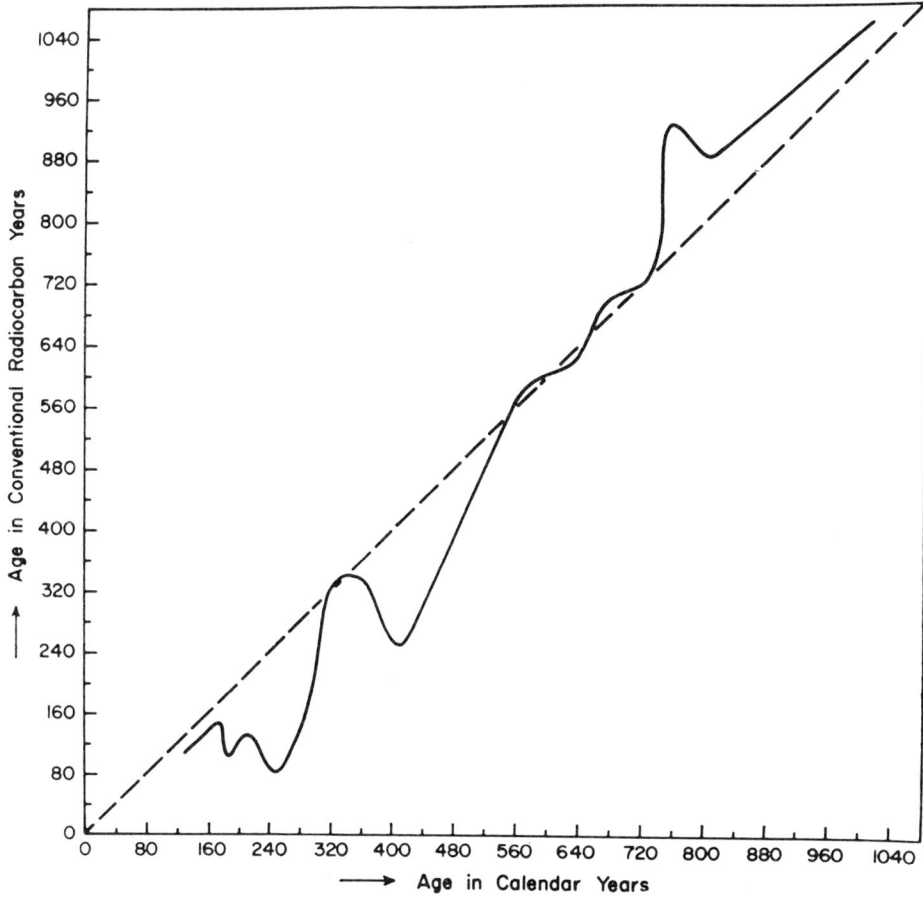

FIG. 11. Relationship of carbon-14 date to calendar date (after Stuiver and Suess, 1966).

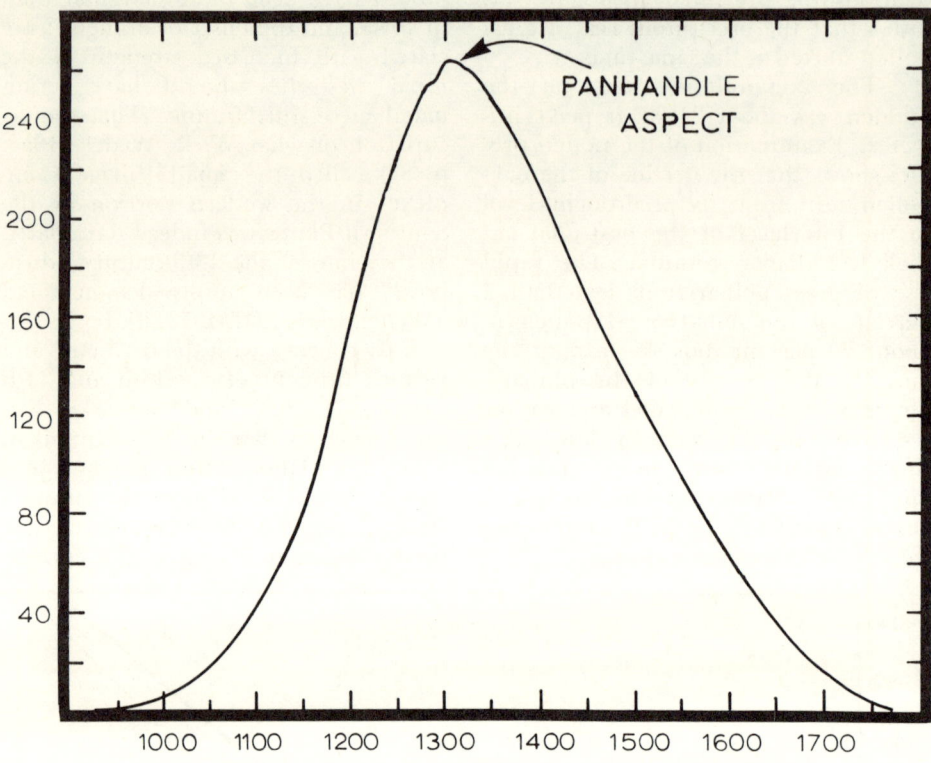

Fig. 12. Probability density distribution calculated from 37 Panhandle Aspect carbon-14 dates plotted according to radiocarbon dates on the abscissa. Arbitrary ordinate.

ture might have become possible after A.D. 1200 in the Panhandle region of Texas and Oklahoma. This raises the question of whether the sedentary occupation of the Panhandle region might be a response to improved environmental conditions *after* the climatic change of A.D. 1160-1200.

Puebloan trade sherds found in various sites of the Panhandle Aspect are mostly from the late 14th century, but with some indicating contact a century earlier. Radiocarbon dates range from A.D. 1120 to A.D. 1620. Neither clearly indicates when the sites were first occupied, for trade contracts with the Pueblo area appear never to have been intense, and it is unlikely that the first settler, linked to eastern or northeastern sources, immediately undertook trade with the Southwest. The latest radiocarbon dates are clearly suspect, which makes one doubt the other extreme as well (Baerreis and Bryson, 1966).

It is possible to treat this question of initial occupation of the Panhandle in terms of probability, however. Each radiocarbon date is based on a sampling of the rate of radioactive decay of carbon-14 atoms in the material under test. When a carbon-14 date is assigned, it means that the sampled decay rate is that which would be expected if the sample had decayed for the indicated length of time, but the "standard deviation" indicated in years is in reality a measure of the standard error of the *sampling* of the frequency of radioactive disintegrations. Each date is thus the modal or most probable value of a "normal distribution," and the probability is approximately two to one that the true value of the disintegra-

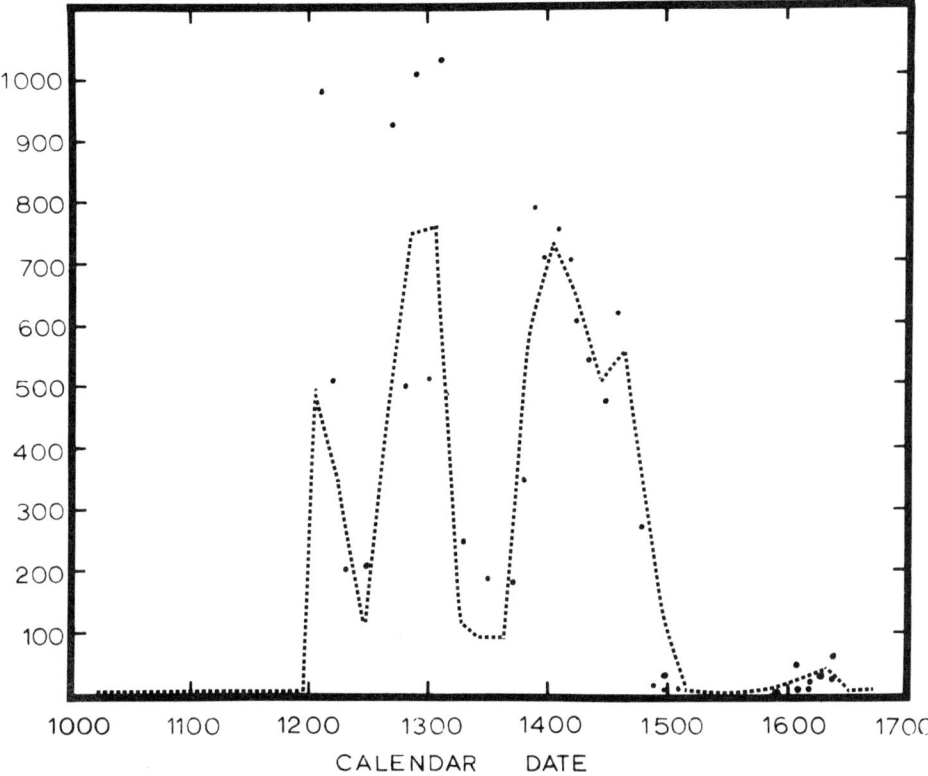

FIG. 13. Probability density distribution of 37 Panhandle Aspect carbon-14 dates by calendar date derived from Figure 12 by transformation of coordinates. Calendar dates on the abscissa. Arbitrary ordinate.

tion count lies within the range which, converted to years, is given by the indicated standard deviation. An array of dates, then, is an array of probability distributions that may be combined into a probability density function. The probability distribution derived from the 37 available Panhandle Aspect dates is given in Figure 12. The probability of occupation of the sites is not represented by this distribution, however, unless the date equivalents of the sampled disintegration rates are linearly related to the calendar date. If the Stuiver-Suess curve of calendar equivalents is correct, then transformation of Figure 12 to a linear calendar-date scale on the abscissa also distorts the ordinates to keep the total area at unity. The result is shown in Figure 13. No longer does the probability slowly increase between A.D. 950 and A.D. 1300, but it rises abruptly at A.D. 1200 by a factor of 100. With considerable statistical certainty one can say that the people were *not* there between A.D. 1180 and 1200 and that they *were* there by A.D. 1220. Evidently the sudden climatic change was followed by a rapid immigration of sedentary peoples into the Panhandle—probably not as rapid as the "Boomers" of 1889, but also probably not a slow drift from site to site that finally dribbled into the Panhandle.

The preceding paragraphs have been concerned with the most recent of the significant environmental changes given in Table 1, its rapidity and its importance. The data presented indicate that the climatic change occurred within decades, changed the ecology and economy of western Iowa sufficiently to affect the lifeways of the Mill

Creek people, terminated agricultural occupation of a large area farther west on the plains, and opened the Panhandle region to agriculture. In terms of paleoecology and archaeology such a change is both sudden and significant.

THE CLIMATIC CHANGE THAT TERMINATED THE PLEISTOCENE

Between about 13,000 and 10,500 BP the southern edge of the Laurentide ice sheet oscillated within a range of a few hundred miles (Bryson and Wendland, 1967a). Then began a rapid retreat, which, except for minor halts, essentially cleared North America of glacial ice (see Fig. 2). Because glaciers integrate the climate, the zero of the derivative of the glacial mass must be taken as the time of climatic change, that is, the time of onset of the rapid retreat (Fig. 14). For the purposes of the present discussion we will call the time prior to 10,500 BP Pleistocene and the following time Holocene.

In Late Pleistocene time the boreal forest and the boreal-forest climate extended southward into the sand hills of Nebraska and into northeastern Kansas and, from this plains area, probably northwestward to the Rockies and east to the Atlantic. By the beginning of Atlantic time the plains-forest border had receded northeastward to central Minnesota (Wright, 1968) and northward to near the ice margin in the prairie provinces (Bryson and Wendland, 1967b).

According to Wright (1968), pollen profiles show that change was directly from spruce forest to prairie in Nebraska, Kansas, and the Dakotas, but to mesic deciduous forest in southern Minnesota and in Illinois. He stated that the grasslands reached their maximum extent about 7000 BP (in early Atlantic time).

Computer analysis of regionally distributed surface-pollen spectra and climatic data indicates that about 40 percent of the surface-pollen spectral variance is covariance with macroclimate (Cole and Bryson, 1968). The eigenvectors of the regional covariance matrix correspond quite closely in the pollen portions with the eigenvectors of the stratigraphic pollen spectra in Minnesota, indicating that space and time are to a certain extent exchangeable in the interpretation of pollen profiles. This, quantitatively, is what is done qualitatively when a palynologist seeks modern analogues of a fossil pollen assemblage. Multiple vector regression of the fossil assemblages (stratigraphic eigenvectors) on the regional pollen *cum* climate eigenvectors then yields estimates of past climates in quantitative terms (Cole and Bryson, 1968).

This type of analysis indicates that in northwestern Minnesota the duration and frequency of inflow of Pacific air increased drastically (about 40 percent) from Late Pleistocene to Atlantic time. Both Borchert (1950) and Bryson (1966) have emphasized that it is the prevalence of this kind of air that characterizes the prairies. Estimates of the climate derived from the pollen profiles cannot change faster than the pollen assemblages themselves. Assuming that the plant communities must change more slowly than the climate, the rate of change of the climate as indicated by the pollen of the plant communities must be a minimum rate. Let us now consider what this rate was at the Pleistocene-Holocene transition.

Ogden (1967), commenting on this sudden change of climate in the Great Lakes region, cited a decline of spruce pollen from 55 percent to 18 percent and a rise of pine pollen from 3 percent to 52 percent in an estimated 170 years at Glacial Lake Aitken in Minnesota. The average of his dates for this sort of change is 10,400 BP. He also cited a figure of about 1100 years

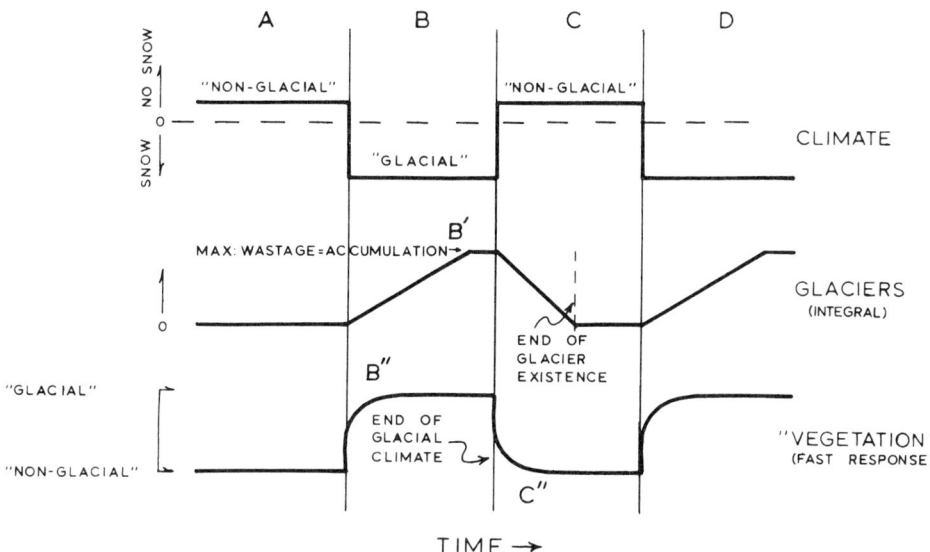

FIG. 14. Schematic glacial and vegetative response to abrupt temperature changes with time (Bryson and Wendland, 1967b).

for the transition from 50 percent spruce to 50 percent oak at Silver Lake, Ohio. To assess which of these rates is more representative, other similar calculations may be made.

Using the data of McAndrews (1966) one may calculate that at Terhell Pond, Mahnohmen County, Minnesota, the sedimentation rate was perhaps 7.8 years per centimeter in gyttja and 11 years per centimeter in silty gyttja. The rapid decline of spruce on McAndrews' diagram is from 65 percent to less than 5 percent in 50 centimeters, whereas pine rises from 2 percent to 45 percent. However, he indicated that this is the transition to a Graminae-*Artemisia* zone, whereas most of the transition to *Populus, Betula, Pinus,* and *Quercus* occurs in approximately 10 centimeters. Thus the decline of the spruce forest might have taken as little as 77 years, and the complete transition to the Graminae-*Artemisia* zone, 550 years. As a test of the method, we may compare the transition into the modern *Ambrosia* zone attributed to the advent of European settlers. The figure one obtains is 78 years (one sampling interval), a rather reasonable figure for Minnesota.

Because many of the transitions appear to follow exponential curves asymptotic to the new pattern, it is difficult to say exactly when the new pattern has been achieved. However, one may define a "half-life" in a manner analogous to the exponential decay of carbon-14, using the interval in which a pollen type changes halfway from its earlier level to its later level. The "half-life" of the spruce decline at Terhell Pond is estimated at 77 years. Using McAndrews' data for Bog D Pond, Hubbard County, Minnesota, one obtains a "half-life" of 88 years for the transition from spruce to pine dominance in the pollen rain, and of 55 years for the pine to grass transition at the beginning of the Atlantic. Of course, if drought-abetted fire were a major immediate agent of vegetational change, such declines might be even faster—one year, for example.

West's data (1961) for Wisconsin sites gives "half-lives" of the spruce decline at the Pleistocene-Holocene boundary of from less than 90 to 175 years, and for pine-oak transition of 140-200 years. These rates and those from McAndrews' data cited above are quite compatible with the data cited for the change in northwestern Iowa near A.D. 1200.

After the initial global climatic change that terminated the Pleistocene, it is likely that there was a slower regional climatic change in the Northern Plains as the Laurentide ice retreat opened a low corridor to the Arctic, allowing the free flow of intensely cold Arctic air into the plains (Bryson and Wendland, 1967a). The effects of this outflow probably spread gradually over the entire plains as the corridor widened. Reference to Figure 1 will show that the corridor was quite wide by the beginning of Atlantic time. It is suggested that the frequent severe northers of the Pre-Boreal and Boreal times might have been a more significant factor in the extinction of the megafauna, such as the mastodon, than overhunting by man. The climatic change that replaced good browse with short grass over vast areas must have provided the *coup de grâce* for the large browsers in what is now the Great Plains area.

SUMMARY

We have attempted in the previous pages to indicate that the major climatic changes must be global and synchronous and that they occur with considerable rapidity. The periods between these changes appear to constitute quasi-stable episodes, which are broken into subepisodes by minor changes. This is consonant with the modern view of the multistable character of the general atmospheric circulation.

We have implied that significant climatic changes are those which produce regional ecological changes of more than a subtle character, but we must reiterate that the direction of the change will not be the same everywhere and that there must be many climatic-biotic core areas with very little change.

Modern analogues may be found or constructed that appear to be useful in the interpretation of climatic change, particularly in elucidating the areas where changes of opposite sign may be associated with a single change of the general circulation, and that suggest optimum locales for field work to verify hypotheses as to the nature of the change. This implies that within the Holocene, climate has changed in "quantity" rather than in "kind." It appears that relatively small changes in the frequency of occurrence of weather patterns and events that occur today can produce significant ecological changes if the changed frequency persists long enough to affect plant competition, prevailing ground-water levels, and so forth.

It is suggested that the evidence is adequate to conclude that a series of quasi-stable climatic episodes separated by rather rapid transitions is a better working hypothesis than that of a steady post-glacial rise of temperatures to an "optimum" followed by a steady decline. Since the timing of these episodes appears to be globally consistent and to coincide with the Blytt-Sernander divisions of the Holocene, it is recommended that paleoecologists and paleoclimatologists adopt the modified Blytt-Sernander terminology for the climatic episodes of the last 10,000 years. This does not mean adopting the European biotic connotations as globally applicable.

Freed from a scheme that is tied to a single climatic parameter, or at best two, which must vary the same way over whole continents, perhaps scholars of past climates can get on with the pressing business of what climates actually existed in each area during each episode of the past. Only the closest of interdisciplinary cooperation can make this goal possible.

ACKNOWLEDGMENTS

This research was supported by the Atmospheric Sciences Division, National Science Foundation (GP-5572X1).

LITERATURE CITED

Antevs, E.
  1948. Climatic changes and pre-white man. Univ. Utah Bull., 38:168-191.
  1952. Climatic history and the antiquity of man in California. Univ. California Arch. Surv. Rep., 16:23-31.
  1955. Geologic-climatic dating in the West Amer. Antiquity, 20:317-335.
Baerreis, D. A., and R. A. Bryson
  1965. Climatic episodes and the dating of the Mississippian cultures. Wisconsin Archeol., 46:203-220.
  1966. Dating the Panhandle Aspect cultures. Bull. Oklahoma Anthro. Soc., 14:105-116.
  1967. Climatic change and the Mill Creek culture of Iowa. Arch. Archeol., Soc. Amer. Archaeol., 29:1-673 (eds.).
Blasing, T. J.
  1968. Patterns of climatic anomaly over the upper Midwest in summer. M.S. thesis, Dept. Meteor., Univ. Wisconsin, Madison, 45 pp.
Borchert, J. R.
  1950. Climate of the central North American grassland. Ann. Assoc. Amer. Geogr., 40:1-39.
Bryson, R. A.
  1966. Air masses, streamlines, and the boreal forest. Geogr. Bull., 8:228-269.
Bryson, R. A., W. M. Irving, and J. A. Larsen
  1965. Radiocarbon and soils evidence of former forest in the southern Canadian tundra. Science, 147:46-48.
Bryson, R. A., and J. F. Lahey
  1958. The march of the seasons. AFCRC-TR-58-223, Final Rep. AF 19 (604) 992, Dept. Meteor., Univ. Wisconsin, Madison, 41 pp.
Bryson, R. A., and W. M. Wendland
  1967a. Radiocarbon isochrones of the retreat of the Laurentide ice sheet. Tech. Rep. 35, Nonr. 1202 (07), Dept. Meteor., Univ. Wisconsin, Madison, 38 pp.
  1967b. Tentative climatic patterns for some late-glacial and post-glacial episodes in central North America. Tech. Rep. 34, Nonr. 1202 (07), Dept. Meteor., Univ. Wisconsin, Madison, 32 pp.
Cole, H. S., and R. A. Bryson
  1968. The application of eigenvector techniques to the climatic interpretation of pollen diagrams: an initial study. Unpubl. M.S. thesis, Dept. Meteor., Univ. Wisconsin, Madison, 52 pp.
Fultz, D.
  1959. Studies of thermal convection in a rotating cylinder with some implications for large-scale atmospheric motions. Amer. Meteor. Soc., Meteor. Monogr., 4 (21):1-104.
Johnson, D. R.
  1966. Non-linear parameter estimation for the partial collective model of air mass analysis. Pp. 63-73, in Appendix, Tech. Rep. 24, Contract 1202 (07) and GP-444, Dept. Meteor., Univ. Wisconsin, Madison.
Küchler, A. W.
  1964. Potential natural vegetation—United States. Amer. Geogr. Soc., Spec. Publ., 36 (map), New York.
Lamb, H. H.
  1966. The changing climate. Methuen & Co., Ltd., London, 236 pp.
McAndrews, J. H.
  1966. Postglacial history of prairie, savanna and forest in northwestern Minnesota. Mem. Torrey Bot. Club, 22 (2):1-72.
Malde, H. E.
  1964. Environment and man in arid America. Science, 145:123-129.
Nichols, H.
  1967a. Central Canadian palynology and its relevance to northwestern Europe in the Late Quaternary period. Rev. Palaeobot. Palynol., 2:231-243.
  1967b. The post-glacial history of vegetation and climate at Ennadai Lake, Keewatin, and Lynn Lake, Manitoba (Canada). Eiszeitalter und Gegenwart, 18:176-197.
Nilsson, T.
  1964. Standartpollendiagramme und C-14 Datierungen aus dem Ageröds Mosse in Mittleren Schonen. Publ. Inst. Mineral. Palaeontol., 124:52.
Ogden, J. G., III
  1967. Radiocarbon and pollen evidence for a sudden change in climate in the Great Lakes region approximately 10,000 years ago. Pp. 117-127, in Quaternary paleoecology (E. J. Cushing and H. E. Wright, Jr., eds.), Yale Univ. Press, New Haven, Connecticut, vii+433 pp.
Sawyer, J. S.
  1966. Possible variations of the general circulation of the atmosphere. Pp. 218-229, in Proc. Internat. Symp. on World Climate, 8000 to 0 B.C., Royal Meteor. Soc., London.
Stuiver, M., and H. E. Suess
  1966. On the relationship between radiocarbon dates and true sample ages. Radiocarbon, 8:534-540.

Wahl, E. W.
  1953. Singularities and the general circulation. Jour. Meteor., 10:42-45.
Wedel, W. R.
  1937. Dust bowls of the past. Science (suppl.), 86:8-9.
  1941. Environment and native subsistence economies in the Central Great Plains. Smithsonian Misc. Collections, 51(3):1-29.
  1953. Some aspects of human ecology in the Central Plains. Amer. Anthropol., 55:499-514.
  1956. Changing settlement patterns in the Great Plains. Pp. 81-92, *in* Prehistoric settlement patterns in the New World (G. R. Willey, ed.), Viking Fund Publ. in Anthropol., no. 23.
  1961. Prehistoric man on the Great Plains. Univ. Oklahoma Press, Norman, 355 pp.
West, R. G.
  1961. Late- and postglacial vegetational history in Wisconsin, particularly changes associated with the Valders readvance. Amer. Jour. Sci., 259:766-783.
Wright, H. E., Jr.
  1968. History of the Prairie Peninsula. Pp. 78-88, *in* The Quaternary of Illinois, Spec. Publ. Univ. Illinois Coll. Agric., 14:1-179.

ANTHROPOLOGY
(Consulting Editor, Alfred E. Johnson)

Geochronology of Man-Mammoth Sites and Their Bearing on the Origin of the Llano Complex

C. Vance Haynes

ABSTRACT

The dispersal of Clovis artifacts throughout central North America in less than 1000 years represents the New World's first technological revolution. Remarkable parallels exist between campsites of mammoth hunters in both the Old and New Worlds during the late glacial period, and mammoths became extinct in the Soviet Union at the same time they did in North America. Early or Mid-Wisconsin tool industries are few in number and do not appear to be ancestral to Clovis tool assemblages.

It appears unlikely that the Clovis progenitors migrated from Siberia to central North America before the Late Wisconsin glacial stages because of the indicated shortness of the period between the time when the Bering land bridge became emergent and the time when Canada became blocked by glacial ice. The Clovis "population explosion" of 11,000 to 12,000 years ago seems best explained by a Late Wisconsin migration of mammoth hunters out of Alaska as soon as passage through Canada became possible during the Two Creeks glacial retreat of 12,000 to 13,000 years ago.

INTRODUCTION

The discovery of the Dent site near Denver, Colorado, in 1932 provided conclusive evidence that man had hunted mammoths in the New World (Wormington, 1957, p. 43). The distinctive fluted projectile points used to kill mammoths became known as Clovis points after significant discoveries of them with mammoth carcasses were made on the Llano Estacado near Clovis, New Mexico (Wormington, 1957, p. 47). Subsequent discoveries on the Llano Estacado led Sellards (1952) to name the mammoth-hunting complex the Llano Complex, which is now defined on the basis of many sites and thousands of single Clovis points found from the Atlantic Coast to the Pacific and from central Canada to central Mexico. It is the earliest well-defined and the most widespread cultural complex known in the New World, and it represents our first technological revolution. But its antecedents are unknown. My objective here is to review the geochronological aspects of man-mammoth associations in America and the Soviet Union in order to evaluate current hypotheses of prehistoric migrations to the New World.

CLOVIS SITES AND THEIR AGE

Since the survey of Clovis sites dated by the radiocarbon method was made several years ago (Haynes, 1964), five more buried sites have been found (Table 1—Lm, E, M, We, MG), one of which—Murray Springs—has yielded

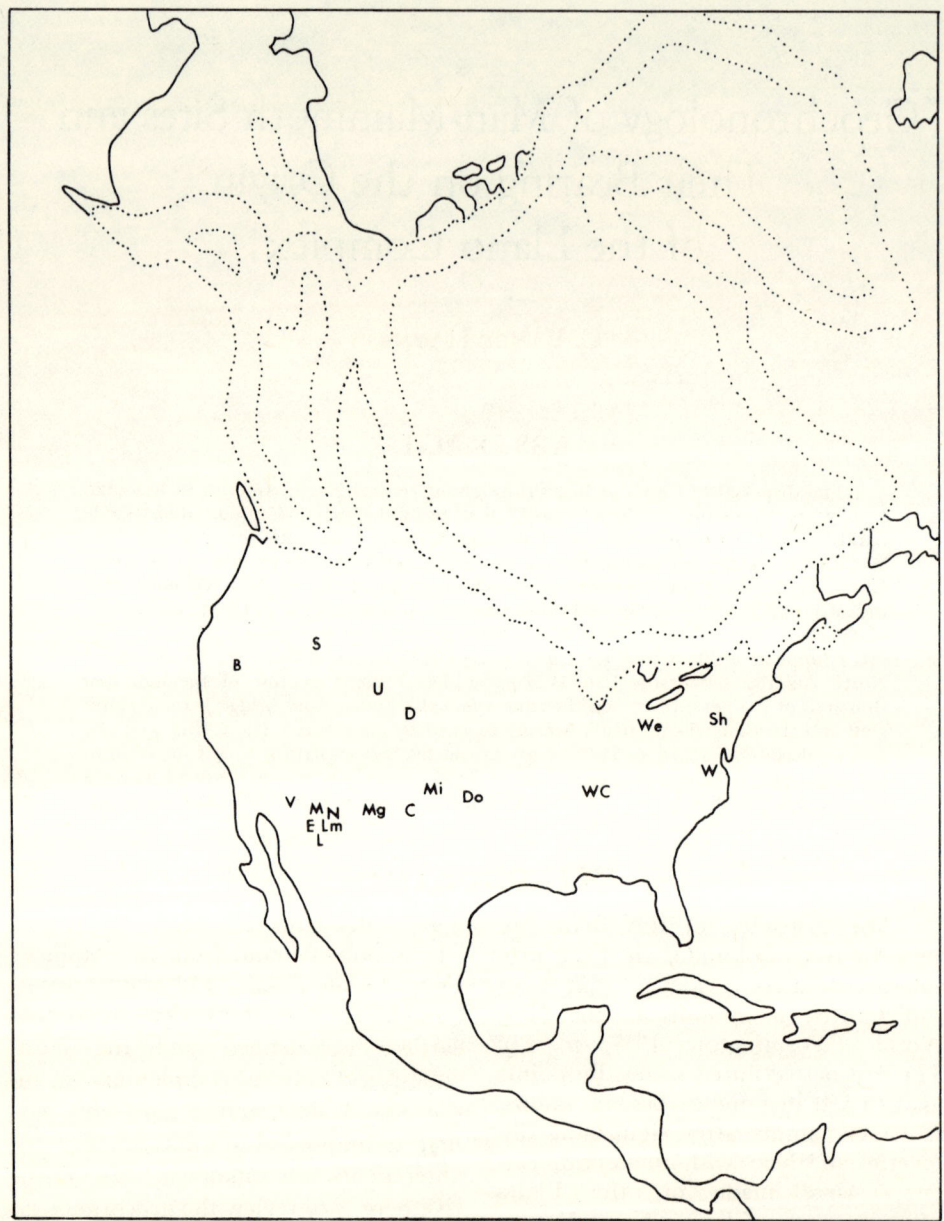

FIG. 1. Outline map of North America showing possible locus of ice borders of 12,000 BP and Clovis sites indexed in Table 1.

another radiocarbon date for the Clovis culture. These finds bring the total number of known Clovis sites to 18 (Fig. 1), of which five had been carbon-14 dated (Table 1). If we include the Union Pacific site and Ventana Cave in the Clovis category, it would raise the number of carbon-14-dated sites to seven, all of which are between 11,000 and 11,500 years old. In fact, statistical treatment of six dates, including that of the Union Pacific site (Irwin et al., 1962), but excluding the Ventana Cave date because of its large statistical error, shows that all six dates meet the chi-square test for a single event at 11,240±140 BP. If we double that statistical error, there is 96 percent prob-

TABLE 1. Clovis Data by Site

Site	Symbol	Type	Game	Deposit	Date in years BP
Dent, Colorado	D	kill	mammoth	overbank alluvium	11,200±500 (1-622)
Clovis, New Mexico	C	hunting camp	mammoth bison	springlaid sand	11,310±240 (av.)
Miami, Texas	Mi	kill	mammoth	pond clay	
Shoop, Pennsylvania	Sh	camp	?	terrace soil	
Williamson, Virginia	W	camp	?	upland soil	
Borax Lake, California	B	camp	?	alluvial fan	
Ventana Cave, Arizona*	V	camp	?	volcanic debris	11,290±1000 (A-203)
Naco, Arizona	N	kill	mammoth	channel alluvium	
Lehner, Arizona	L	hunting camp	mammoth bison	channel alluvium	11,260±360 (av.)
Union Pacific (UP), Wyoming*	U	kill	mammoth	channel alluvium	11,280±350 (1-449)
Domebo, Oklahoma	Do	kill	mammoth	channel alluvium	11,160±500 (av.)
Simon, Idaho	S	cache (?)	?	prairie soil	
Leikem, Arizona	Lm	kill	mammoth	channel alluvium	
Welling, Ohio	We	quarry camp	?	terrace soil	
Murray Springs, Arizona	M	hunting camp	mammoth bison	channel alluvium	11,230±340 (A-805)
Escapule, Arizona	E	kill	mammoth	channel slope	
Wells Creek, Tennessee	WC	camp	?	upland soil	
Mockingbird Gap, New Mexico	MG	camp	mammoth	channel slope	

* Questionable Clovis site.

ability that all of these sites were occupied between 10,960 and 11,520 years ago.

The radiocarbon dates are consistent with the stratigraphy and with correlations between sites based upon archaeological, paleontological, and geological sequences (Fig. 2). Furthermore, undated sites correlated on these bases show the Clovis horizons to be of comparable age at four of the yet undated sites—Miami, Naco, Leikem, and Escapule—and probably of the same age at three others—Borax Lake, Simon, and Mockingbird Gap. The four remaining sites—Shoop, Williamson, Wells Creek, and Welling—are in ancient soils of the plowed zone of 30- and 60-foot terraces or on uplands in the eastern United States and could well be the same age as the western sites. Similarities between western and eastern Clovis artifact assemblages have been described by Hester (1966).

From archaeological sequences and radiocarbon dates known from the eastern United States (Mason, 1962; Coe, 1964), it is clear that there are no places in the sequences for Clovis points after approximately 10,000 BP. The 10,600-year-old Debert points (Stuckenrath, 1966) along with the Bull Brook points (Byers, 1959) are probably derived from Clovis.

Some archaeologists speculate that certain forms of eastern fluted points may be as old as 17,000 years (Griffin, 1965, p. 655) on the basis of the distribution of surface finds with respect to glacial moraines and glacial lake margins. In Wisconsin and Michigan numerous fluted-point finds appear to be related to late Cary, Mankato, or Valders features (Mason, 1958; Quimby, 1958, 1963) of 13,000 to 10,500 years ago. Quimby pointed out that some of these finds can be no older than Valders maximum, or 10,700 BP. The fact that the "Mason-Quimby line," as Martin (1967, fig. 2) called the northern

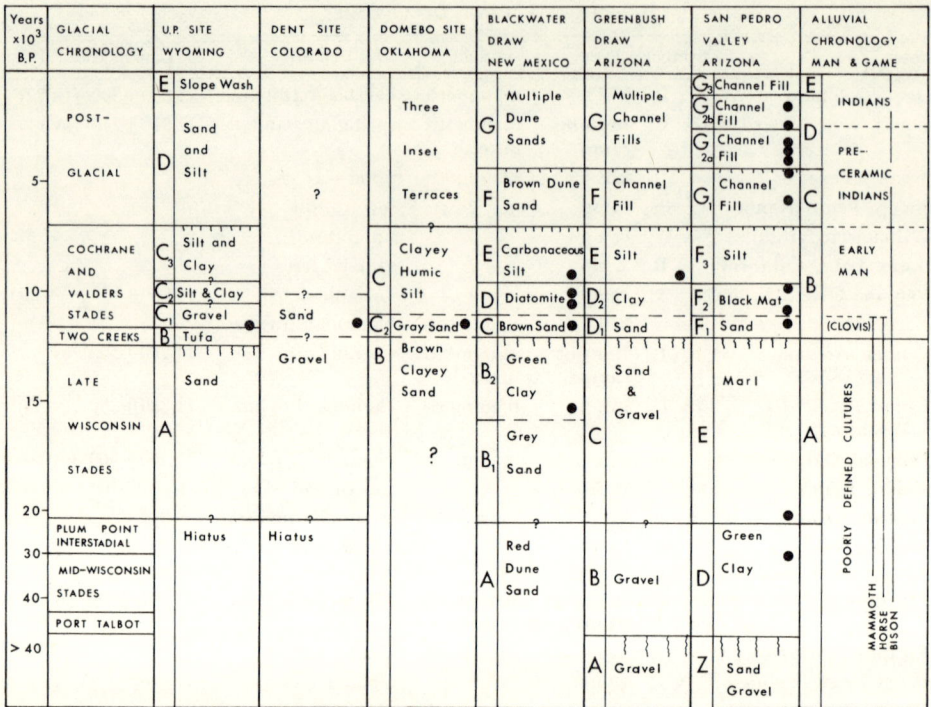

FIG. 2. Correlation chart of radiocarbon-dated mammoth-kill sites in the United States. Glacial chronology after Leighton (1960) and Hughes (1965). Site stratigraphy after Haynes (1970a) for Union Pacific; Malde (1954) for Dent; Albritton (1966) for Domebo; Haynes (1970b) for Blackwater Draw; Haynes and Johnson (1970) for Greenbush Draw; and Haynes (1968b) for San Pedro Valley. Alluvial chronology after Haynes (1968a). Dots indicate position of radiocarbon sample. According to Dreimanis et al. (1966), the Port Talbot interstade began more than 48,000 years ago and ended 24,000 years ago. It is now considered to represent Mid-Wisconsin time, the final part of which includes the Plumb Point interstade.

limit of fluted-point finds, crosses late Cary features indicates that the first occupation of the Lower Michigan peninsula was no earlier than 13,000 BP.

Prufer and Baby (1963, pp. 53-55) have plotted the distribution of more than 1500 fluted points in Ohio. Two of these from Pleistocene beaches near Lake Erie are believed to have been water-worn by the wave action of glacial lakes Warren and Lundy. Wayne and Zumberge (1965) presented two geochronological schemes for these lakes. According to their figure 6, these stages are post-Two Creeks, or between 12,000 and 11,000 years old; but the scheme of their figure 7 places these stages before the Two Creeks interstade and after the Mankato stade, or between 13,000 and 12,500 BP. Thus there is the possibility that some fluted points in Ohio could be pre-Two Creeks, but one of the water-worn points is typologically closer to points of the Plano Complex than to fluted points. This type is referred to as a Holcombe point (Prufer, personal communication), which at the type site is believed to be of main Lake Algonquin age (Fitting et al., 1966, p. 120)—11,000 BP according to the one scheme, or 10,400 BP according to the other. On the basis of present evidence an early age for Holcombe points of 10,000 to 11,000 BP is reasonable, whereas an age of 12,500 to 13,000 is not reasonable. Therefore (1) the pre-Two Creeks age of lake stages Lundy to Warren is incorrect, or (2) the water-worn projectile points have been moved from the original location of wave action, or (3) the wear on the points is not due to water, but to blasting by windblown sand.

As both Roosa (1965, p. 99) and Fitting et al. (1966, p. 133) pointed out, there are several variations of fluted-point types in the eastern United States. Some of these undoubtedly have temporal significance; until the relative ages of specific types are known and taken into account, distribution maps of fluted points may be misleading. This is probably the reason why the Mason-Quimby line does not correspond to glacial features of any one age. From the specific locations of numerous fluted points in Ohio, Michigan, and Wisconsin, it appears that there are relationships to features of glacial lakes Warren, Grassmere, Lundy, and early Algonquin that indicate ages of 13,000 to 12,000 BP, or 12,000 to 10,500 BP, depending upon which interpretation of Great Lakes chronology is correct. The latter interpretation is more consistent with the precise dating of western Clovis sites as discussed previously. The other interpretation is supported by Dreimanis (1966), but allows no temporal overlap between the ages of eastern and western Clovis points.

Of the known Clovis sites where faunal associations exist, all contain mammoth remains, three (Clovis, Lehner, and Murray Springs) contain bones of bison, and this year at Murray Springs a highly probable association with horse remains was found. Although other game was taken as opportunity presented itself, the Clovis hunters were specialists in taking mammoths.

Unlike sites of mammoth hunters in the Old World, Clovis sites have yielded only meager assemblages of tools. In addition to specialized fluted projectile points, Clovis tool kits included: end scrapers, commonly spurred; large unifacial side scrapers; keeled scrapers on large blades; flake knives; some backed, worked blades, gravers, and perforators; bone points; foreshafts; and shaft straighteners (Haynes and Hemmings, 1968). Recently at Murray Springs we found a burinated side scraper, and blade cores are known from the Williamson site in Virginia.

Ovate or leaf-shaped bifacial forms are known from four sites, but those from the Simon site in Idaho demonstrate a logical stepwise sequence from large, crude, bifacial preforms to finished Clovis points. Striking-platform preparation by edge-grinding is a typical trait and is clearly shown in a broken, bifacial preform and associated flakes at Murray Springs.

Until the discovery of the Murray Springs site, the appearance of an undisturbed Clovis living "floor" was unknown. Here partly or wholly dismembered carcasses lay strewn about, apparently near where the game animals were felled (Fig. 3). Broken bones and bone fragments littered certain areas, and in one area there was an artificial stacking of bones. In and around bone concentrations lay Clovis points, broken and whole; a few tools in various states of disrepair; and tens of thousands of flakes, some concentrated into distinct piles, each of distinctive lithology. Shallow depressions of variable size, some containing charcoal and burned bone, occurred throughout the floor, and are readily distinguishable from numerous other elephant-foot-sized depressions, which were concentrated in the southeastern part of the site and which are thought to be mammoth tracks.

OLD WORLD MAMMOTH HUNTERS

Some similarities of the Murray Springs site to mammoth-hunters' camps known from the upper Paleolithic of the Old World have led me

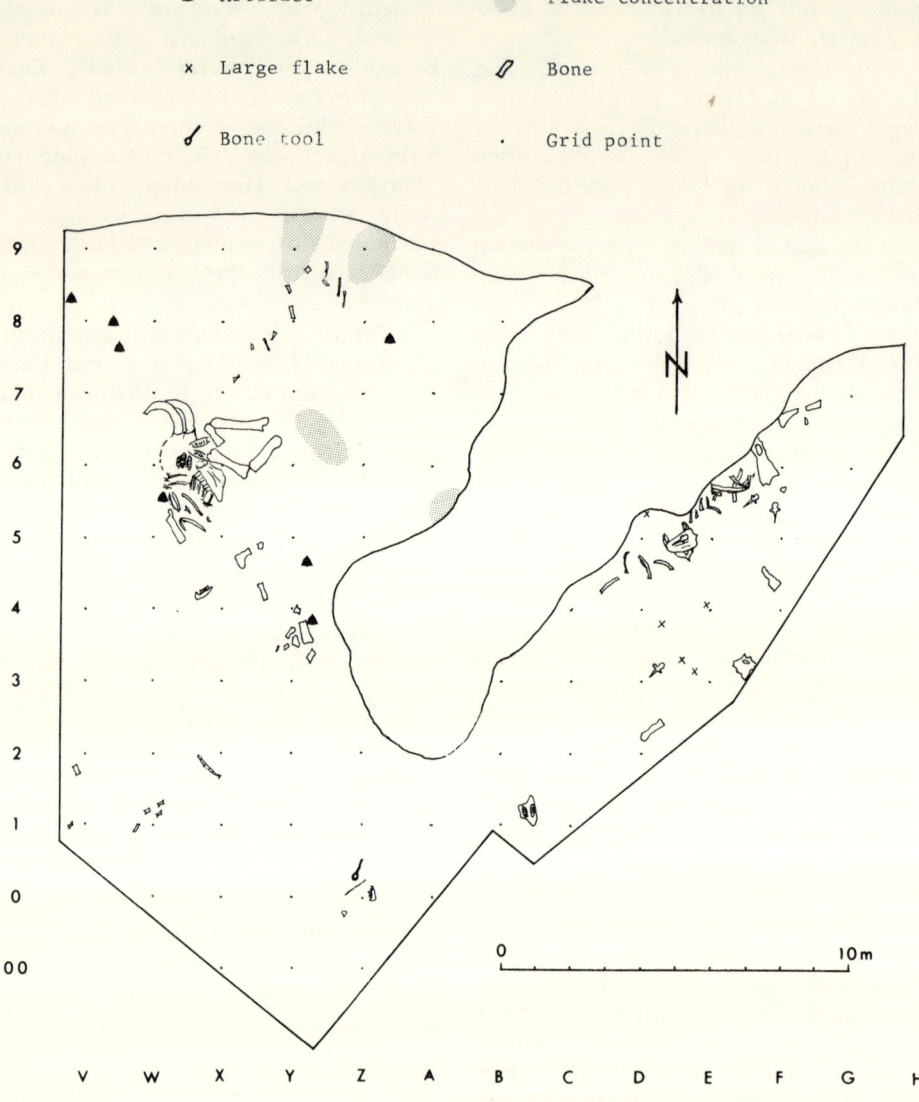

Fig. 3. Excavation map of Murray Springs, locality 1, for 1967.

to examine some of these sites more closely (Fig. 4). Fortunately, through the efforts of Dr. Richard G. Klein, some of the more pertinent Russian literature has been interpreted for English-speaking scholars (Klein, 1969). The site of Pavlov in Czechoslovakia has long been famous and is probably the largest mammoth-hunter camp yet found. Farther east, in Russia, the sites of Molodova V and those of the Kostenki area and Sungir' are most in- structive. Bâtons from Molodova are similar to the Murray Springs shaft wrench (Fig. 5). Characteristic are scattered mammoth bones, some stacked around dwellings and in some cases representing several score individuals; numerous small depressions, some containing art carvings and tools, others containing charcoal and burned bone; thousands of flakes and hundreds of tools, including scrapers, knives, perforators, bifacial points, bone awls,

Fig. 4. Map of U.S.S.R. showing location of radiocarbon-dated sites (P—Pavlov; M—Molodova; Y—Yeliseyevichi; K—Kursk I; Ki—Kostenki area; S—Sungir'; Ko—Kokorevo area; A—Afontova Gora area; T—Taimyr Mammoth; Ma—Mal'ta; C—Chekurova Mammoth; Kh—Kuraanakh; U—Ushki) (Commission for the Study of the Quaternary Period, Academy of Sciences, U.S.S.R., 1968).

needles, and shaft straighteners. Farther east, in Siberia, similar but smaller Upper Paleolithic sites occur at Kokorevo, Afontova Gora II, and Mal'ta. The most outstanding difference between these Old World sites and Murray Springs is size. The latter is small by Old World standards, and no art objects have been found.

Now that geochronological data for the Quaternary of Russia and Siberia are available (Kind, 1967; Ravskii and Tseitlin, 1968; Aksenov and Medvedev, 1968), we can, for the first time, make comparisons with the New World. There are radiocarbon dates from eight Paleolithic sites in Russia and Siberia (Fig. 6), and many more apply to late Quaternary geological sites, including some of the well-known frozen-mammoth discoveries such as Taimyr and Chekurovka (Commission for the Study of the Quaternary Period, Academy of Sciences, U.S.S.R., 1968).

In Siberia the late Quaternary is divided into two main glacial periods, the Zyrianka (early) and the Sartan (late), separated by the Karginsky interstadial and followed by the postglacial period. The Zyrianka glaciation is correlated with the Wurm II period of Europe, the Karginsky interstadial to the Paudorf interstadial, and the Sartan glaciation to Wurm III. Most of the radiocarbon dates are from loess-mantled terraces in major river basins, within which local stratigraphic sequences are reasonably well understood. At most of the sites, post-glacial deposits occur beneath the modern flood plain and on the first terrace above the modern flood plain. Deposits of Sartan and Karginsky age make up the loess mantle on higher terraces, and at some sites they overlie eroded deposits of Zyrianka age.

In the Kostenki area, loess-like loam of the second terrace (15 to 20 meters) is partly colluvial, having been redeposited from higher surfaces, and the first terrace is an alluvial inset terrace of late Sartan age (Fig. 7). At

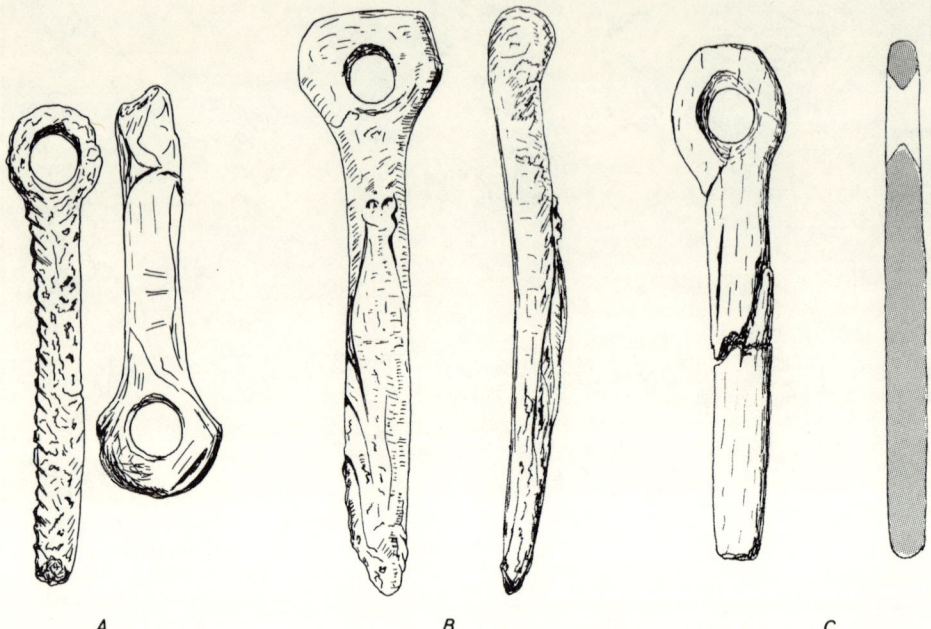

Fig. 5. Shaft wrenches or "bâtons" from (A) Pekárna Cave, Moravia (no scale); (B) Molodova V, western Russia (length 31 cm.); and (C) Murray Springs, Arizona (length 26 cm.). (*A* after Augusta J. and Z. Burian, 1960, pl. 39; *B* after Abramova, 1967; *C* after Haynes and Hemmings, 1968.)

Sungir', Sartan loess-like loam overlies a greater than 35,000-year-old soil in the third terrace, and post-glacial deposits make up two lower inset terraces. At Afontova Gora II in Siberia, Sartan loess-like loam makes up the second terrace, whereas it mantles the third terrace at Mal'ta farther east. Two buried soils that are correlated with the Alleroed and Boelling intervals of Europe occur near the top of Sartan loess-like loam both at Afontova Gora II and at Kokorevo I and II in south-central Siberia.

At all of these sites Upper Paleolithic mammoth-hunting industries are found in deposits of Sartan age and are dated between 11,000 and 23,000 BP. At some of the sites flint tools made of flakes predominate over blades; at others the reverse is true. At Molodova V, farther west, early Upper Paleolithic flake artifacts occur in deposits 23,000 to 30,000 BP and are correlated with the Paudorf Interstadial; whereas to the east, in Siberia, both blades and leaf-shaped bifaces are found in Late Pleistocene deposits. Stone blades and bone needles, awls, and points are known from Mal'ta (Griffin, 1960), in deposits believed to be of Zyrianka age in spite of a radiocarbon date of 14,750 ±110 BP (Aksenov and Medvedev, 1968) on bone, which is commonly contaminated by younger carbonaceous matter. Leaf-shaped bifacial forms are known from the Irkutsk Military Hospital site, also thought to be of Zyrianka age (Aksenov and Medvedev, 1968).

So far, the latest that mammoths have been found to occur in archaeological sites is between 11,000 and 13,000 BP, and the Taimyr Mammoth from the second terrace of the Mamontova River in extreme northern Siberia has been dated at 11,500 BP. Thus it appears that mammoths became extinct in Siberia at the same time as in North America. They were hunted to the bitter end on both continents (Vereshchagin, 1967; Martin, 1967).

CLOVIS ORIGINS

Returning now to America, excavations of the past few years have shown that there are four sites that may be older than the known Clovis sites (Fig. 8). These sites, near Valsequillo and Tlapacoya in Mexico, range in age from 21,000 to possibly more than 25,000 BP. Stratigraphic correlations are at present uncertain, and the archaeological validity of some of these sites has not gone unchallenged; nevertheless they constitute the most convincing evidence yet offered that man had entered the New World before the last major advance of continental glaciers.

The question now arises: Does the Llano Complex owe its origin to indigenous earlier complexes, as hypothesized by Wormington (1962), Bryan (1965), Wendorf (1966), and Müller-Beck (1967), or does it represent a separate late-glacial migration (Green, 1963; Haynes, 1964)?

Wormington's (1962) model has early immigrants bringing simple leaf-shaped bifacial points to the New World south of Canada before Late Wisconsin time. These were the beginning of a developmental sequence that led to Sandia points, fluted Sandia points, and eventually to Clovis points. What has merit about this hypothesis is not only its logic, but the distinct possibility that Sandia points are older than Clovis points. Unfortunately, precise dating of the Sandia Complex has not yet been attained. Bryan's (1965) model differs from Wormington's mainly in that he attributed all

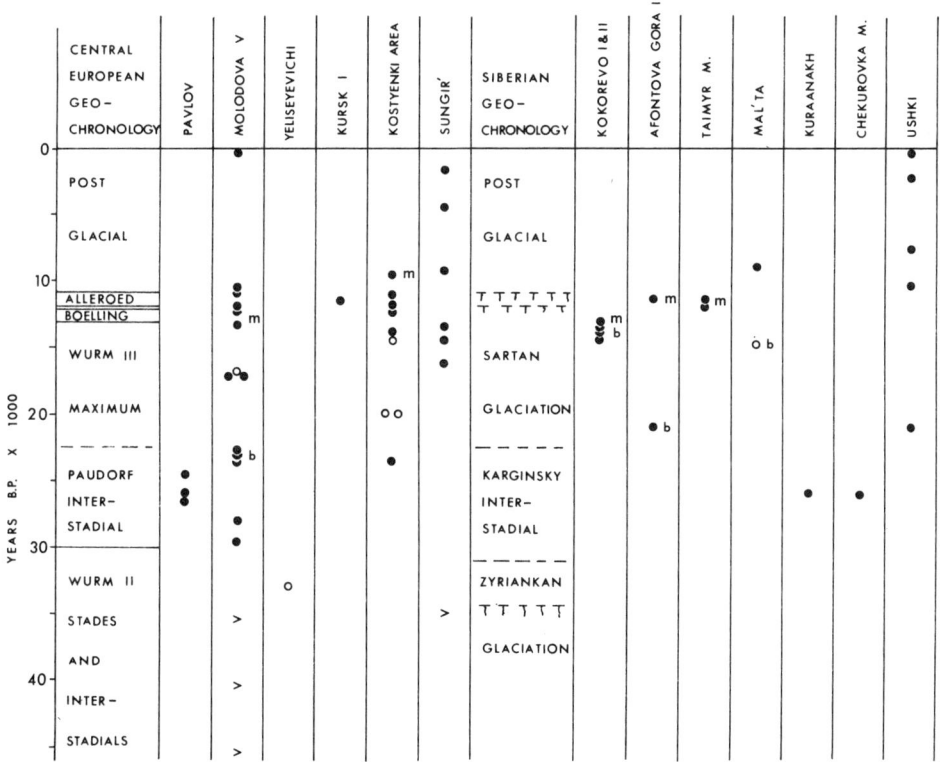

FIG. 6. Correlation chart of radiocarbon-dated Paleolithic sites in the U.S.S.R. Solid dots indicate chronologic position of radiocarbon samples; open circles indicate dates of questionable value; > indicates date in excess of limits of detection; b indicates date applicable to "bâtons"; and m indicates minimum date for mammoths. Geochronology after Kind (1967).

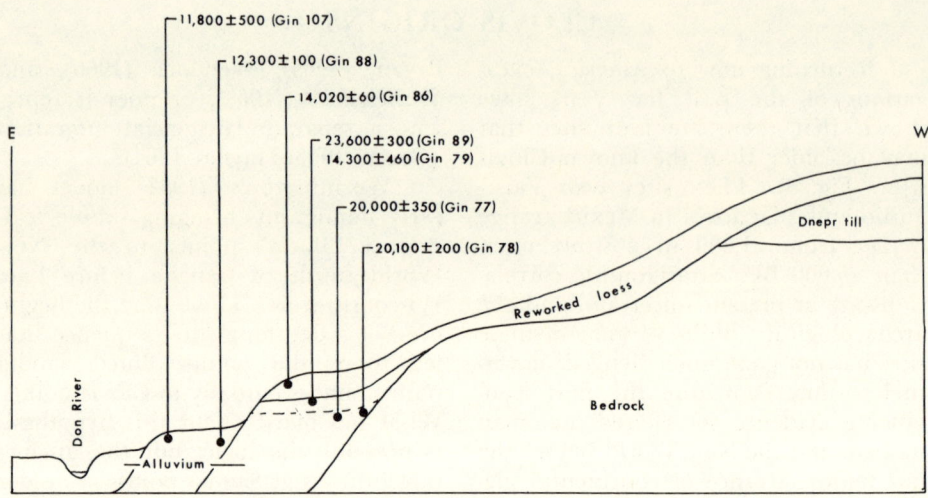

Fig. 7. Generalized cross section of the terraces of the Don River in the Kostenki-Borshevo region (after Klein, 1969), showing relative positions of radiocarbon dates.

New World early-man complexes to Early Wisconsin or pre-Wisconsin migrations of pre-projectile-point complexes from Siberia. The evidence for such a migration is tenuous indeed. Wendorf (1966) and Müller-Beck (1966), finding no similarities between Clovis artifact assemblages and the late Upper Paleolithic blade assemblages of eastern Asia, derived the Llano Complex from the middle or lower Upper Paleolithic flake industries of Siberia. But, as emphasized here, it is now apparent that in Eurasia the blade industries overlap in time the flake industries, and it now seems probable that both industries coexisted throughout the Upper Paleolithic of Siberia as well as the Russian Plain.

In 1964 I proposed the possibility that Clovis sites might represent a late-glacial migration distinct from earlier migrations, because three factors became apparent at that time: (1) re-evaluation of the dating of the Two Creeks glacial interstage showed this significant glacial retreat to have occurred immediately before 11,900 BP, not 11,400 BP as previously thought (Broecker and Farrand, 1963); (2) geochronological work concentrated on Clovis sites showed them to postdate the Two Creeks retreat (Haynes, 1964); and (3) progenitors for Clovis points were not known in underlying deposits (Haynes, 1967). The logical implication is that the sudden appearance of the Llano Complex throughout all of North America south of the ice border was related to a relatively rapid natural event, namely the separation of the Cordilleran and Laurentide ice sheets to form a trans-Canadian passage for man and animals sometime between 12,000 and 13,000 years ago.

An alternative model offered by Hester (1966) differs mainly in that he places the opening of an ice-free corridor between 14,000 and 15,000 BP.

In considering migrations, we must examine events in Beringia, where the land bridge was the gateway to America, and in Canada, where glaciers at times blocked passage to the south. The results of a recent symposium on the Bering Land Bridge (Hopkins, 1967a) provide the latest estimates of when the bridge was passable and when the ice-free corridor through Canada was open. According to these, it is estimated that the Bering Land Bridge was open from approximately 25,500 to 15,000 years ago, again between 13,500 and 12,300 years ago, and finally between 11,800 and 10,000 years ago. Considering the sea-

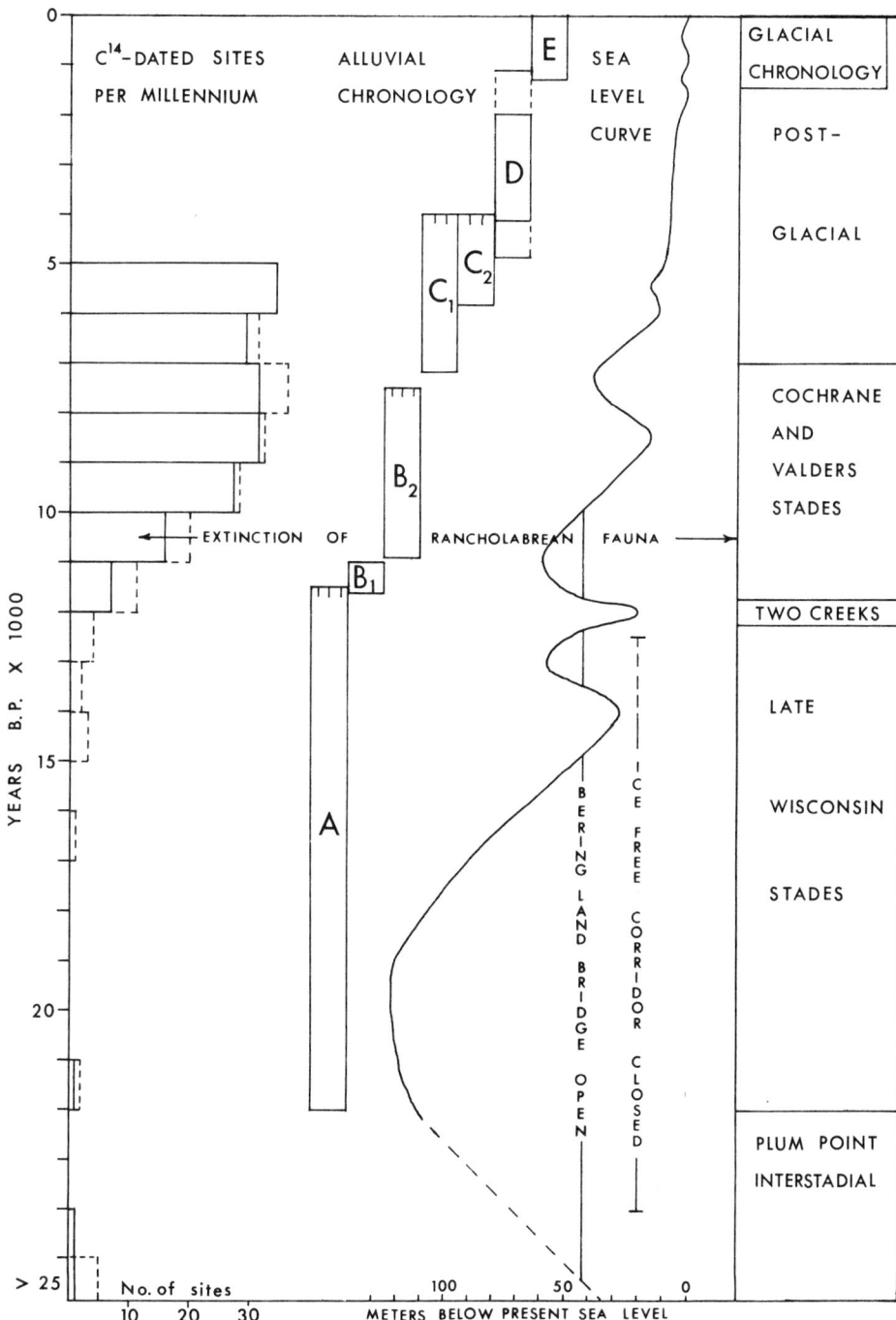

FIG. 8. Correlation chart showing histogram of radiocarbon-dated sites in the Americas compared to sea-level fluctuations and alluvial and glacial chronologies (after Haynes, 1967 and 1968a, and Hopkins, 1967b).

level curve of Milliman and Emery (1968), the bridge was open 9000 to 21,000 years ago. Passage through Canada was believed to have been blocked by Cordilleran and Laurentide ice from 12,500 BP to approximately 24,000 BP (Hopkins, 1967b, fig. 4), although in other hypotheses it is suggested that there was no major obstruction in Late Wisconsin time (MacNeish, personal communication) or that passage was not unblocked until as late as 9800 years ago (Bayrock, personal communication).

The main difficulty with Hester's (1966) model is that his estimate of when a Canadian passage opened falls within what is now thought to be the most intense period of Late Wisconsin glaciation of the Cordillera (Armstrong et al., 1965). Furthermore, the Two Creeks retreat is the one most marked during deglaciation of the mid-continent and is matched by a similar retreat of the southern Cordilleran ice during the Everson interstade between 11,000 and 13,000 BP (Armstrong et al., 1965).

Some of the pre-Clovis sites in America could derive from migrations during 1500 years or so when both the land bridge and the trans-Canadian Corridor may have been simultaneously open, between 24,000 and 25,500 BP. Müller-Beck (1967) has suggested this approximate period for the entry of a "Mousteroid" tradition that developed into Clovis in North America. But from the new knowledge of the Upper Paleolithic radiocarbon dates from Kostenki, Sungir', and other sites, it is now clear that these sites, from which Müller-Beck derived the migrants, are of Sartan age and therefore too late to make a trans-Canadian passage possible before the Late Wisconsin glacial maximum. Migrants so derived could have made the passage during deglaciation, 12,000 to 13,000 years ago, after 10,000 years of diffusion and mammoth-hunting across Siberia and Beringia. Thus they could have led to the Llano Complex.

The most severe limitation to this model is with the differences in the lithic industries between the Llano Complex and the potential Old World antecedents as presently known. As both Müller-Beck (1966) and Wendorf (1966) have pointed out, blade and burin industries are predominant in the Upper Paleolithic of Siberia as at Afontova Gora II and Mal'ta, whereas flake tools and bifacial projectile points predominate in Clovis assemblages. But, because of radiocarbon dating, it is now apparent that flake industries with bifacial points are contemporary with blade industries in Russia as at Kostenki and Sungir', and bifacial leaf-shaped forms occur at the Irkutsk Military Hospital site in Siberia. Elements of both industries overlap on the Russian Plain, and during 10,000 years it is possible that blades were either not emphasized or de-emphasized by some groups before reaching central North America. Furthermore, Müller-Beck's belief that there is nothing "Aurignacoid" in the Llano Complex is not valid, considering the blades, burins, bone points, foreshafts, and the "bâton" that are known from Clovis assemblages.

It should also be mentioned that all of the known Clovis assemblages put together yield only the merest glimpse of what made up their tool industry. As yet we do not have the type of Clovis habitation site that would yield a more complete picture of their industry. New world sites are kill and quarry sites, whereas most of those in the Old World are habitation sites at or near kills.

Microblade industries, which are probably unrelated to Clovis origins, may be as old as 13,000 to 14,000 BP in Kamchatka (Dikov, 1968) and Japan (Hayashi, 1968), and Alaska as well, if suggested extrapolations below radiocarbon-dated levels are correct (Anderson, 1968; McKennan and Cook, 1968). There is now evidence that cultural diversity had occurred in Alaska at least by 9000 BP (Laughlin,

1967, p. 447), which is not surprising when the period of existence of the land bridge is considered; but datable sites for the Clovis points known to be from Alaska or for Clovis progenitors have not yet been discovered in Alaska.

If the hypothesis of a Late Wisconsin migration of Clovis progenitors is correct, then there should exist in Alaska mammoth-hunter sites of a flake and bifacial-point industry that coexisted with core and blade, and possibly microblade—industries of people who did not emphasize mammoth-hunting. As Laughlin (1967, p. 422) noted: "The settlement patterns and the patterns of annual movements of the interior herbivore hunters of historical time are different from those of contemporary coastal marine hunters, and comparable differences must have characterized the coastal and interior land-bridge peoples of 15,000 years ago."

New World cultures older than 25,000 years must be derived from Bering crossings during a lower sea-level stand that predates the last glaciation. This would be prior to the Woronzofian Transgression of sea level more than 30,000 years ago (Hopkins, 1967b). Such migrations would be derived from the Middle Paleolithic of Siberia, which is as yet unknown. It is possible that all pre-Clovis-aged migrations had this minimum age, because after the Bering Land Bridge emerged approximately 25,500 years ago, there may have been less than 1500 years in which to diffuse into central North America before the trans-Canadian corridor became blocked. Considering the 18,000 years or so that early man has had to develop in America before Clovis, it is amazing that the site frequency and cultural diversity seen after 11,000 BP in the New World did not take place much earlier (Haynes, 1967, fig. 7). That it did not is apparent from the paucity of sites known from beds that are immediately pre-Clovis (Fig. 2). That ecological conditions were not unfavorable then is attested to by the abundant remains of big game animals in deposits equivalent to Unit A of the alluvial chronology (Fig. 7), dated 11,500 to 22,000 BP (Haynes, 1968a).

Tool assemblages from American sites earlier than Clovis are too meager for adequate comparison with Clovis assemblages; but from what we do see, there is no tradition represented that is as closely related to Clovis as are traditions in the Upper Paleolithic of the Russian Plain between 15,000 and 23,000 BP, when bifacial points, bone points, end scrapers, blade tools, and bone-shaft wrenches were used by mammoth hunters.

CONCLUSION

In conclusion, it appears from what little data are available that poorly defined hunting cultures entered central North America more than 30,000 years ago, but did not prosper and develop as well as later migrants. As Wendorf (1966) has suggested, descendants of these earliest people may have taken refuge in warmer climates of Latin America during the Late Wisconsin glaciation, but it now appears that the main peopling of the New World took place during deglaciation, when something close to a population explosion occurred between 11,000 and 11,500, by mammoth hunters entering from Alaska and finding abundant, relatively untapped resources (Haynes, 1966). Considering the 10,000 years during which Siberia and Alaska were joined while the development and spread of the Upper Paleolithic mammoth hunters occurred, it is obvious that they would have found their way to Alaska several millenia before the opening of a trans-Canadian corridor. In Alaska they would have further evolved mammoth-hunting skills inherited from their Siberian ancestors, while others continued the blade and microblade traditions inherited from eastern Asia. Upon deglaciation, the mammoth hunters would find their way into central North America, which abounded in big game and was sparsely populated at best. That the Clovis mammoth hunters were supremely successful is attested to by their rapid expansion throughout the continent and by the diversity of cultures derived from Clovis. The phenomenal dispersal of Clovis sites in less than 1000 years is more compatible with a distinct migration from the north in response to removal of the ice block than with a sudden outgrowth from meager indigenous cultures after 12,000 years of sluggish development for which a continuum has not been demonstrated. For such an outgrowth to have fortuitously coincided with opening of an ice-free corridor seems unlikely.

Testing of the hypothesis emphasized here would be aided by the systematic excavation of any and all sites of extinct game animals in Late Quaternary deposits of Siberia and America. Only thus will the negative evidence for man's absence from the New World become statistically significant. In Alaska there must be evidence of at least 10,000 years of progress in mammoth-hunting, and in Canada the geochronology of ice-free corridors is being sought with vigor.

ACKNOWLEDGMENTS

I am indebted to Richard G. Klein for making available to me portions of his forthcoming book on the Kostenki-Borshevo area, and to Alfred E. Johnson for constructive criticism of the manuscript. The helpful comments of A. Dreimanis and W. F. Farrand are appreciated, but I assume all responsibility for errors and omissions. The financial assistance of the National Geographic Society in supporting the Murray Springs excavations and of the National Science Foundation for grant GA-1288 is especially appreciated. I also wish to acknowledge the cooperation of the landowners at Murray Springs, Arizona, Kern County Land Company, and Mr. Andrea Cracchiolo, for allowing us to conduct scientific excavations on their property.

LITERATURE CITED

Albritton, C. C.
  1966. Stratigraphy of the Domebo site. Pp. 11-13, in Domebo: a Paleo-Indian mammoth kill in the prairie-plains (F. C. Leonhardy, ed.), Contrib. Mus. Great Plains, 1:1-53.

Aksenov, M. P., and G. I. Medvedev
  1968. New data on the Pre-Neolithic period of the Angara region. Arctic Anthro., 5 (1):213-223.

Anderson, D. D.
  1968. A Stone Age campsite at the gateway to America. Sci. Amer., 218:24-33.

Armstrong, J. E., D. R. Crandell, D. J. Easterbrook, and J. B. Nobel
  1965. Late Pleistocene stratigraphy and chronology in southwestern British Columbia and northwestern Washington. Bull. Geol. Soc. Amer., 76:321-330.

Broecker, W. S., and W. R. Farrand
  1963. Radiocarbon age of the Two Creeks forest bed, Wisconsin. Bull. Geol. Soc. Amer., 74:795-802.

Bryan, A. L.
  1965. Paleo-American prehistory. Occas. Papers Idaho State Univ. Mus., 16:1-247.

Byers, D. S.
  1959. Radiocarbon dates from the Bull Brook site, Massachusetts. Amer. Antiquity, 24:427-429.

Coe, J. L.
  1964. The formative cultures of the Carolina Piedmont. Trans. Amer. Philos. Soc., 54 (5):1-130.

Commission for the Study of the Quaternary Period (Academy of Sciences, U.S.S.R.)
  1968. Radiocarbon dates from Soviet laboratories, 1 January 1962–1 January 1966. Radiocarbon, 10:417-467.

Dikov, N. N.
 1968. The discovery of the Paleolithic in Kamchatka and the problem of the initial occupation of America. Arctic Anthro., 5 (1):191-203.

Dreimanis, A.
 1966. Lake Arkona Whittlesey and post-Warren radiocarbon dates from "Ridgetown Island" in southwestern Ontario. Ohio Jour. Sci., 66:582-586.

Dreimanis, A., J. Terasmae, and G. D. McKenzie
 1966. The Port Talbot interstade of the Wisconsin glaciation. Canadian Jour. Earth Sci., 3:305-325.

Fitting, J. E., J. DeVisscher, and E. J. Wahla
 1966. The Paleo-Indian occupation of the Holcombe Beach. Papers Mus. Anthro., Univ. Michigan, 27:1-159.

Green, F. E.
 1963. The Clovis blades: an important addition to the Llano Complex. Amer. Antiquity, 29:145-165.

Griffin, J. B.
 1960. Some pre-historic connections between Siberia and America. Science, 131:801-812.
 1965. Late Quaternary prehistory in the northeastern woodlands. Pp. 655-667, in The Quaternary of the United States (H. E. Wright, Jr., and D. G. Frey, eds.), Princeton Univ. Press, Princeton, New Jersey, x+922 pp.

Hayashi, K.
 1968. The Fukui microblade technology and its relationships in Northeast Asia and North America. Arctic Anthro., 5 (1):128-190.

Haynes, C. V., Jr.
 1964. Fluted projectile points: their age and dispersion. Science, 145:1408-1413.
 1966. Elephant hunting in North America. Sci. Amer., 214:104-112.
 1967. Carbon-14 dates and early man in the New World. Pp. 267-286, in Pleistocene extinctions: the search for a cause (P. E. Martin and H. E. Wright, Jr., eds.), Yale Univ. Press, New Haven, Connecticut, x+453 pp.
 1968a. Geochronology of Late Quaternary alluvium. Pp. 591-631, in Means of correlation of Quaternary successions (R. B. Morrison and H. E. Wright, Jr., eds.), Univ. Utah Press, Salt Lake City, xi+631 pp.
 1968b. Preliminary report of the late Quaternary geology of the San Pedro Valley, Arizona. Pp. 79-96, in Southern Arizona guidebook III, Arizona Geol. Soc.
 1970a. Geological stratigraphy of the Union Pacific mammoth site, Carbon County, Wyoming. Unpubl. manuscript.
 1970b. Pleistocene and Recent Stratigraphy of Blackwater Draw, New Mexico, and Rich Lake, Texas. In press, in Paleoecology of the Llano Estacado, 2 (F. Wendorf and J. J. Hester, eds.), Fort Burgwin Res. Center, Taos, New Mexico.

Haynes, C. V., Jr., and E. T. Hemmings
 1968. Mammoth-bone shaft wrench from Murray Springs, Arizona. Science, 159:186-187.

Haynes, C. V., Jr., and A. E. Johnson
 1970. The Leikem mammoth site, Arizona. Unpubl. manuscript.

Hester, J. J.
 1966. Origins of the Clovis culture. 36th Congreso Internacional de Americanistas, 1:129-142.

Hopkins, D. M.
 1967a. The Cenozoic history of Beringia: a synthesis. Pp. 451-484, in The Bering land bridge (D. M. Hopkins, ed.), Stanford Univ. Press, Stanford, California, xiii+495 pp.
 1967b. Quaternary marine transgressions in Alaska. Pp. 47-90, in The Bering land bridge (D. M. Hopkins, ed.), Stanford Univ. Press, Stanford, California, xiii+495 pp.

Hughes, O. L.
 1965. Surficial geology of part of the Cochrane District, Ontario, Canada. Geol. Soc. Amer. Spec. Paper, 84:535-565.

Irwin, C., H. T. Irwin, and G. A. Agogino
 1962. Ice Age man vs. mammoth in Wyoming. Nat. Geogr., 121:828-837.

Kind, N. V.
 1967. Radiocarbon chronology in Siberia. Pp. 172-192, in The Bering land bridge (D. M. Hopkins, ed.), Stanford Univ. Press, Stanford, California, xiii+495 pp.

Klein, R. G.
 1969. Man and culture in the Late Pleistocene: a case study. Chandler Publishing Co., San Francisco, California, in press.

Laughlin, W. S.
 1967. Human migration and permanent occupation in the Bering Sea area. Pp. 409-450, in The Bering land bridge (D. M. Hopkins, ed.), Stanford Univ. Press, Stanford, California, xiii+495 pp.

Leighton, M. M.
 1960. The classification of the Wisconsin glacial stage of North Central United States. Jour. Geol., 68:529-552.

McKennan, R. A., and J. P. Cook
 1969. Prehistory of Healy Lake, Alaska. Paper presented at 8th Internat. Cong. Anthro. Ethnol. Sci., Tokyo-Kyoto, Japan, in press.

Malde, H. E.
 1954. Memorandum concerning the Dent site. U.S. Geol. Surv., Denver Fed. Center, 4 pp.

Martin, P. S.
 1967. Prehistoric overkill. Pp. 75-120, in Pleistocene extinctions: the search for a cause (P. S. Martin and H. E. Wright, Jr., eds.), Yale Univ. Press, New Haven, Connecticut, x+453 pp.

Mason, R. J.
  1958. Late Pleistocene geochronology and the Paleo-Indian penetration into the Lower Michigan Peninsula. Papers Mus. Anthro., Univ. Michigan, 11:1-48.
  1962. The Paleo-Indian tradition in eastern North America. Current Anthro., 3:227-278.

Milliman, J. D., and K. O. Emery
  1968. Sea levels during the past 35,000 years. Science, 162:1121-1123.

Müller-Beck, H.
  1966. Paleohunters in America: origins and diffusion. Science, 152:1191-1210.
  1967. On migration of hunters across the Bering land bridge in the Upper Pleistocene. Pp. 373-408, in The Bering land bridge (D. M. Hopkins, ed.), Stanford Univ. Press, Stanford, California, xiii+495 pp.

Prufer, O. H., and R. S. Baby
  1963. Paleo-Indians of Ohio. Ohio Hist. Soc., Columbus, 68 pp.

Quimby, G. I.
  1958. Fluted points and geochronology of the Lake Michigan Basin. Amer. Antiquity, 23:247-254.
  1963. A new look at geochronology in the Great Lakes region. Amer. Antiquity, 28:558-559.

Ravskii, E. I., and S. M. Tseitlin
  1968. Geological periodization of the sites of the Siberian Paleolithic. Arctic Anthro., 5 (1):76-81.

Roosa, W. B.
  1965. Some Great Lakes fluted point types and sites. Michigan Archaeol., 9:44-48.

Sellards, E. H.
  1952. Early man in America. Univ. Texas Press, Austin, 211 pp.

Stuckenrath, R., Jr.
  1966. The Debert archaeological project, Nova Scotia: radiocarbon dating. Quaternaria, 8:75-80.

Vereshchagin, N. K.
  1967. Primitive hunters and Pleistocene extinction in the Soviet Union. Pp. 365-398, in Pleistocene extinctions: the search for a cause (P. S. Martin and H. E. Wright, Jr., eds.), Yale Univ. Press, New Haven, Connecticut, x+453 pp.

Wayne, W. J., and J. H. Zumberge
  1965. Pleistocene geology of Indiana and Michigan. Pp. 63-84, in The Quaternary of the United States (H. E. Wright, Jr., and D. G. Frey, eds.), Princeton Univ. Press, Princeton, New Jersey, x+922 pp.

Wendorf, F.
  1966. Early man in the New World: problems of migration. Amer. Nat., 100:253-270.

Wormington, H. M.
  1957. Ancient man in North America. Denver Mus. Nat. Hist., Pop. Ser., 4:1-322.
  1962. A survey of early American prehistory. Amer. Sci., 50:230-242.

Early Ceramic Environmental Adaptations

MARVIN F. KIVETT

ABSTRACT

Archaeological sites assignable to the Plains Woodland pattern have been recognized by researchers for some 30 years. Sites of the Western and Central Plains are believed to fall into the general time period between A.D. 1 and 900. Dates for Hopewellian remains are generally earlier, but the prcise time span has not been fully determined. Variations in site locations in relation to stream valleys may suggest climatic or seasonal controls. It is during this period that the first evidence of domesticated plants appears. Sufficient variations occur in the artifact inventory to suggest several distinct foci, which may have occurred at different time periods. The small village sites may represent an adaptation to a hunting and gathering economy, whereas the communal burials suggest some overall organization, perhaps based on religious rituals.

The terms Woodland "period," "pattern," "horizon," "culture," and "complex" all have general significance and meaning to archaeological researchers. As a time indicator, generally accepted dates cover the period of A.D. 1 to approximately A.D. 900 and include a number of culturally related complexes in the Central Plains that have sufficient diagnostic traits to be considered related. It should be noted that dates earlier and later have been reported for certain sites (Neuman, 1967). In some cases, however, these dates are based on shell rather than charcoal.

In contrast to earlier complexes, the presence of distinctive pottery in Woodland complexes suggests semi-sedentary habits. When found in stratigraphical relationship to other pottery complexes, however, materials of the Woodland culture have consistently been at the lowest level of pottery content (Champe, 1946, p. 52).

One of the early references to excavations at an archaeological site of this category in the general area under consideration was by Sterns (1915, pp. 121-127) at the Walker-Gilmore Site in eastern Cass County, Nebraska. In his paper, a complex now recognized as assignable to the Woodland pattern was recognized but not assigned to a particular pattern or time horizon. Fifteen years later, however, Strong (1933, pp. 281-283) continued investigations at the site, referred the materials to the Sterns Creek culture, and suggested its assignment to the "Algonkian" or Lake Michigan culture. At the same time a major point made by Strong (1933, p. 286) in his review of the area was that "the Plains area generally has produced or supported a considerable variety and succession of culture types, indicating that its environmental limitations are not so drastic as has often been believed."

Wedel (1935, pp. 188-189) appears to have been among the first to recognize variations in ceramics of the general Woodland culture and to extend the range of such sites westward into south-central Nebraska.

During the late 1930's and early 1940's, W.P.A.-assisted excavations in Nebraska (Hill and Kivett, 1940) and intensive field work in Kansas by the U.S. National Museum (Wedel, 1959, pp. 542-551) greatly expanded the knowledge of Woodland sites. Geo-

Fig. 1. Walker-Gilmore Site, Sterns Creek, eastern Cass County, Nebraska. Cultural levels are exposed in the terrace cut.

graphically, such remains now have been reported from all areas of the Central and Northern Plains. Although they have certain similarities, the remains are widely distributed and vary sufficiently to suggest their assignment to several distinct complexes or foci. Thus, some 30 years of sporadic field work have served to substantiate the judgment of researchers of the

Fig. 2. Kelso Site, Sandhills of Hooker County, Nebraska.

1930's that these were the earliest pottery-bearing complexes in the Central Plains. The relatively recent use of carbon-14 dating methods has suggested a general time sequence for these complexes.

The earliest reports of Woodland sites came as a result of stream-bed cuts that exposed cultural zones from six to 25 feet below the present surface (Wedel, 1940, p. 304). In the 1930's the people responsible for these sites were commonly referred to as "bog trotters." This may be significant in terms of environment, or it may only suggest seasonal villages (Fig. 1). Camp or village sites may have been on the flood plains in the winter and at higher elevations during the summer, when the lower elevations were more subject to flooding.

There have been certain studies relating to the nature of the overburden on such sites. This varies somewhat in relation to the elevation of the site above the stream bed. Creek-valley sites, such as Walker-Gilmore, immediately adjacent to stream beds, may be buried in and beneath a considerable depth of alluvial material designated as T° to the most recent fill at the site, representing the flood-plain fill or its equivalent (Champe, 1946, p. 71). West of the one-hundredth meridian the overburden more often consists of loess or other wind-laid deposits (Kivett, 1950, pp. 88-89). The Kelso Site in the Nebraska Sandhills, dated at A.D. 800±200, had an overburden of sterile sand up to nine feet in depth (Fig. 2). It should also be noted that the greatest depth of overburden at both the loess- and sand-covered sites was on the northwestern side (Kivett, 1952, p. 35).

Wedel (1959, pp. 381-412) has reported on the stratified Pottoroff Site in Lane County, western Kansas. The lower occupation level, B, was assigned to the Woodland pattern. The nonoccupation level immediately above the Woodland zone is described as buff to light gray in color without visible bedding lines. It may also be of interest to

Fig. 3. Valley cord-roughened vessel, 25VY1.

note that Woodland village levels are characteristically dark gray to black in color when compared to the village level of the later earth-lodge dwellers. Whether this represents a more intensive use of a limited area, a longer occupation, or has other meaning is not clear at this time (Kivett, 1952, p. 35; Wedel, 1959, pp. 390-391).

Wedel (1959, pp. 542-557) has provided detailed as well as summary descriptions of Woodland complexes with particular reference to the Kansas and Nebraska areas. It is not the purpose of this paper to provide detailed comparisons of artifacts, but rather to consider the adaptation of these early ceramic groups to a Central Plains environment.

Pottery remains vary in detail; but with the exception of that assignable to the Hopewellian complex, the ves-

FIG. 4. Habitation floor, 25VY1, Valley County, Nebraska.

sels tend to have elongated bodies, conoidal bottoms, and direct rims with flat or rounded lips. Vessel walls are usually thick and cord-roughened for the Keith and Valley foci (Fig. 3).

Village sites are relatively small in area, but for the most part they have sufficient village refuse to suggest an occupation of some duration. Thus they cannot be considered temporary campsites. House remains are not well defined, but certain sites have revealed irregular post patterns and, in some cases, circular basins ranging from eight to 18 feet in diameter and excavated up to two feet into the subsoil (Fig. 4). Hearth areas often occur near the central portion of the basin (Hill and Kivett, 1940, pp. 153-161). Such remains with limited posts suggest a structure of some permanency as compared to the later skin tipi. It may be significant that sites with basin-house remains have been reported primarily from the central and western parts of Nebraska (Hill and Kivett, 1940; Kivett, 1952). Such semisubterranean house floors may represent an initial adaptation that was to be more fully developed in the later substantial plains earth-lodge. On the other hand, house remains from the later Sterns Creek focus, best known from the Walker-Gilmore Site of eastern Nebraska, seem to lack this basin floor. Wedel (1940, p. 305) has described the remains as including masses of thatch and small post molds, indicating that the probable type of habitation was an eastern rather than a western form.

As indicated, village sites are small, seldom exceeding a few acres in extent. Several, particularly those west of the one-hundredth meridian in Nebraska, cover a very limited area and consist of one to four small house sites, perhaps representing single-family groups. These sites are often on the first terrace immediately above the flood plains at an elevation generally below that of the later earth-lodge sites assignable to the Upper Republi-

Fig. 5. Location of Site 25VY1, immediately above the stream bed.

can Aspect (Fig. 5). Whether this placement reflects climatic difference, an attempt at concealment, or some other basis cannot now be determined (Kivett, 1949, p. 282). Wedel (1953, p. 506) summarized the present thinking: "On the whole, most Woodland manifestations in the Central Plains look primarily like a creek valley hunting and gathering economy, characteristically with small population aggregates—perhaps limited family groups. Even where these remains are found far out in the short-grass plains, the picture is much the same—a mode of life, one thinks, that was probably carried west out of a forest margin habitat into the tree-fringed creek valleys of the plains."

It is significant that on the basis of present research the first limited evidence of maize horticulture appears in the Woodland pattern. The Hopewell village sites near Kansas City, with their maize and bean remains, abundant storage pits, and deep village refuse (Wedel, 1943), suggest more stable communities than do those sites to the west in Kansas and Nebraska. Corn has also been reported from a Woodland village site in Platte County, Nebraska (Kivett, 1952, pp. 57-58).

It has been indicated on the basis of limited research (Strong, 1935, p. 268; Kivett, 1952, p. 39) that animal-bone refuse suggests a greater reliance on deer, small mammals, and birds than on bison. Perhaps this statement should be reexamined in the light of the additional sites that now have been excavated within the range of the bison, as compared to those in the eastern forested areas of the Central Plains that were the primary subject of earlier Woodland reports. Relatively few detailed studies of the faunal remains have been made, and in many cases deer and pronghorn antelope have been classified in a single category.

The range of artifacts from western Woodland sites is extensive when compared with earlier, nonpottery complexes, but somewhat limited in comparison with the late prehistoric complexes. The bison shoulder-blade hoe and other artifacts utilized as gardening tools are rare or absent. Exceptions include a possible bison bone-digging stick tip from the Valley focus (Hill and Kivett, 1940, p. 167).

Present evidence suggests the use of both the atlatl, or throwing stick, and the bow and arrow during much of the Woodland period. This suggestion is based on the variation in the size of projectile points (Fig. 6) and the excavation of atlatl weights from a

FIG. 6. Chipped stone from 25VY1, showing the range in size of projectile points.

number of Woodland sites in the Central Plains (Neuman, 1967, pp. 36-53).

As indicated, the faunal list includes a wide range of birds and large and small game animals. Other remains have been identified as dog. In addition, most sites yield a considerable amount of fresh-water mussel remains, suggesting that this was an important food source. The shells were also utilized as tools and ornaments.

Although chipped stone tools made from local chert outcrops or from glacial and stream deposits are common, polished and ground stone tools are rarely present in the western Woodland sites.

Bone was utilized to provide nearly the full range of artifacts found in the later complexes with the exception of the specialized agricultural tools. Awls and needles that were used for skin-working and perhaps for weaving fabrics are present. Fish remains as well as fishhooks are rare or absent.

Pipes, giving evidence of the habit of smoking, are rare or absent in the western Woodland complexes, although present in the Hopewell sites of Missouri and Kansas. Clay pipes are also present in the later Sterns Creek focus in the eastern Central Plains.

Of interest in the inventory of

Fig. 7. Burial from Woodruff Ossuary 14PH4, with shell disk beads. Photograph by the Smithsonian Institution.

artifacts is the presence of marine and fresh-water shell, often fashioned into beads and other ornaments (Fig. 7). The bulk of such ornaments has come from burials (Kivett, 1953). In contrast to the small, family village sites, which may suggest an independent society lacking community organization, certain of the burial sites do suggest some overall cooperation and organization. It is evident that certain of the earthern mounds in Kansas and the ossuaries of northern Kansas and southern Nebraska cannot be assigned to a single-family village. Such ossuaries may contain the secondary buri-

FIG. 8. Excavations at end of ridge mark 25HK13, one of the numerous small Woodland sites in southwestern Nebraska, located adjacent to a small tributary.

als of 20 to more than 60 individuals. The abundance of grave goods, largely in the form of shell beads fashioned from fresh-water mussels, represents a considerable expenditure of time and effort (Kivett, 1953, p. 137). There can be little doubt that such burial grounds were utilized by a number of the small villages, perhaps ranging up to several miles apart (Fig. 8). The presence of some native copper, conch shell, and other marine forms also suggests trade routes and at least some overall organization. Such remains may indicate burial and religious rituals much more complex than those suggested by the village remains.

One may cite the small villages of family size as evidence of a lack of stability. On the other hand, such small groups may represent an adaptation to the game and fuel supply. Lacking strongly developed gardening practices, the Woodland people were largely dependent upon a hunting and gathering economy. Such small groups would be able to occupy one or two villages, perhaps on a seasonal basis, without exhausting the food and fuel supply. Their greater reliance on a stream-valley economy would lessen the impact of periodic droughts and game shortages and enable them to expand more readily up the stream valleys to the western Central Plains without exhausting available resources. It is to be expected, however, that certain of the ceremonial and religious rituals might diminish as the need increased for greater effort in securing a food supply.

One of the final tests of whether a people adapt to their environment would seem to be the length of time that they subsist in a particular area. Our present evidence suggests that the Woodland culture, in varying but recognizable form, lasted nearly 1000 years. In contrast, our present culture has adapted to the Central Plains for a period of slightly more than 100 years, and during the drought of the 1930's it was often doubtful whether certain areas would continue to be occupied.

LITERATURE CITED

Champe, J. L.
    1946. Ash Hollow Cave. Univ. Nebraska Studies, n.s., 1:ix+1-104.

Hill, A. T., and M. F. Kivett
    1940. Woodland-like manifestations in Nebraska. Nebraska Hist., 21:147-243.

Kivett, M. F.
    1949. Archeological investigations in Medicine Creek Reservoir, Nebraska. Amer. Antiquity, 14:278-284.
    1950. Archeology and climatic implications in the Central Plains. Proc. Sixth Plains Archeol. Conf., 1948, Anthro. Papers, Univ. Utah, Dept. Anthro., 11:88-89.
    1952. Woodland sites in Nebraska. Nebraska State Hist. Soc., Publ. Anthro., 1:1-102.
    1953. The Woodruff ossuary, a prehistoric burial site in Phillips County, Kansas. Bull. Bur. Amer. Ethnol., 154:103-141.

Neuman, R. W.
    1967. Atlatl weights from certain sites on the Northern and Central Great Plains. Amer. Antiquity, 32:36-53.

Sterns, F. H.
    1915. A stratification of culture in eastern Nebraska. Amer. Anthro., 17:121-127.

Strong, W. D.
    1933. Plains culture area in light of archeology. Amer. Anthro., 35:271-287.
    1935. Introduction to Nebraska Archaeology. Smithsonian Misc. Collections, vol. 93 (10):vii+1-323.

Wedel, W. R.
    1935. Reports on field work by the Archeological Survey of the Nebraska State Historical Society, 1934. Nebraska Hist., 15:132-255.
    1940. Culture sequence in the Central Great Plains. Smithsonian Misc. Collections, 100:291-352.
    1943. Archeological investigations in Platte and Clay counties, Missouri. Bull. U.S. Nat. Mus., 183:viii+1-284.
    1947. Culture chronology in the Central Great Plains. Amer. Antiquity, 12:148-156.
    1953. Some aspects of human ecology in the Central Plains. Amer. Anthro., 55:499-514.
    1959. An introduction to Kansas archeology. Bull. Bur. Amer. Ethnol., 174:xvii+1-723.

Aspects of Adaptation Among Upper Republican Subsistence Cultivators

Richard A. Krause

ABSTRACT

Many recent archaeological and ethnological studies have been devoted to showing the adaptive value of various social institutions. Such inquiries usually stress the interaction among technology, social practices, and environmental variables within an analytical framework, which may be called ecological in orientation and perspective. Patterned relations between elements of the cultural, social, and technological orders are often clear enough among simple hunting-gathering peoples. But can the ecological approach be as useful when applied to agricultural societies whose subsistence techniques are more efficient and allow greater latitude for borrowing and divergent internal development?

It is suggested in this paper that selected features of the social life of Upper Republican subsistence cultivators may be usefully analyzed in an ecological frame of reference. Others have argued that the density and distribution of a population along with the size, composition, distribution, and degree of permanency of settlements have verifiable adaptive significance in most environmental contexts. To these factors may be added the role of the sexes and the family, and the composition and size of communal groups formed for such economic enterprises as hunting, gathering, fishing, or farming. I will attempt to generate the data necessary for an examination of these demographic and cultural factors and assess the adaptive role of each insofar as this is possible through a study of the archaeological data.

INTRODUCTION

The 19th-century archaeologist attempted to document an evolutionary succession of social and technological orders. The operational concept developed for that study was one of sequent stages of technological complexity, and analysts concentrated on the cumulative aspect of the evolutionary process. As success was achieved in this inquiry, many archaeologists turned to the study of individual cultures, correlating them in time and space and tracing their genetic relationships. These studies focused upon cultural history, and the basic conceptual unit was the individual culture. The relationship between cultures was studied in terms of the development and spread of artifacts or artifact complexes. Environment was considered a limiting factor; trait mobility, rate of diffusion, and innovation or invention were seen as the active or creative agents in social development.

More recent research has centered upon cultural evolution as a process involving the active interaction among elements of technology, social practice, and environment. Julian Steward, the first anthropologist to make substantial use of this approach, labeled it "cultural ecology." According to Steward, "cultural ecology" deals with "the adaptive processes by which the nature of society and an unpredictable number of features of culture are affected by the basic adjustment through which man utilizes a given environment" (*in* Tax, 1953, p. 243). In retrospect it is evident that Steward's work marked a turning point in archaeological theory. As Geertz (1963, p. 2) pointed out, Steward's analysis goes beyond the traditional approaches of anthropogeography and possibilism by selecting only those features of environment and cultural pattern that are functionally interrelated. We need no longer accept either a gross determinism or passive environmental limitation in our analyses of specific cultural practices. In sum, then, cultural ecology is an important concept for the archaeologist, because it directs attention to specifying "the human relations between selected human activities, biological transactions, and physical processes by including them within a single analytical system, an ecosystem" (Geertz, 1963, p. 3).

Patterned relations among elements of the ecosystem are clearest among technologically simple hunting and gathering peoples. Steward (1955, pp. 122-150), for example, was able to convincingly demonstrate the adaptive advantage of small localized patribands, in the presence of a bow-spear-club technology and scattered nonmigratory game source. But is the ecological approach as useful when it is applied to agricultural societies whose subsistence techniques are more efficient and allow greater latitude for borrowing and divergent internal development? As Steward (1938, p. 257) himself pointed out: "When, however, ecology allowed latitude in subsistence activities non-economic factors such as warfare, festivals, ceremonies, etc. became determinants of socio-political patterns." Unless technology and environment alone determine the rest of culture, there is latitude for variation in the social forms associated with a pattern of adaptation to local environment. One problem confronting the ecological study of agricultural peoples is, then, the selection of those aspects of social life which are functionally related to environmental variables. Another, perhaps more important, problem lies in determining which of these relations have demonstrable adaptive significance.

Steward (1938), Netting (1965, p. 87), and others have suggested that population density and distribution, together with the size, composition, distribution, and degree of permanency of villages, have verifiable adaptive significance in most environmental contexts. To these factors may be added the role of the sexes and the family, and the composition and size of communal groups formed for such economic enterprises as hunting, gathering, fishing, or farming. All of these do, indeed, seem to be interrelated aspects of adaptation. Taken together they form the complex of factors in Steward's (1955, p. 37) "culture core" and the "social instrumentalities" discussed by Netting (1965, p. 86). But the possibility of independent variation among factors is strong enough, and the interfactor relationship complex enough, to make separate treatment an analytical desideratum.

A few examples of the complexity of interfactor relationships may be instructive. A region's natural resources, for instance, may set an upper limit to, but do not determine, population density. While it is true enough that in technologically simple hunting-gathering societies the food yield per area is often small and the population density is low, even a cursory inspection of population maps will show that agriculture is not invariably accompanied by greater population densities. Factors other than technology and environment must be called upon to explain such situations. The general tendency for agriculture to result in population increases may, for example, be balanced by cultural practices affecting health and fertility, morbidity, or fecundity. Then, too, effective use of technological capabilities may be hampered by the accuracy and extent of traditional knowledge about, or attitudes toward, climatic, edaphic, and biotic factors.

Still other cultural and economic

forces may shape significant social alignments. Endemic warfare may influence the areal distribution of a population as well as alter the age and sex composition of villages, households, and communal work groups. Involvement in trade may put a premium on the drawing together of interdependent craftsmen, may induce the participants to settle along trade routes, or may reshape the traditional economic role of the sexes. Finally, historical factors must not be neglected. Patterns of social life, which evolved in one area, may be transferred to another and only slowly modified to suit conditions in the new region. In short, each and every factor bears close examination in each and every case.

In this paper I suggest that selected features of the social life of Upper Republican subsistence cultivators may be usefully analyzed in an ecological frame of reference. First, however, I must clarify several theoretical points associated with the organization and analysis of data. As I hinted above, adaptation is my major concern, not trait origin nor the mechanism of trait transmission. This is not to say that trait mobility and rate of diffusion are unimportant, merely that they must be considered a subproblem of ecologically oriented research. More important for my purpose is the existence of ecological boundaries beyond which a trait, or set of traits, may gain or lose adaptive significance.

It must also be understood that what is to be studied is neither a teleological nor a unidirectional process. Traits that have adaptive value do not necessarily arise as a response to need, and there is no question of causality in my analysis. Hence, I hope to stay within the bounds of a theory of evolution that provides functional explanations for the fixation or loss of random events within a defined system.

Then, too, it must be emphasized that the basic unit of my study will be society, defined as a spatially bounded human population, whose members may be characterized by a specific range of biological and cultural traits. Society in this sense is a statistical concept and does not imply a unit in any way greater than the sum of its parts. It is a system understood in terms of its parts and statable relations among them. I reserve the term "culture" to denote traits shared by a significant number of individuals and transmitted through the learning process. This distinction is standard in social anthropology and should facilitate the archaeological study of social, cultural, and environmental variables as interacting factors in the adaptive process.

In keeping with the above provisions, no discussion of the origins of Upper Republican life will be offered. Instead, I will confine myself to Upper Republican complexes in north-central Kansas and south-central Nebraska in an attempt to generate data for a study of the demographic and cultural factors discussed previously. Finally, the adaptive role of various factors will be assessed, insofar as this is possible through a study of the environmental and archaeological data to which I shall now turn.

SUBSTANTIVE DATA

ENVIRONMENT

The most intensive Upper Republican occupation of the Central Plains lay east of the 100th meridian in the vast area classed as tall-grass prairie (in effect, the southern half of Nebraska and the north-central portion of Kansas). This region is blanketed with a layer of loess originally covered with dense sod formed under grass one to three feet high (bluestem, prairie, and spear grasses were the most abundant species). Generally speaking, the rivers flow through broad valleys having extensive bottomlands and flood plains, which support mixed stands of cottonwood, ash, elm, willow, box elder, and black walnut, as well as herbaceous undergrowth. The soils of the riparian area are rich and fertile, and the interfluvial areas, for the most part, are open and rolling, although steep-walled ravines dissect large areas in the west. The climate is harshly continental with long, cold winters and hot, dry summers. Today the growing season averages 120 days and annual rainfall about 20 inches.

The tall-grass prairie and adjacent wooded bottomland supported a relatively rich and varied fauna. The grass-covered uplands were the heart of the old bison range. Large herds of these animals previously inhabited the area, which they shared with a number of symbiotes, among them jackrabbit, badger, prairie dog, and prairie chicken. The predators—cougar, bobcat, gray wolf, and coyote—lived primarily in the wooded bottoms but made forays into the grasslands to scavenge and hunt. White-tailed deer, elk, and mule deer foraged in the grasslands but often used the bottomland stands of timber and undergrowth in which to bed down. Black bear, beaver, raccoon, and porcupine were more perennial inhabitants of the wooded bottomlands and river banks. Waterfowl nested on the rivers; both the main streams and their perennial tributaries contained a variety of fish, turtles, and fresh-water clams.

By A.D. 1200, perhaps earlier, the river and creek valleys of the Central Plains seemingly supported a network of Upper Republican hamlets and homesteads. People bearing an Upper Republican culture apparently settled the major streams in Kansas first and spread from here to the smaller perennial tributaries and nearby terracelands in north-central Kansas and south-central Nebraska. Unfortunately, the details of population growth and dispersal have yet to be worked out on an area-wide scale. Hence, while I may draw upon Upper Republican archaeological complexes in several areas for comparative materials, I must, by necessity, restrict my intensive analysis to a single locale. The Glen Elder locality of the Solomon River Valley was chosen because it is the area I know best. There were at least three different kinds of Upper Republican settlement in the locality, which can be placed in an interesting developmental sequence when combined with radiocarbon dates and an assessment of trends toward change in material culture and settlement pattern. Hopefully, this sequence is representative of events and processes that were replicated in other Central Plains riverine settings.

THE UPPER REPUBLICAN COMPLEXES OF NORTH-CENTRAL KANSAS

1. The earliest and most impressive of the Glen Elder Upper Republican settlements were small farming hamlets, composed of from six to about 10 large lodges covered with timber, grass, and earth. These lodge clusters were situated upon prominent, windswept first terraces of the North or South Solomon River, in full view of any incidental passerby and about one-half mile to a mile from the junction of a feeder creek with the river. There was usually a special work-storage area along one edge of the hamlet—a place where food was stored in underground chambers and a place where at least some tools were manufactured and other tasks were performed. While these hamlets varied in size, a population range of 60 to 120 individuals does not seem unreasonable in view of the size and apparent permanence of the lodges within them.

The houses in these early hamlets were substantial rectangular structures with a relatively massive four- to six-centerpost superstructure. Lodges were 25 to 35 feet long, and 20 to 30 feet wide, with long, broad, rectangular, covered entranceways. From two to nine deep, wide, subterranean storage chambers usually were dug beneath the lodge floor, along the side walls, and back wall of the lodge, away from the central domestic hearth. The artifact content of such lodges was rich and varied, indicating a degree of residential permanence. Fifteen persons could be comfortably sheltered within a single lodge, and an estimate of 10 to 15 inhabitants would, I suspect, be reasonable, if somewhat low. Cylindrical pits filled with fine, whitish wood ash accompanied several houses, as did carefully prepared caches of raw materials and stored tools. Both practices suggest periodic abandonment with reoccupation intended.

The economic basis for hamlet life seems to have been provided by mixed hunting-collecting and gardening, with a surprising emphasis upon collecting. Walnut-stained stones for crushing and grinding nuts, some of elaborate manufacture, are a relatively frequent and integral part of hamlet debris, as are sizable heaps of fresh-water mussel shells, and more modest deposits of hackberry and sunflower seeds. Bison and more solitary game animals were hunted, probably with lance and bow and arrow. But the scarcity of butchered animal bone in the hamlet detritus and the presence of sizable stores of unmodified bison scapulae, elk antlers, and deer racks point to out-hamlet butchering. Bison-scapula hoes, shelled maize, and the charred remains of beans and squash are direct evidence for the cultivation of river-bottom garden plots.

The burial practices of these hamlet dwellers suggest a hitherto unsuspected degree of ceremonial elaboration with an interesting emphasis upon dedicatory reburial. The single excavated cemetery lay upon the same terrace as the associated hamlet and contained two distinct clusters of burial pits. The majority of these pits were devoid of skeletal material or cultural debris. Pottery fragments were, however, recovered from some pits, as were finger and toe bones; and one pit contained the flexed burial of an aged male. These two pit clusters were spread around a common burial chamber, which contained the disarticulated remains of an estimated 50 to 60 individuals and two complete pottery jars whose size, shape, and contents suggested a ceremonial function. The two associated vessels, the two discrete clusters of individual burial pits, and definable concentrations of bone in the common chamber all support the idea that this cemetery was in use over a long period of time. The dead were evidently first placed in individual bell-shaped chambers and later exhumed to be ceremoniously reinterred in a common pit together with special grave goods.

2. The second kind of Upper Republican settlement consisted of three or four households, which occupied smaller, less substantial lodges hidden away along the timbered banks of one of the Solomon's feeder creeks. These houses were not clustered together, as were their counterparts in the earlier farming hamlets, but instead were built 100 to 200 yards apart in a rough line paralleling the creek bank. These settlements also lacked special work-storage areas and seem to reflect a less substantial clustering of the local populace.

The houses scattered along feeder creeks were built on the same basic plan as those in the hamlets, but were smaller and of lighter and less regular construction. They were only 12 to 15 feet wide and 12 to 15 feet long and had narrower, rectangular, covered entranceways. The superstructure was built of less substantial timbers, and the walls were of smaller poles, less carefully and closely placed. These structures seemingly reflect far less expenditure of labor than their hamlet analogues and would be crowded indeed by as many as 15 occupants.

Again, the earlier emphasis upon collecting appears to have waned, to judge by the far fewer nutting stones and less substantial amounts of mussel shell in and around small lodge remains, although this change may be more apparent than real. The same variety of environmental resources was exploited, but perhaps less intensively. Finally, the artifact content was more sparse and less varied than in earlier lodges; storage pits were smaller, and stores of raw material and other evidence of periodic abandonment and intended reoccupation were lacking.

3. The third type of Upper Republican settlement was situated on a high first terrace, or old natural levee, immediately overlooking the Solomon River. From all appearances these were seasonal fishing or hunting camps. They are characterized by sizable heaps of clam shells, scattered post holes, and fire basins, but there are no definable houses or permanent domestic structures. The artifact content is relatively sparse, consisting of pottery fragments, and bone and stone tools spread unevenly through heaps and layers of shell. The storage chambers are small and are usually packed with clam shells rather than stored tools, raw materials, or the usual run of domestic garbage.

The available carbon-14 dates and an inspection of changes in domestic architecture and trends in ceramic change seem, at present, to support the relative chronological order developed above: hamlets early, creek-bank settlements and seasonal hunting-fishing camps later. The seven carbon-14 determinations on charcoal from farming hamlets suggest that an occupation between the 9th and 13th centuries A.D. is a reasonable estimate. The ceramics also suggest a relatively early placement—a high incidence of flared, cord-roughened rims with plain or incised lips and a correspondingly low proportion of incised collared rims with plain or incised lips. The relatively low proportion of collared rims with finger-pinched lower collars and incised upper collars also indicates an early placement.

The carbon-14 dates from creek-bank settlements and seasonal camps suggest a later placement, but are too few to serve as guidelines for assessing the span of occupation. Nevertheless, an examination of domestic architecture, ceramics, and other features of artifact content place these sites within the temporal range assigned to Upper Republican settlements in the Republican River drainage. In ceramics there is a relatively higher proportion of collared, incised, and pinched and incised collared rims—a situation with close parallels in hunting camps in western Nebraska and eastern Colorado. The domestic architecture of the creek-bank settlement is more like that in Nebraska sites, and the reduction in the number of nutting stones and in the size of shell heaps also points to conditions farther north and west where these features are rare. On the basis of this information, a time span encompassing the 11th to 15th centuries A.D. is practicable.

THE UPPER REPUBLICAN COMPLEXES OF SOUTH-CENTRAL NEBRASKA

The Upper Republican inhabitants of the Republican River drainage apparently lived in much the same way as did their homestead-dwelling counterparts in the Solomon Valley. According to Wedel (1961, p. 95), Nebraska's Upper Republican groups "lived in single houses, randomly scattered at intervals of a few yards to several hundred feet, or in clusters of two to four lodges, similarly separated from other small clusters or single units." Wood (1969) has characterized this spread-out pattern of settlements as one of neighborhoods composed of semi-independent homesteads or homestead aggregates. The occupants of these homesteads apparently utilized the full range of natural resources within the reach of their knowledge and technology. Corn, beans, and squash were cultivated in river-bottom gardens and stored in underground chambers; and both herd and solitary game mammals were hunted, as were a surprisingly wide range of smaller local animals. Falk (1969, pp. 44-51), for example, listed 29 species of small animals, among them mammals, birds, reptiles, amphibians, and crustaceans, from a single lodge.

The intense utilization of local fauna led Falk (1969, p. 51) to argue that "a number of small semi-independent groups strung out along stream valleys, would soon over-exploit local resources on a year round basis." Similar considerations led Wood (1969) to suggest that Nebraska's Upper Republican groups engaged in major communal hunting forays in the adjacent plains. There are, for example, campsites that indicate hunting parties ranged far to the west of their homes and gardens. Pottery and other Upper Republican artifacts occur in Ash Hollow Cave, Signal Butte, and other sites in the Nebraska Panhandle; in southeastern Wyoming; in the Agate Bluff rock shelters in northern Colorado; and in the South Platte, Republican, and Arkansas drainages in eastern and northeastern Colorado (Bell and Cape, 1936, pp. 357-359; Champe, 1946, pp. 47-50; Strong, 1935, p. 229; Irwin and Irwin, 1957, pp. 15-33; Withers, 1954, pp. 1-3; and Morton, 1954, pp. 30-41).

Ossuary burial was still practiced by Nebraska's homestead-dwelling population, but with several noteworthy modifications. Ossuaries were no longer situated on river terraces adjacent to lodge sites, but instead were positioned upon prominent heights overlooking the neighborhood. Then, too, empty pits for primary deposition of the dead have not been found surrounding the common burial chamber, and their lack led Wedel (1961, p. 96) to posit exposure as the first stage in the cycle of dedicatory reburial. Exposure or, for that matter, primary interment in pits not surrounding the common chamber would seem to be significant alterations of the burial pattern, as would the relocation of the common burial chamber from occupied terrace land to prominent points.

DISCUSSION

Both the overall density and the areal distribution of Upper Republican populations seem directly related to factors of environment and technology. The available bone agricultural tools made the sod-covered, grassy uplands an effective barrier to cultivation and limited arable land to easily turned alluvial creek- and river-valley soils. Thousands of acres thus lay beyond the means of the Upper Republican cultivator. Hence, population growth and the development of those social forms functionally related to agriculture were dependent upon success in utilizing a rather limited sector of the total environment. Settlements of any size and permanence were literally tied to the river bottoms and adjacent terracelands, and a combination of environment and technology is sufficient to explain the concentration of sites in these regions. But while factors of environment and technology may define arable soils and the areal distribution of populations practicing agriculture, these two factors do not alone determine soil use or overall population density.

The available evidence indicates that Upper Republican peoples were shifting cultivators who did not use many of the techniques within the range of their technical capabilities. Among these were bench terracing, which would allow the use of otherwise unarable tree-covered slopes, and ridging techniques, which would prevent runoff, hinder erosion, and conserve moisture in dry years or in perennially dry fields. Ditching procedures for the reclamation of potentially rich but waterlogged areas and irrigation practices that utilize the potential of perennial springs and small streams apparently were yet other unrealized potentialities. Lacking, too, was heavy annual use of compost enriched with human or collectable animal dung, although the stores of fine, white wood ash in some early houses may have been a step in the direction of more intense fertilization. In Africa, for example, groundnuts and sweet potatoes are fertilized with carefully saved household ash (Netting, 1965, p. 86). Admittedly, the plains environment with its annual temperature extremes and seasonal rainfall may limit the productive potential of some of these practices, but their absence is in no way enjoined by the environment.

Similar considerations may apply to the use of faunal and floral populations. While we may assume that the location of stands of walnut, sunflowers, wild turnips, and the like were known and that these plants were used by Upper Republican peoples, we can adduce no evidence of attempts to manage them. Purposeful planting or tending, the planned elimination of competitors, or careful environmental manipulations, which would alter growth and maturation rates or encourage the spread of desirable species, all seem to have been lacking. Likewise, the creation of artificial environments for fish and shellfish and the selective cropping of resident bison and pronghorn antelope populations were seemingly not practiced. Intensive farming and successful management of natural resources both require heavy investments of time and labor and a substantial degree of residential permanence during the spring, summer, and autumn when hunting, fishing, and collecting were most productive. Residential stability could have been achieved through careful resource management, but judging by the archaeological record, it was not. Instead, Upper Republican peoples ranged considerable distances in search of game and collectable plant foods and in so doing, abandoned their dwellings and gardens for appreciable periods. All of this is perhaps an overly complex way of saying that the use made of the environment is dependent upon the user's knowledge of climatic, edaphic, floral, and faunal potentialities as well as factors of technology and environment.

In effect, then, the actual population density permitted by Upper Republican economic practices should lie between that characteristic of hunter-gatherers and that possible for shifting cultivators in temperate regions. Earlier I noted the seeming concentration of sites along virtually every stream with arable bottomland and pointed to environment and technology as responsible. The large number of sites superficially suggests a teeming population, but given Upper Republican economic practices, these many sites are probably the consequence of short-lived dwelling units. On the basis of the faunal and floral remains at the Mowry Bluff site, Wood (1969) suggests a duration of five years per dwelling unit, and I suspect that by far the greater number of Upper Republican peoples were thinly scattered over the terrain, with mobility and short-lived settlements responsible for the apparent number of sites—perhaps a tenth or less of which may have been occupied at the same time.

Hamlet life and the posited shift from hamlet to homestead also need careful examination from an ecological point of view. The size and permanence of local settlements permitted by shifting cultivation have been the center of considerable controversy. Most authorities agree that low population densities and shifting cultivation are concomitant phenomena. But although shifting cultivation may limit the population density, it by no means precludes the gathering together of the agricultural populace into villages of considerable size and permanence. One has but to look to the fortified prehistoric towns of the Dakotas to measure the possibilities of village size in a plains riverine environment. Shifting cultivation obviously does not prohibit, but neither does it encourage, the growth of sizable population aggregates. Where they are found, factors of trade or of warfare (or both) may be responsible. Elsewhere I have argued that war and trade shaped the density and distribution of protohistoric and early-historic populations along the Missouri River in South Dakota (Krause, 1967). Similar arguments have been used to explain the size, permanence, and distribution of historic Iroquois settlements (Trigger, 1963); and the long-range demographic effects of war and trade have been measured against the extensive archaeological sequence in Egyptian Nubia (Trigger, 1965).

Warfare and trade, however, are not reasonable explanations for the formation of early Upper Republican population aggregates. Indeed, exotic, or extra-local, trade items have been found in Upper Republican sites; marine shells of ultimate southeastern provenience accompany both burial and dwelling sites, copper-covered discs were recovered from a single ossuary, and carved stone pipes of exotic design have been found in several houses. These items, however, can represent no more than a trickle. Certainly they do not suggest the volume of trade necessary to induce Upper Republican peoples to concentrate along trade routes. Nor is there evidence of extensive warfare or raiding. Fortifications and archaeologically determinable incidences of violence are notably lacking in the Upper Republican sequence, and one certainly could not argue that Upper Republican hamlet or house sites were chosen with defensibility in mind.

At this point I wish to suggest that historical factors were responsible for hamlet living. It is an anthropological axiom that socio-cultural adaptations indigenous to one area may be transferred to another where they are slowly modified to suit new conditions. That this may be the case in the Glen Elder locale is suggested by the following admittedly speculative evidence. The ceramics accompanying Glen Elder hamlets show considerable evidence of struggle and experimentation: there are numerous instances of warping, spalling, and shattering and of the use of diverse vessel forms, tempers, and clays. Evidently early Upper Republican potters worked within the framework of a well-developed manufacturing tradition, but had not fully adjusted their manufacturing practices to the possibilities and limitations of local raw materials. Then, too, the number of species of small animals utilized by hamlet dwellers seems restricted in view of the wide range of such forms used by their cultural heirs and relatives in south-central Nebraska. This latter judgment may, of course, reflect sampling error as well as actual practice. It also seems unusual that there was an apparent emphasis upon extensive walnut collecting and the use of communal work-storage areas, which may likewise be examples of newly introduced traits. Wherever the ultimate origin of these practices, it seems fairly evident that together with hamlet dwelling they were an integral part of a new way of life in the Central Plains.

Farming hamlets, however, did not become the predominant mode of Upper Republican settlement, and a combination of factors may be responsible for this. One important limitation to shifting cultivation is the amount of land necessary per capita. Hamlets must have been surrounded by the fields upon which their inhabitants were dependent, and heavy use of unimproved land and unmanaged wild faunal and floral resources would almost invariably be accompanied by declining yields. When combined with problems of transportation, declining agricultural yields and depletion of reserves of wild food would force either the periodic relocation of hamlet units or the redistribution of individual lodges. The latter seems to have been the preferred, if not the only, solution; and a reasonable explanation can be tendered by examining the adaptive value of the two modes of settlement.

I have already suggested that population density was low. There should have been enough tillable land and bison for all in most localities, not to mention herds of pronghorn antelope and more modest resources—rodents, fish, shellfish, and crayfish included. Agricultural produce and wild foods could be stored against periods of want, and meat could be dried, but a continuing need for fresh meat and skins for blankets, robes, and clothing could not thus be satisfied. Nor could the local reserves of collectable wild foods be efficiently utilized by a concentrated population. The local herd and solitary game animals and the collectable wild foods would soon be exhausted, and trips farther and farther afield could be anticipated by a hamlet-dwelling population. The effort required in hunting and gathering would increase with the distance one had to go from a settlement, and transportation would be a paramount problem, especially in the winter months when dog traction would be difficult if not impossible. Economic and social inducements for the continuation of hamlet life, trade, warfare, access to improved land, or managed natural resources, all seem to have been absent; under these circumstances a spread-out pattern of isolated homesteads or homestead aggregates would seem to be of significant adaptive value. In short, a spreading out of the local populace would minimize the problems of transportation and maximize the efficiency of shifting cultivation, hunting, and collecting. Generally speaking, a dispersed pattern of settlement is an adaptation reflecting dependence on hunting and fishing.

Baerreis and Bryson's (1965) data on climatic shifts suggest that additional pressures were brought to bear upon hamlet-dwelling populations. According to these authors, from about A.D. 800 to A.D. 1250 subtropical anticyclones brought moist tropical air and abundant summer rainfall into the Central Plains. The abundant rainfall apparently supported the westward expansion of maize horticulture by Upper Republican peoples, and it was during this favorable climatic period that hamlet dwelling seems to have flourished in the Glen Elder locality. About A.D. 1250, however, a sharp change in atmospheric circulation seemingly took place, with moist, warm summers giving way to cooler, drier conditions. Baerreis and Bryson (1965) posited catastrophic consequences for plains horticulturists, but the intensity and extent of climatic change have yet to be accurately measured in the full range of Central Plains physiographic and environmental contexts. Nevertheless, the posited change in the Upper Republican settlement pattern corresponds fairly well with the suggested change in climatic conditions. It seems reasonable, then, to suggest that climatic change be considered a deviation-amplifying process (Maruyama, 1963, p. 164), working with the previously discussed factors to stimulate a shift from hamlet to homestead dwelling.

To carry the analysis further is admittedly speculative, but I don't think the integrative and socializing functions of group co-residence and communal work-storage areas should be neglected. The posited shift from hamlet to homestead would almost certainly be accompanied by significant alterations in the intensity and scheduling of day-to-day, face-to-face social interaction. The performance of common domestic tasks and socialization of the young in communal contexts should develop social bonds that tend to mute the divisive pull of separate domestic group interests. With the lapse of hamlet life and the spreading out of local populations, the pattern of social interaction must have been modified. Such suprahousehold integrative and socializing mechanisms as existed would lose at least some of the reenforcing influence of residential proximity and common performance of daily domestic tasks.

The ethnography of the plains contains many examples of the social and ceremonial modifications that seemingly accompanied increased mobility and population dispersal. It is in this light that the inferred modification of Upper Republican burial practices seems most significant. It should, for example, be reasonable to assume that dedicatory reburial played a significant role in promoting social solidarity for both hamlet and homestead dweller. With the shift from hamlet to homestead, the change of burial milieu from hamlet vicinity to prominent hilltop can be seen as an adaptation serving to focus the attention of dispersed domestic groups upon a wider, suprahousehold network of social relations. In sum, I suggest that the continuation of common burial and the modification of burial rites that accompanied population dispersal were adaptive in the sense that they provided a centripetal force countermanding the centrifugal pull of increased mobility and geographic separation.

The structure and composition of domestic groups will be my final concern. In either hamlet or homestead the domestic group must be adapted to a seasonal round of horticulture, hunting, and gathering. How such a unit recruited members and what residence rules it had are problematical, but a core of related males or females, their affines, and offspring would seem highly productive in terms of certain necessities. The need for periodic expenditures of large amounts of labor, for example, could best be handled by an extended household composed of several potentially independent polygynous families. When shifting cultivation is practiced, the extended household apparently mobilizes needed labor most effectively. A large domestic unit with several males and their wives and children guarantees the presence of sufficient labor for cultivating extensive tracts of land or collecting wild foods at opportune moments during the growing season. The periodic allocation of large amounts of labor makes the addition of personnel highly desirable, for they enhance productive capacity.

How such a unit might combine with other forces to facilitate a shift from hamlet to homestead living or how to explain the colonization of new lands requires an additional bit of speculation. When farms and collecting territories are limited, as they must have been (if hamlets had any reasonable temporal duration, or if local homestead populations began to press hard upon available resources), additions to the extended-household labor force would bring no corresponding increase in productivity. Under these circumstances, fission along family lines could be expected as a necessary by-product of demographic strain. We should then see the budding off of independent family units and their physical removal from the parent hamlet or homestead neighborhood. This process would, for instance, explain the smaller size of creek-bank dwellings in the Glen Elder locale and the flimsy houses occurring far from household aggregates in south-central Nebraska. At this point I should stress that my arguments are not offered as a thoroughly documented or proven explanation. All I claim is that domestic group fragmentation and population growth in the context of the environmental, social, and cultural forces discussed here provide an archaeologically testable model for exploring the evolutionary implications of the Upper Republican occupation of the Central Plains.

In conclusion, I see the Upper Republican occupation of the Central Plains as something akin to an adaptive radiation in much the same sense as the term is often used in biological evolution. It is often pointed out that the major difficulty in applying evolutionary concepts to socio-cultural phenomena is the fact that culture traits are not bound to genetic mechanisms; hence different rules govern the rate, direction, and mode of their transmission. But how important is this difference if adaptation is the major concern? The trait-environment relationship may have the same adaptive effect whether the trait be biological or cultural. Archaeological studies that focus on the adaptive level of populations measured in terms of population density, size, and distribution within specific environmental contexts would seemingly make a real contribution to our understanding of process in cultural evolution.

ACKNOWLEDGMENTS

Early drafts of this paper were read by W. Raymond Wood, David Evans, and Carl Falk, whose comments and suggestions were of considerable value. As usual, however, the author alone is responsible for the contents.

The discussion of Upper Republican remains in the Glen Elder locality is a summary statement of data obtained by the University of Nebraska's Summer Field School for Archaeology during the 1965, 1966, and 1967 field seasons. The research was supported by funds from the National Park Service. A similar summary was presented to the 25th Annual Plains Conference.

All carbon-14 dates are from Gakashuin Laboratory and are on file at the Laboratory of Anthropology, University of Nebraska, Lincoln.

LITERATURE CITED

Baerreis, D. A., and R. A. Bryson
 1965. Climatic episodes and the dating of the Mississippian cultures. Wisconsin Archaeol., 46:203-220.
Bell, E. H., and R. E. Cape
 1936. The rock shelters of western Nebraska. Chapters in Nebraska Archaeol., 1:357-400.
Champe, J. L.
 1946. Ash Hollow Cave: a study of stratigraphic sequences in the Central Great Plains. Univ. Nebraska Studies (n.s.), 1: ix+1-130.
Falk, C. R.
 1969. Faunal remains. Pp. 44-51, in Two house sites in the Central Plains: an experiment in archaeology (W. R. Wood, ed.). Mem. Plains Anthro., 6:x+1-132.
Geertz, C.
 1963. Agricultural involution: the process of ecological change in Indonesia. Univ. California Press, Berkeley, xx+176 pp.
Irwin, C., and H. Irwin
 1957. The archeology of the Agate Bluff area, Colorado. Plains Anthro., 8:15-38.
Krause, R. A.
 1967. Arikara ceramic change: a study of the factors affecting stylistic change in late 18th and early 19th century Arikara pottery. Unpubl. Ph.D. dissertation, Yale Univ., New Haven, Connecticut.

Maruyama, Magorah
 1963. The second cybernetics: deviation-amplifying mutual causal processes. Amer. Sci., 51:164-179.
Morton, H. C.
 1954. Excavation of a rock shelter in Elbert County, Colorado, Southwestern Lore. Colorado Archaeol. Soc., 20:30-41.
Netting, R. McC.
 1965. A trial model of cultural ecology. Anthro. Quart., 38:81-95.
Steward, J. H.
 1938. Basin-plateau aboriginal sociopolitical groups. Bull. Bur. Amer. Ethnol., 120: xii+1-346.
 1955. Theory of culture change. Univ. Illinois Press, Urbana, 244 pp.
Strong, W. D.
 1935. An introduction to Nebraska archaeology. Smithsonian Misc. Collections, 93 (10):vii+1-323.
Tax, S., and others
 1953. An appraisal of anthropology today. Univ. Chicago Press, Chicago, Illinois, xiv+395 pp.
Trigger, B. G.
 1963. Settlement as an aspect of Iroquoian adaptation at the time of contact. Amer. Anthro., 65:86-101.
 1965. History and settlement in Lower Nubia. Yale Univ. Publ. Anthro., 69: viii+1-224.

Wedel, W. R.
 1961. Prehistoric man on the Great Plains. Univ. Oklahoma Press, Norman, xviii+355 pp.

Withers, A. M.
 1954. Reports of archaeological fieldwork in Colorado, Wyoming, New Mexico, Arizona, and Utah in 1952 and 1953. Univ. Denver Archaeol. Fieldwork. Southwestern Lore, 19:1-3.

Wood, W. R.
 1969. Ethnographic reconstructions. Pp. 102-108, in Two house sites in the central plains: an archaeological experiment (W. R. Wood, ed.). Mem. Plains Anthro., 6: x+1-132.

Climate and Culture History in the Middle Missouri Valley

Donald J. Lehmer

ABSTRACT

The close correspondence between the dates of certain widespread climatic episodes defined by Reid A. Bryson and others and a sequence of episodes in the history of the native cultures of the Missouri Valley in the Dakotas suggest a close correlation between climatic and cultural changes. The beginning of Bryson's Neo-Atlantic episode, when influxes of moist tropical air produced favorable conditions for corn agriculture, correlates with the first appearance of horticultural villages in South Dakota around A.D. 900. The Pacific I episode, beginning around A.D. 1250, was a time of lowered temperatures and decreased precipitation, and it correlates with a drastic reduction in the extent of the occupation of the area by the village tribes. More favorable conditions during the Pacific II episode, which lasted from about A.D. 1450 to 1550, saw a marked increase in the number and geographic extent of occupied villages. The Neo-Boreal episode was a time of cool summers that began about the middle of the 16th century. Many of the villages occupied between A.D. 1550 and 1675 were small and temporary affairs, which suggests a marginal economy. Moderation of the Neo-Boreal conditions during the first half of the 18th century appears to be reflected in the development of larger and more permanent villages in South Dakota.

Reid A. Bryson and co-workers have defined and characterized a series of climatic episodes for the period since the end of the Pleistocene (Bryson and Wendland, 1967; Baerreis and Bryson, 1965; Bryson et al., this volume). There is a remarkable correspondence between the dates that Bryson has given for his later episodes and the dates that have been established for a series of cultural episodes in the history of the native villages of what is generally called the Middle Missouri Region. This consists of the Missouri Valley in North and South Dakota between the mouths of the White and Yellowstone rivers. It should be emphasized that the dates for the two sets of episodes were arrived at independently and that they are based on different bodies of data. In view of this, the correspondence between the dates for the climatic and cultural episodes strongly suggests a close interrelation between the cultural and climatic changes. This suggestion is reinforced by the fact that the culture changes themselves are the sort that would be expected as responses to the climatic conditions that obtained during Bryson's several episodes.

The villagers of the Middle Missouri Valley are usually classified as sedentary horticulturalists. They lived in reasonably permanent villages, and the corn and other crops from their gardens were a major element in their food supply. They also depended heavily on hunting big game animals, especially bison, which provided them with meat and with hides and bones as raw materials for their manufactures. The enormous quantities of animal bones found in the village sites are incontrovertible evidence of the importance of hunting in the economy. I would estimate that, under optimum climatic conditions, the food supply for the villages was drawn in roughly equal parts from the gardens and the hunt.

Summaries of the characteristics of the native cultural traditions of the region have already been published (Lehmer, 1954), and a classification of their subdivisions has also been presented (Lehmer and Caldwell, 1966). It should be noted that the term "variant," which is used in this paper, is approximately synonymous with the "horizon" of the Lehmer and Caldwell classification.

There were three aspects of the cultural ecology of the Middle Missouri villages that would have been particularly sensitive to climatic changes. Successful corn growing was obviously dependent on a favorable combination of precipitation, summer temperatures, and length of growing season. A large supply of game animals would have been in part a function of adequate pasturage, which would again have depended on favorable climatic conditions. The third aspect involves the timber supply. The villagers used great quantities of posts in building their houses and also stockades, which often surrounded the settlements. They also used enormous amounts of wood for fuel, especially during the Dakota winters. The importance of the timber supply was aptly summed up by an observer at the beginning of the 19th century. The trader, Pierre Antoine Tabeau, wrote: ". . . the Ricaras . . . cultivate only new lands being forced to change their habitation often for want of wood which they exhaust in five or six years" (Abel, 1939, p. 69).

Horticulture, hunting, and timber all were affected, for better or for worse, by climatic changes. Through those effects, the climate influenced the whole way of life of the village tribes.

The Neo-Atlantic episode, which began about A.D. 900, was the earliest one to influence the Middle Missouri villages. It has been described as a period during which there was an influx of moist tropical air into the Great Plains region. This situation encouraged the westward extension of corn agriculture and was responsible for the prairies expanding westward at the expense of the steppe. The beginning of the Neo-Atlantic episode seems to correlate directly with the first movement of the village tribes into the Middle Missouri Valley.

The Initial Variant of the Middle Missouri Tradition is the oldest known village culture in the region. From the time it first appeared, it was a fully integrated complex with a horticultural and hunting economy supporting substantial villages that show every evidence of having been occupied for considerable periods of time. There are more than 35 village sites known to represent this complex. A few lie east of the Missouri in the valleys of the James and the Big Sioux. The rest are concentrated in the southern part of the Middle Missouri region—occurring on both banks of the river in the section between the White River and Chapelle Creek, and on up the right bank as far as the mouth of the Cheyenne (Fig. 1).

All of the available evidence suggests that the Initial Middle Missouri Variant was an intrusive complex, which was carried to the Missouri Valley from southwestern Minnesota and northwestern Iowa at the beginning of the 10th century. This movement coincided with the beginning of the Neo-Atlantic episode. It seems reasonable to conclude that climatic conditions of this episode were favorable for corn growing in the west and must also have resulted in the growth of lush pasturage for native game animals. It is probably no coincidence that the earliest villages in the Central Plains seem to have appeared at about the same time, presumably in response to the same climatic stimulus.

Around A.D. 1100 another, and closely related, cultural complex moved into the southern reaches of the Middle Missouri Valley. This is the Extended Middle Missouri Variant. The

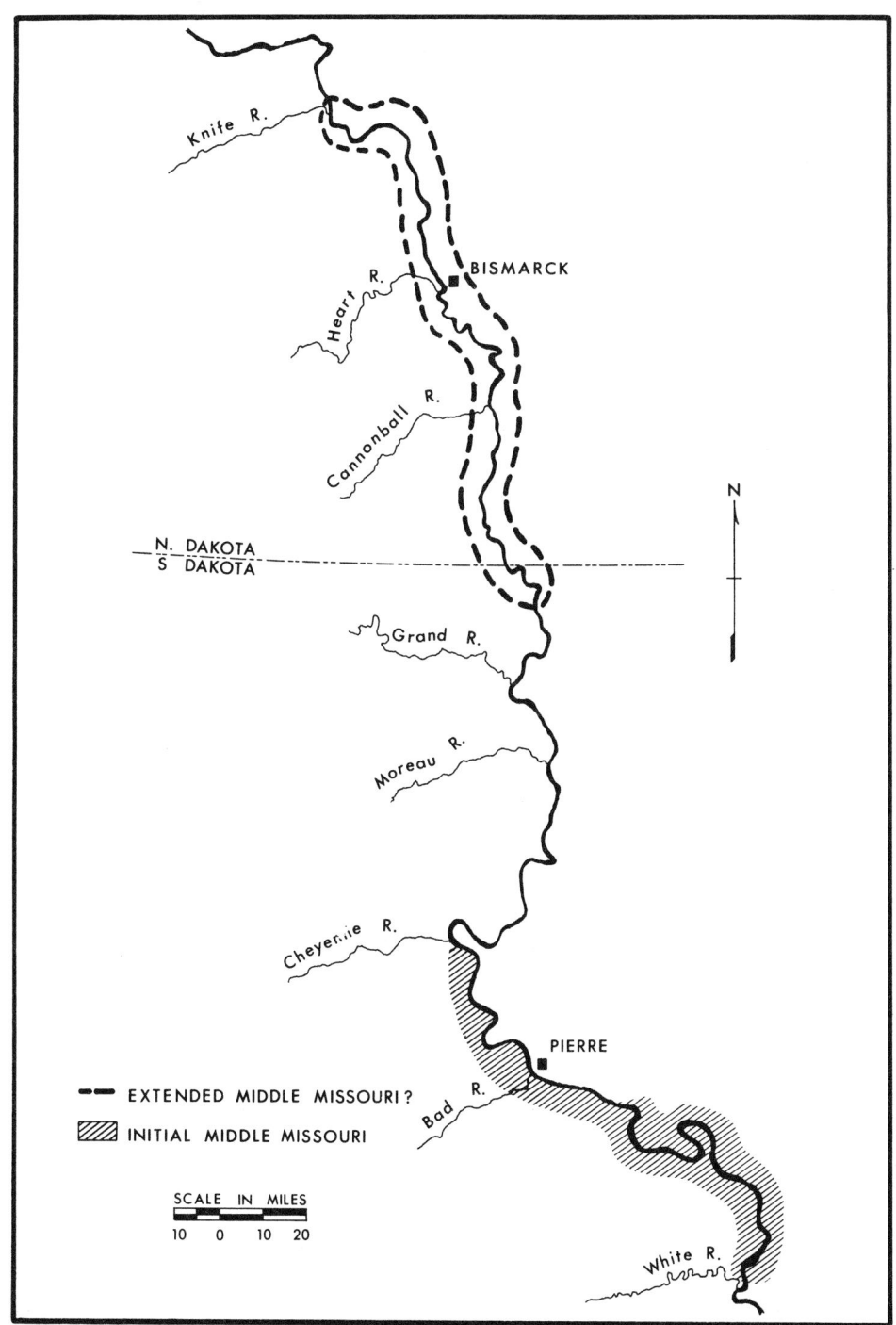

FIG. 1. Village distribution, A.D. 900-1100.

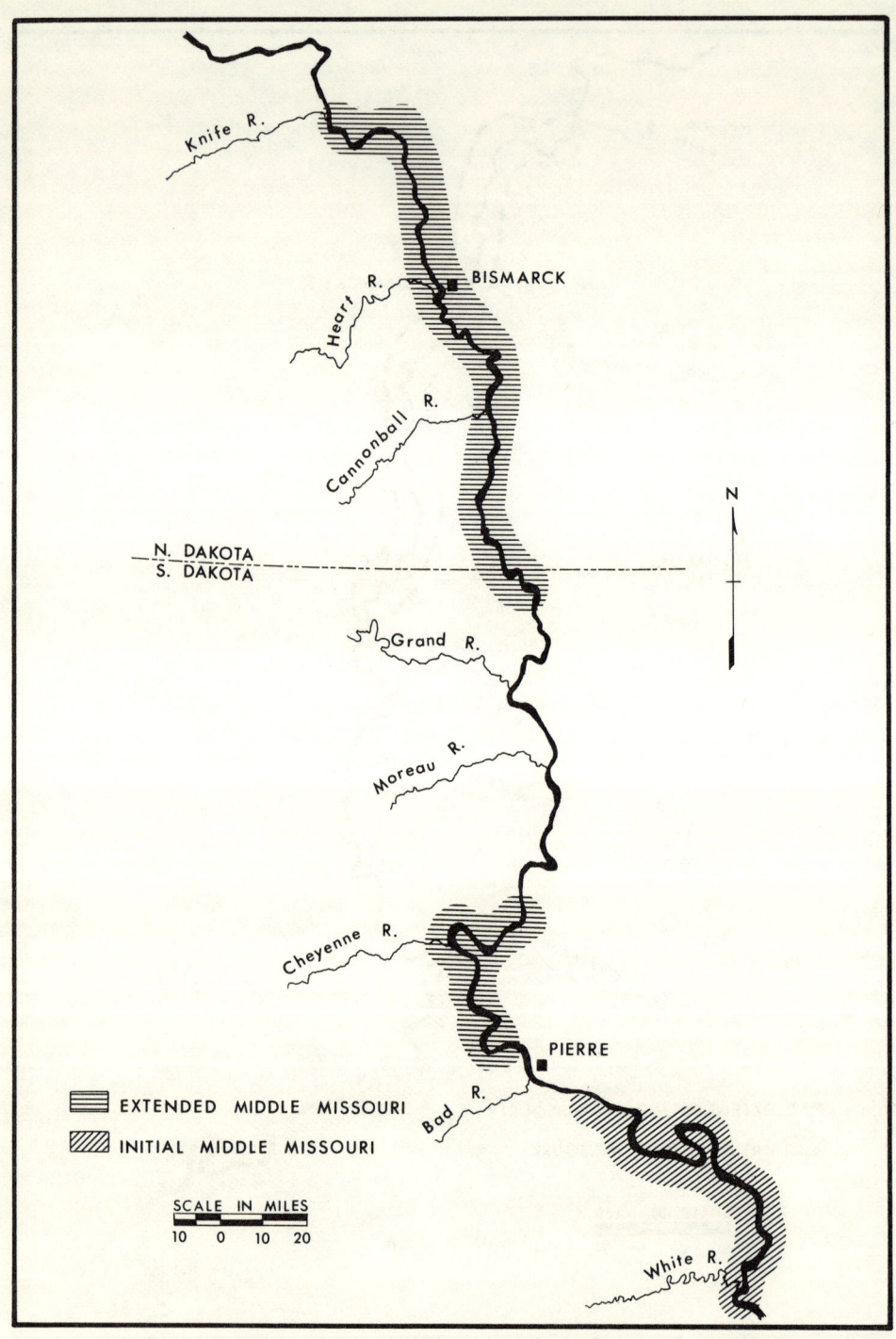

FIG. 2. Village distribution, A.D. 1100-1250.

Extended Middle Missouri groups seem to have displaced the older Initial Middle Missouri population from the area upstream from the Bad River, but there is every indication that the two peoples occupied adjacent sections of the valley for the next century or so (Fig. 2).

The southern Extended Middle Missouri sites were the products of a culture complex that is represented by a much larger group of villages in North Dakota, and it seems likely that the southern occupation was the result of a downriver movement. The northern and southern Extended Middle Missouri sites show a basic similarity to the Initial Middle Missouri Variant, and there is every reason to suppose that both were derived from a common ancestral form.

Unfortunately, we have no data that will allow a correlation of the early Extended Middle Missouri movements with climatic changes. Chronological controls are completely inadequate for an assessment of the time of the first Extended Middle Missouri movement into the northern sections of the valley. I am not aware of any recognized climatic factor that might have a bearing on the appearance of Extended Middle Missouri groups in central South Dakota at the beginning of the 12th century.

The next major population shift among the village tribes correlates closely with one of Bryson's climatic changes—the end of the Neo-Atlantic and the beginning of the Pacific I episode. This change has been variously dated at A.D. 1200 (Bryson and Wendland, 1967), A.D. 1250 (Baerreis and Bryson, 1966), and A.D. 1300 (Baerreis and Bryson, 1965). I am using the A.D. 1250 date here, partly because of its median position in the series and partly because it fits best with the choronological evidence from the Middle Missouri Valley.

The Pacific I episode seems to have been the product of changed atmospheric circulation patterns involving an increased flow of the westerlies across the Northern Plains. This resulted in the introduction of greater amounts of cool dry air, lowered temperatures, and decreased precipitation. These conditions, which lasted until the mid-15th century, would have been unfavorable for corn agriculture, and they may also have diminished the game supply by cutting back the grass cover available for pasturage. They also presumably inhibited tree growth through decreased precipitation and increased evaporative stress.

These climatic conditions appear to be directly reflected by the distribution of villages in the Middle Missouri Valley during Pacific I times (Fig. 3). Settlements representing the Initial Middle Missouri Variant seem to have been limited to a few pockets on the right bank of the river between the mouths of the White and the Cheyenne. A lack of any acceptable dates from southern Extended Middle Missouri villages for the period A.D. 1250 to 1450 strongly suggests that these settlements were abandoned completely. Inadequate dating of the northern Extended Middle Missouri sites makes it impossible to say anything regarding population distributions in North Dakota at this time. However, in South Dakota the relatively unfavorable Pacific I climate seems to be directly reflected by the abandonment of most of the Missouri Valley and the concentration of the remaining villages in a few small areas.

The Pacific I climate began to moderate sometime after the beginning of the 15th century, and a change was fully apparent by the beginning of the Pacific II episode. Pacific II covered the period A.D. 1450 to 1550. Its climate is described as having reverted in part to the character of Neo-Atlantic times, and Pacific II conditions correlate with a much more extensive occupation of the Missouri Valley in South Dakota by the village tribes.

A new cultural complex, the Initial Variant of the Coalescent Tradi-

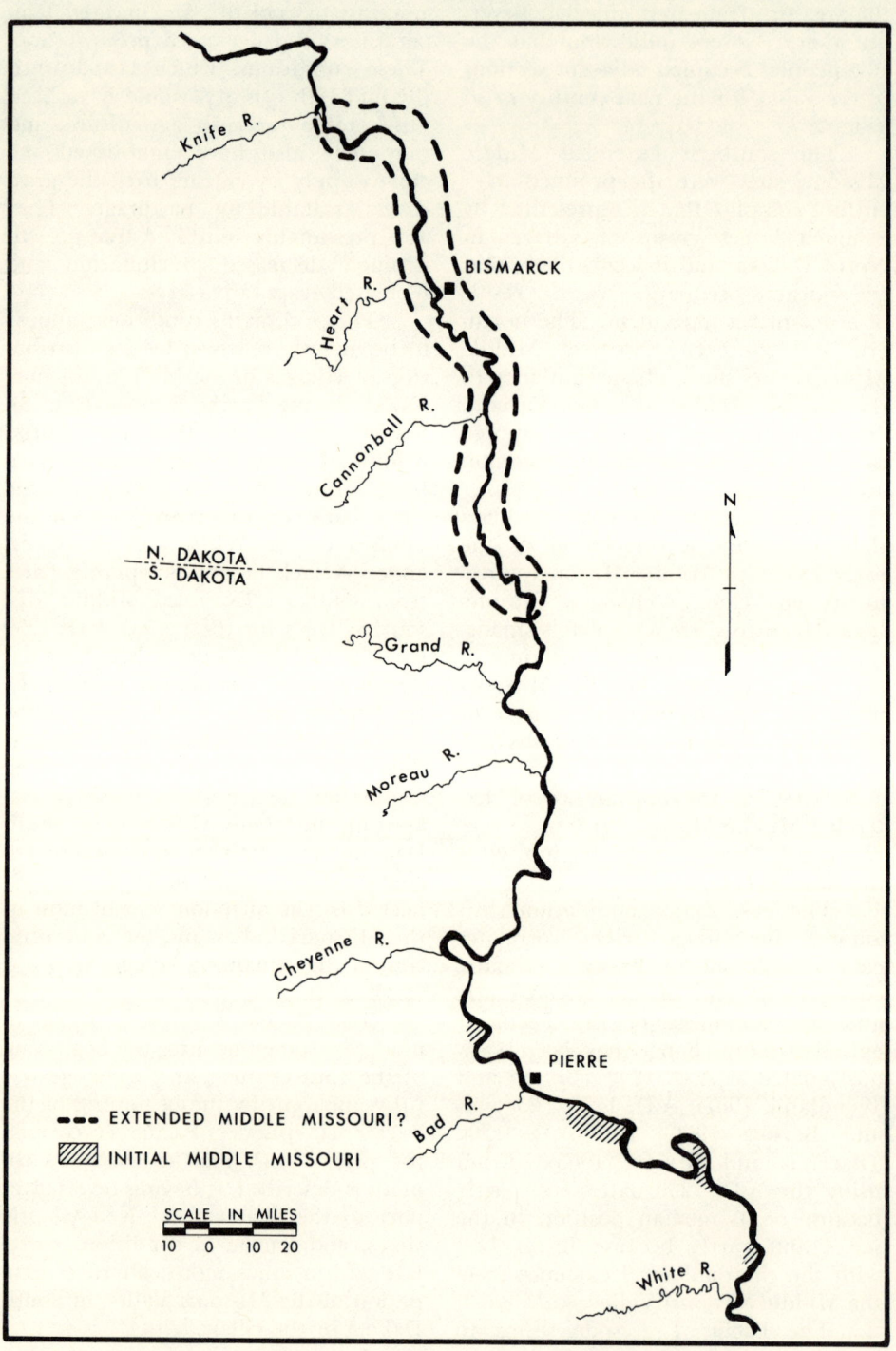

FIG. 3. Village distribution, A.D. 1250-1400.

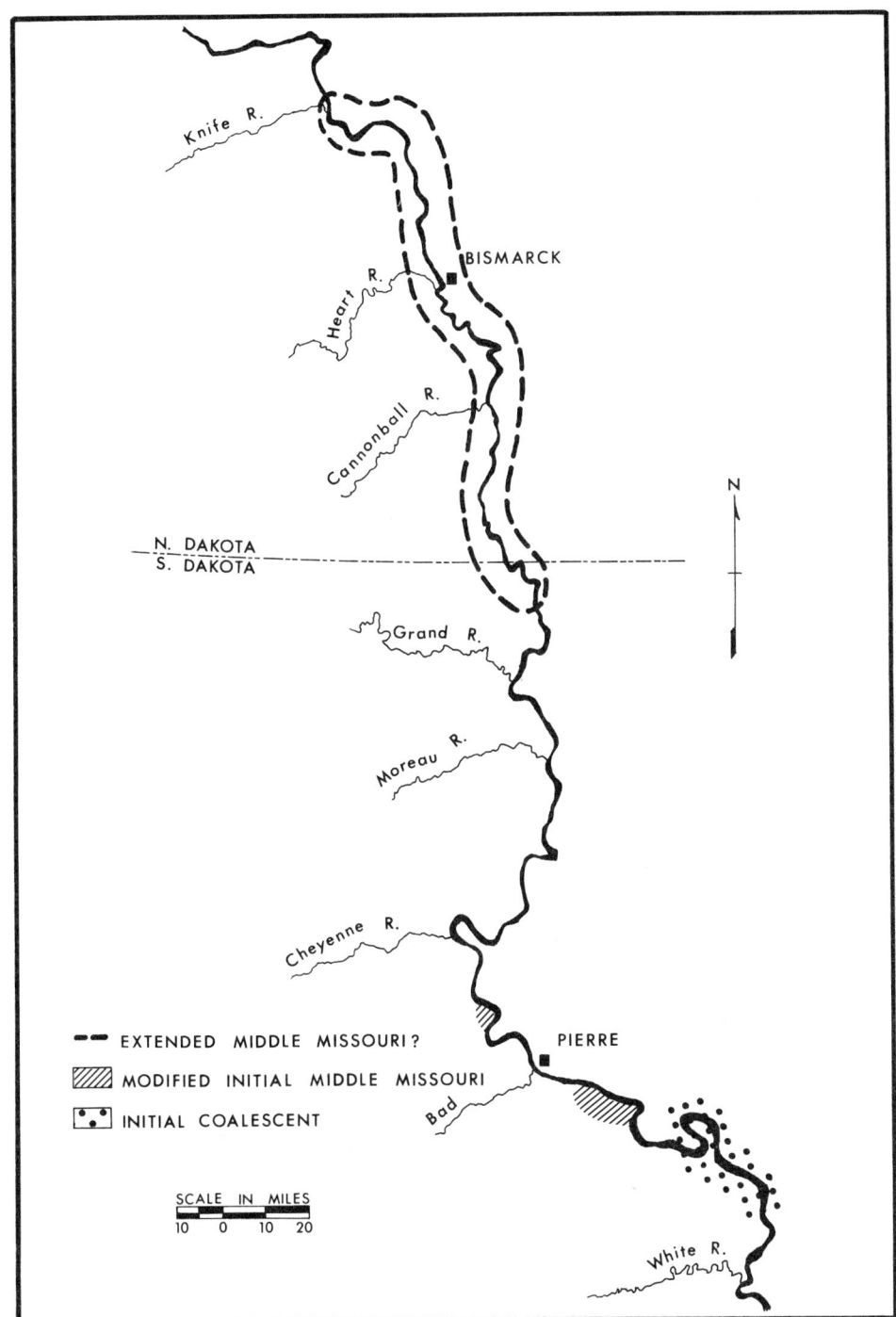

FIG. 4. Village distribution, A.D. 1400-1450.

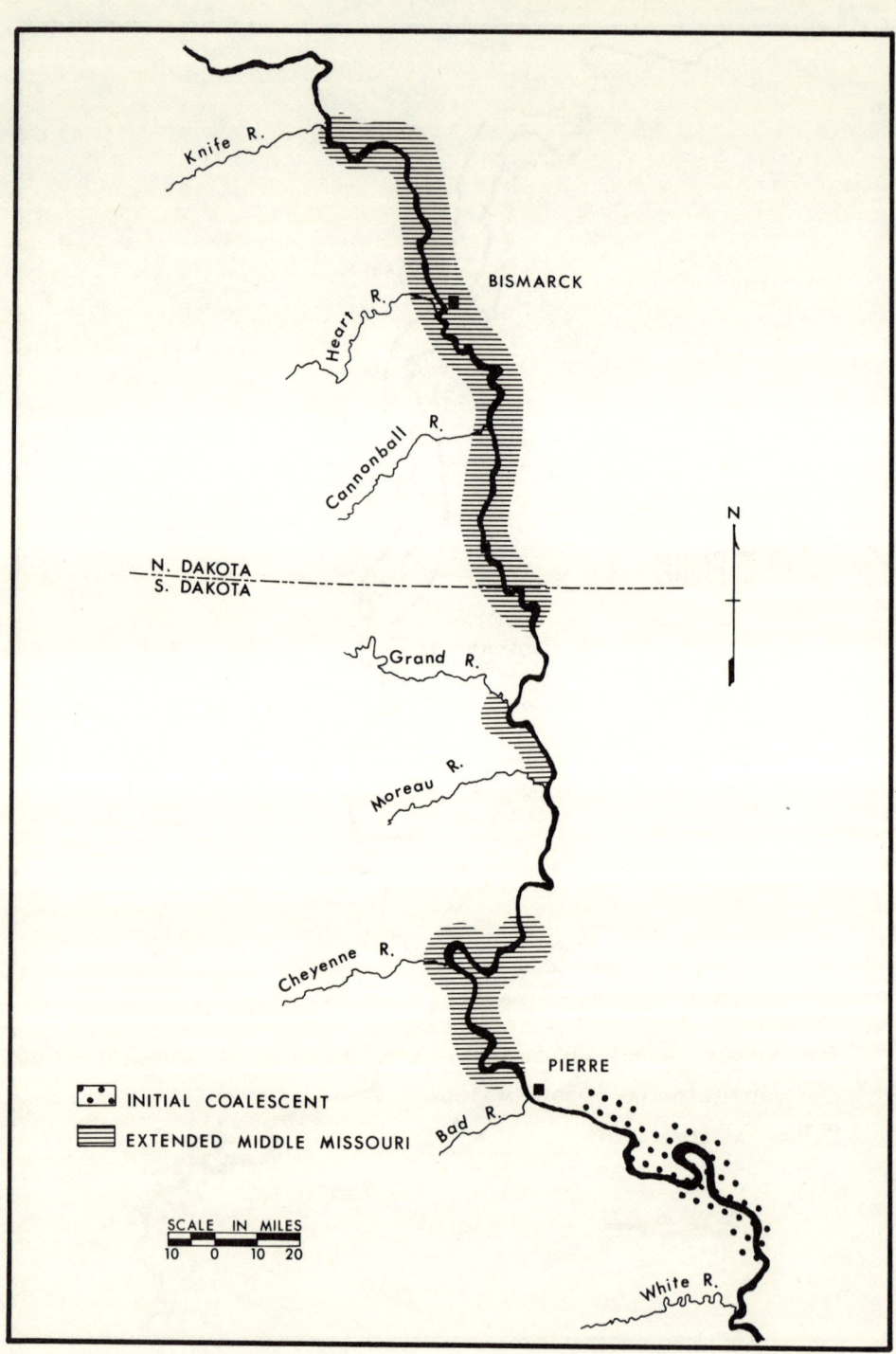

Fig. 5. Village distribution, A.D. 1450–1550.

tion, appeared in the region about the beginning of the 15th century. It is known from a group of less than a dozen villages on both banks of the Missouri, mostly in and below the Big Bend (Fig. 4). The cultural complex represented shows close similarity to that of the Central Plains Tradition, which had its geographic center in Nebraska and northern Kansas, whereas it differs significantly from the Middle Missouri Tradition. There is every reason to assume that the Initial Coalescent Variant reflects a population influx from the Central Plains. The new arrivals appear to have displaced the southernmost of the remnants of the Initial Middle Missouri population, but there are suggestions that the Initial Middle Missouri settlements just downstream from the Bad River continued to be occupied until the mid-15th century.

About A.D. 1450 there was another movement into the Missouri Valley in central South Dakota, one that correlates with the beginning date of the Pacific II episode. This consisted of a reoccupation of the old southern Extended Middle Missouri area by people having essentially the same culture as their predecessors. These groups presumably also moved down the valley from the main Extended Middle Missouri center in North Dakota. The southern Extended Middle Missouri people and the Initial Coalescent groups appear to have occupied adjacent sections of the valley throughout the Pacific II episode (Fig. 5).

The middle of the 16th century saw the beginning of a new climatic episode, the Neo-Boreal. This was another period of reduced summertime penetration of tropical air northward across the United States. As a result, summers were appreciably cooler in the Midwest. The lower summer temperatures must, in turn, have reduced crop yields in the Middle Missouri, which is at best a marginal region for corn agriculture. This situation certainly seems to be reflected in the native cultures that existed during the period A.D. 1550-1675.

The southern Extended Middle Missouri groups abandoned their towns in central South Dakota about A.D. 1550. There are indications that they established a few villages on the right bank between the Grand and the Moreau, but moved still farther upstream after a few years. The final manifestation of the Middle Missouri Tradition is the Terminal Variant. It is represented by a number of large and heavily fortified towns that show every indication of having been occupied for considerable periods of time. Almost all of them are concentrated in the relatively short stretch of the Missouri Valley between the mouth of the Cannonball River and Painted Woods Lake, about 35 miles north of Bismarck. Three Terminal Middle Missouri sites that were located well below the Cannonball were likely abandoned at a relatively early date (Fig. 6).

The large and stable communities of the Terminal Middle Missouri Variant appear, at first glance, to present a decided anomaly with the climatic conditions under which they were occupied. There are indications, however, that the part of the Missouri Valley where they occur is decidedly more favorable for agriculture than the areas either upstream or downstream. Certainly, this section of the valley and the adjacent uplands are farmed more productively today than any comparable part of the Middle Missouri Region. This appears to reflect the fact that this section of the valley has the lowest evaporative stress in the Middle Missouri Region (Bryson and Wendland, 1967, fig. 77).

In South Dakota the effects of the Neo-Boreal episode on the village cultures can be clearly seen, although they took a somewhat different form from those of earlier periods of unfavorable conditions. The earlier responses involved the abandonment of large sections of the valley and a sharp reduction in the number of occupied vil-

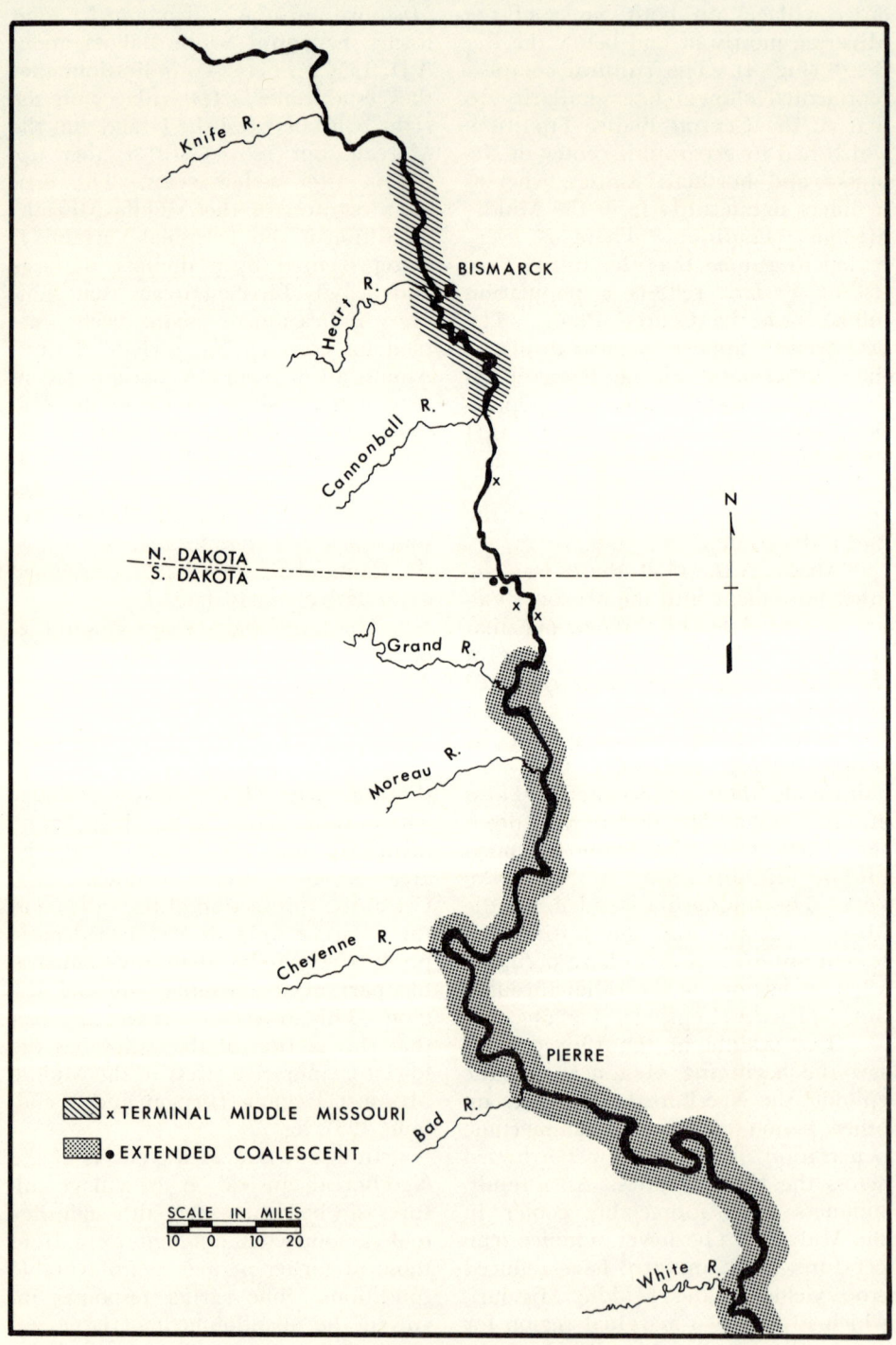

Fig. 6. Village distribution; A.D. 1550-1675.

lages. During the early Neo-Boreal times, the cultural response was in terms of reducing both the size of the individual villages and the length of time that they were occupied.

The first century and one-quarter of the Neo-Boreal episode saw the emergence and spread of the Extended Variant of the Coalescent Tradition. Extended Coalescent was a direct outgrowth from the earlier Initial Coalescent configuration. About the middle of the 16th century it seems to have begun a rapid expansion upstream

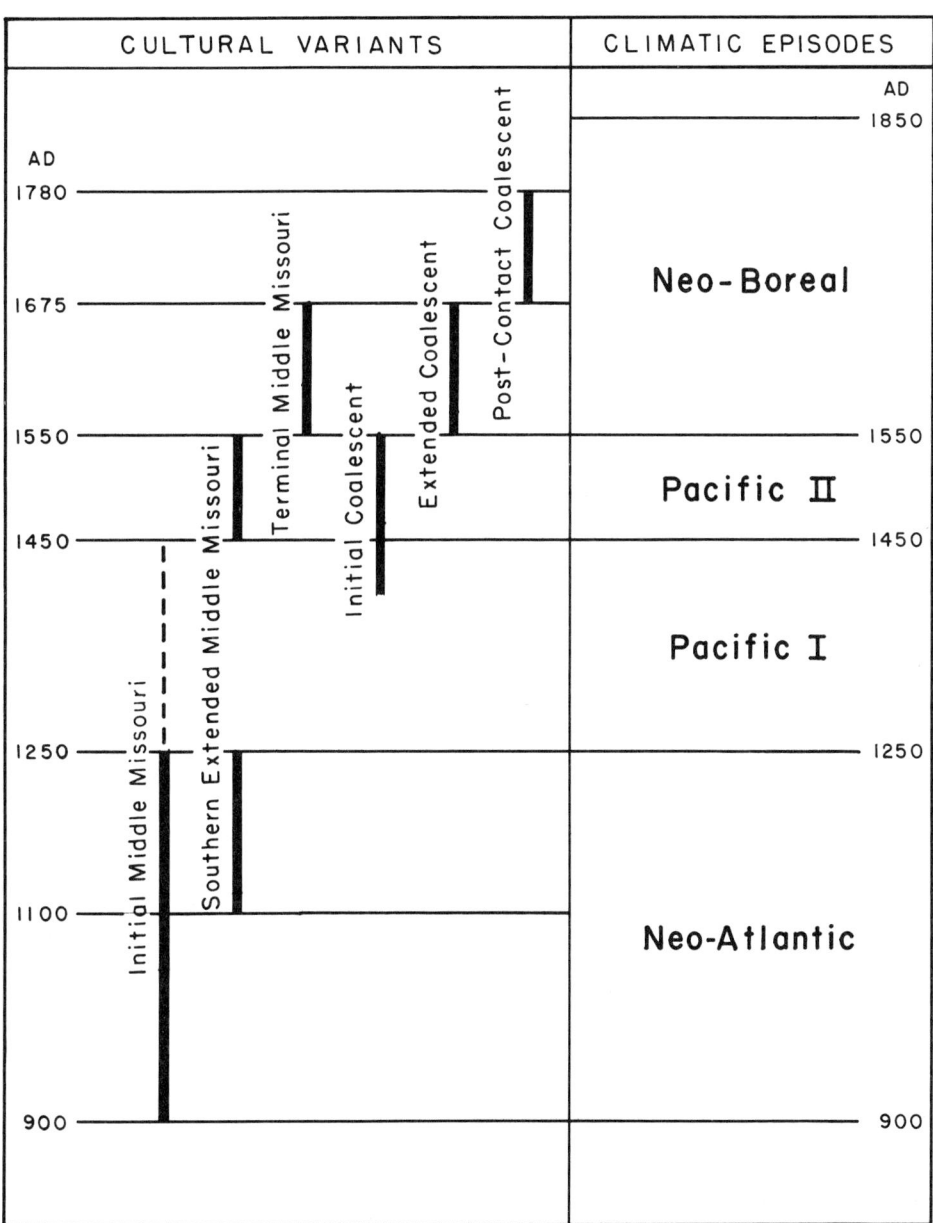

FIG. 7. Chart showing the correlation between the climatic and cultural episodes in the Middle Missouri Valley.

from the old Initial Coalescent heartland. Eventually, the Extended Coalescent groups established more than 150 villages between the White River and the North Dakota border. But the character of the villages differed from those established earlier and later in the region.

Most Extended Coalescent sites contain relatively few houses, although some are fairly large. The houses are characteristically scattered over a considerable area. The scarcity of cultural debris found in the settlements offers a sure indication that they were occupied for only a few years. It is also worth noting that many of the Extended Coalescent villages occupy a unique topographic situation.

A few sites have been found on the Missouri flood plain, presumably winter villages built to take advantage of the protection of the tree cover and a readily available supply of wood for fuel. The great majority of all of the villages in the Middle Missouri region were built on the river edge of the first terrace above the flood plain, and many of the Extended Coalescent sites are located there. However, a substantial number of villages of this complex are situated on flat-topped ridges high up in the breaks that form the border of the Missouri trench. Their locations are well removed from the wood, water, and garden lands of the river bottoms. They might have been easier to defend than the lower villages, but there is almost always easy access from the high ground on one side, and there are no indications of fortifications. The upland villages were built in situations that gave them a wide view of the lower portions of the valley, and it seems possible that this was done so that an almost continuous watch for game animals could be kept during a period when poor pasturage had forced most of the large bison herds out of the region.

The available evidence suggests that the great majority of the Extended Coalescent people lived a hand-to-mouth existence in typically small communities that had a high degree of geographic mobility. This pattern contrasts sharply with that of the other village complexes of the region, and there seems to be every reason to assume that it represented a response to the unfavorable climatic conditions of the Neo-Boreal episode.

After A.D. 1675 a new set of factors began to exert strong influences on the Middle Missouri villages. These were cultural rather than climatic factors, involving the effects of the White penetration of North America. Native reactions to innovations, which included the horse, the fur trade, and European epidemic diseases, overshadow cultural changes that can be attributed to climatic factors. There is one that might be mentioned, however.

Baerreis and Bryson (1965) have written: "... the character of the Neo-Boreal was relatively uniform except for a break in the first half of the eighteenth century." This may be reflected in the settlement pattern of the southern villages that represent the Post-Contact Variant of the Coalescent Tradition. These villages had an extensive distribution along the Missouri up until the end of the 18th century. Unlike the Extended Coalescent towns, they were fairly large and appear to have been occupied for several decades. It seems possible that the return to larger and more permanent towns was made possible by improved climatic conditions during the break in the Neo-Boreal episode.

To summarize, favorable climatic conditions of the Neo-Atlantic and Pacific II episodes correlate in time with known movements of sedentary village populations into the Middle Missouri region and with numerous and reasonably permanent settlements. Unfavorable conditions during Pacific I time correlate with a marked contraction in the area occupied and with a sharp reduction in the number of villages. The unfavorable Neo-Boreal conditions equate with a pattern in

South Dakota of small settlements that were lived in for short periods of time, presumably because of a rapid local exhaustion of resources such as timber. This situation was less noticeable in North Dakota, where the Missouri Valley is subject to less evaporative stress. Larger and more permanent villages seem to have reappeared in South Dakota during the 18th century, at a time when Neo-Boreal conditions are believed to have moderated (Fig. 7).

Climatic changes in the Middle Missouri region still have to be documented by pollen analysis, studies of changes in the microfauna, and similar investigations. At present it is only possible to say that there is a correspondence there between culture changes and the widespread climatic episodes that have been postulated. However, that correspondence is such a remarkably close one that it strongly suggests an interrelationship between climatic and cultural change.

LITERATURE CITED

Abel, A. H. (ed.)
    1939. Tabeau's narrative of Loisel's expedition to the Upper Missouri. Univ. Oklahoma Press, Norman, xi+272 pp.

Baerreis, D. A., and R. A. Bryson
    1965. Climatic episodes and the dating of the Mississippian cultures. Wisconsin Archaeol., 46:203-220.
    1966. Dating the Panhandle Aspect cultures. Bull. Oklahoma Anthro. Soc., 14:105-116.

Bryson, R. A., and W. M. Wendland
    1967. Tentative climatic patterns for some late glacial and post-glacial episodes in central North America. Pp. 271-298, in Life, land and water (W. J. Mayer-Oakes, ed.), Occas. Papers Dept. Anthro., Univ. Manitoba, xvi+414 pp.

Lehmer, D. J.
    1954. Archeological investigations in the Oahe Dam area, South Dakota, 1950-51. Bull. Bur. Amer. Ethnol., 158:xi+1-190.

Lehmer, D. J., and W. W. Caldwell
    1966. Horizon and tradition in the Northern Plains. Amer. Antiquity, 31:511-516.

Some Environmental and Historical Factors of the Great Bend Aspect

WALDO R. WEDEL

ABSTRACT

The Great Bend Aspect comprises archaeological sites and associated materials in central Kansas. Archaeological and documentary evidence dates these in the late 15th to 17th centuries, and points to their probable affiliation with Wichita-speaking groups of the earliest period of Spanish contact. Historically, the local culture apparently represents a fusion of earlier cultures to north and south, and is comparable to the Coalescent tradition in the Northern Plains.

Native subsistence featured both hunting and horticulture. The artifact inventory indicates successful and intensive exploitation of the resources of the local environment. Contacts with other groups and localities, notably with the Puebloan peoples of the upper Rio Grande, are manifested in the utilization of certain raw materials and finished products of nonlocal origin.

Living near the western limit of optimum corn-growing conditions, the Great Bend people nevertheless enjoyed substantial crop surpluses. About A.D. 1700 their communities in central Kansas were abandoned, and a shift southward is indicated. Although the reasons for this must await much more intensive study, possible factors include climatic deterioration plus hostile action by tribes from the east and west. Thus, by the second quarter of the 18th century, the centers of Wichita culture and influence were south of Kansas on the lower Arkansas and Red rivers, where French trade contacts stimulated a temporary affluence and a notable florescence of culture.

The archaeological materials here included under the designation of Great Bend Aspect are largely from central and south-central Kansas. Sites center on the middle Arkansas River drainage between approximately 97° and 99° west longitude and between 37° and 38°30′ north latitude, but are not wholly restricted to this area (Fig. 1). They are known only from limited excavations and inadequate surveys; much of the work has been done by nonprofessionals, and its results are not in the scientific literature. Many interpretations to date are impressionistic and tentative. The materials are attributable to a semisedentary Plains Village Indian complex dating from roughly the 15th to late 17th centuries. Assignment to this time level is strongly supported by the inclusive presence of limited White contact items and of datable glaze-paint–decorated potsherds from contemporaneous Pueblo towns of the upper Rio Grande district in New Mexico.

Identification with a historic tribal entity is feasible, and the evidence points toward affiliation with one or more of the southern Caddoan-speaking groups ancestral to the Wichita. The likelihood is strong that some of these sites were inhabited when the first European expeditions under Coronado (1541) and Oñate (1601) set foot on the soil of future Kansas. Thus, the revelations of archaeology can be amplified by comparison with the documents from the earliest entradas to provide an important base line from which other native cultures, both earlier and later, can be more precisely

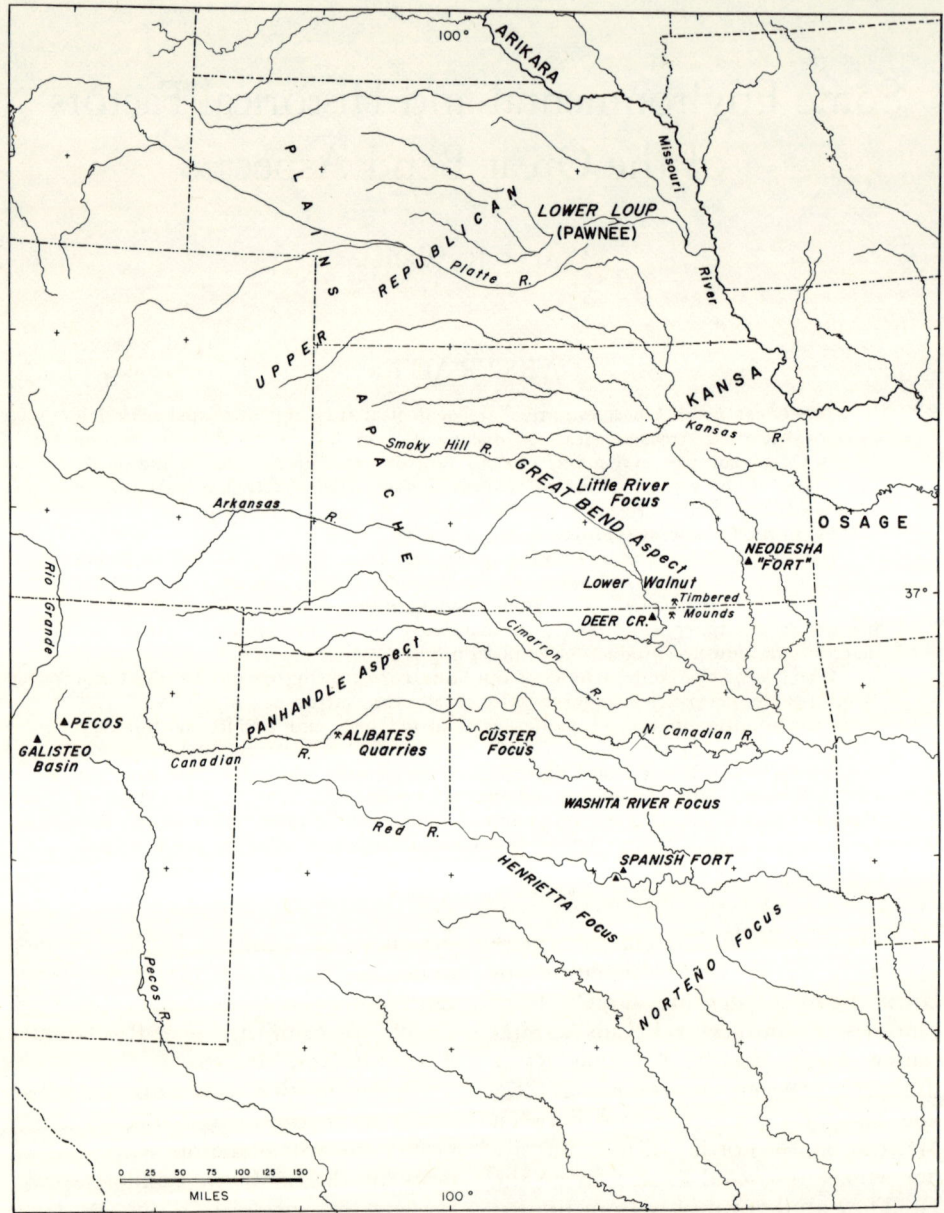

Fig. 1. Index map to show location of certain late-prehistoric (post-A.D. 1000) archaeological sites and culture complexes, historic tribes, and other features discussed in the text.

dated and their significance more perceptively appraised in terms of the cultural history of the plains region.

The materials involved originate in village and camp sites (Wedel, 1942, 1959). The principal known sites consist of trash-littered areas from an acre or two up to 100 or more acres in extent. Before modern cultivation, low refuse mounds were often visible, as their traces still are, here and there. Among these mounds, and evidently under many of them, there are large subsurface storage pits last used by the Indians for deposition of refuse. At some sites, such as the Udden site on

Paint Creek, the Tobias site on Little Arkansas River, and the Malone site on Cow Creek, these caches are present by the hundreds (Udden, 1900; Wedel, 1959). They are usually bell-shaped, that is, have undercut sides flaring out and down from a constricted neck, and they range in size up to eight feet or more in depth by as much as nine feet in bottom diameter. Their total storage capacity must have aggregated many hundreds of bushels at some sites.

House remains are usually absent or still unrecognized, although subcircular posthole configurations around fireplaces on the Pawnee River near Larned suggest dome-shaped pole-and-grass structures, probably like those reported by the Spanish discoverers of the 16th century. There are no indications of defensive works, such as ditches, palisades, or earthen walls surrounding the village sites.

At five sites in Rice and McPherson counties, a notable feature is a shallow ditch or series of oblong depressions arranged end to end around a low central earthen mound. Limited excavations at three sites have shown that each of these features consists of four semisubterranean oblong pithouses, originally with a superstructure of poles and grass, earth-covered wholly or in part, which seem to have been destroyed by fire. In two, the fill contained large numbers of sandstone boulders from nearby streamside ledges. Two have yielded human bones, either as complete skeletons or as disarticulated elements, or both. From the plan, location, and content of these so-called council circles within the village communities, it is inferred that they were special-purpose structures. Still imperfectly understood are their alignment and orientation along lines that intersect the horizon at solstitial points (Wedel, 1967).

The village sites occur in a variety of topographical locations. In Rice and McPherson counties, where they have been tentatively assigned to the Little River focus, sites are found on streamside terraces and on gently sloping hillsides and narrow prairie ridges. On the ridges they may develop a strongly linear arrangement, with refuse areas scattered at 20- to 50-yard intervals along the flat-to-rounded, and often sloping, ridges. None is directly on the banks of the larger streams, such as the Smoky Hill and Arkansas rivers. Farther south, in Cowley County, closely related but slightly deviant sites, designated Lower Walnut focus, lie close to the Walnut River banks. Here some of the sites consist of loose groups of trash heaps and cache pits, with other similar groupings a few hundred yards away, suggesting a settlement pattern like that described by some members of the Oñate party in 1602 (Scholes and Mera, 1940, p. 274). In nearly all cases, the sites are near good springs or other sources of permanent water. Apparently there was a strong preference for locating villages on the more elevated bluestem-prairie benches and ridges, away from the tree-fringed streamside flood-plains.

From the village sites come abundant cultural materials, including quantities of broken pottery, a wide and varied assortment of chipped and ground stone, bonework, and other materials. These are generally far more plentiful and more varied in nature than are the comparable materials from older sites in the area. Except for the pottery, which can be described as drab, lackluster, and largely utilitarian in character, the products of the local native arts and crafts compare favorably with those produced at the same time level among the contemporary Pawnee in east-central Nebraska or among the Arikara and Mandan on the Middle Missouri.

The subsistence economy was a dual one, featuring both production and collection of food items. Domesticated plants included maize, beans, squash, and sunflower. The size and quality of some of the charred ears that have been collected by archaeologists show that these people possessed well-developed and productive crop varieties, evidently reflecting a long and deeply rooted tradition of crop-raising. The bison-scapula hoe was the usual gardening tool, and use of the digging stick may be inferred. Large milling stones for grinding corn are present. The size and number of erstwhile storage pits is good evidence of a highly successful horticultural economy, even though this locality is near the western margin of what is generally considered the corn belt. The crops were presumably grown in the bottomlands along the small streams on the banks of which the communities stood. There is no evidence of irrigation, nor do we know what, if anything, in the way of fertilizers may have been applied. The maize evidently included several varieties, among them 10-, 12-, and 14-rowed ears, along with a smaller number of 8-rowed ears that possibly represent Maiz de Ocho, the variable and highly productive late-prehistoric Mexican variety that is thought by some to have been a major factor in the florescence of native American culture after about A.D. 700 (Galinat and Gunnerson, 1963).

Hunting was evidently of considerable importance. Bison bones, especially ribs, limb elements, and skulls, are common at the larger sites. Pronghorn antelope, elk, deer, and smaller species are present in lesser amounts. There are great numbers of chipped end-scraper blades, as well as numerous small, well-made projectile points and a wide variety of cutting tools, all suggesting a strongly developed skin-working industry at a time when the White traders were not yet on hand. There is no convincing evidence, either in the bone refuse or in the inventory of material-culture items, that the horse was known; and the horse motif, or anything that can be reasonably so identified, is uncommon among the petroglyphs of the region. Despite the inferred absence of the horse, the hunting methods of the local people—on foot, with bow and arrow—seem to have produced ample game to supplement the yield from the gardens. The juxtaposition of open grasslands teeming with bison and pronghorn, and timbered stream valleys well populated with deer, elk, bear, and other mammals is undoubtedly reflected in the archaeological record. At sites in southern Kansas, where the flood-plain forest connects with the upland scrub oak, wild turkey bones are found in the sites; these have been reported sparingly from central Kansas also. There is neither documentary nor archaeological evidence that the turkey was domesticated. Fish bones are scarce, and unlike the older sites of the Central and Southern Plains, those of the Great Bend Aspect have yielded no bone fishhooks or other devices clearly designed for taking fish. Dog bones indicate large, powerful beasts, presumably of draft-animal type.

For their various arts and industries, the local craftsmen drew on a wide range of raw materials. Projectile points, end scrapers, knives, and other chipped objects were commonly fashioned from the attractive and distinctively banded or fossiliferous (fusulinid), or both, Florence flint, which was taken from extensive aboriginal quarries—the Timbered Mounds of Gould (1898, 1899)—in Permian formations in the southern Flint Hills of Cowley County, Kansas, and Kay County, Oklahoma, 50 to 150 miles distant. Blue-gray cherts, also of Permian age, from the Flint Hills farther north in Marion and Chase counties, were likewise drawn on. Still another popular material was the less tractable dense yellow-to-brown "jasper" found in seams and layers in a limestone matrix on many streams in the Smoky Hill and Republican drainages of northern Kansas and southern Nebraska. The locally abundant and easily obtainable Dakota sandstone was extensively used for arrowshaft abraders, awl sharpeners, milling stones, and other tools requiring an abrasive surface. Grooved mauls and other heavy-duty tools were fashioned from harder sandstones and from Sioux quartzite, presumably collected in the glacial drift of the Kansas River Valley. A fine-grained reddish sandstone, not yet localized as to origin, was commonly used in pipe-making. Cone-in-cone calcite from the Smoky Hill Valley served for arrowshaft straighteners.

Other raw materials are evidently of more distant origin. From southwestern Minnesota came catlinite for pipes and ornaments. Alibates agatized dolomite from the Texas Panhandle near Amarillo was made into weapon points and other chipped tools. Large tubes, probably intended for personal adornment, were made from marine shell, probably the columella of the Gulf conch; and *Olivella* shell beads were received as finished products in trade from the Southwestern Indians in the upper Rio Grande drainage. Fresh-water mussel shells of local origin found limited use as pottery tempering, for ornaments, and as occasional domestic tools.

Of particular interest and significance are the indicated contacts between these early historic Kansans and the Pueblo Indians in New Mexico. The evidence includes turquoise beads and pendants; obsidian for arrowpoints; ridged mica-schist arrowshaft-straighteners, made in Rio Grande Pueblo form and of New Mexico stone (sometimes clumsily imitated in cone-in-cone calcite or other local stone); malachite or azurite in clay, for pigment; and, most important, pottery fragments of glaze-paint–decorated and polychrome wares from the Rio Grande Pueblo peoples.

The Puebloan sherds from central Kansas represent at least a dozen named and dated Southwestern wares that range in time from Gila polychrome of the 14th century to such late and short-lived wares as Kotiyiti and Pojoaque polychrome. They have been recovered at most of the principal known sites of the local complex in central and southern Kansas, and examples occur in many local collections. Most are surface finds, some are from roadcuts or other soil disturbances; but they are supported by systematically excavated pieces that range in time from about A.D. 1500 to 1675. Besides their usefulness in dating, these specimens are evidence of the contacts that existed between the central Kansas villages and the flourishing Tano, or southern Tewa, towns in the Galisteo Basin south of Santa Fe for a century or more after, as no doubt for some time before, the Coronado entrada. It is not yet clear whether these contacts were direct or, alternatively, involved such nomadic intermediaries as the Plains Apache bison hunters. The Rice County sites specifically known to be associated with these trade materials lie mainly within a few miles of the 19th-century route of the Santa Fe trail near its crossings of Cow Creek and the Little Arkansas River, some 550 miles from Santa Fe via the wagon trail as measured by Gregg in the 1830's (1954, p. 217).

The origins of the Great Bend Aspect culture can only be conjectured at the moment, pending a great deal more field research and laboratory analysis, as well as the establishment of a firm chronology among the sites comprising the complex itself. Identification of the manifestation as Wichita-affiliated suggests a southerly beginning; this is further indicated by certain items in the material culture, which, on present knowledge, appear earlier in the Oklahoma-Texas region than they do in central Kansas. Included here, among other items, are metapodial digging-stick tips, transversely scored ribs, certain pottery traits, a stone pipe form, and perhaps deer-mandible sickles. Specifically, it seems probable that the Great Bend Aspect is rooted in part in the late prehistoric cultures of central and western Oklahoma (Pillaert, 1963, p. 46; Wedel, 1968a, pp. 58-60), such as the Washita River, Custer, Optima, and

Antelope Creek (Panhandle Aspect) foci, with other increments from contemporary groups to the north. These southern complexes, interestingly enough, all lie mainly at or beyond the western margin of the Western Cross Timbers, in a geographical setting strongly remindful today of the Great Bend region of central Kansas.

I have elsewhere suggested that Great Bend may be a sort of southern "coalescent" culture comparable to the taxonomically sanctified "Coalescent" in the Northern Plains. Unlike the latter, however, where strong leads connect the Lower Loup to Pawnee, and Coalescent to Arikara and Mandan, there are at present few firm threads tying the 16th- and 17th-century materials from central Kansas to the historic tribal entity to which they have been ascribed, namely the Wichita. Indeed, at the moment, wide and impressive differences are apparent between the Great Bend and the later Wichita materials found at Deer Creek and at Spanish Fort, in northern and southern Oklahoma, respectively. Thus, the exact time of departure from central Kansas and the mechanics by which these northerners were transformed into the Wichita when they became the objects of concentrated interest to Spanish and French in the 18th century remain to be worked out.

The consistent presence of limited quantities of European trade materials inclusively at these sites is well established. Udden (1900, pp. 66-67) long ago recorded the discovery of chain-mail fragments, since lost, as well as the surface finding of "one or two perforated beads of blue glass" on a midden at the Paint Creek (Udden, or 14MP1) site. Subsequently, chain mail has been unearthed at the C. F. Thompson site (14RC9) and at Saxman (14RC301) (Terry, 1961). Pea-sized blue glass beads have been excavated at the Tobias (14RC8), K. Hayes (14RC13), and Paul Thompson (14RC12) sites; and, so far as my information goes, only this type of bead has come to light from subsurface findspots at the central Kansas locations. Metal goods also appear; they include an iron axhead, a butcher-knife blade, an awl-like object and fragments, and scraps of iron and brass that are not further identifiable. It is important to note that materials of European origin are proportionately much less common in Great Bend Aspect sites than at later 18th-century Wichita sites at Deer Creek in northern Oklahoma and at Spanish Fort on the Red River (Bastian, 1966). It is probably safe to infer that at the latter locations the Indians were in direct and regular contact with French traders, whose activities, as recorded in Du Tisne's and La Harpe's time, were too late to be reflected in the archaeology of the central Kansas communities.

The Great Bend Aspect sites may be said to represent the high-water mark in the Wichita Indian penetration northward into the Central Plains. None is reported north of the Smoky Hill River or west of Larned. Outside the Arkansas River drainage, several sites once were situated in the upper Cottonwood River drainage, near Marion and perhaps as far east as Cottonwood Falls. Whether the "Neodesha Fort" on the Verdigris River is a related complex, as has been suggested elsewhere, is not yet certain. From present indications, the region between these central Kansas sites and those of the Lower Loup Pawnee in east-central Nebraska was not certainly inhabited by semisedentary people at this period.

In geographical terms, the known Great Bend Aspect sites are mainly in, or immediately peripheral to, the Great Bend Lowland, with a probable extension eastward into the Flint Hills Upland (Schoewe, 1949, p. 292). Basically, they are in a tall-grass prairie seamed with hardwood-fringed small streams. Currently, it has a normal 10-inch summer rainfall, a 180- to 200-day frost-free growing season, and an abundance of groundwater that is near enough to the surface to support numerous springs and seeps; and so the district seems as well suited environmentally to permanent Indian occupancy as any to the north or south. It seems reasonable to ask: Why did not Indians of the Great Bend Aspect extend their settlements northward beyond the Smoky Hill, and more important, why and under what circumstances did they abandon this district and withdraw, or so it appears, toward the south?

The historic documents give us little help in this problem. Nothing that has come down to us from the Coronado expedition of 1541 or from the Oñate entrada of 1601 enlightens us much as to contemporary climatic and environmental conditions. The Spaniards complained of the bitter cold on the Rio Grande during the winter of 1540, of the shortage of fuel and winter clothing, and of inadequate drinking water on the plains en route from the Rio Grande to Quivira. But once they had arrived in the Wichita habitat at Quivira, there were only laudatory remarks on the quality of the soil, on the various native fruits and berries, and on the well-watered character of the land by contrast with that toward the Rio Grande. Coronado described Quivira as having the most suitable soil "that has been found for growing all the products of Spain" (Hammond and Rey, 1940, p. 189). Jaramillo, who also made the trip to Quivira and so qualifies as an eyewitness, says that the land "has a fine appearance, the like of which I have never seen anywhere in our Spain, Italy, or part of France, nor indeed in other lands where I have traveled in the service of his majesty" (Hammond and Rey, 1940, p. 305).

Sixty years later and 150 miles or so down the Arkansas River, Oñate was impressed by the cornfields and gardens of the natives. The corn, he observed, was as tall as that in New Spain, or taller; the land was so fertile that having harvested one crop of corn, the Indians had started a second, which was already six inches high; and there were many beans and calabashes. All of this was growing without irrigation, the crops depending entirely on seasonal rains, which the Spaniards judged "must be very regular in that land, for in the month of October it rained as in New Spain in August" (Hammond and Rey, 1953, p. 755). In none of these firsthand accounts is there any indication of climatic adversities, of droughts, or of other problems bearing on the production of domestic crops. On the contrary, there is indication that these people had mastered the problems of food production and were probably enjoying substantial crop surpluses.

In central Kansas, by the 18th century, the Arkansas Valley was no longer the locus of settled semihorticultural Indians but had become, as Kroeber (1939, p. 75) aptly put it, "back country of the Osage and Kansas, the latter a small tribe." At this period and later, the maize-growing Indians of the plains lived primarily in the tall-grass prairies of eastern Nebraska, eastern Kansas, eastern and central Oklahoma, and farther north on the Missouri mainstem in the Dakotas. The western limits of aboriginal corn-growing at this period were somewhere along the 98th meridian, a little farther west on the Missouri River. In the 12th and 13th centuries, corn was being grown as far west as the Colorado-Kansas line, 200 miles or more beyond the 18th-century limit, and farther south, in the Oklahoma and Texas panhandles. This earlier maize production seems to have been a much less rewarding one, at least north of the Arkansas River, if the smaller and less plentiful storage pits at Upper Republican and other early sites are indicative.

That far-reaching population shifts occurred in late-prehistoric times in the western Nebraska-Kansas-Oklahoma region has been recognized by archaeologists for some time. Recently, in the light of tree-ring and radiocarbon-dating studies, together with cross-finds of Puebloan sherds in plains sites from Texas to Nebraska, the picture has become somewhat clearer. In briefest outline, there are now available some 35 radiocarbon dates for Upper Republican and Central Plains Phase sites, ranging from A.D. 465 to "modern." Among these are seven sites represented by two to five dates each. Site averages developed from these group determinations suggest that the principal Upper Republican occupation in southwestern Nebraska dates between about A.D. 1050 and 1250. If this dating approach is valid, it suggests further that at least this section of the Upper Republican habitat may have been largely abandoned long before the mid–15th-century drought indicated by tree-ring evidence and sometimes regarded as responsible for driving these prehistoric farmers from the western part of the plains (Wedel, 1959, p. 570).

Baerreis and Bryson (1966) have suggested from site averages that the Texas and Oklahoma Panhandle Aspect sites were occupied mainly between A.D. 1200 and 1450. Thus, they fall in general after the Upper Republican sites, from which, it has been proposed, they may have derived their culture if not, indeed, their population as well. The Great Bend Aspect sites, to judge from cross-finds of Puebloan pottery and from unsatisfactory radiocarbon determinations, apparently date mainly after about A.D. 1450. If this is so, both their material culture and their population could have been drawn in part from the Panhandle culture, as this could have been drawn in part from the still older Upper Republican culture.

It has long been suspected that drought may have played an important role in the inferred late-prehistoric exodus from the western part of the plains and the movement of its erstwhile residents in the 15th or 16th century into the better-watered eastern tall-grass prairies (Wedel, 1941). One observer (Griffin, 1961, p. 711) has asserted that the later, more easterly plains villages were located "along the major stream valleys with sources in the Rocky Mountains," the implication here being, I suppose, that streams heading in the Rockies were drought-proof and so afforded asylum to refugees from the drought-plagued Western Plains. Whatever merit this idea may have for the Middle Missouri mainstem settlements in the Dakotas, I doubt its relevance for Nebraska and Kansas. Most of the Great Bend Aspect villages occur not on the banks of the major through-flowing streams, which have low sandy banks that are subject to overflow, but rather on the tributaries, near localities where water was obtainable from outcrops of Dakota sandstone or at contacts between this and the underlying impervious Kiowa shale. Likewise, in the Pawnee district of east-central Nebraska, the early historic (Lower Loup) communities occur mostly not on the through-flowing Platte River, with its Rocky Mountain spring water, but on the banks of the Loup River, whose various branches have their sources in the Nebraska Sandhills some hundreds of miles east of the Rocky Mountains. It is oversimplification to suggest that the mountain-based rivers pulled the Central Plains Village Indians through the postulated droughts in late-prehistoric times.

The most recent and stimulating contribution to climatology and human prehistory in the Southern Plains stems from the work of Baerreis and Bryson (1965). Pertinent to the present discussion is their comment on the Neo-Boreal episode, dated about A.D. 1550 to 1880, where we read (1965, p. 217; see also Bryson and Wendland, 1967, p. 296), in part:

> After about A.D. 1450 the climate seems to have reverted in part to the character of Neo-Atlantic times, but a definite change to stronger circulation and a new climatic episode occurred about 1550. Apparently, the westerlies and polar storm tracks shifted southward and intensified, carrying

wet cloudy summers deep into Europe. . . . Glaciers . . . once again were found in the U.S. Rockies as far south as New Mexico. . . . It is likely that the growing season was shortened, and that the summers were cooler in the upper Midwest. . . .

Whether there is anything in the climatological setting for the Southern and Central Plains, and specifically for the middle Arkansas drainage, that will offer an explanation for the inferred withdrawal of the Wichita-speaking groups from central Kansas by the opening of the 18th century, I am not prepared to say. The North Platte tree-ring sequence (Weakly, 1943) records a 20-year drought, 1688-1707. The much-criticized Bismarck, North Dakota, tree-ring sequence also is interpreted as showing a long drought at about this same time, specifically during 39 years from 1663 to 1702 (Will, 1946). I know of no comparable data for the central Kansas region, but this approximates the time period in which the principal Great Bend Aspect occupation almost certainly ended.

As Baerreis and Bryson (1965, p. 217) have further noted: "Neo-Boreal began just at the time of European contact, obscuring the relative roles of cultural and climatic impacts on the lifeway of the Indian. It is an intriguing possibility that deteriorating climatic conditions making an agricultural economy precarious may have been more influential in producing a shift to a fur trading economy than the magnet of European trade goods."

In the same vein, and without denying the possibility of a climatic factor being involved, I think it a strong probability that cultural-historical factors may have been at least equally influential on the Middle Arkansas. The Coronado documents suggest that the Indians of Quivira had enemies among their contemporaries to the northeast, who were probably Siouan and may well have been early Kansa or Osage. We know that these tribes were mortal foes of the Wichita in the 18th century. In the High Plains west of Quivira, the emergence of the Plains Apache as typical nomadic bison-hunters and their acquisition of the horse early in the 1600's boded ill for the semisedentary Wichita tribes. By Oñate's time, the Wichita and the Plains Apache were deadly foes, with each tribe openly and avowedly practicing cannibalism toward the other as opportunity offered. There are strong indications at several Great Bend Aspect sites of extended conflagration that may well have been incendiary in nature, as well as indications of violent death and hasty interment of the only skeletons so far found that reasonably can be attributed to the local population (Wedel, 1968b). All this seems to be further evidence of hostile action against the Indians of the Great Bend Aspect by an outside force. Such action could have been stimulated, of course, by natural phenomena of adverse character, such as droughts and the resultant local disappearance of game animals or other food resources. The nomads of the Western Plains must have found the corn-filled caches of the Village Indians a tempting prize at any time, and particularly so when because of drought or for other reasons, game and wild vegetal foods were scarce or difficult to get.

With the information we have today, it seems probable that the questions posed earlier regarding the northward spread and subsequent southward withdrawal of Wichita peoples in the Middle Arkansas River drainage must remain unanswered for the present. We urgently need much more field data on the nature of Great Bend Aspect culture and its variations through time and space. Further archival search for possible contemporary documents bearing on the Indians represented by this complex might be rewarding. If these researches in field and library have as one of their prime goals the acquisition of environmental data in depth, including among others such matters as soil and microfaunal analyses, settlement patterns, and site distributions through time and space, they may one day carry us measurably nearer the answers we seek.

ACKNOWLEDGMENTS

The archaeological field work on which this paper is partly based was supported by National Science Foundation grant GS-556, by Smithsonian Research Foundation award no. 3301, and by previous grants from the Smithsonian Institution.

LITERATURE CITED

Baerreis, D. A., and R. A. Bryson
  1965. Climatic episodes and the dating of the Mississippian cultures. Wisconsin Archeol., 46:203-220.
  1966. Dating the Panhandle Aspect cultures. Bull. Oklahoma Anthro. Soc., 14:105-116.
Bastian, T.
  1966. Initial report on the Longest site. Great Plains Newsletter, 3(1), Lawton, Oklahoma.
Bryson, R. A., and W. M. Wendland
  1967. Tentative climatic patterns for some late glacial and post-glacial episodes in central North America. Pp. 271-298, in Life, land and water (W. J. Mayer-Oakes, ed.), Occas. Papers, Dept. Anthro., Univ. Manitoba, xvi+414 pp.
Galinat, W. C., and J. H. Gunnerson
  1963. Spread of eight-rowed maize from the prehistoric Southwest. Bot. Mus. Leaf., Harvard Univ., 20:117-160.
Gould, C. N.
  1898. The timbered mounds of the Kaw Reservation. Trans. Kansas Acad. Sci., 15:78-79.
  1899. Additional notes on the timbered mounds of the Kaw Reservation. Trans. Kansas Acad. Sci., 16:282.
Gregg, J.
  1954. Commerce of the prairies. M. L. Moorhead, ed., Univ. Oklahoma Press, Norman, xxxviii+469 pp.
Griffin, J. B.
  1961. Some correlations of climatic and cultural change in eastern North American prehistory. Ann. New York Acad. Sci., 95:710-717.
Hammond, G. P., and A. Rey
  1940. Narratives of the Coronado expedition, 1540-1542. Univ. New Mexico Press, Albuquerque, xii+413 pp.
  1953. Don Juan de Oñate, colonizer of New Mexico 1595-1628. Coronado Cuarto Centennial Publ., 1540-1940, Univ. New Mexico Press, Albuquerque, 5:xvi+1-584, 6:xv+585-1187.
Kroeber, A. L.
  1939. Cultural and natural areas of native North America. Univ. California Publ. Amer. Archaeol. Ethnol., 38:xii+1-242.
Pillaert, E. E.
  1963. The McLemore site of the Washita River focus. Bull. Oklahoma Anthro. Soc., 11:1-114.
Schoewe, W. H.
  1949. The geography of Kansas. Part II. Physical geography. Trans. Kansas Acad. Sci., 52:261-333.
Scholes, F. V., and H. P. Mera
  1940. Some aspects of the Jumano problem. Publ. Carnegie Inst. Washington, 523:265-299.
Terry, K., and I. Terry
  1961. Chain mail and other exotic materials from south central Kansas. Plains Anthro., 6:126-129.
Udden, J. A.
  1900. An old Indian village. Lutheran Augustana Book Concern, Rock Island, Illinois, 80 pp.
Weakly, H. E.
  1943. A tree-ring record of precipitation in western Nebraska. Jour. Forestry, 41:816-819.
Wedel, W. R.
  1941. Environment and native subsistence economies in the Central Great Plains. Smithsonian Misc. Collections, 101(3):1-29.
  1942. Archaeological remains in central Kansas and their possible bearing on the location of Quivira. Smithsonian Misc. Collections, 101(7):1-24.
  1959. An introduction to Kansas archeology. Bull. Bur. Amer. Ethnol., 174:xvii+1-723.
  1967. The council circles of central Kansas: were they solstice registers? Amer. Antiquity, 32:54-63.
  1968a. Some thoughts on Central Plains—Southern Plains archeological relationships. Great Plains Jour., 7:53-62.
  1968b. After Coronado in Quivira. Kansas Hist. Quart., 34:369-385.
Will, G. F.
  1946. Tree-ring studies in North Dakota. Bull. Agric. Exp. Sta., North Dakota Agric. Col., 338:1-24.

BOTANY
(Consulting Editor, Philip V. Wells)

Pollen Analysis of Pre-Wisconsin Sediments from the Great Plains

RONALD O. KAPP

ABSTRACT

This paper summarizes the results of pollen analysis of sediments from the Kingsdown Formation in southwestern Kansas and northwestern Oklahoma. The fossiliferous beds are assigned to Illinoian glacial and Sangamon interglacial ages. Fossil mollusks and vertebrates have been recovered from the nine sites treated in this paper; these paleontological interpretations have been previously published by Hibbard and his co-workers.

Three of the sites that have previously been assigned to the Illinoian glacial age on the basis of animal fossil remains have yielded pollen diagrams dominated by spruce *(Picea)* and pine *(Pinus)* pollen. This fossil assemblage suggests that forests of the Rocky Mountain Front Range extended far into the plains during the Illinoian age.

Late Illinoian and Sangamon (interglacial) sites have yielded sediments with little spruce pollen, less pine than earlier glacial-age sediments, and sufficient grass and composite pollen to indicate open vegetation. Pollen preservation is poor at some of the Sangamon interglacial sites, but abundance of pine pollen suggests that pine trees persisted in local stands on the plains. While conclusive interpretation of the Sangamon interglacial vegetation and paleoecological conditions must await additional pollen data, it seems evident that prairie vegetation in the central United States is not unique to the post-glacial period, whether the chief cause of the restriction of trees is climate, fire, soils, or some other factor or combination of factors.

This paper summarizes studies of pre-Wisconsin pollen analyses from the plains and relates this evidence to the paleoecologic implications of the associated faunas and rock units. Some of the sites mentioned cover virtually the entire plains, but most of the detailed data are from the Southern High Plains where Claude Hibbard and his associates have worked for several decades (Hibbard, 1958, 1960; Hibbard and Taylor, 1960; Taylor, 1960; Kapp, 1962, 1965).

INTRODUCTION

Geologic sediments and their entombed fossils are the primary source of our understanding of the genesis of the existing biotic assemblages, biogeographic patterns, and habitats of the Great Plains. These attempts at paleoecologic reconstruction require that we invoke uniformitarian comparisons between Recent biogeographic or ecological patterns and the fragmentary fossil evidence. Such inferences, based on modern biogeographic patterns, Recent biotic communities, and ecosystems, may be somewhat misleading. Ideally we should reconstruct the history of these environments genetically, beginning from well-documented Early Pleistocene biotic communities. Our full understanding of process and change in the origin, evolution, and extinction of ecological systems—that is, a full and accurate history of plains ecology —must, therefore, await the discovery and analysis of dozens of additional fossiliferous sites.

At the present time the Tertiary ancestry of the Pleistocene plains biota, and the Quaternary record itself, are fragmentary. The available data from Pleistocene fossil faunas and pollen analysis do, however, permit a glimpse of the ancestral ecosystems of the Great Plains.

At only a few localities has the entire fossiliferous assemblage—vertebrates, mollusks, pollen, and so forth—been studied. A full reconstruction of the environments of the past awaits analysis of all fossil groups by teams of taxonomic experts. This paper emphasizes certain sites where both faunal and pollen evidence are available. In addition, it attempts to evaluate the meaning of this incomplete record as it relates to the history and origin of the mid–North American grassland biome.

SEDIMENTARY CYCLES AND FAUNAL PROGRESSION

In the High Plains, especially in southwestern Kansas and adjacent Oklahoma (Hibbard, 1958; Hibbard and Taylor, 1960; Taylor, 1960), it has been possible to reconstruct major parts of the Pliocene and Pleistocene sedimentary record (Fig. 1). While such reconstructions are partly based on long-distance correlations and include discontinuities, there are some rock units that serve as stratigraphic markers. The widespread Pearlette Ash, which has been petrographically described and identified from nearly the entire plains region (Swineford, 1949), has been correlated with the glacial section in the Central Great Plains (Frye et al., 1948). The actual age of this ash is uncertain; recently reported and rather ancient K-Ar dates require further study before acceptance (Dreeszen, this volume). Caliche beds are useful for correlation within local areas in the Southern High Plains; these strata are considered to be indicators of arid climates (Bretz and Horberg, 1949). The rock sequence for southwestern Kansas (Fig. 1) gives evidence of three or four cycles of deposition during the Pleistocene. The widespread sheet deposits of the Ballard and Crooked Creek formations reveal sediments that vary from coarse basal gravels to fine upper silts. This appears to reflect climatic cycles that gave rise to variations in precipitation, erosion, and stream flow. The three episodes of caliche formation apparently represent the most arid phases of these glacial-interglacial (pluvial-interpluvial) cycles. Correlation of this sedimentary sequence with the stages of eastern or cordilleran glacial geology has been difficult. The tentative correlations proposed in Figure 1 are inferred from the sediments and from the character of the vertebrate and molluscan faunas from the various strata. Analysis of large numbers of specimens from faunal assemblages has resulted in the tentative placement of several dozen local faunas into the Pleistocene sequence. As mentioned, it is frequently necessary to use inferential evidence of climatic requirements of the fauna itself to assign it to a position in the glacial-interglacial sequence, thus there is the ever-present danger of circular reasoning in such stratigraphic analyses.

Three generalizations about the sequence of local faunas seem to be well established:

(1) There are cyclic fluctuations from "glacial-pluvial faunas" to those indicating "interglacial-interpluvial" conditions.

(2) These contrasts between glacial and interglacial faunas become more pronounced later in the Pleistocene.

(3) Throughout long periods of the Early and Middle Pleistocene (both during glacials and interglacials), the faunas indicate more equable, less continental, climates than at present. The early interglacial faunas, especially the croco-

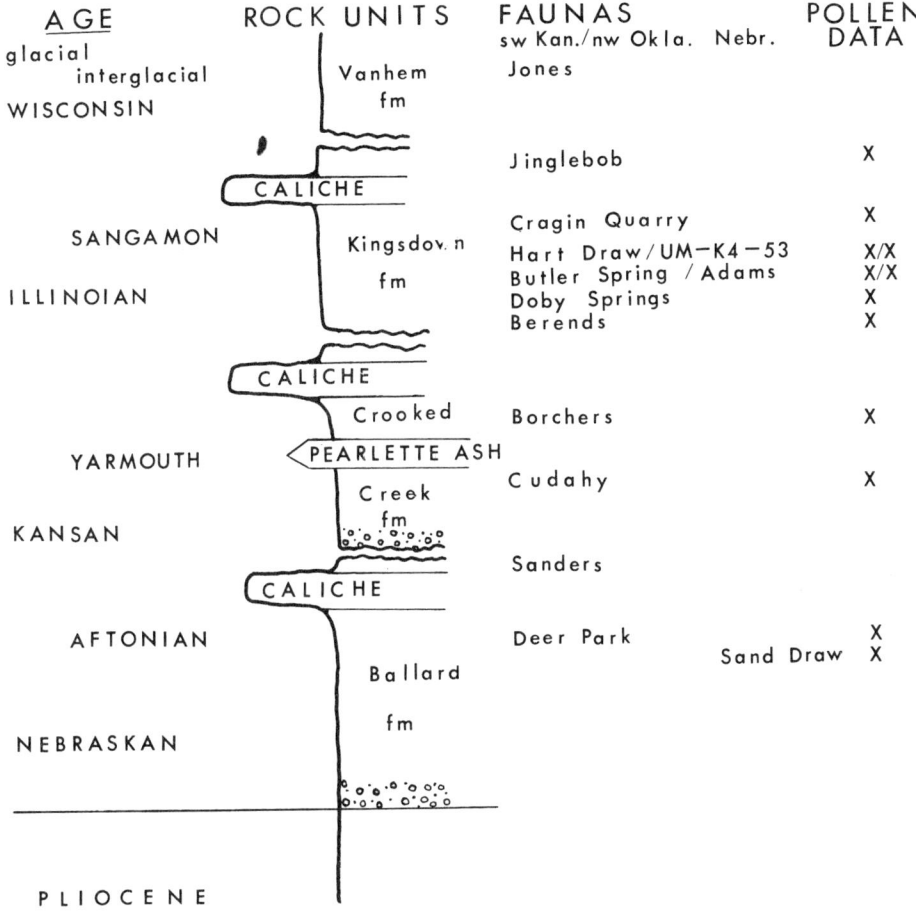

FIG. 1. Sequence of rock units for southwestern Kansas and adjacent Oklahoma, with suggested correlations with glacial and interglacial stages and fossil faunas. Sites at which pollen analysis has been attempted are marked (X). Modified from Hibbard and Taylor, 1960, p. 7.

dilian and giant land tortoise components are considered by Hibbard (1960 and this volume) to reflect milder winters than at present. The Early and Middle Pleistocene vertebrate and molluscan faunas generally required more equable temperatures and more effective precipitation than at present, except during periods of caliche formation. The progression of the faunas from Early to Late Pleistocene seems to indicate substantial deterioration of climate in the Late Pleistocene, especially during the Wisconsin glacial age.

POLLEN ANALYSIS

TECHNIQUES

Pollen analysis of Pleistocene sediments from the Great Plains has been difficult because a large number of the sites, even those rich in molluscan or vertebrate fossils, have not yielded well-preserved fossil pollen in quantities suitable for analysis. Furthermore, it has been impossible to consistently predict whether a certain type of sediment will be suitable for analysis. Generally, fine silts and clays are more productive than coarse silts or sandy horizons, but there are numerous exceptions. Interestingly, many of the promising dark brown or black strata in alluvial and lacustrine deposits, although rich in finely divided plant remains, have generally been devoid of pollen. Even the polleniferous sediments generally have a low pollen density when compared with lake and bog sediments from forested regions; this requires the collecting and processing of large-volume samples, a tedious and time-consuming procedure.

All samples for these studies were collected from open pits and gully banks. At each sample site, quantities of overburden or weathered surface sediments were removed to expose fresh, unfissured beds, which lacked evidence of color changes associated with weathering. Usually, bulk samples, 500 to 1000 cubic centimeters in volume, were collected and sealed in plastic bags to prevent contamination. The single attempt to collect these consolidated sediments with motor-driven coring equipment was unsuccessful because good stratigraphic control could not be maintained, and the risk of mixing or contaminating sediments was great. It is probable that hollow-core drilling techniques would yield unweathered polleniferous sediments from certain sites.

The quality of preservation of the pollen from these Pleistocene sediments is variable. In some instances, especially when pollen is present in abundance, it is well preserved and is readily identified as post-glacial or fresh material. In many instances, however, the pollen and spores were poorly preserved. Using the terminology and definitions of Cushing (1967), broken and crumpled palynomorphs were frequent in some samples. This suggests that some of the pollen spectra show an inaccurate representation of thin-walled pollen types such as *Juniperus* and *Populus*. In many cases the palynomorphs were corroded (exine etched or pitted, the corroded spots usually with scalloped margins) or variously degraded (structural elements of exine homogenous, fused). Corrosion and degradation are probably indications of biological decay or weathering, or both.

Laboratory processing techniques include two main series of steps: (1) removal of mineral sediments and (2) acetolysis to remove extraneous organic material and enhance microscopic analysis of pollen exines. Mineral sediments usually were removed by use of the following sequence (some steps eliminated, where appropriate): (1) flooding with six percent HCl to remove carbonates; (2) "swirling" to remove sand (coarse sediments only); (3) covering sample with twice its volume of HF (concentrated) to remove silicates; (4) flotation with $ZnCl_2$ (saturated) to concentrate the organic fraction. An ultrasonic generator was used in some instances to improve particle dispersion.

Pollen counting of each sample was continued until the prepared material was exhausted or approximately 200 grains had been tallied. Pollen percentages and histogram-spectra were prepared when the pollen count approximated 200 grains (50-150 in a few instances). At some sites, a sufficient number of stratigraphically-ordered pollen spectra were studied to permit construction of a standard pollen diagram; in other instances only a few isolated pollen spectra are available.

EARLY AND MIDDLE PLEISTOCENE SITES

No sites have yet been discovered from Early or Middle Pleistocene beds that have yielded fossil pollen in quantities suitable for extensive analysis or the calculation of pollen spectra.

A preliminary survey of sediments from the Early Pleistocene Sand Draw locality (including *Stegomastodon* Quarry) from Brown County, Nebraska, has yielded small quantities of degraded pollen. Until more satisfactory data are available, it may be worthy of note that these fragmentary records are dominated by coniferous (spruce and pine) pollen. Hibbard (1958) correlated the Sand Draw fauna with the Nebraskan glacial age, but subsequently (1960) considered it to be

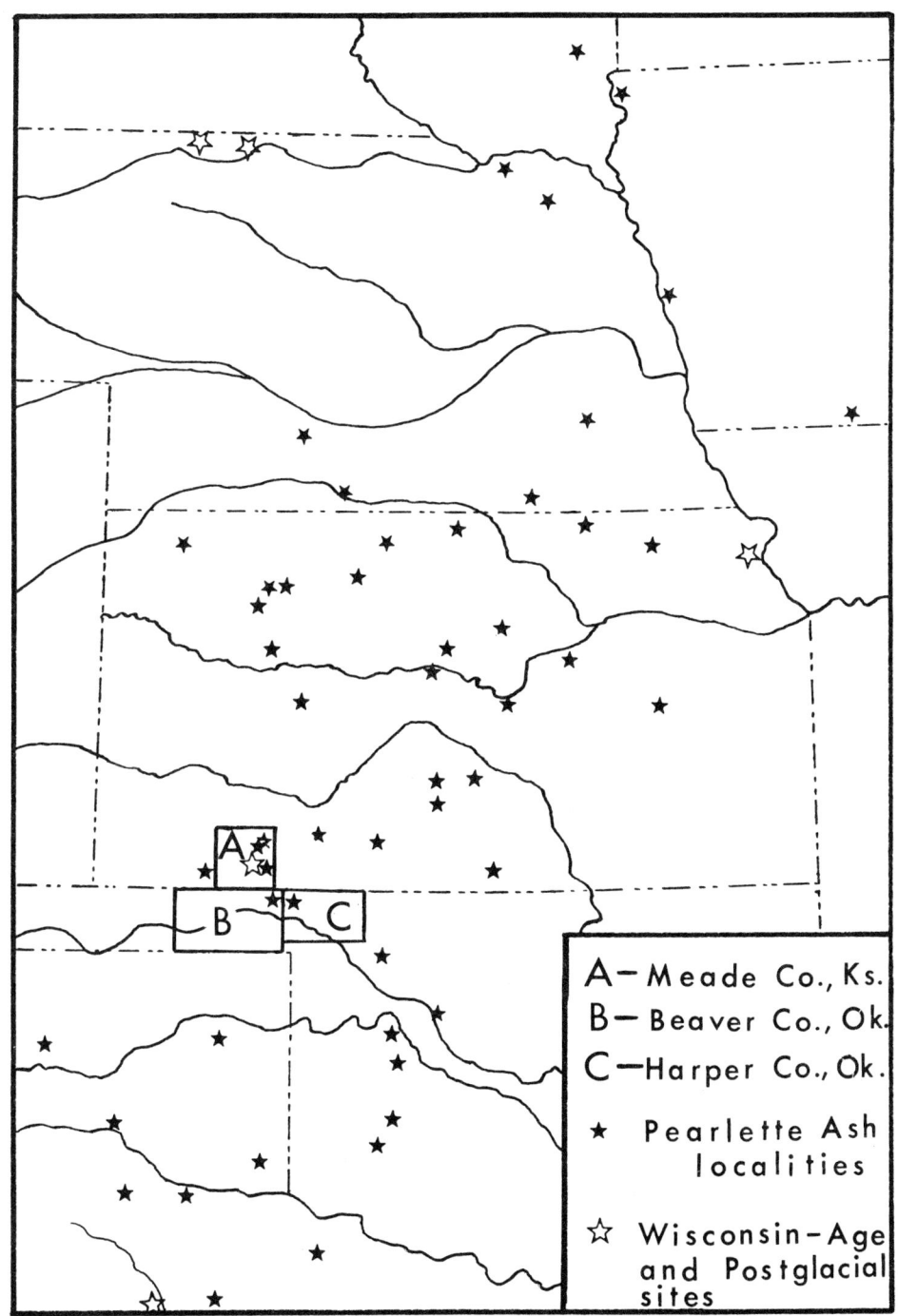

Fig. 2. Localities in the Central Plains for which pollen analytical data are available.

Aftonian. Whatever its exact glacial correlation, the deposit is clearly from the Early Pleistocene, containing *Plesihippus,* the fossil horse that typifies Blancan faunas.

The sediment matrix of the Deer Park local fauna (Aftonian) in Meade County, Kansas, has not yielded pollen.

A rather exhaustive analytical pollen survey has been completed on the Middle Pleistocene sediments associated with the late-Kansan Pearlette Ash. This ashfall was widespread (Fig. 2), and collections for pollen analysis were made at nearly 50 sites in six states. The Cudahy vertebrate fauna has been recovered immediately beneath the ash at sites in Nebraska, Kansas, Oklahoma, and Texas. In addition, mollusks have been studied at other localities (Frye *et al.,* 1948). In spite of numerous collections, variable sediment types, and repeated attempts to recover pollen, not a single pollen spectrum has been counted from sediments associated with the Pearlette Ash. At a few sites in Kansas occasional degraded pine, oak, and grass pollen grains have been seen. It seems probable that the physical-chemical conditions associated with the volcanic ash deposits are unsuitable for pollen preservation. The wide distribution of this dependable time-stratigraphic marker in the plains region warrants continued attempts at pollen recovery from sediments with Pearlette Ash lenses, however. Several attempts to recover pollen from sediments associated with the Borchers local fauna (Yarmouth, above Pearlette Ash beds) have been similarly unsuccessful.

LATE PLEISTOCENE SITES

Detailed pollen studies have been completed on sediments associated with eight local faunas taken from the Kingsdown Formation in Meade County, Kansas, and Beaver and Harper counties, Oklahoma (Fig. 2); these are believed to span a considerable segment of Illinoian and Sangamon ages.

The detailed pollen diagrams and descriptions of the individual sites have been published earlier (Kapp, 1962, 1965). The following is a summary of those data and their possible implications for paleoecology of the plains.

Pollen diagrams and spectra from the several Kingsdown Formation sites have been somewhat simplified and plotted together in Figure 3. The stratigraphic correlation was derived from published interpretations based on direct field correlation of rock units, faunal similarity, and faunal progression. Thus, these pollen data constitute an independent variable and have not influenced the stratigraphic ordering of the sites. It should be added, however, that the palynological data do not conflict with other evidence and that the stratigraphic ordering is further substantiated. For example, C. L. Smith (1958) and Stephens (1960) analyzed the Doby Springs vertebrate fauna and concluded that it was a close correlative of the Berends local fauna and similar to, but apparently younger than, the Butler Springs (including Adams) faunas.

Criteria for interpretation.—Attempts to reconstruct the nature of vegetation require cautious interpretation of pollen data. In undertaking such a reconstruction of the Illinoian-Sangamon vegetation, available data on modern pollen rain in the prairie and Rocky Mountain regions were reviewed. The objective is an attempt to determine the relationships between the frequencies of pollen of certain types and the actual proximity of certain trees or plant communities to the depositional basins. Studies of modern pollen deposition in windmill tanks on the Llano Estacado (an extension of the High Plains in Texas, south and southwest of these study sites) by Hafsten (1961) indicate the contribution of *Picea* (spruce) and *Pinus* (pine) pollen from stands of trees at varying distances from depositional basins in treeless areas. More detailed studies by Potter and Rowley (1960) of the

POLLEN ANALYSIS OF PRE-WISCONSIN SEDIMENTS 149

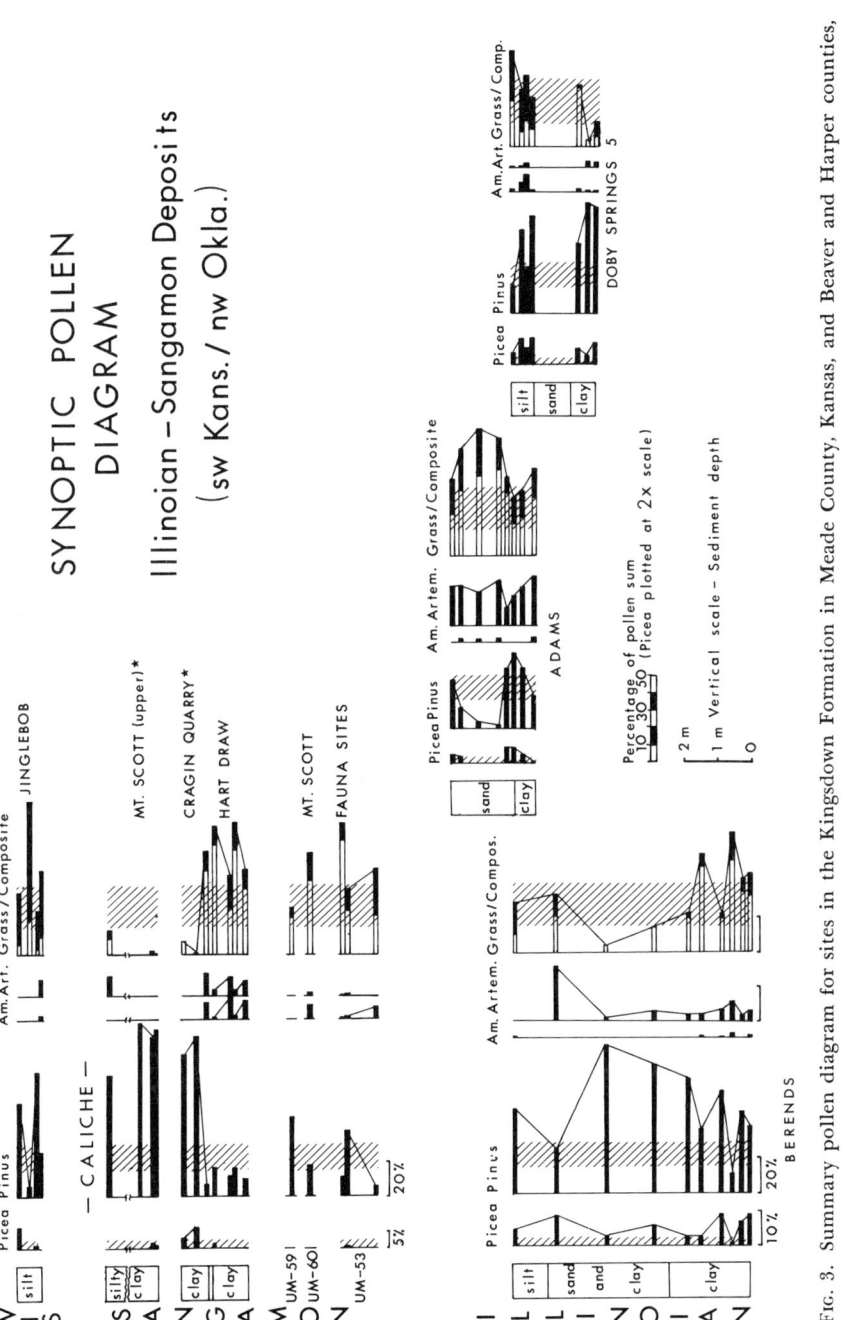

FIG. 3. Summary pollen diagram for sites in the Kingsdown Formation in Meade County, Kansas, and Beaver and Harper counties, Oklahoma. The stratigraphic sequence of the sites is based on previous geologic and paleontological studies by Hibbard and others. The Mt. Scott fauna sites are probably transitional between the Illinoian glacial and the Sangamon interglacial (Illinoian late glacial). The vertical extent of each of the pollen diagrams is drawn to scale; site names in capital letters refer to the designations of the fossil local faunas. Aquatic pollen types and those of low incidence have not been plotted in this diagram. Stars (Cragin Quarry and Mt. Scott) indicate that the diagrams are considered to be of dubious validity because of poor pollen preservation and a limited number of pollen types.

relation between local vegetation and pollen deposition on the San Augustin Plains of New Mexico are helpful in interpreting the meaning of conifer pollen percentages in fossil spectra. My studies of modern pollen rain from cattle-watering tanks extend across the prairie from Colorado to the Mississippi River (Kapp, 1965). They have permitted an assessment of the contribution of the montane forests of the West and deciduous forests of the East to recent pollen rain in the grasslands.

Detailed studies of modern pollen rain by Bent and Wright (1963) in western New Mexico and by Maher (1963) in Colorado provide additional data that have contributed to the development of the criteria described below. These latter studies are not directly applicable to considerations of plains vegetation, because they were conducted in mountainous regions where conditions for pollen dissemination and deposition are quite different from those in the flat, open High Plains.

While quantification in the interpretation of fossil pollen spectra is fraught with risks of oversimplification and misjudgment, it seems useful to describe as precisely as possible the criteria upon which these interpretations are based. There are essentially two kinds of quantitative information in a pollen diagram or in groups of temporally ordered pollen diagrams (see Fig. 3). First, the trends or gradual changes in pollen frequency in successive strata are considered to reflect the direction of changes occurring in the regional vegetation. A trend of this kind is noted in Figure 3; *Picea* pollen frequencies decline from Berends to Adams to Mt. Scott sites. Second, the pollen analyst often attempts to relate percentages of each pollen type in the modern pollen rain to the composition of the specific plant communities that produced the pollen mixture. This may take the form of proposing correction values (r values) for adjusting the pollen frequencies to reflect some direct measure of vegetational abundance or dominance. This method requires intensive studies of many vegetational types and their pollen production before fossil pollen spectra can be interpreted as representative of specific plant communities; furthermore, such interpretations have decreasing reliability in increasingly ancient sediments.

There has been no attempt to estimate r values for the various pollen types in Figure 3. Instead, modern pollen rain data (see above) have been used to establish percentages believed to indicate absence or presence of certain vegetation. Diagonal bars have been drawn over certain pollen profiles in Figure 3 to indicate the critical levels of pollen frequency. If the pollen frequencies fall below (left of) the minimum indicated by the diagonal lines, *Picea* and *Pinus* trees were probably not present in the region. If the pollen frequencies rise above (right of) the diagonal lines, the trees probably were nearby. If the percentage is within the hachured zone, the interpretation is uncertain.

Picea.—Spruce pollen is particularly important in these interpretations, because it is not disseminated far from the source trees. Furthermore, in these latitudes and altitudes, spruce trees seem to respond sensitively to climatic changes. The latitudinal and altitudinal limits of spruce distribution seem to follow the 70°F (21°C) July average isotherm. Potter and Rowley (1960, p. 20) stated that "any appreciable amount of spruce pollen in a sedimentary profile would surely represent the near presence of spruce." Similarly, Hafsten (1961, pp. 71-72) found only occasional grains of spruce or fir *(Abies)* pollen at distances of 200 miles or more from montane spruce-fir forests. He concluded (p. 79) that "only the windmill tanks situated very close to the present distribution areas of these trees (cf. WT-1 and WT-2 at

50-100 miles) show a higher incidence of these pollen types." I have not found *Picea* pollen frequencies above 1 percent or *Abies* above 0.5 percent in the modern pollen rain of the central states—Kansas, Nebraska, southeastern South Dakota, southern Minnesota, Iowa, and northwestern Missouri (Kapp, 1965, p. 187). It is evident that *Picea* percentages of two percent or greater provide conclusive evidence that spruce trees were near the depositional site or within the region that contributed the regional pollen rain. Interpretation of spruce-pollen percentages below two percent is uncertain. Thus, in Figure 3, spruce trees were certainly near the Berends and Doby Springs depositional sites in the Oklahoma Panhandle and near the Adams site in Meade County, Kansas, during Illinoian glacial time.

Pinus.—Hafsten (1961, pp. 78-79) found 35 percent pine pollen in windmill tanks situated 15 to 20 miles from the westerly pine areas; the percentages drop to 13 and 16 at a distance of 150 miles, and to as low as 9.5 at a site nearly 200 miles east of the nearest pine stands. On the San Augustin Plains of New Mexico where the pinyon pine–juniper association occupies 31.5 percent of the area and the ponderosa pine association 16 percent, pine-pollen contribution to the total pollen rain averaged 26 percent. None of my surveys (Kapp, 1965) shows pine-pollen percentages above eight in the Central Plains area. On the basis of these data it seems sound to estimate that if pine-pollen percentages are above 30, pine trees grew nearby; whereas percentages below 15 suggest that the source may be quite distant. Accordingly, the hachured zone on the pine-pollen profiles in Figure 3 indicate the 15 to 30 percent range of uncertain interpretation. Higher percentages indicate local pine stands, and lower percentages suggest long-range transport as the source.

Grass and composite pollen.— There are many uncertainties in estimating the openness (treelessness) of vegetation by means of pollen analysis. For example, in the San Augustin Plains, where at least 51 percent of the vegetational cover is grama grassland and the saltbush-grama association, arboreal pollen at eight stations comprised 71 to 99 percent of the pollen rain (average 93 percent). In their study of that area, Potter and Rowley (1960, p. 22) concluded that "if shrubby vegetation is involved at the site of sampling, a percentage of arboreal pollen exceeding 40-50% of the total collected must indicate a forest border within a few miles. In grassland areas, however . . . arboreal pollen percentages may exceed 90% even when forested areas are farther away." These results and the results of similar studies in mountainous areas have little similarity in vegetational structure or pollen rain to open grasslands. The data are not of direct aid in interpreting the pre-Wisconsin pollen diagrams from the High Plains. My surveys of modern pollen rain across the Central Plains (Kapp, 1965, pp. 186-189) suggest that the sum of grass and composite (excluding *Ambrosia* and *Artemisia*) pollen may be a useful indicator of the prairie vegetational formation. (Weedy chenopods and amaranths grow abundantly in the area of trampling near the cattle-watering tanks; these pollen types are doubtless overrepresented in the modern pollen deposition in this study. Computation of the pollen percentages excluding chenopod-amaranth pollen is believed to most accurately reflect the presettlement pollen rain of the native grasslands; this exclusion was made in computing the grass-composite percentages in the modern pollen rain that served as the basis for the 15 to 40 percent uncertainty range suggested below.) The range of 15 to 40 percent (diagonal lines, Fig. 3) of grass plus composite pollen (excluding *Ambrosia* and *Artemisia*) is tentatively suggested as the range of uncertain interpretation.

When percentages are lower than 15, the local region may have been largely forested; when above 40, the vegetation was probably open—either savanna, scrub, or grassland.

Illinoian glacial age.—The Berends diagram is dominated by pine *(Pinus)* pollen, generally above 50 percent; spruce *(Picea)* pollen ranges from four to 10 percent; deciduous tree pollen is nearly absent; and nonarboreal pollen types (NAP) are low, restricted primarily to sagebrush *(Artemisia)* and grass *(Gramineae)* pollen. The probability that the abundant conifer pollen was carried from great distances to this basin by streams is minimal, because the Berends sediments are not alluvial, but were deposited in a local sink. Pollen preservation at this site is excellent, further suggesting that the pollen is autochthonous. In the early Berends record, the presence of substantial quantities of grass pollen probably indicates an open woodland, or perhaps a mosaic of woodland and grassland communities.

The pollen diagram from the Doby Springs 5 site is similar to the Berends diagram. Small numbers of *Pseudotsuga* (Douglas-fir) pollen grains have been recovered at this site. The Doby Springs and Berends sites had been previously considered to be correlative on the basis of the fish and mammal fossils.

The pollen diagram from the Adams local fauna site in Meade County, Kansas, is dominated by conifer pollen in the lower clay sediments, thus resembling the Berends and Doby Springs pollen records. Pollen spectra from the overlying sands appear to reflect a retreat of spruce and pine to greater distances with the prairie or sagebrush-prairie formation prevailing.

Apparently the Illinoian glacial-age vegetation of southwestern Kansas and the Oklahoma Panhandle included significant stands of spruce and pine trees. These trees may have been restricted to "scarp woodlands" such as those of the modern plains (Wells, 1965 and this volume), although I have previously suggested (Kapp, 1965) that the trees might have grown in gullies and streambeds.

Pollen is sparser and its preservation poorer in sediments associated with the Butler Springs local fauna (not shown in Fig. 3), the Mt. Scott local fauna sites, Hart Draw, and especially the Cragin Quarry local faunas. From the pollen spectra available (Fig. 3), however, certain trends are discernible. Spruce *(Picea)* and pine *(Pinus)* pollen percentages drop to minima, which mostly represent long-range drift from western mountains. There is a slight increase in pollen representation of deciduous trees. This may represent an "unmasking" of the pollen production of trees present at an earlier stage or it may reflect an actual influx of some trees, perhaps near escarpments, as at present. The increase in species of open communities is most significant, however. *Ambrosia* (ragweed) and *Artemisia* (sagebrush) percentages are generally higher than in Illinoian sediments. Compositae (aster-sunflower family) frequency is above 20 percent; grass pollen generally exceeds 40 percent. The spectra from the Hart Draw site compare favorably with modern pollen rain deposited in cattle-watering tanks in the Central Plains of Kansas and Nebraska (Kapp, 1965, pp. 188-189). One major exception to this similarity is the relatively low representation of the Chenopodiaceae and Amaranthaceae (goosefoot-pigweed-tumbleweed) families in Sangamon sediments. While it is likely that localized pine stands persisted in the region, spruce pollen is most likely long-range drift. It is clear that woodland contracted and that open plant communities expanded during this episode.

The Butler Spring local fauna (pollen data inadequate and not plotted in Fig. 3) and the Mt. Scott local fauna have been considered late Illinoian or early interglacial in age (Hibbard, 1963; G. R. Smith, 1963; Miller, 1961). Vertebrate and molluscan fossils from the Cragin Quarry and upper Mt. Scott exposure are interglacial in character (Hibbard and Taylor, 1960).

Interpretation of the Cragin Quarry and upper Mt. Scott pollen spectra is uncertain. The deposits are immediately below the late Sangamon caliche; in fact some lime nodules are present in the sediments that yielded the fossil fauna. Pollen preservation is poor, but pine predominates, totaling 100 percent in one spectrum from Mt. Scott. The reduction in variety of pollen types and severe degradation make any firm interpretation risky; it is tentatively concluded that differential degradation resulted in overrepresentation of the resistant and easily recognized coniferous types. It might be noted that there is no ambiguity about stratigraphic position, for the UM-K4-53, UM-K1-60, and UM-K2-59 (Mt. Scott local fauna sites), Hart Draw, and Cragin Quarry sites are located within a few hundred yards of one another on the Big Springs Ranch (Meade County, Kansas).

During Sangamon times, spruce trees probably were absent from the region, suggesting a warmer climate than in Illinoian. Pine trees probably were of more restricted distribution, although they likely persisted at favored sites. Prairie plant communities expanded.

Pollen is virtually absent from the caliche horizon, and no palynological or paleoecological appraisal is possible of the standard interpretation that caliche formation represents increased aridity.

The Jinglebob local fauna was originally described by Hibbard (1955) as interglacial, probably of Sangamon age. Since that publication, the Kingsdown Formation is better known; it is clear that the Jinglebob strata postdate the caliche. Fossils of the southeastern rice rat *(Oryzomys)*, a northern shrew *(Sorex cinereus)*, and the meadow vole *(Microtus)* suggest a moist, equable climate with mild summers. Preliminary pollen analysis by Kathryn Clisby (Hibbard, 1955, p. 200) noted the abundance of pine pollen. Indeed the recent pollen analyses shown in Figure 3 also reveal significant percentages of spruce pollen. The total pollen assemblage differs from Illinoian glacial pollen diagrams (for example, Berends) in that composite percentages are higher. The phytogeographic shifts that explain these pollen spectra are uncertain. The increased moisture required by the Jinglebob fauna and cooler-moister climates suggested by the abundant conifer pollen may reflect climatic events associated with the earliest phases of the Wisconsin glacial period. Clear definition of an interglacial-to-glacial biotic or paleoecologic break is probably not possible. Inasmuch as this was a period of climatic transition, the question of Late Sangamon as opposed to Early Wisconsin placement of the Jinglebob biota is not worthy of controversy. High conifer pollen percentages, especially spruce, are sufficiently similar to the Illinoian glacial pollen diagrams to warrant tentative assignment to the Early Wisconsin.

CONCLUSION

During the Illinoian glacial age in southwestern Kansas and adjacent Oklahoma, it appears that the coniferous forests of the Rocky Mountain Front Range and foothills extended to the Southern High Plains. During glacial time, forest trees clearly advanced at the expense of open, treeless prairie. The existence of woodlands on the Plains is also well documented for Wisconsin and post-glacial time in north-central Nebraska and South Dakota (Sears, 1961; Watts and Wright, 1966; Wright, this volume—for locations see Fig. 2), northeastern Kansas (Horr, 1955; Wright, this volume), and the Llano Estacado of New Mexico and Texas (Wendorf, 1961 and this volume).

During the glacial ages there were undoubtedly phytogeographic shifts that provided east-west migrational routes along forest corridors at least across the Northern Great Plains. Distributional patterns for Recent mammals (Hoffmann and Jones, this volume) and small mammals in the fossil faunas of Illinoian glacial age (Stephens, 1960; Kapp, 1962) undoubtedly date from glacial ages in which trees were more extensive on the Central Plains.

Pollen analytical evidence from the Sangamon interglacial sediments, especially the absence of *Picea*, suggests a warmer summer climate; the abundance of nonarboreal pollen in some pollen spectra (especially Hart Draw) reflects the expansion of prairie vegetation. Pine trees may have persisted in localized stands. While conclusive interpretations of the Sangamon vegetation await additional pollen data, it seems evident that prairie vegetation in the Central United States is not unique to the post-glacial period, whether the chief cause of the restriction of trees be climate or fires.

ACKNOWLEDGMENTS

Financial support for this research was provided by grants from the National Science Foundation, specifically G-10689 (to the University of Michigan) and GB-1023 (to Alma College). Invaluable aid was provided by Professor Claude Hibbard in locating sites, discussing results, and reviewing the manuscript. Professor W. S. Benninghoff provided important assistance during initial stages of the research. Numerous property owners cooperated by permitting field collections on their property; Stephen Bushouse served as field and laboratory assistant; and the facilities of the Departments of Botany of the University of Michigan and Alma College were used throughout the study. I gratefully acknowledge each of these sources of aid.

LITERATURE CITED

Bent, A. M., and H. E. Wright, Jr.
   1963. Pollen analysis of surface materials and lake sediments from the Chuska Mountains, New Mexico. Bull. Geol. Soc. Amer., 74:491-500.

Bretz, J. H., and L. Horberg
   1949. Caliche in southeastern New Mexico. Jour. Geol., 57:491-511.

Cushing, E. J.
   1967. Evidence for differential pollen preservation in Late Quaternary sediments in Minnesota. Rev. Palaeobot. Palynol., 4:87-101.

Frye, J. A., A. Swineford, and A. B. Leonard
   1948. Correlation of Pleistocene deposits of the Central Great Plains with the glacial section. Jour. Geol., 56:501-525.

Hafsten, U.
   1961. Pleistocene development of vegetation and climate in the Southern High Plains as evidenced by pollen analysis. Pp. 59-91, in Paleoecology of the Llano Estacado (F. Wendorf, ed.), Mus. New Mexico Press, Santa Fe, 144 pp.

Hibbard, C. W.
   1955. The Jinglebob interglacial (Sangamon?) fauna from Kansas and its climatic significance. Contrib. Mus. Paleontol., Univ. Michigan, 12:179-228.
   1958. Summary of North American Pleistocene mammalian local faunas. Papers Michigan Acad. Sci., Arts, Letters, 43:3-32.
   1960. An interpretation of Pliocene and Pleistocene climates of North America. Rep. Michigan Acad. Sci., Arts, Letters, 62:5-30.
   1963. A Late Illinoian fauna from Kansas and its climatic significance. Papers Michigan Acad. Sci., Arts, Letters, 48:187-221.

Hibbard, C. W., and D. W. Taylor
   1960. Two Late Pleistocene faunas from southwestern Kansas. Contrib. Mus. Paleontol., Univ. Michigan, 16:1-223.

Horr, W. H.
   1955. A pollen profile study of the Muscotah marsh. Univ. Kansas Sci. Bull., 37:143-149.

Kapp, R. O.
1962. Pollen analytical investigation of Pleistocene deposits on the Southern High Plains. Unpubl. Ph.D. thesis, Univ. Michigan, 226 pp.
1965. Illinoian and Sangamon vegetation in southwestern Kansas and adjacent Oklahoma. Contrib. Mus. Paleontol., Univ. Michigan, 19:167-255.

Maher, L. J., Jr.
1963. Pollen analysis of surface materials from the southern San Juan Mountains, Colorado. Bull. Geol. Soc. Amer., 74:1485-1504.

Miller, B. B.
1961. A Late Pleistocene molluscan faunule from Meade County, Kansas. Papers Michigan Acad. Sci., Arts, Letters, 46:103-125.

Potter, L. D., and J. Rowley
1960. Pollen rain and vegetation, San Augustin Plains, New Mexico. Bot. Gaz., 122:1-25.

Sears, P. B.
1961. A pollen profile from the grassland province. Science, 134:2038-2040.

Smith, C. L.
1958. Additional Pleistocene fishes from Kansas and Oklahoma. Copeia, 1958:176-180.

Smith, G. R.
1963. A Late Illinoian fish fauna from southwestern Kansas and its climatic significance. Copeia, 1963:278-285.

Stephens, J. J.
1960. Stratigraphy and paleontology of a Late Pleistocene basin, Harper County, Oklahoma. Bull. Geol. Soc. Amer., 71:1675-1702.

Swineford, A.
1949. Source area of Great Plains Pleistocene volcanic ash. Jour. Geol., 57:307-311.

Taylor, D. W.
1960. Late Cenozoic molluscan faunas from the High Plains. U.S. Geol. Surv. Prof. Paper, 337:1-94.

Watts, W. A., and H. E. Wright, Jr.
1966. Late-Wisconsin pollen and seed analysis from the Nebraska Sandhills. Ecology, 47:202-210.

Wells, P. V.
1965. Scarp woodlands, transported grassland soils, and concept of grassland climate in the Great Plains region. Science, 148:246-249.

Wendorf, F. (ed.)
1961. Paleoecology of the Llano Estacado. Mus. New Mexico Press, Santa Fe, 144 pp.

Vegetational History of the Central Plains

H. E. WRIGHT, JR.

ABSTRACT

Enough pollen diagrams have been completed for the central and northern parts of the Great Plains to sketch the history of this prairie region. During the time of maximum Wisconsin glaciation, about 20,000 to 14,000 years ago, spruce forest covered most of eastern and central United States as far west as northeastern Kansas. West of there it may have been interrupted by a vast area of sand dunes, produced as a result of intensive wind action occasioned by the presence of the ice sheet nearby to the northeast. When the ice lobes retreated and the winds decreased, the dunes became stabilized, and the spruce forest spread rapidly over them, perhaps extending as far west as the Black Hills.

The climatic change that accelerated the ice retreat in Late Wisconsin time also caused the spruce forest to fail. In the western part of the area, it was replaced directly by prairie. Farther east, closer to the edge of the present prairie, the spruce forest gave way to birch and alder in the south and to pine in the north, and these in turn were succeeded by temperate deciduous trees. These transformations were abrupt, and they occurred sooner in the south (about 12,500 years ago) than in the north (9500 years ago). The temperate forest in the east was soon interrupted by grassy openings, and by 8000 years ago a fully developed prairie covered the region and extended into central Minnesota as well. Reversal of the climatic trend about 6000 to 7000 years ago caused the prairie to withdraw to the west, and groves of deciduous trees have reoccupied the slopes of depressions far out into the modern prairie.

INTRODUCTION

The Central Great Plains are a great grassland, and it is difficult to visualize them otherwise. This region is characterized by dry weather and strong winds; and the great areas of sand dunes and loess deposits, although not now active, seem in a sense to fit into the modern environment. Indeed, during the droughts of the 1930's the dunes were locally reactivated, and the great "black blizzards" attested to distant transport of dust.

Yet from the edge of Kansas northward, the presence of glacial drift forces the realization that conditions were not always as they are today. What were these conditions? The former existence of glacial ice in the area, unfortunately, does not automatically indicate the nature of the environment peripheral to it: the North American ice sheet of the Pleistocene has no modern analogue, and the situation around the European ice sheets of the Pleistocene was different because the climate and morphology of Europe are different. Thus the field is open for objective study of the periglacial environment and of the development of the vegetation during the time of ice retreat and subsequently up to the present time.

Vegetational history is best deciphered by stratigraphical pollen analysis of lake or wetland sediments. Permanent wetness is required to prevent decomposition of pollen grains. Paucity of suitable sites for such investigations in the Central Plains has discouraged the necessary studies for many years, but a few recent projects have provided a beginning. They suggest that the grasslands are indeed young, and that much of the Pleistocene periglacial landscape, rather than being a vast tundra (as in Europe) or a steppe, may have been covered with a boreal forest instead. This picture seems incompatible with that implied by the sand dunes and loess deposits, so we are faced particularly with the need for more data, with both paleoecologic and chronologic control, to resolve the apparent conflict. The present paper will describe what is known thus far about the vegetational history of the region and will present a hypothesis to explain juxtaposition of sand dunes and spruce forest.

EXTENT OF THE LATE-WISCONSIN SPRUCE FOREST IN THE GREAT PLAINS

Although the area is large and the localities investigated are mostly confined to the periphery of the Central Plains (Fig. 1), they all point to the existence of a widespread cover of boreal spruce forest during retreat of the Wisconsin ice sheet. In the north, the Hafichuk site in the modern prairie of southern Saskatchewan (Ritchie and deVries, 1964) shows that spruce was dominant at least until about 10,000 years ago (Fig. 2). In the Riding Mountain area of southern Manitoba, an outlier of spruce forest midst the aspen parkland only about 100 miles from a large reentrant of continuous prairie (Ritchie, 1964), the spruce-pollen zone at the base of the sections gave way directly to a thick zone dominated by herbs. The carbon date on spruce wood from a nearby locality is 9570 BP (S-129).

Although there are no sites in Montana, several localities in the prairie of North Dakota indicate the presence of spruce forest during late-glacial time. Remains of white spruce were collected from the base of a pond deposit near Tappen, in central North Dakota (Moir, 1958), and were dated as 11,480 BP (W-542). The Woodworth site in the same area (McAndrews et al., 1967) shows that the basal spruce-pollen zone is overlain first by a zone with pollen of elm, oak, and hazel before the herb zone that continues to the top of the section. Farther east, at the Seminary site near Fargo (McAndrews, 1967), peat with dominant spruce pollen that is dated as 9900 BP (W-993) is overlain by the sediments of Glacial Lake Agassiz II.

In the modern prairie region of western and southern Minnesota, basal spruce-pollen zones are overlain by zones dominated by pollen of deciduous trees, although herb-pollen types are common and increase upward until they prevail. At Thompson Pond (McAndrews, 1966), the spruce-pollen zone terminated soon after 11,000 years ago, according to a carbon date at a correlative site in the forested area farther east. At nearby Qually Pond (Shay, 1965, 1967), spruce was dominant before 11,740 years ago (Y-1327), and at Madelia (Jelgersma, 1962), a spruce forest gave way gradually to deciduous forest during the interval 12,000 to 10,000 BP (Wright, 1968a). In adjacent northeastern South Dakota, at Pickerel Lake (Watts and Bright, 1968), today about 75 miles into the prairie from the edge of the deciduous forests of Minnesota, a spruce-pollen assemblage marks the base of the sediment. A carbon date of 10,670 BP (Y-1361) indicates the time of abrupt demise of this forest, with replacement for a short time by deciduous elements like birch, elm, and oak, already with a major component of herbs, as indicated by the relatively high pollen percentages for grasses. The increased warmth and dryness that brought about this change continued, and the deciduous forest was soon completely replaced by prairie,

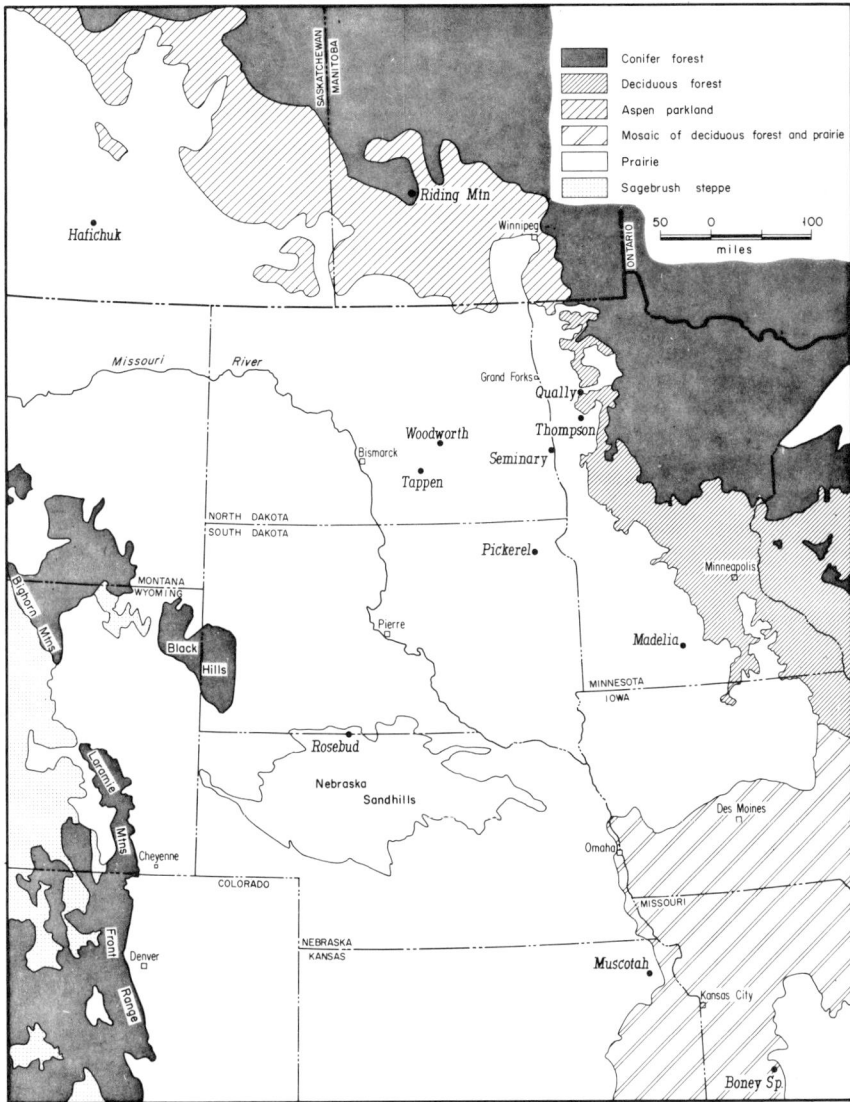

FIG. 1. Vegetation map of the Great Plains and adjacent areas, showing location of pollen sites mentioned in the text. Simplified from Küchler (1964).

probably by about 8000 to 9000 years ago.

A site called Rosebud on the South Dakota-Nebraska state line, at the northern edge of the Sandhills, shows the same kind of spruce-pollen domination in the basal sediments (Watts and Wright, 1966). Here the change to a strictly prairie assemblage, dominated by ragweed, chenopods, sage, and grasses, came soon after 12,600 years ago (Y-1359 and Y-1360).

If we skip to west-central Missouri, at the northern edge of the Ozark Highlands, in the vegetation mapped by Küchler (1964) as a mosaic of prairie and oak-hickory forest, a spruce-pollen assemblage was found in a spring bog (Boney Spring) in association with bones of an extinct Pleistocene mammal—the pollen count does not differ appreciably from that for

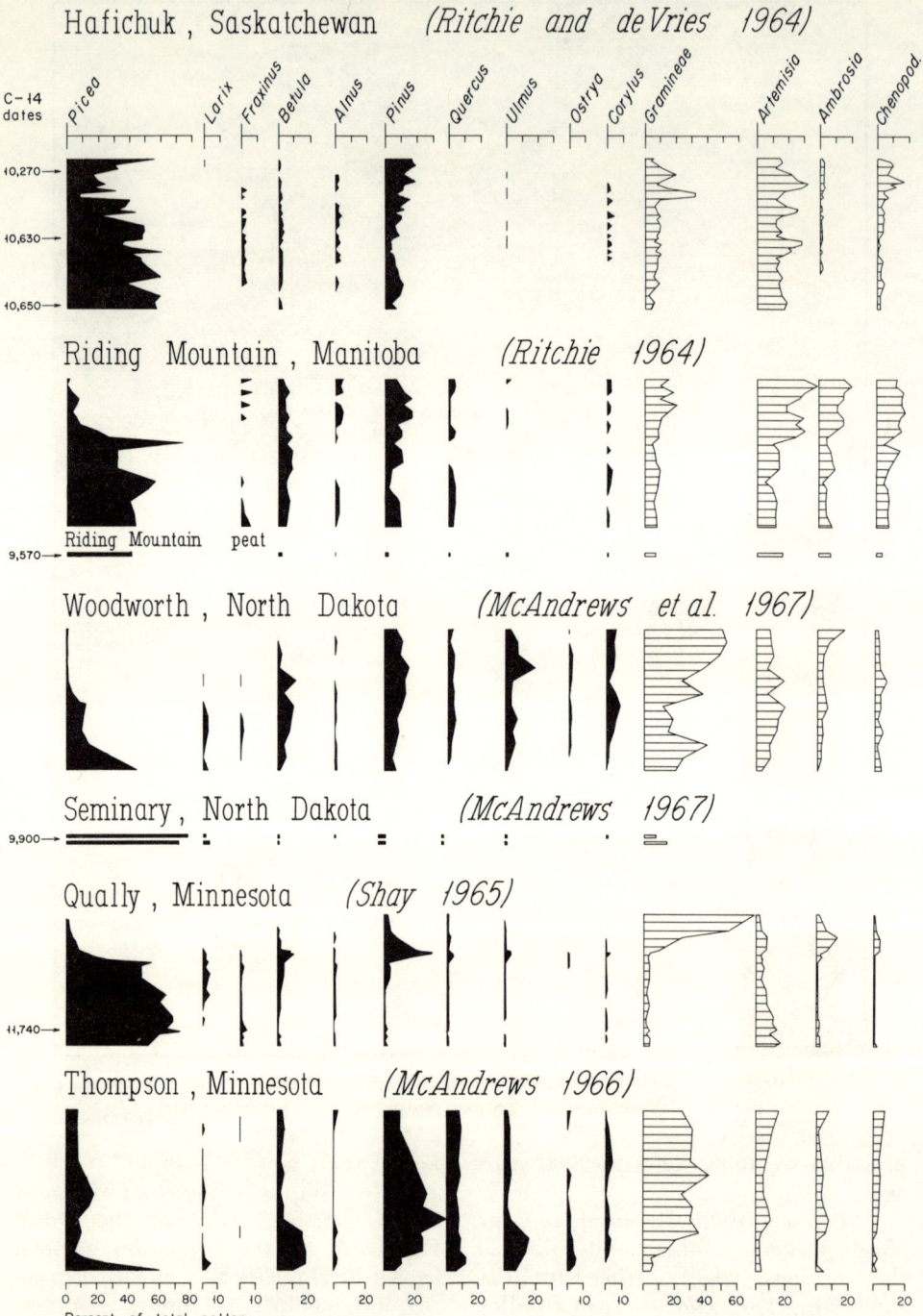

Fig. 2. (See also facing page.) Pollen diagrams of late-glacial and early post-glacial lake and bog sediments in localities at the northern and eastern edge of the prairie. See Fig. 1 for location. The pollen sum is generally total pollen excluding aquatics; Cyperaceae pollen (not shown) is included in the sum only at Madelia and Pickerel; at Madelia it is abundant in the spruce zone.

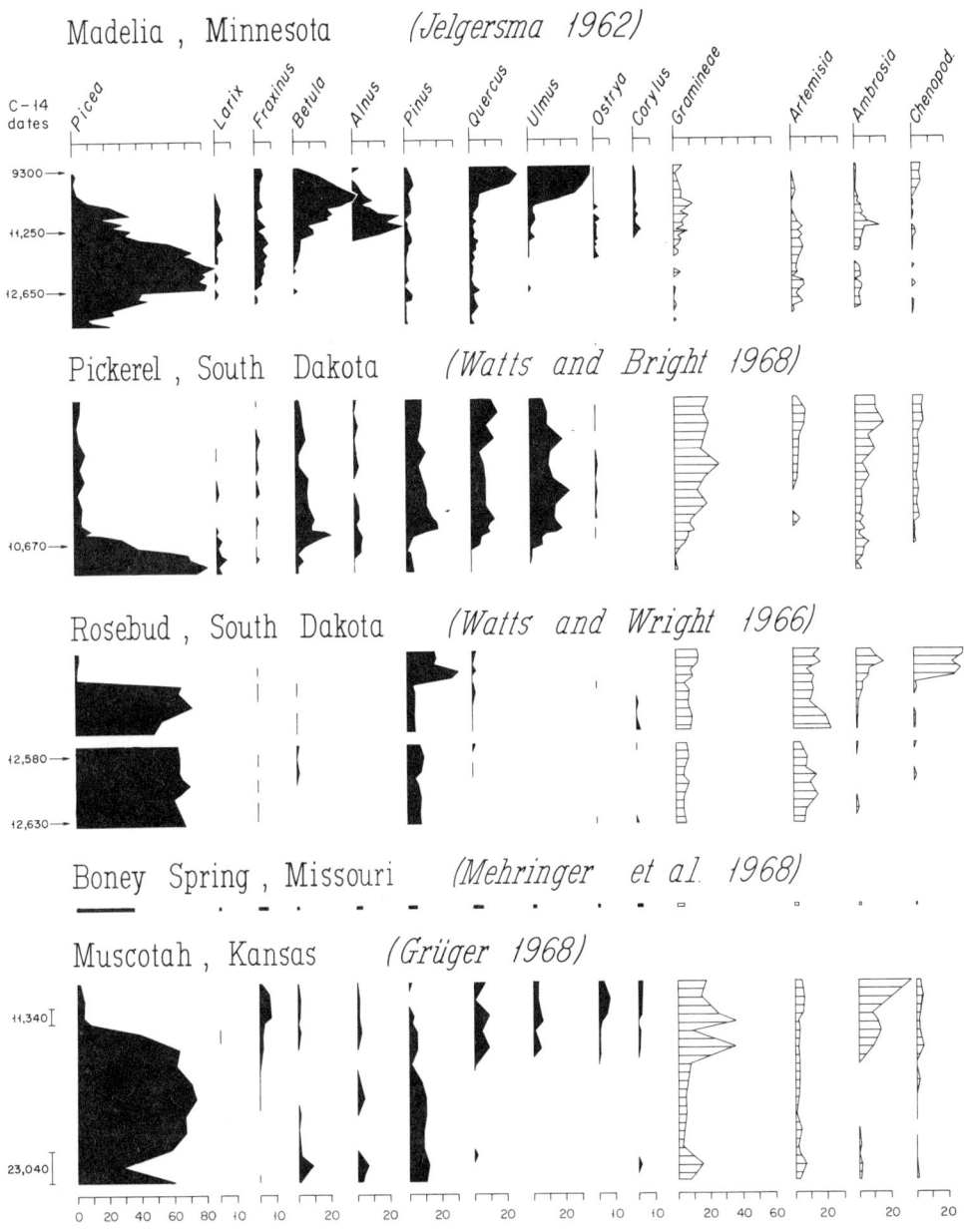

late-glacial sediments 1000 miles to the north in Minnesota (Mehringer et al., 1968 and this volume).

Except for Boney Spring, all of the pollen sites so far mentioned are in the region of Wisconsin glaciation or, in the case of the Nebraska Sandhills, a region of active dune formation within Wisconsin time. The sites, therefore, did not originate until late-glacial time, and the radiocarbon dates, where available, show only the time of destruction of the boreal spruce forest that prevailed throughout the area 10,000 to 12,000 years ago.

In northeastern Kansas, however, in the region of Kansan glaciation on the side of the Delaware River flood plain near Muscotah, artesian-spring marshes provide a longer record (Grüger, 1970). This locality is at the eastern edge of continuous prairie. It has the familiar pollen sequence, with a spruce zone near the base, overlain by a zone dominated by pollen of deciduous trees, and this by an herbaceous-pollen zone. The spruce zone terminated between 15,880 (W-2206) and 11,340 years ago (W-2149). Samples from two marshes provide dates for the base of the spruce zone of 23,040 (W-2150) and 24,500 (W-2205) years ago.

It seems clear from this summary of pollen sites located throughout most of the Northern Plains and the eastern edge of the Central Plains that a boreal spruce forest dominated the landscape until about 12,000 years ago in the south and about 9500 years ago in the north. Although the coverage of sites has many gaps, especially in the western half of the plains, the presence of white spruce in the Black Hills of southwestern South Dakota implies that the boreal forest covered the western part of the plains as well. Its southern limit on the Great Plains is not known. Eastward, it stretched far across the country, probably all the way to the Atlantic Coast, from New England (Davis, 1968) to Georgia (Watts, 1970).

COMPOSITION OF THE SPRUCE FOREST

Throughout most of this vast extent, the boreal forest seems to have had a remarkably uniform composition. Birch and alder were probably present, and tamarack was generally a constituent of the forest except in the south and west. Among the expectable herbs, Cyperaceae was the most common, and *Artemisia* generally the next. These are the two herb types that dominate the herb-pollen zone underlying the spruce zone in sites in the Late Wisconsin periglacial region of northeastern Minnesota.

A puzzling feature in practically all of the diagrams is the relatively high pollen percentages of the temperate deciduous trees ash, oak, elm, ironwood, and hazel. They are generally as high in the northern sites as in the southern, with a total of five to 15 percent. Today the deciduous genera range barely into the boreal forest of southern Manitoba and adjacent Ontario, and they are represented only in a minor way in the modern pollen rain of the boreal forest (Lichti-Fedorovich and Ritchie, 1965).

This problem has been discussed previously with regard to several different parts of the country, for deciduous pollen types are conspicuous in the spruce zone from Minnesota (Wright, 1968a) to New England (Davis, 1968). Because the percentages reach such relatively high values and show no consistent increase to the south (toward large refuges that should occur south of the boreal forest), I do not believe that long-distance transport of these pollen types, from the south is the correct explanation. Rather, I favor the explanation that minor stands of these tree genera actually occurred throughout the spruce forest, during the ice retreat of the Late Wisconsin, to a greater extent than they do today in the boreal forest, and that they thereby provided a local pollen source.

The opposite problem exists in the case of pine, which shows less than three percent pollen for the spruce zone in most sites west of the Appalachian Mountains, despite the fact that jack pine is now found today throughout the boreal spruce forest, except on the northern fringe (Wright, 1968a). Apparently jack pine was not common in late-glacial time in the Great Lakes region, nor was either red pine or white pine. The refuges for these trees must have been in the Appalachian Mountains, and indeed jack pine pollen and needles have been identified in quantity as far south as northern Georgia, in association with spruce pollen and the seeds of numerous northern aquatic and marsh plants (Watts, 1970). Some pine species may have occurred in the south-central or southwestern United States, and stands of these trees may have contributed the five to eight percent pine pollen in the southern sites described here.

The spruce-pollen zone ends as abruptly at sites in the Central Plains as it does in areas to the east. The timing, according to a few radiocarbon dates, probably ranges from about 12,000 years ago in Kansas and Nebraska to 9500 years ago in southern Canada. This progression was started by the same climatic change that accelerated the wastage of the Wisconsin ice sheet, which was retreating during this period from central Wisconsin and Minnesota into western Ontario and southern Manitoba. The ice-sheet margin fluctuated in position during its retreat, although the different ice lobes did not always advance and retreat in unison. Comparable fluctuations cannot be detected within the spruce zone, however, presumably because the vegetation several hundred miles from the ice front was not sensitive enough to minor climatic changes to be recorded in the pollen sequence—if indeed the glacial fluctuations were caused directly by climatic change at all and not by some nonclimatic factor such as surging.

THE SPRUCE FOREST AND PERIGLACIAL ENVIRONMENT

The widespread distribution of a closed spruce forest through at least the eastern and northern parts of the plains, as well as vast areas to the east, at a time when the ice sheet was still in the Great Lakes region, raises some questions about other aspects of the periglacial landscape. Pollen and plant-macrofossil studies from northeastern Minnesota (Watts, 1967) indicate that a narrow belt of tundra bordered the ice in that region until 11,500 to 10,500 years ago. Was there also tundra farther south and west when the Wisconsin ice sheet was at its maximum in Iowa and Illinois, 18,000 to 20,000 years ago, or did the spruce forest come to the edge of the ice sheet, as it apparently did in southern Minnesota during the retreatal phases of Wisconsin glaciation?

Only one pollen site provides information on this question—Muscotah, in northeastern Kansas, where the spruce-pollen zone is as old as 24,500 BP. This site is about 150 miles southwest of the edge of the Des Moines lobe (Fig. 3), from the Tazewell (20,000 BP) to the Cary (14,000 BP) phases of Wisconsin glaciation (Ruhe et al., 1968). The Muscotah diagram shows no indications of tundra during this interval.

The interval 20,000 to 14,000 years ago includes the time of maximum loess deposition in Illinois (Frye et al., 1968) and Iowa (Ruhe et al., 1968). If the upland at this time was covered by a closed spruce forest, it could not have served as a source for loess. This is unlike the situation in much of Europe, where frost disturbance of tundra soils broke the mineral particles down to silt size and exposed them to wind action (Dücker, 1937). The principal sources of loess therefore must have been the flood plains of glacial meltwater streams, which were kept bare of vegetation by intermittent flooding during summer. Between floods, strong winds could dry out and deflate deposits of silt. Loess can be seen to blow today into the forest from the Tanana River flood plain in Alaska, which drains the glaciers of the Alaska Range (Péwé, 1951). In Illinois the clear decrease in thickness, particle size, and carbonate content of loess away from major rivers indicates that flood plains were the major source, and furthermore that summer winds strong enough to carry loess were mostly from the west (Frye *et al.*, 1968).

For eastern Iowa, Ruhe *et al.* (1968) indicated that some of the more sandy loess was derived from smaller local streams that were dissecting areas of older till, presumably without the aid of glacial meltwaters. It is difficult to visualize flood-plain conditions suitable for loess deflation in a heavily forested landscape without the intermediation of glacial meltwaters to keep the flood plains bare. Perhaps the winds in the periglacial climate were sufficiently strong to utilize even minor source areas for loess; perhaps the loess did in fact come from more distant glacial flood plains; or perhaps our concept of the climatic, geomorphic, and vegetational conditions of this time are so inadequate that we cannot answer the questions satisfactorily.

West of the Missouri River, the loess of Wisconsin age is as extensive as it is in Iowa (Fig. 3). In northeastern Kansas, the source of the loess is said to be the Missouri River, and the molluscan fauna of the loess implies a wooded terrain (Leonard and Frye, 1954)—a conclusion substantiated by the findings at Muscotah. The rapid decrease in loess thickness westward from the Missouri River is attributed to the dissipation of strong easterly surface winds by the tree cover, and hence the rapid deposition of loess borne from flood plains.

Farther west in Kansas, the principal source of loess is attributed by Frye and Swineford (1951) to the Platte River Valley, which carried glacial outwash from the Rocky Mountains. Because variations in the thickness and grain size of loess away from the Platte River and other river sources are much more gradual than is the case with the Missouri River, the vegetation is pictured by them as prairie (much as today), which does not obstruct the wind transport of silt. The molluscan fauna of the loess in this region also implies prairie, with local riverine woodlands to supply habitats for nonprairie types (Leonard and Frye, 1954). These two evidences for prairie are opposed if the spruce forest of the Muscotah area extended very far westward.

For Nebraska, an additional source for loess was available—the Nebraska Sandhills, which cover almost the entire northwestern third of the state (Lugn, 1962). This remarkable physiographic region was formed by Pleistocene periglacial winds in an environment that is pictured by Smith (1965) as first a desert (with large transverse dunes built by northerly winds) and later an area of patchy vegetation (with longitudinal dunes superimposed on the older dunes by northwesterly winds). The thickness and grain size of the loess decreases southeastward from the Sandhills toward (rather than away from) the Platte River. The loess was therefore not derived from the Platte, but rather it represents the fines blown by northwesterly winds, perhaps in part the same winds that shaped the longitudinal series of dunes in the Sandhills, although the age relations of the dunes and the loess are not certain (Smith, 1965).

At this time, the James River Lobe of the ice sheet stood in southern South Dakota, less than 100 miles northeast of the Sandhills. A spruce forest existed in northeastern Kansas (250 miles southeast of the Sandhills) and probably covered most of Iowa peripheral to the ice sheet. Is a spruce forest in eastern Kansas compatible with a semidesert area of active dunes in central Nebraska?

The hypothesis may be presented that a treeless landscape developed in the western two-thirds of Nebraska as an indirect result of strong periglacial winds from the north. Sand was deflated from alluvium that was derived in turn from poorly consolidated Tertiary sediments. In the beginning, the area of deflation could have been the dissected surface of the Ogallala Group

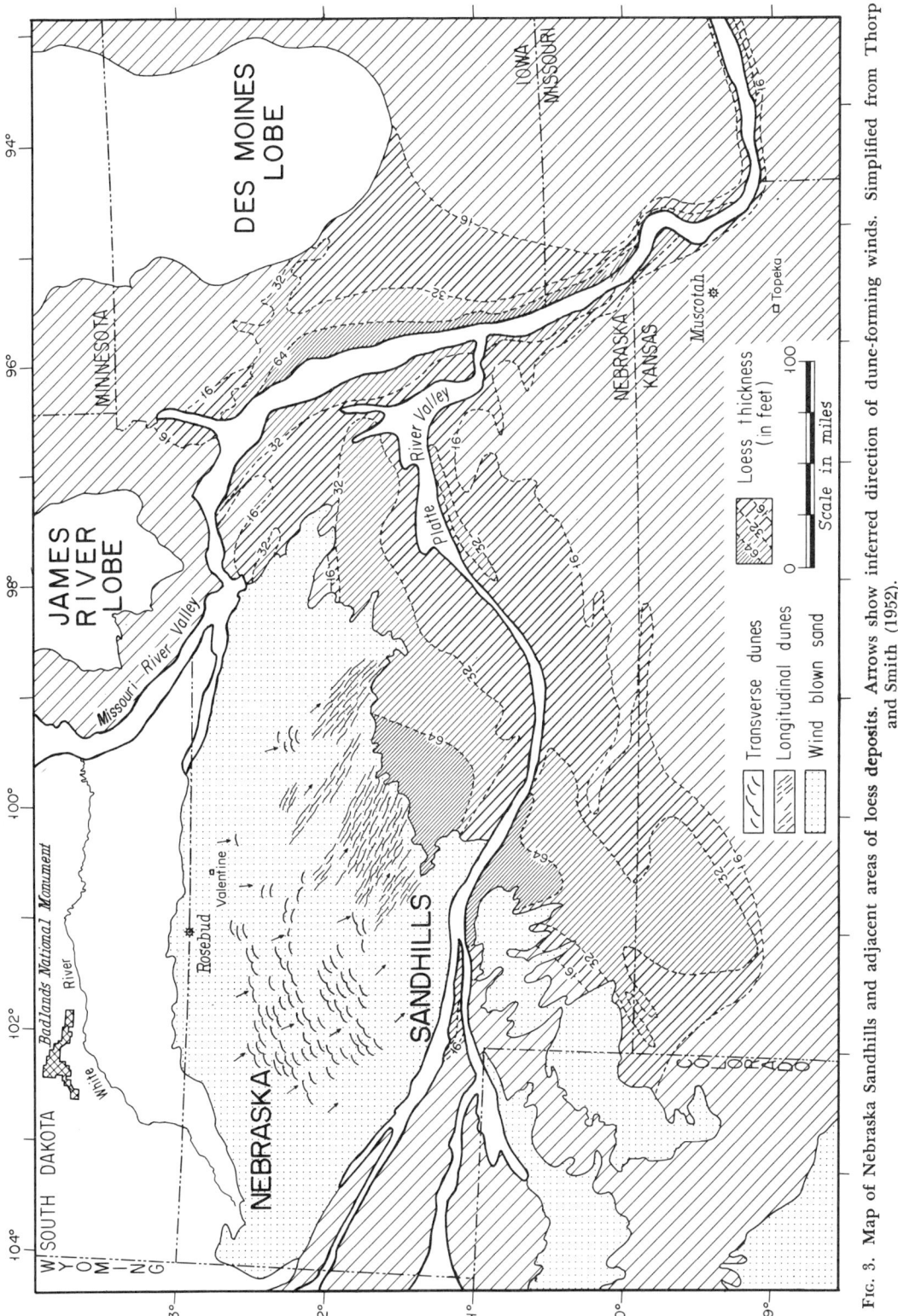

FIG. 3. Map of Nebraska Sandhills and adjacent areas of loess deposits. Arrows show inferred direction of dune-forming winds. Simplified from Thorp and Smith (1952).

in the High Plains of northern Nebraska. But in time this source must have become largely buried in sand, and thereafter the sources must have been in part north and northwest of the buried Pine Ridge escarpment that bounds the High Plains on the north, that is, the dissected terrain of the White River drainage, including the South Dakota Badlands, where rapid stream erosion even today supplies bare hill slopes and bare alluvial trains that are susceptible to wind deflation. Recent work in the Badlands area points to local occurrence of both sand dunes and loess deposits and also to abundant unconsolidated alluvium, which is susceptible to wind deflation (Harksen, 1968). The northwest-trending small streams in much of western South Dakota may reflect some control by wind action (White, 1961).

The abundant sand and the strong winds overwhelmed whatever vegetation might have been rooted in the sand, and a vast area of massive transverse dunes was created, resembling an African desert landscape (Smith, 1965), even though the rainfall and temperature in the area may have been suitable for tree growth. In certain respects the features might be compared to the sand dunes that interrupt a heavy forest cover along the coasts of many humid regions—Oregon, for example —where persistent on-shore winds and an abundant supply of shore sand prevent the growth of vegetation (Cooper, 1958).

The time of formation of the main transverse dunes in the Sandhills is placed as Early Wisconsin by Smith (1965). Possibly the dunes may have accumulated progressively during times of earlier glaciation as well. Some of the loess south of the Sandhills may have been deposited at these times. Later during the Wisconsin glaciation, when the winds shifted more to the northwest and diminished somewhat in strength, sparse herbaceous vegetation gained a foothold on the sand, and the transverse dunes were partially reshaped to longitudinal dunes. The rest of the loess southeast of the Sandhills must have been deposited at this time. Some of it may have been derived from further winnowing of fine particles during reworking of the sand dunes, but most of it may have come from sources farther northwest, which could supply silt as well as sand. The loess was carried southeastward until the general wind velocity decreased (as a result of weakening of the periglacial influence) and as the surface wind velocity decreased (because of the tree cover, which may have extended up to the edge of the dunes).

With further decrease in periglacial wind action and the stabilization of the sand dunes, the spruce forest rapidly spread westward across the Sandhills to Rosebud and perhaps to the Black Hills. The climatic change that finally terminated the Pleistocene then brought about the rapid destruction of the spruce forest, and ultimately the development of the prairie of today.

This hypothesis can best be tested by further pollen investigations in Nebraska and Kansas, along with radiocarbon dating for chronologic control. Unfortunately, the abundant lakes and marshes of the Nebraska Sandhills may all be too young to record the critical period. The lake sediment examined by Sears (1961), for example, proved to be only 5040 years old (Y-912) at the base. Perhaps other spring marshes in the loess area, in addition to Muscotah, will yield longer pollen records.

POST-GLACIAL FOREST SUCCESSION

The vegetation that succeeded the deteriorating spruce forest in the Great Plains varied from east to west as well as from south to north. The variations largely reflected the different proximities of refuges of various tree types that could stock the areas newly made available by the demise of the spruce forest. In the eastern part of the plains, as represented by Muscotah, Pickerel, Madelia, and Woodworth, the spruce forest was succeeded by temperate deciduous forest, including oak, elm, ironwood, and hazel. These trees may already have existed in minor numbers within the spruce forest, so local seed sources may have been available for rapid spreading. The percentage of herbs (especially grasses) rose at the same time, however, and we must postulate the presence of sizable openings in the deciduous forest. This type of parkland exists at present in a rather restricted way in the prairie-forest border region from Kansas to Minnesota, and on into Canada as the Aspen Parkland. Woodlands are confined to sheltered slopes and the flanks of depressions, and prairie covers much of the upland. A savanna vegetation, with scattered rather than clumped trees, would give a similar pollen rain, but among the common trees present, only bur oak develops a savanna structure.

Farther north, at sites also along the eastern side of the modern prairie, birch and alder showed a short interval of prominence immediately after the spruce fall and before the rise in the temperate deciduous types. At Madelia, the major but partial fall in the spruce-pollen curve occurred about 12,000 years ago and was followed by a pause marked by the dominance of alder. Then, as the spruce curve resumed its fall about 10,000 years ago, birch reached a maximum; birch trees had been common in the spruce forest, so seed sources were close at hand. The steep rise of elm and oak followed about 9500 years ago. This sequence is matched (with less detail) at Pickerel Lake, where the rise of alder and birch preceded that of the temperate deciduous tree types. The same can be seen at Woodworth.

Still farther north along this eastern side of the modern prairie, at Thompson and Qually, the spruce fall is not well dated but must have occurred about 10,000 years ago, according to results at Seminary. It is matched by a rise in pine pollen, which, however, does not reach levels high enough for us to be certain that pine trees were actually in the area, because pine pollen is easily dispersed from distant pine stands. The pollen sequence farther east shows that jack or red pine (or both) rapidly invaded east-central Minnesota from the east about 10,500 years ago, but probably reached only about as far west and south as they are today.

At the northern edge of the modern prairie in Manitoba, the spruce forest, which fell after 9500 years ago, was replaced directly by prairie—without an interval of deciduous forest.

The expansion of the prairie may be followed throughout the region. First detectable in the Central Plains soon after 12,600 years ago (Rosebud), it followed the spruce forest northward, reaching Manitoba about 9500 years ago. Its eastward spread was slowed by the intervention of deciduous forest, which, however, contained parklike openings with grasses. Continuous prairie reached its present extent on the southwest (Muscotah) as early as 9900 BP (W-2202), as the deciduous forest withdrew in turn to the east.

These vegetational changes, which reflected the climatic warming that also brought the end of the glacial period, did not stop at this point, however. They continued with the same trend for another thousand years or so. Prairie expanded farther into the deciduous forest, reaching about 75 miles northeast of its present limit in Minnesota (Wright, 1968b) and perhaps a similar distance into the spruce forest of Manitoba (Ritchie, 1964). Within the prairie itself the changes are recorded by changing proportions of herb types, as well as by the decrease in

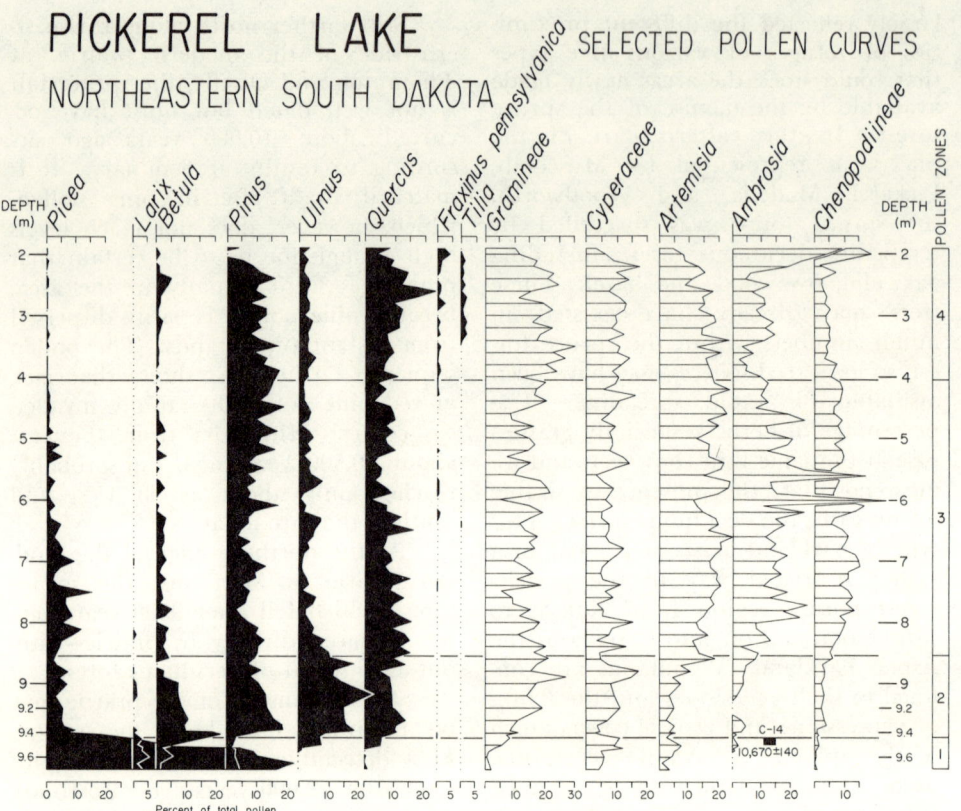

Fig. 4. Summary pollen diagram for Pickerel Lake, northeastern South Dakota. Extracted from Watts and Bright (1968).

percentages of deciduous-tree pollen. At Pickerel Lake, for example, the pollen percentages of *Ambrosia, Artemisia,* and chenopods increase at the expense of grasses (Fig. 4), and minor prairie pollen types such as *Iva ciliata, Petalostemum prupureum,* and *Amorpha* are regularly present (Watts and Bright, 1968). Probably the fringe of woodland that occurs around the basin today (and did also before about 8000 years ago) was absent in mid-post-glacial time. Of equal significance at Pickerel Lake for this interval is the increased incidence of seeds of upland-prairie plants and also of wet-ground annual weeds, which imply occasional drying of portions of the lake, presumably as a result of a drier climate (Fig. 5).

The record of this mid-Holocene dry interval is a conspicuous feature of all the pollen diagrams of Minnesota, even deep in the coniferous forest. It can be detected as far east as Ohio and even New England by the increased percentages of ragweed pollen for the period of about 8000 to 4000 years ago (Wright, 1968b)—although in New England, especially, the record may result from distant transport of *Ambrosia* pollen from the prairie as much as from local spread of prairie plants.

The reversal of the climatic trend eventually resulted in the change once again of the composition of the prairie and in the readvance of deciduous trees into the prairie border. At Muscotah, a decrease in pollen of *Ambrosia* occurred 5100 years ago (W-2203), whereas in the prairie border of Minnesota the conspicuous advance of the forest border occurred at a later date—again a manifestation of slow migra-

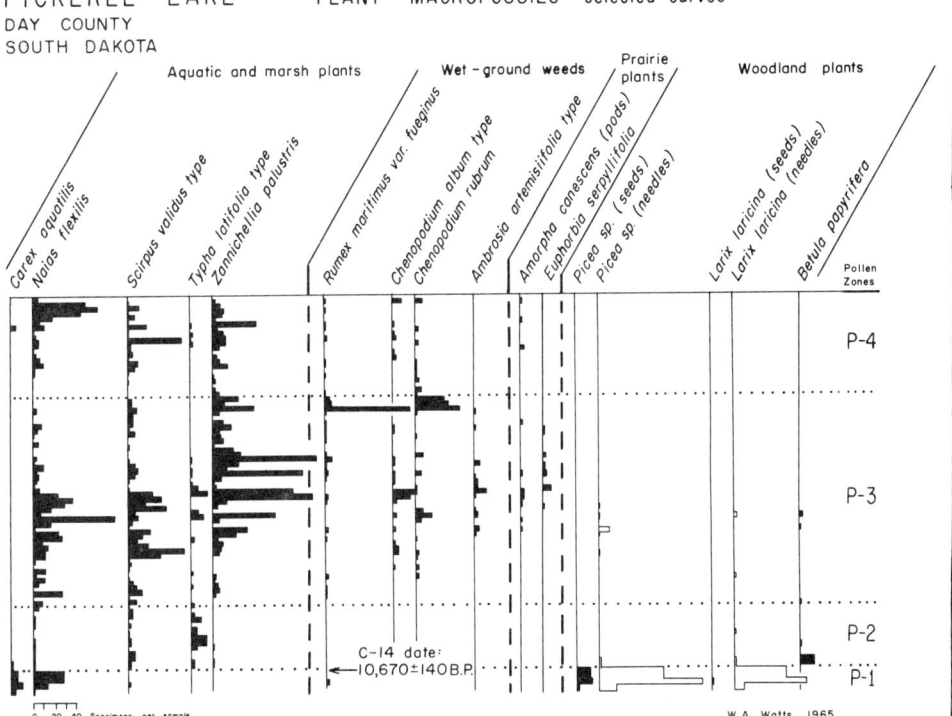

Fig. 5. Summary seed diagram for Pickerel Lake, northeastern South Dakota. Extracted from Watts and Bright (1968).

tion of major vegetational belts. The composition of the tree component in the prairie in subsequent time was different from that of the earlier interval. At Pickerel Lake, for example, where elm was perhaps the dominant deciduous tree in the early Holocene, it was minor in the late Holocene. On the other hand, green ash and basswood, absent before, became common in the later period. Oak occurred at both times. The lake is now rimmed by a woodland with all these trees, but the proportions of the several species must have been different in the early part of the Holocene.

SUMMARY AND CONCLUSIONS

The Late Wisconsin and Holocene pattern of vegetational and climatic changes that has been established by pollen and seed analysis in the forested areas of Minnesota can be extended to the west and southwest into the Great Plains with little change in concept. The entire area from southern Canada to Kansas and Missouri in Late Wisconsin time was marked by a boreal spruce forest, which extended eastward to the Appalachian Mountains. The new pollen diagram from Muscotah indicates that in eastern Kansas, and perhaps elsewhere, the spruce forest prevailed from 24,000 to at least 14,000 years ago, during the time of main Wisconsin glaciation, when the ice was as close as 150 miles away.

Although evidence is still lacking for the immediate periglacial area in Iowa and Illinois, there is no indication of tundra in the Muscotah diagram for this time range, and the forest probably grew to the ice edge. Farther west, however, in northern Nebraska, the periglacial winds were apparently so strong, and the sand supply from eroded Tertiary sediments so abundant, that a vast area of desert-like dunes developed, with loess deposits to the southeast. But when the winds decreased in late-glacial time, the spruce forest advanced over the stabilized dunes.

Post-glacial climatic changes in the central grasslands followed the pattern established for the forested area to the east—a mid-post-glacial maximum of warmth and dryness allowed eastward expansion of the prairie into the forest and a change in composition of the prairie to more xeric types, with fewer groves of trees. The changes were slow and progressive from west to east and from south to north. The prairie can first be detected as early as 10,000 years ago in Kansas, and it reached eastern Minnesota as prairie openings by 7000 years ago. Reversal of the climatic trend closed the forest again and brought groves to the prairie border. The trend towards forest expansion—and of conifers within the forest—has continued to the present.

Thus, after the relatively steep climatic gradient about 12,000 to 11,000 years ago that accelerated glacial retreat and brought about deterioration of the almost continent-wide boreal forest, the climatic curve for subsequent time seems to have been a smooth and gentle curve, with its culmination about 7000 years ago. Minor embellishments on the basic trends (such as the drought of the 1930's) may well have occurred, but they have not been detected by pollen analysis.

The mid-post-glacial Altithermal interval is therefore a reality at least at the eastern edge of the plains, as seen from the full-length diagrams of Muscotah, Pickerel, Thompson, and Riding Mountain. Its chronological limits must be selected arbitrarily, for its effects on the vegetation reached certain thresholds at different times in different places. Thus, the change from forest to prairie ranged from about 10,000 years ago in the south to 8000 in the north, whereas the reverse occurred about 4000-5000 years ago in various areas.

Although more dates are needed to work out the time relations carefully, it seems clear that vegetation belts migrated at rates measurable by radiocarbon dating, and that the array of dominant species changed during the migration. Pollen analysis, in revealing proportions among many of the major plant types, provides a sense of the dynamic character of the vegetation under the stress of long-term climatic change. Even in the prairie regions, where the geographic zones of vegetation are less conspicuous, the method is sensitive enough to detect shifts in the proportion of grasses to more xeric herbs and shrubs. With more investigations farther west in the prairie, the vegetational and climatic sequence may be traceable to the western border against the Rocky Mountains, where the tradition of the Altithermal is stronger. When a network of radiocarbon dates is available for pollen-stratigraphic horizons throughout the prairie and its bordering areas, it may be possible to recognize regional trends and their chronological relations.

ACKNOWLEDGMENTS

The work reported in this paper is an outgrowth of studies supported by a grant from the National Science Foundation (GB-3401). It is contribution no. 70 of the Limnological Research Center, University of Minnesota. I am obliged to Johanna Grüger and Meyer Rubin, respectively, for unpublished pollen data and radiocarbon dates of the Muscotah site, and to H. T. U. Smith for comments on the manuscript.

LITERATURE CITED

Cooper, W. S.
1958. Coastal sand dunes of Oregon and Washington. Mem. Geol. Soc. Amer., 72: 1-169.

Davis, M. B.
1968. Climatic changes in southern Connecticut recorded by pollen deposition at Rogers Lake. Ecology, 50:409-422.

Dücker, A.
1937. Über Strukturböden im Riesengebirge. Ein Beitrag zum Bodenfrost- und Lössproblem. Zeitschr. Deutsch. Geol. Ges., 89:113-129.

Frye, J. C., and A. Swineford
1951. Petrography of the Peoria loess in Kansas. Jour. Geol., 59:306-322.

Frye, J. C., H. B. Willman, and H. D. Glass
1968. Correlation of Midwestern loesses with the glacial succession. Pp. 3-22, in Loess and related eolian deposits of the world (C. B. Schultz and J. C. Frye, eds.), Univ. Nebraska Press, Lincoln, 369 pp.

Grüger, J.
1970. A new pollen investigation for Muscotah Marsh, northeastern Kansas. Unpubl. manuscript.

Harksen, J. C.
1968. Red Dog loess named in southwestern South Dakota. South Dakota Geol. Surv. Rep. Inv., 98:1-17.

Jelgersma, S.
1962. A late-glacial pollen diagram from Madelia, south-central Minnesota. Amer. Jour. Sci., 260:522-529.

Küchler, A. W.
1964. Potential natural vegetation of the conterminous United States. Amer. Geogr. Soc. Spec. Publ., 36:v+1-38.

Leonard, A. B., and J. C. Frye
1954. Ecological conditions accompanying loess deposition in the Great Plains region of the United States. Jour. Geol., 62:399-404.

Lichti-Federovich, S., and J. C. Ritchie
1965. Contemporary pollen spectra in central Canada. II. The forest-grassland transition in Manitoba. Pollen et Spores, 7:63-87.

Lugn, A. L.
1962. The origin and sources of loess. Univ. Nebraska Studies, n. s., 26:xi+1-105.

McAndrews, J. H.
1966. Postglacial history of prairie, savanna and forest in northwestern Minnesota. Mem. Torrey Bot. Club, 22 (2):1-72.

1967. Paleoecology of the Seminary and Mirror Pool peat deposits. Occas. Papers Dept. Anthro., Univ. Manitoba, 1:253-269.

McAndrews, J. H., R. E. Stewart, Jr., and R. C. Bright
1967. Paleoecology of a prairie pothole: a preliminary report. Pp. 101-114, in Glacial geology of the Missouri Coteau and adjacent areas (L. Clayton and T. F. Freers, eds.), North Dakota Geol. Surv. Misc. Ser., 30:1-170.

Mehringer, P. J., Jr., C. E. Schweger, W. R. Wood, and R. B. McMillan
1968. Late-Pleistocene boreal forest in the western Ozark highlands? Ecology, 49: 567-568.

Moir, D. R.
1958. Occurrence and radiocarbon date of coniferous wood in Kidder County, North Dakota. North Dakota Geol. Surv. Misc. Ser., 10:108-114.

Péwé, T.
1951. An observation on wind-blown silt. Jour. Geol., 59:399-410.

Ritchie, J. C.
1964. Contributions to the Holocene paleoecology of westcentral Canada. I. The Riding Mountain area. Canadian Jour. Bot., 42:181-196.

Ritchie, J. C., and B. deVries
1964. Contributions to the Holocene paleoecology of westcentral Canada: a late-glacial deposit from the Missouri Coteau. Canadian Jour. Bot., 42:677-692.

Ruhe, R. V., W. P. Dietz, T. E. Fenton, and G. F. Hall
1968. Iowan drift problem, northeastern Iowa. Iowa Geol. Surv. Rep. Inv., 7:1-40.

Sears, P. B.
1961. A pollen profile from the grassland province. Science, 134:2038-2040.

Shay, C. T.
1965. Postglacial vegetation development in northwestern Minnesota, and its implications for prehistoric man. M.S. thesis, Univ. Minnesota.

1967. Vegetation history of the southern Lake Agassiz basin during the past 12,000 years. Occas. Papers Dept. Anthro., Univ. Manitoba, 1:231-252.

Smith, H. T. U.
1965. Dune morphology and chronology in central and western Nebraska. Jour. Geol., 73:557-578.

Thorp, J., and H. T. U. Smith, et al.
1952. Pleistocene eolian deposits of the United States, Alaska, and parts of Canada. Geol. Soc. Amer., map, scale 1: 2,500,000.

Watts, W. A.
1967. Late-glacial plant macrofossils from Minnesota. Pp. 59-88, in Quaternary paleoecology (E. J. Cushing and H. E. Wright, Jr., eds.), Yale Univ. Press, New Haven, Connecticut, vii+433 pp.

1970. The full-glacial vegetation of northwestern Georgia. Ecology, 50: in press.

Watts, W. A., and R. C. Bright
1968. Pollen, seed and mollusk analysis of a sediment core from Pickerel Lake, northeastern South Dakota. Bull. Geol. Soc. Amer., 79:855-876.

Watts, W. A., and H. E. Wright, Jr.
1966. Late-Wisconsin pollen and seed analysis from the Nebraska Sandhills. Ecology, 47:202-210.

White, E. M.
1961. Drainage alignment in western South Dakota. Amer. Jour. Sci., 259:207-210.

Wright, H. E., Jr.
1968a. The roles of pine and spruce in the forest history of Minnesota and adjacent areas. Ecology, 49:937-955.
1968b. History of the Prairie Peninsula. Pp. 78-88, in The Quaternary of Illinois, Spec. Publ. Univ. Illinois Coll. Agric., 14:1-179.

A Record of Wisconsin-Age Vegetation and Fauna from the Ozarks of Western Missouri

PETER J. MEHRINGER, JR., JAMES E. KING, AND EVERETT H. LINDSAY

ABSTRACT

Excavations of two spring deposits in western Missouri resulted in the first radiocarbon-dated associations of Pleistocene biota from the Ozark Highlands. Although the pollen record is not complete, three distinct assemblages are recognized. From about 32,000 to 13,500 years ago pollen dominance changed from NAP-pine, to spruce dominance, to spruce with deciduous elements. Mastodon (*Mammut americanum*) was the most abundant mammal recovered. It was associated with horse and muskox in sediments dominated by NAP-pine pollen, and with giant beaver and ground sloth in deposits with spruce and deciduous tree pollen, and spruce macrofossils.

INTRODUCTION

The effect of past climatic change in expansion or contraction of the plains biota should be reflected in areas bordering the Great Plains. This paper reports the preliminary studies of Wisconsin-age deposits that are ideally located to reveal the history of the biota of the eastern border of the Central Plains and the western Ozark Highlands.

The sites, Boney and Trolinger springs (Fig. 1), are located along the Pomme de Terre River in the oak-hickory forest near the eastern edge of the prairie (Kucera, 1961, p. 226). Fenneman (1938, p. 652) placed the area within the Springfield Plateau, physiographically a part of the Ozark Highlands. Bretz (1965, p. 103) placed the Springfield Plateau in the Prairie Plains of the Central Lowlands. The Springfield Plateau dips gently to the west with major relief along the rivers.

Paleontological investigations of spring deposits of the Pomme de Terre River Valley have a long history. According to Mehl (1962, pp. 30-31), in 1806 B. S. Barton wrote Baron Georges Cuvier, the founder of vertebrate paleontology, describing the abundant mastodon remains of the region. In the 1830's, A. K. Koch collected hundreds of specimens of mastodon and other fossils from Missouri. It is particularly interesting that he reported tropical plants with mastodon from a spring approximately two kilometers from Trolinger Spring where mastodon remains occur with boreal plant species (Fig. 1). Koch was noted for his imaginative interpretations and promotional zeal.

Interest in the deposits of the Pomme de Terre River Valley was renewed in 1963 with the initiation of excavations at Rodgers Shelter (Fig. 2) by W. Raymond Wood and R. Bruce McMillan of the University of Missouri. Rodgers Shelter contains an exceptionally rich cultural sequence spanning the last 10,000 years (Wood and McMillan, 1967). In an attempt to find a record of vegetational history during the period of occupation of Rodgers Shelter, we investigated nearby spring deposits.

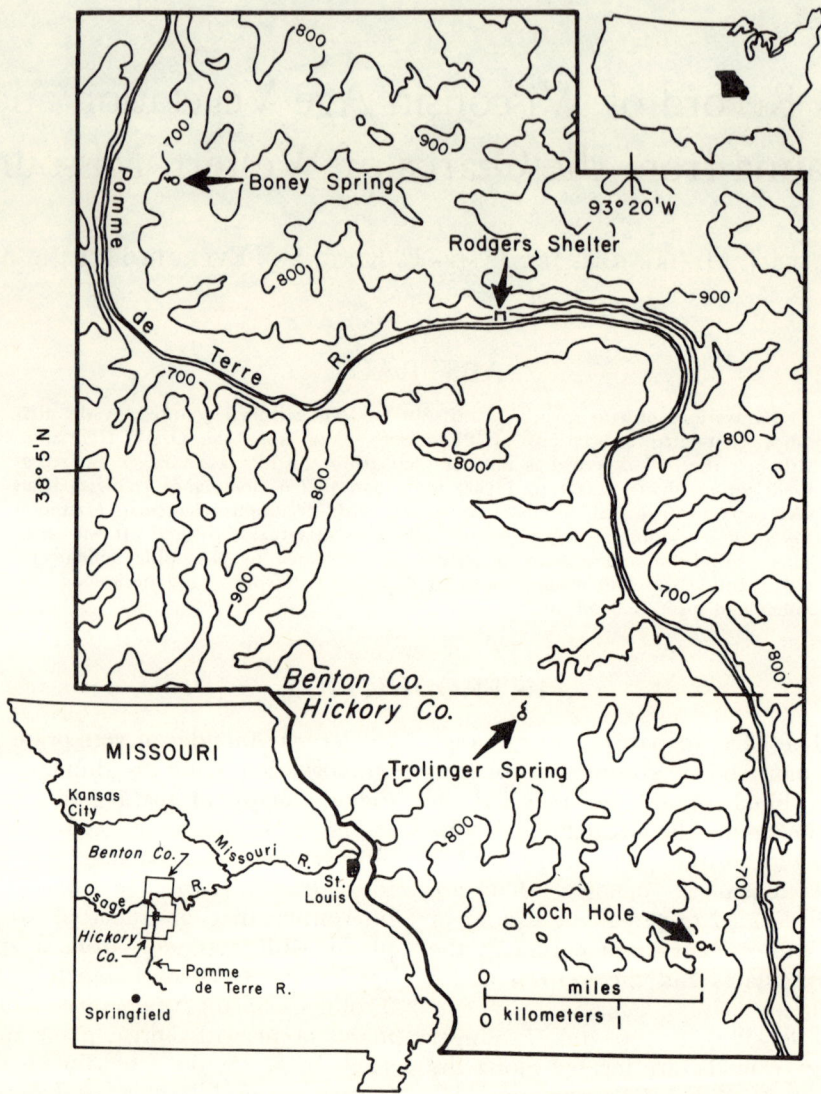

Fig. 1. Site locations.

Core samples were obtained from Boney Spring in 1966. These contained a spruce-dominated pollen record, spruce and larch macrofossils, and bone and tusk fragments (Mehringer et al., 1968). Excavations of Boney Spring and Trolinger Spring were started in 1967 under the direction of W. R. Wood and R. B. McMillan. Geological studies of the spring deposits and valley alluvium by C. V. Haynes are in progress.

Excavation of Boney Spring (Fig. 3) is incomplete, partly because unusually heavy rains in August, 1968, added to the difficulty of excavating a live spring. Neither the lateral nor vertical extent of the bone concentration was established. Only the spring feeder conduit and a small area adjacent to the conduit (between four and five meters depth, where the bone concentration was overlain by gray clay) were excavated. We assume the major part of the deposit is still in place.

Trolinger Spring excavations

FIG. 2. Rodgers Shelter. (Photograph, R. B. McMillan)

(Figs. 4, 5) were completed in 1968. The entire bone concentration was exposed, insuring an adequate paleontological sample. Some of the deposits were excavated and mapped by archaeological methods; stratigraphic cross-sections and other geological data were gathered. The excavation of Trolinger Spring was less difficult than that of Boney Spring because the deposits were shallower and the spring ceased to flow in 1964.

POLLEN AND RADIOCARBON DATES

Analysis of pollen and plant macrofossils from Boney and Trolinger springs is not complete. Preliminary analyses of selected samples (Fig. 7) are included to show the major palynological features of the two sites. Although these results may be modified by further studies, they clearly illustrate a marked change in Ozark vegetation during the Wisconsin.

Two radiocarbon dates are available from Boney Spring (Fig. 3). Plant debris (twigs, seeds, and mosses) from the pulp cavity of tusk A (Figs. 6, 7) was dated at $13,700 \pm 600$ BP (M-2211). A spruce log (identified by R. C. Koeppen) recovered from the bone concentration contained 65 to 70 annual rings. Ten annual rings, representing years 20 to 30 in the growth of the tree, were dated at $16,580 \pm 220$ BP (I-3922). These two dates establish a late-Wisconsin age for the bone concentration at Boney Spring.

Fig. 3. Boney Spring. The top of the bone concentration is visible at the base of a gray clay, immediately above water standing in the spring feeder conduit. The uppermost dark unit is a post-glacial peat containing plant macrofossils of oak-hickory forest species. Artifacts were recovered from the base of the peat, indicating the use of the spring by prehistoric inhabitants of the valley.

Fig. 4. Trolinger Spring at conclusion of the 1967 excavations. Depth and location of Pollen Profile III samples (Fig. 7) are indicated by the horizontal lines. (Photograph, W. R. Wood)

Four pollen spectra from pulp cavities (Fig. 6) of mastodon tusks from Boney Spring are similar to that previously reported with spruce and larch macrofossils (Mehringer et al., 1968). They differ, in their low pine values, from modern pollen assemblages from eastern North America. Values for spruce pollen and thermophilous deciduous elements are suggestive of the southern boreal forest. However, pines are present in these forests, and pine pollen is important in their modern pollen rain. The pine pollen (four percent or less) in the Boney Spring samples is probably the result of long-distance transport.

It is probable that the spruce forest represented by the Boney Spring samples contained minor deciduous elements, but it is unlikely that pines were present. This is particularly interesting in view of the similarity with late-glacial deposits of the western Great Lakes region. Wright (1968) reviewed the present and past distributions of pine and spruce pollen and their biogeographic significance in eastern North America. Other relevant occurrences of midwestern boreal pollen assemblages south of the Wisconsin ice margin are discussed by Wright (this volume).

The Trolinger Spring pollen profiles are marked by a significant change from NAP-pine in organic mud sediments, to spruce dominance in the overlying blue-gray clay (Figs. 4, 5). Two radiocarbon dates from the lower NAP-pine zone ($32,200+1900,-1600$ BP, I-3599, and $25,650\pm700$ BP, I-3537) are in proper stratigraphic sequence. They suggest a mid-Wisconsin interstadial age for NAP-pine dominated pollen samples. The change to spruce dominance may reflect the onset of late-Wisconsin full-glacial conditions.

Three other dates, stratigraphically above the first two, range from

FIG. 5. Trolinger Spring Pollen Profile V. Depth and location of the pollen samples (Fig. 7) are indicated by the horizontal lines. The lower dark unit is an organic clayey silt with dispersed sand. All of the extinct megafaunal remains were from the lower part of this unit. Some small mammals were recovered by wet-screening of the overlying blue-gray clay.

Fig. 6. Plant debris being removed from a *Mammut americanum* tusk pulp cavity (Boney Spring, tusk A; Fig. 7). The fill of tusk pulp cavities served as a source of radiocarbon dates, pollen samples, and plant macrofossils.

29,340 to 14,450 BP. These dates are apparently not in proper stratigraphic sequence and are not acceptable within the framework of our present understanding. More samples will be dated, and pollen profiles analyzed. These studies, together with Haynes' continuing geological investigations should lead to a clearer understanding of the geochronology of the deposits. At this time, the most that can be suggested is that the transition to spruce dominance is probably younger than about 25,000 and older than 16,500 years ago, the age of the spruce log from Boney Spring.

The oldest samples from Trolinger Spring are dominated by NAP, mainly sedge, with less than 30 percent pine pollen. Unlike Boney Spring, with the exception of oak (less than six percent), pollen of thermophilous deciduous trees is rare. The only other possible tree species of importance are willow and birch.

Some of the pine pollen could be the result of long-distance transport, from distant pine forests, but with at least part of the pine pollen produced by local trees. The pine pollen percentages do not indicate forest, but rather open vegetation with scattered pines, possibly forming a parkland. While there are similarities between these spectra and the modern pollen rain of the aspen parkland of southern Manitoba (Lichti-Federovich and Ritchie, 1965, p. 75), we hesitate to suggest specific comparisons until the macrofossils are examined and there is a thorough study of possible pine species represented by pollen.

Within the transition zone to spruce dominance (see Fig. 7, Profile V, 1.70 meters), the pollen spectra more closely resemble modern samples from the northern boreal forest (Wright, 1968, fig. 2, table 1). However, above this transition, pine percentages decrease and spruce pollen is dominant. The spruce pollen values (60 to 91 percent) might indicate either forest-tundra transition or northern boreal forest (Wright, 1968, fig. 2). The pine values (less than 14 percent) suggest that pine trees were absent or were

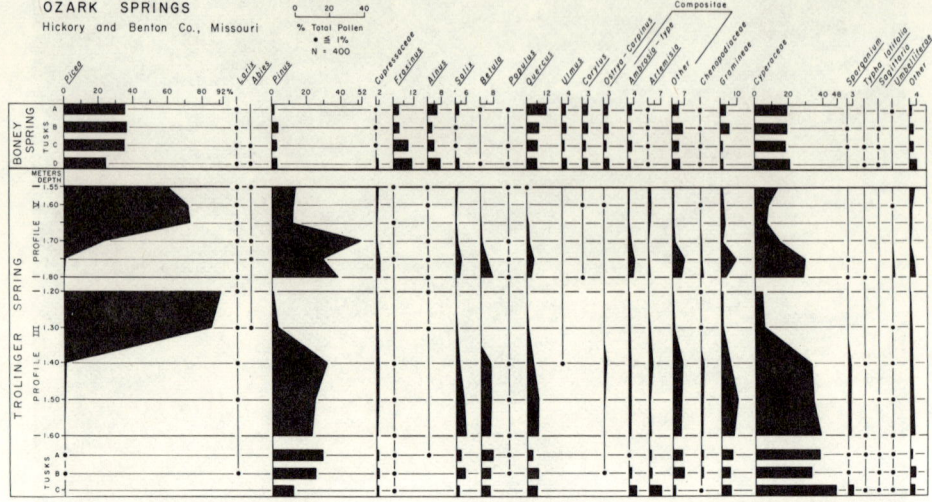

Fig. 7. Pollen profiles and spectra from mastodon tusk pulp cavities. Pollen spectra from the Trolinger Spring tusk samples are shown in their proper stratigraphic position within the deposits. The same arrangement was attempted for the Boney Spring tusk samples; however, because of the mixed sediments and bones in the spring conduit, the stratigraphic relationship of these samples is questionable.

only a minor element of the Wisconsin Ozark spruce forest.

Our present lack of confidence in assigning an age to the NAP-pine to spruce transition at Trolinger Spring does not alter the conclusions concerning the directly observable association of faunal remains and pollen samples. All of the megafaunal remains were recovered from organic muds stratigraphically below the spruce-dominated pollen spectra (Fig. 7). This is further reflected in the pollen content of the Trolinger Spring tusk samples. They contain less than one percent spruce. By comparison, spruce is the most abundant pollen type (26 to 36 percent) from the Boney Spring tusks.

A hiatus of unknown time exists between the youngest Trolinger Spring samples and the oldest Boney Spring samples. Pollen preservation was too poor for analysis above the uppermost samples shown in the Trolinger Spring pollen profiles (Fig. 7). Future excavations of nearby spring deposits may yield data to fill this hiatus.

FAUNA

The faunal remains (Table 1) are concentrated in the organic mud at Trolinger Spring (NAP-pine zone) and at the base and below the gray clay at Boney Spring (Fig. 3).

Mammut americanum, the only species found at both sites, is represented by 81 specimens (seven individuals) from Trolinger Spring, and 102 specimens (eight individuals) from Boney Spring. We estimate that as many as 1000 specimens of *Mammut* remain *in situ* at Boney Spring. *Mammut* was widespread in North America during the Pleistocene, and is the most common mammal in spring deposits of the Ozark Highlands. *Mammut* was a browser, characterized as an inhabitant of boreal forest or woodland (Dreimanis, 1968).

A badly tumbled cheektooth fragment of *Mammuthus* was discovered by W. D. Frankforter in an excavation wall at Trolinger Spring. The speci-

TABLE 1. Faunal List

Trolinger Spring	Boney Spring
INSECTIVORA	EDENTATA
soricid	*Paramylodon* cf.
RODENTIA	*harlani*
Peromyscus sp.	RODENTIA
Synaptomys sp.	*Castoroides* sp.
PROBOSCIDEA	PROBOSCIDEA
Mammut	*Mammut*
americanum	*americanum*
*Mammuthus sp.	ARTIODACTYLA
PERISSODACTYLA	*Odocoileus* sp.
Equus sp.	
ARTIODACTYLA	
Symbos sp.	

* *Mammuthus* is not associated with the other Trolinger Spring fauna in spring deposits (see text).

men was found at the top of an alluvial deposit adjacent to, and truncated by, spring deposits. Truncation of the alluvium indicates that the *Mammuthus* record predates the spring deposits containing other mammal remains. *Mammuthus* remains are common in Middle to Late Pleistocene deposits of North America, but *Mammuthus* apparently was never abundant in the Ozark Highlands.

Equus and *Symbos* were the only other large mammals recovered from Trolinger Spring. *Equus* is represented by an upper incisor and scapula fragment. Although *Equus* is well represented in Pleistocene deposits elsewhere in Missouri (Mehl, 1962, p. 60), fossil equids have not been recorded previously from Hickory or Benton counties.

Symbos is represented by three incisors, five cheekteeth, and a cannon bone fragment. Initial identification was made by C. E. Ray of the National Museum. *Symbos* was probably a browser, inhabiting the boreal forests of eastern North America. *Ovibos*, the extant grazing muskox, is well adapted for survival on the Arctic tundra.

All small mammal remains (soricid, *Peromyscus*, and *Synaptomys*) were collected from Trolinger Spring by wet screen-washing. The soricid and *Peromyscus* came from sands in the spring conduit, and their stratigraphic relationship is questionable. The soricid specimen, a fragmentary lower incisor, is too poorly preserved to warrant generic identification. Several shrews (for example, *Blarina brevicauda* and *Cryptotis parva*) inhabit this area today.

Peromyscus is represented by an upper and a lower anterior molar. The specimens were badly corroded, leaving only an enamel cap. Measurements of the teeth are: M1 length 1.95 mm, width 1.12 mm; m1 length 1.51 mm, width 1.00 mm. Dimensions of the lower molar are similar to teeth of *Peromyscus maniculatus* and *Peromyscus leucopus*, species that presently inhabit the Ozark Highlands. The upper molar is larger than the normal range of these species, suggesting more than one species of *Peromyscus* may be represented. *Peromyscus* exhibits a wide range of habitat preferences and is ubiquitous in North American Pleistocene deposits.

Enamel from an incisor of *Synaptomys* was collected from the blue-gray clay (spruce zone) at Trolinger Spring. The anterior surface of this incisor has a lateral shallow groove, as seen in *Synaptomys*. During the Pleistocene, *Synaptomys* ranged as far south as Nuevo Leon, Mexico, but its present range is north of the 35th parallel (Lundelius, 1967, p. 305). *Synaptomys cooperi* presently inhabits Benton and Hickory counties, Missouri. Enamel plates from prismatic microtine cheekteeth and enamel caps from incisors of rodents other than *Synaptomys* were collected from the organic mud and overlying blue-gray clay. The prismatic enamel plates could be the remains of *Synaptomys* or another microtine rodent.

Along with the mastodons, *Odocoileus*, *Paramylodon*, and *Castoroides* were recovered from Boney Spring. *Odocoileus* is represented by a deciduous upper premolar, a cannon bone fragment, and a fragment of antler.

Odocoileus presently inhabits the Ozarks.

A mylodont ground sloth *(Paramylodon* cf. *harlani)* was represented by five isolated cheekteeth. The specimens are lobate and asymmetrical in occlusal outline, thereby distinguishing them from megalonychid ground sloths whose cheekteeth are transversely elongate in occlusal outline. Mylodont remains from Benton County, described by Harlan (1843) as *Orycterotherium missouriense,* were later referred to *Mylodon harlani* by Owen (1843). The habitat of mylodontids, especially *Mylodon,* was characterized as open grassland by Stock (1925, p. 27).

Castoroides is represented by two incisors. These large rodent incisors have strong longitudinal ridges and grooves on the enamel surface. *Castoroides* is recorded from Florida to Oregon in Pleistocene deposits of North America, but it is more commonly found east of the Mississippi River and north of the 37th parallel. Lundelius (1967, p. 298) characterized *Castoroides* as an inhabitant of "forested areas with a cool humid climate."

DISCUSSION AND CONCLUSIONS

The megafaunas from both Trolinger and Boney springs show similarity in the dominance of mastodon, but they share no other genera. At Trolinger Spring, about 32,000 to 25,000 years ago, a mastodon–horse–muskox fauna occupied an open vegetation, possibly a pine parkland, lacking spruce. Whereas at Boney Spring, about 16,500 to 13,500 years ago, a mastodon–ground sloth–giant beaver fauna occupied a spruce forest with minor deciduous elements.

The boreal nature of the plant and pollen record from Trolinger and Boney springs is not surprising in view of the Pleistocene boreal small mammals known from the Ozarks. These include showshoe hare *(Lepus americanus)* from Conard Fissure in northern Arkansas (Brown, 1907), and the more recent finds of *Microtus xanthognathus, Synaptomys borealis, Sorex arcticus,* and others (Hawksley, 1965; Oesch, 1967).

A few Ozark insects and plants show boreal affinities. Ross (1965) described existing disjunct populations of caddisflies and stoneflies, relict in cold-water springs in Missouri, that have their main distribution in Canada and the northern Great Lakes region. Plant species having northern affinities include *Liparis loeselii, Galium boreale* var. *hyssopifolium,* and *Ilex verticillata* var. *padifolia* (Steyermark, 1959).

The many endemic plant species of the Ozarks have been cited in support of long vegetational isolation and stability (Steyermark, 1959, p. 104). The majority of these have distributions to the south, southeast, and southwest (Steyermark, 1959, figs. 1-10, p. 106). The dynamic nature of the Wisconsin vegetational change, reported here, makes explanations of endemism based on assumptions of long vegetational stability seem unlikely. An alternative interpretation is that most of the Ozark endemics are of post-glacial age, resulting from northward migration and isolation within the last 12,000 years.

Many more deposits, similar to the two described in this paper, will be inundated upon completion of Kaysinger Bluff Dam. Destruction of these, including the remaining deposits at Boney Spring, will result in the loss of their rich Pleistocene biota and archaeological sites associated with the springs. We hope to complete excavation of Boney Spring and to salvage others before they are lost to the rising waters.

ACKNOWLEDGMENTS

The research on which this paper is based was supported by National Science Foundation grant GS-1185 to W. Raymond Wood. The paper is contribution no. 183 of the Program in Geochronology, the University of Arizona, Tucson, Arizona 85721. We thank Austin Long, Laboratory of Isotope Geochemistry, Department of Geochronology, the University of Arizona, for furnishing a radiocarbon date.

LITERATURE CITED

Bretz, J. H.
 1965. Geomorphic history of the Ozarks of Missouri. Missouri Div. Geol. Survey and Water Resources, XLI (ser. 2), 147 pp.
Brown, B.
 1907. The Conard Fissure, a Pleistocene bone deposit of northern Arkansas: with descriptions of two new genera and twenty new species of mammals. Mem. Amer. Mus. Nat. Hist., 9:155-208.
Dreimanis, A.
 1968. Extinction of mastodons in eastern North America: testing a new climatic environmental hypothesis. Ohio Jour. Sci., 63:257-272.
Fenneman, N. M.
 1938. Physiography of the eastern United States. McGraw-Hill Book Co., New York, 714 pp.
Harlan, R.
 1843. Description of the bones of a new fossil animal of the order Edentata. Amer. Jour. Sci., 44:69-80.
Hawksley, O.
 1965. Short-faced bear (Arctodus) fossils from Ozark caves. Bull. Nat. Speleological Soc., 27:77-92.
Kucera, C. L.
 1961. The grasses of Missouri. Univ. Missouri Press, Columbia, 241 pp.
Lichti-Federovich, S., and J. C. Ritchie
 1965. Contemporary pollen spectra in central Canada, II. The forest-grassland transition in Manitoba. Pollen et Spores, 7:63-87.
Lundelius, E. L.
 1967. Late-Pleistocene and Holocene faunal history of central Texas. Pp. 287-319, in Pleistocene extinctions: the search for a cause (P. S. Martin and H. E. Wright, Jr., eds.), Yale Univ. Press, New Haven, Connecticut, x+453 pp.
Mehl, M. C.
 1962. Missouri's ice age animals. Missouri Div. Geol. Survey and Water Resources, Educ. Ser., 1:1-104.
Mehringer, P. J., Jr., C. E. Schweger, W. R. Wood, and R. B. McMillan
 1968. Late-Pleistocene boreal forest in the western Ozark Highlands? Ecology, 49:567-568.
Oesch, R. D.
 1967. A preliminary investigation of a Pleistocene vertebrate fauna from Crankshaft Pit, Jefferson Co., Missouri. Bull. Nat. Speleological Soc., 29:163-185.
Owen, R.
 1843. Letter from Richard Owen . . . on Dr. Harlan's notice of new fossil Mammalia. . . . Amer. Jour. Sci., 44:341-345.
Ross, H. H.
 1965. Pleistocene events and insects. Pp. 583-596, in The Quaternary of the United States (H. E. Wright, Jr., and D. E. Frey, eds.), Princeton Univ. Press, Princeton, New Jersey, x+922 pp.
Steyermark, J. A.
 1959. Vegetational history of the Ozark forest. Univ. Missouri Studies, 31:1-138.
Stock, C.
 1925. Cenozoic gravigrade edentates of western North America. Publ. Carnegie Inst. Washington, 331:206 pp.
Wood, W. R., and R. B. McMillan
 1967. Recent investigations at Rodgers Shelter, Missouri. Archaeology, 20:52-55.
Wright, H. E., Jr.
 1968. The roles of pine and spruce in the forest history of Minnesota and adjacent areas. Ecology, 49:937-955.

Vegetational History of the Great Plains: A Post-Glacial Record of Coniferous Woodland in Southeastern Wyoming

Philip V. Wells

ABSTRACT

Accumulated evidence indicates that the extensive treeless grasslands now occupying the plains of central North America have not had a continuous existence there since the mid-Tertiary. An increasing number of Pleistocene pollen and macrofossil records have documented the existence of coniferous woodland or forest vegetation at several sites on the Great Plains from South Dakota and Minnesota south to Kansas and western Texas, and even in the most arid reaches of the adjacent Chihuahuan Desert.

There appears to have been a shift to nonconiferous vegetation, usually interpreted as more xerophytic, at most of the fossil-pollen sites in the plains region at the close of the Wisconsin. However, dated evidence on the nature and history of post-glacial vegetation is relatively scanty. Here reported is a remarkable record of post-glacial coniferous vegetation from the floor of the Laramie Basin, which lies in the rain-shadow of high mountains and is more arid than the treeless grasslands of the High Plains to the east of the Laramie Range in southeastern Wyoming. The record consists of plant macrofossils ranging in size from massive logs of western red cedar (*Juniperus scopulorum* Sarg.) up to four feet in diameter, exposed at the surface or partly buried in eolian sand deposits, to leaves, cones, and seeds of red cedar and western yellow pine (*Pinus ponderosa* var. *scopulorum* Engelm.), preserved in ancient, rock-sheltered woodrat middens. Radiocarbon ages of the outer sapwood of 10 cedar logs range from 205 ± 95 to 1735 ± 80 years. The dates are scattered over the time interval of 1500 years, suggesting a secular trend to more arid climate rather than catastrophic mortality. Radiocarbon ages of two woodrat deposits containing remains of ponderosa pine and red cedar are 1860 ± 80 and 4060 ± 80 years.

If the now arid floor of the Laramie Basin was wooded recently with red cedar and even ponderosa pine, what was the potential vegetation on the less arid grasslands of the Great Plains, where at present red cedar and ponderosa pine are still abundantly and widely distributed as scarp woodlands?

INTRODUCTION

The traditional assumption that the treeless condition of the grasslands in the Central Plains region of North America has a continuous history "since the Miocene at least" was essentially a uniformitarian extrapolation from the present situation, based on interpretations of early ecologists, notably Clements (1920). The widespread occurrence in the High Plains of fossilized fruits of certain grasses and borages in the Ogallala formation of Miocene-Pliocene age was thought to support this view (Elias, 1942). However, evidence has been accumulating that indicates the extensive treeless grasslands now occupying the plains have not had a continuous existence there since the mid-Tertiary.

It has long been known that fossilized fruits of the hackberry tree *(Celtis)* are more abundant and more widely distributed in the Ogallala sediments than are the grass fruits (Chaney and Elias, 1938; Frye *et al.*, 1956). More recently, a well-preserved leaf impression and pollen flora (the Kilgore flora) from the Ogallala formation of western Nebraska has come to light (MacGinitie, 1962). This remarkable record documents the existence of numerous, ecologically diverse tree species in the central part of the Great Plains during Late Miocene time. Among the broad-leaf trees are representatives of existing subtropical genera, for example *Cedrela* (Meliaceae). One of the more temperate elements is *Celtis kansana* Chaney and Elias; the leaf impressions that here complement the widely distributed silicified endocarps of *Celtis* closely resemble the relatively mesophytic existing species, *C. occidentalis* L. Of course, it is likely that the more mesophytic broad-leaf trees occurred only along the streams. On the other hand, the presence of leaf impressions of evergreen, sclerophyllous live-oaks *(Quercus)*, algerita *(Mahonia)*, and relatively abundant pollens of two species of pines *(Pinus)*, and lesser amounts of pollens of grasses, and of *Artemisia, Ambrosia,* and other composites, indicates that open, xerophilous woodland with a ground cover of herbs prevailed in the more xeric upland habitats. The coexistence in nearby fossil faunas of about the same age of browsing and grazing mammals, including a forest-dwelling horse *(Hypohippus)* with low-crowned teeth ill-adapted for grazing, cannot be taken as evidence for a prevalence of treeless grassland. According to MacGinitie (1962), paleobotanists, handicapped by data much scantier than is now available, together with vertebrate paleontologists, "have indulged in a kind of circular reasoning with regard to the prairie problem, each drawing evidence from the findings of the other." He further wrote: "The sum of evidence indicates that the plains vegetation, between the stream valleys, was an open grassy forest of small live-oaks, pines, hackberry, and persimmon, with shrubs of *Mahonia*, currant, hawthorne, sagebrush, and relatively abundant species of composites."

It is significant to note that the Kilgore flora was deposited in a shaly lens intercalated between the usual sandy Ogallala sediments containing silicified fruits of the grass *Stipidium gamma* and of the hackberry. Rather than constituting proof of the existence of treeless grassland, the fragmentary record afforded by the silicified clasts of initiallly very indurated fruits, namely the stony endocarps of hackberry, the bony nutlets of borages, and the hard caryopses of grasses more closely allied to the southerly genus *Piptochaetium* than to *Stipa* (Stebbins, 1950, p. 527), is probably a highly distorted one of erosion-resistant dispersal units sorted out placer-fashion in the coarse, stream-laid sediments of the Ogallala formation. In a review of the literature, MacGinitie (1962) pointed out that there is really no firm evidence for the existence of extensive treeless grassland in North America during the Tertiary or, for that matter, the interglacials of the Pleistocene.

PLEISTOCENE AND POST-GLACIAL HISTORY: THE FOSSIL EVIDENCE

An increasing number of Late Pleistocene macrofossil and pollen records from the Great Plains or plains border (see Wright, this volume) indicate the presence of boreal spruce *(Picea)* forest on the Northern Plains (Fig. 1, Table 1). White spruce *(P. glauca)* grew on high ground of the Coteau in southern Saskatchewan as recently as $10{,}270 \pm 150$ years ago (Ritchie and de Vries, 1964), on morainal deposits in central North Da-

TABLE 1. Pollen and macrofossil records of forest or woodland vegetation during late Wisconsin time, 10,000 to 20,000 years ago, in what is now grassland or desert in the Plains region of central North America

Fossil site	Radiocarbon dates (BP)	Pollen (%) Spruce	Pollen (%) Pine	Macrofossils	Distance (km) from existing stands of Spruce	Distance (km) from existing stands of Pine
Moose Jaw, Saskatchewan	10,270±150 11,650±150	55	20	spruce	160	250
Tappen, North Dakota	11,480±300			spruce	300	
Rosebud, South Dakota	12,580±160	70	10	spruce	200	20
Pickerel Lake, South Dakota	>10,670±140	85	5	spruce, larch	160	160
Muscotah, Kansas	15,500±1500	70	10	spruce	700	320
Boney Spring, Missouri	13,700±600 16,580±220	36	5	spruce, larch	750	100
Rich Lake, Texas	17,400±600	10	90		320	200
*Maravillas Canyon, Texas	14,800±180			pine (pinyon), oak, juniper		50
*Dagger Mountain, Texas	16,250±240 20,000±390			pinyon, oak, juniper		40

* Sites of maximum aridity, at latitude (29°30′ N) and elevations (600 and 880 m) now occupied by desert vegetation.

kota about 11,500 years ago (Moir, 1958), and in the Sandhills of southwestern South Dakota (radiocarbon age about 12,600 years, Watts and Wright, 1966). There were high levels of spruce pollen and macrofossils of spruce and larch (Larix) at the western border of Minnesota (McAndrews, 1966) and in eastern South Dakota (radiocarbon age ca. 10,670±140, Watts and Bright, 1968). Spruce also was present at Muscotah Marsh in northeastern Kansas—radiocarbon age of spruce macrofossils 15,500±1500 years (Horr, 1955; McGregor, 1968, reported as Abies); and spruce and some larch were represented by macrofossils and pollens at the western edge of the Ozark Highlands in Missouri (Mehringer et al., 1968 and this volume).

In the relatively arid southwestern corner of Kansas, deposits of Illinoian, Sangamon (interglacial), and Wisconsin age have yielded high relative percentages of pine pollen, but there were only small amounts of spruce and Douglas-fir (Pseudotsuga) pollens, chiefly during the Illinoian glacial (Kapp, 1965 and this volume). On the still more arid High Plains of the Llano Estacado of western Texas, several pollen profiles from Wisconsin sediments, one at Rich Lake with a radiocarbon age of 17,400 years, record high levels of pine pollen (more than 90 percent) and low levels (10 percent or less) of spruce pollen (Hafsten, 1961).

Unfortunately, it has not been possible to establish the identity of the pines recorded at the fossil pollen sites on the Southern Plains. However, macrofossil evidence has been obtained on the southwestern border, about 250 miles south of Rich Lake. Beautifully preserved plant materials with radiocarbon ages ranging from 11,500 to 20,000 years and older are abundant in numerous rock-sheltered Neotoma deposits at now-desert elevations in southwestern Texas (Wells, 1966). Even in now-arid reaches of the Chihuahuan Desert, pine-dominated vegetation was widespread during the Wisconsin glacial, but the pine was a xerophytic variety of the Mexican pinyon (Pinus cembroides var. remota Little). The fossil evidence from the southwestern edge of the Stockton Plateau, adjacent to the plains, indicates that the pinyon pine formed an open woodland with live-oaks (Quercus pungens Liebm. and Q. grisea Liebm.) and juniper, together with cacti, Agavaceae, and

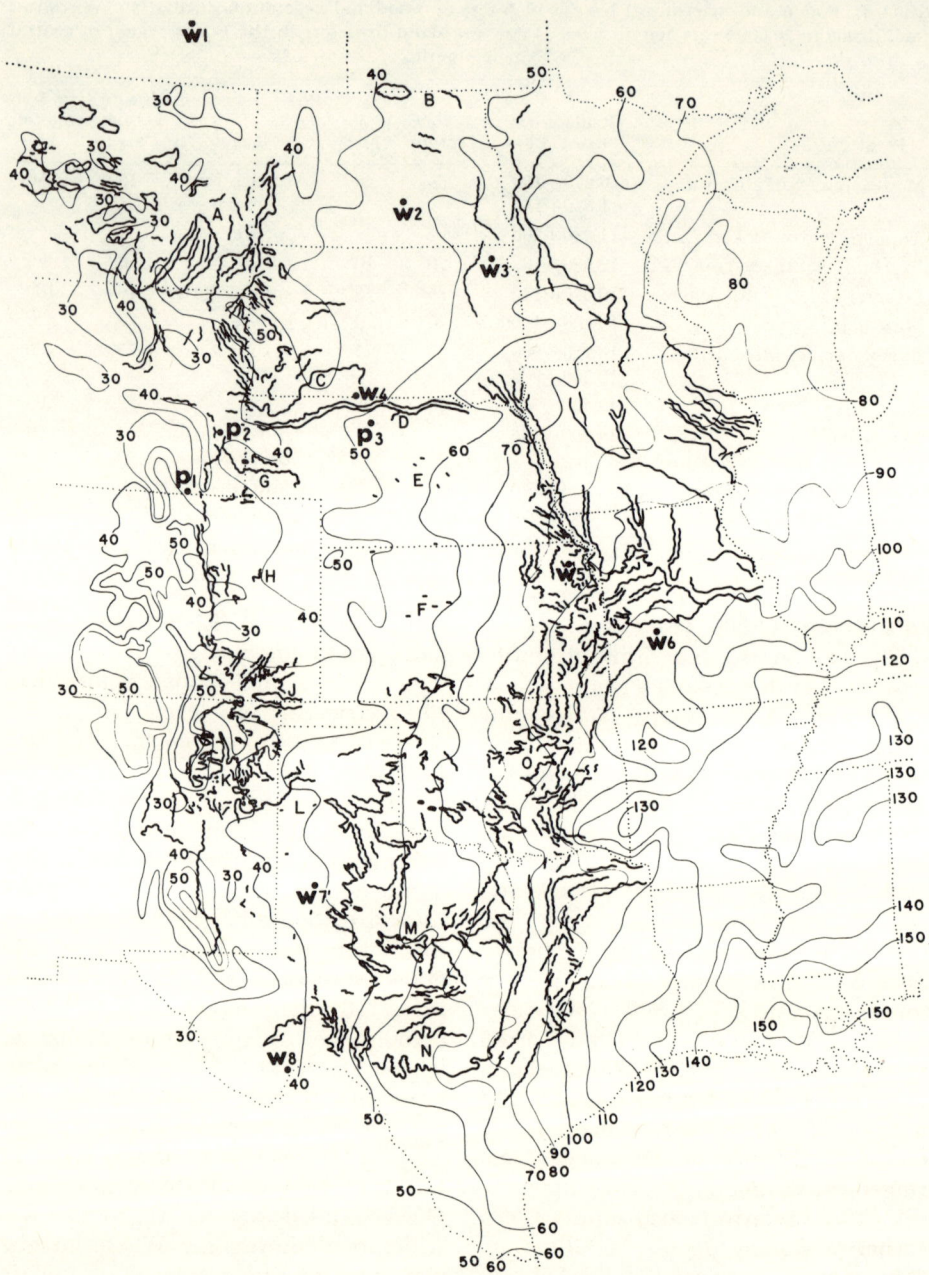

Fig. 1. Location of late-Wisconsin (W) and post-glacial (P) fossil records of forest or woodland vegetation in relation to distribution of existing nonriparian, scarp woodlands (heavy lines) in the Great Plains region. Thin, continuous lines are isohyets at 10-centimeter intervals, ranging from 30 to 150 centimeters of mean annual precipitation. Dotted lines indicate state or other boundaries. W 1, Hafichuk Ranch, southwest of Moose Jaw, Saskatchewan; W 2, 16 km south of Tappen, Kidder Co., North Dakota; W 3, Pickerel Lake, Day Co., South Dakota; W 4, 30 km southwest of Rosebud, Todd Co., South Dakota; W 5, Muscotah Marsh, Atchison Co., Kansas; W 6, Boney Spring, Benton Co., Missouri; W 7, Rich Lake, Terry Co., Texas; W 8, Maravillas Canyon and Dagger Mountain, Brewster Co., Texas; P 1, southwestern corner of Laramie Basin, Albany Co., Wyoming; P 2, 3 km southeast of Guernsey, Platte Co., Wyoming; P 3, Hackberry Lake, Cherry Co., Nebraska. Some dominant upland tree species of existing scarp woodlands

other light-demanding, semidesert plants. Some grass fruits representing four genera *(Bouteloua, Buchlöe, Heteropogon, Tridens)* also entered the *Neotoma* record, documenting their coexistence with woodland conifers then, just as they coexist today in grassy woodland on high mountains within the Chihuahuan Desert province (Wells, 1966). On the other hand, there is no indication that treeless grassland shifted southward into what is now the arid Chihuahuan Desert during the Wisconsin glacial, when much of the Great Plains south of the ice sheet was occupied by coniferous forest, woodland, or possibly parkland.

At most of the fossil-pollen sites in the plains region, there is a dramatic decrease in coniferous pollens in the post-glacial sediments that is usually interpreted as a climatically induced shift to a more xerophytic nonarboreal vegetation similar to the prairies of today. However, dated fossil evidence on the post-glacial history of vegetation and climate in the grassland province is relatively scanty. Sears (1961) has obtained a number of pollen profiles with significant levels of pine from post-Wisconsin lake sediments in the Sandhills of western Nebraska. A dated pollen record from Hackberry Lake, located only 12 miles from existing scarp woodlands of *Pinus ponderosa* along the Snake River breaks south of the Niobrara Valley in the Sandhills, shows that sediment at a depth of five meters, with a radiocarbon age of about 5000 years, contains more than twice the percentage of arboreal pollen (chiefly pine), and less than one-fifth the percentage of grass pollen, found in sediments at the top of the profile.

BIOGEOGRAPHIC EVIDENCE

The extensive migration of boreal or cool-temperate tree species during the Wisconsin glacial has left a record in the existing flora of the plains region. The white spruce *(Picea glauca)* persists today in the midst of the Northern Plains as isolated populations in the Black Hills of South Dakota (McIntosh, 1931) and the Cypress Hills of southern Alberta and Saskatchewan (Breitung, 1954), and the balsam fir, *Abies balsamea* (L.) Mill., has disjunct, relictual populations as far south as northeastern Iowa (Conard, 1938). Some of the boreal deciduous trees that occur in the Black Hills, namely the paper birch *(Betula papyrifera* Marsh) and the aspen *(Populus tremuloides* Michx.), range disjunctly southward in the Plains to the canyons of the Niobrara River in the Sandhills of Nebraska (Pool, 1914).

Post-glacial migration may be implied by the more or less disjunct distributions of temperate deciduous species in the western sector of the Northern Plains. In the Black Hills, for example, the bur oak, *Quercus macrocarpa* Michx., hop-hornbeam, *Ostrya virginiana* (Mill.) Koch, American elm, *Ulmus americana* L., and two species of hazel, *Corylus americana* Walt. and *C. cornuta* Marsh, form an isolated outlier of the eastern deciduous forest (McIntosh, 1931). The disjunct occurrence of a race of the sugar maple *(Acer saccharum* Marsh) in the Caddo Canyons and Wichita Moun-

---

from selected areas of the Plains, indicated by smaller letters A to O on map, are as follows: A,C,D,G,H—*Pinus ponderosa, Juniperus scopulorum;* B—*Populus tremuloides* Michx., *Quercus macrocarpa* Michx.; D,E—*P. ponderosa, J. virginiana* L., *Q. macrocarpa;* F—*J. virginiana;* J,K—*P. ponderosa, P. edulis* Engelm., *J. monosperma* (Engelm.) Sarg., *J. scopulorum, Quercus undulata* Torr.; L—on northwest, similar to J,K, but minus *P. ponderosa;* L—on east (Break of the Plains), *J. pinchotii* Sudw., *Quercus mohriana* Buckl.; M,N—*Q. virginiana* L., *Q. shumardii texana* Ashe, *Q. mohriana, J. ashei* Buchholz, *J. pinchotii;* O—*Quercus stellata* Wang., *Q. marilandica* Muenchh., *Q. muehlenbergii* Engelm., *Q. shumardii, J. virginiana.*

tains of western Oklahoma (Little, 1939) may have a similar history, or may be older.

Evidence of hybridization between the now widely separated eastern bur oak *(Quercus macrocarpa)* and the western Gambel's oak *(Q. gambelii* Nutt.) indicates a relatively recent sympatry in New Mexico and South Dakota (Tucker and Maze, 1966; Maze, 1968). The southern hybrid populations are small and isolated in the canyon of Tramperos Creek, well out on the short-grass steppe of northeastern New Mexico. The hybrids are in contact with the western parent, *Q. gambelii,* a seemingly unstable situation, but they are 250 miles west of the existing range of bur oak in Oklahoma. The disjunction is analogous to that of the isolated sugar maple populations of western Oklahoma, inasmuch as a climate less arid than at present would be sufficient to account for the implied westward migration.

However, extensive introgression of the large populations of bur oak in the Black Hills of South Dakota suggests a massive northward migration of the Gambel's oak. Since the existing northern limits of the less cold-tolerant Gambel's oak are in the central Rocky Mountains, about 250 miles to the southwest, a northeastward migration across the intervening plains to the outlying Black Hills requires a climate both warmer and wetter than at present, thus ruling out a Wisconsin or earlier glacial age for the indicated distribution. Any evidence of older incursions of Gambel's oak into the Black Hills almost certainly would have been eliminated during the now well-documented spruce maximum of the Late Wisconsin glacial in South Dakota, dated as recently as about 11,000 to 12,000 years ago. Hence, a post-glacial incidence of warm-moist climate is implied for the Northern Great Plains.

A POST-GLACIAL MACROFOSSIL RECORD OF CONIFEROUS WOODLAND FROM THE FLOOR OF THE LARAMIE BASIN, WYOMING

There is a remarkable record of coniferous woodland during the latter half of post-glacial time on the floor of the Laramie Basin in southeastern Wyoming. The record consists of plant macrofossils ranging in size from massive logs of western red cedar, *Juniperus scopulorum* Sarg., up to four feet in diameter, exposed at the surface or partly buried in shallow sand deposits, to leaves, cones, and seeds of red cedar and ponderosa pine, *Pinus ponderosa* var. *scopulorum* Engelm., preserved in ancient, rock-sheltered woodrat middens. The Laramie Basin (Fig. 1) is partially enclosed on the east, west, and south by high mountains that create a pronounced local rain-shadow, and the extensive plain on the floor of the basin is now even more arid than the treeless grasslands of the High Plains to the east of the Front (Laramie) Range of the Rocky Mountains in southeastern Wyoming. The mean annual precipitation at Laramie is only 12 inches.

Existing vegetation.—The upland vegetation in the vicinity of the subfossil materials, located in the arid southwestern corner of the Laramie Basin about 20 miles south of Laramie, may be described as a shrub-steppe. The flat plains on the floor of the basin are covered with a short-grass sod of blue grama, *Bouteloua gracilis* H.B.K., and a sparse scattering of dwarf sagebrush, *Artemisia arbuscula* ssp. *nova* (Nels.) Ward and *A. frigida* Willd. Also present are the grasses *Stipa comata* Trin. and Rupr., *Agropyron smithii* Rydb., and *Koeleria cristata* (L.) Pers., the arid-land sedge, *Carex filifolia* Nutt., and the prickly-pear, *Opuntia polyacantha* Haw. On the eolian sand deposits, the vegetation is dominated by the large silver sagebrush, *Artemisia cana* Pursh, with reduced numbers of the above species. Also present on the sands are rabbit brush, *Chrysothamnus viscidiflorus* (Hook.) Nutt., soapweed, *Yucca glauca* Nutt., Indian rice-grass, *Oryzopsis hymenoides* (R. & S.) Ricker, antelope brush, *Purshia tridentata* (Pursh) DC., horse brush, *Tetradymia canescens* DC., four-wing saltbush, *Atriplex canescens* (Pursh) Nutt., and winter-fat, *Eurotia lanata* (Pursh) Moq. Hence, it is clear that the prevalent vegetation at the site is not only desertic in physiognomy, but also bears a strong floristic resemblance to the winter-cold deserts of the Great Basin and Colorado Plateau. On the other hand, the low ridges of sandstone at the site are sparsely vegetated with somewhat less xerophytic deciduous shrubs, principally mountain mahogany, *Cercocarpus montanus* Raf., and squaw bush, *Rhus trilobata* Nutt., with service berry, *Amelanchier utahensis* Koehne, and wax currant, *Ribes cereum* Dougl. At one spot below a particularly massive, gently sloping surface of bare sandstone bedrock, three individuals of western red cedar, *Juniperus scopulorum* Sarg., were surviving in 1968, though they appeared to be dying back in response to the severe droughts of recent decades. The three junipers are isolated here in a microsite uniquely favored by catchment of runoff from an extensive area of bare rock slope, but even so are close to their tolerance limits. The general lower limit of coniferous timber on mountains surrounding the arid south end of the basin is about 1000 feet higher, at an elevation of about 8500 feet. The last surviving junipers provide a living sample of a vegetation that used to be extensively distributed on the floor of the basin, during the past several thousand years, but is now represented mainly by beautifully preserved subfossil plant materials.

Former vegetation.—The numerous sandstone outcrops in the southern part of the Laramie Basin have an extraordinary development of cavities and rock shelters that harbor many ancient *Neotoma* middens. Seven deposits that have been dated range in radiocarbon age from about 1100 to 4000 years. The bulk constituents of the deposits from an area of several square kilometers are uniformly *Juniperus scopulorum* Sarg., represented by leafy twigs and wood. Two of the middens with radiocarbon ages of 1860 ±80 and 4060±80 years also contain two-needled leaf fascicles of another conifer, *Pinus ponderosa* var. *scopulorum* Engelm., which no longer grows at the sites, although it is present on mountain slopes rimming the basin. On the other hand, most of the other species represented in ancient *Neotoma* deposits are still growing near the sites of deposition. Of the 15 species identified in the middens (Table 2), 13 are xerophytic shrubs or herbs that are more or less widely distributed on the floor of the basin, and lend a semi-desert character to the existing vegetation (see description above). In particular, the species of *Artemisia, Atriplex, Chrysothamnus, Eurotia,* and *Oryzopsis* are characteristic of the winter-cold deserts of western North America. The composition of the old *Neotoma* middens corresponds to an open, xerophilous woodland dominated by the western red cedar and fewer trees of ponderosa pine, with a lower synusia of semidesert shrubs and grasses. The recent decline of the woodland conifers to the vanishing point indicates a dramatic shift in the physiognomy of the vegetation, from woodland to semidesert shrubland, but the floristic composition shows remarkably little change during the past 4000 years, aside from the local demise of the conifers.

An outstanding feature of the paleoecological record at this locality is the presence of old wood of *Juniperus scopulorum* at or near the surface of the extensive sandy flats on the floor of the basin. The logs are widely scat-

TABLE 2. Summary of semiquantitative analysis of plant remains in the oldest *Neotoma* deposit (radiocarbon age, 4060±80 years) among the seven dated middens from the southwestern corner of the Laramie Basin[1]

Species	Structures	Relative Abundance
Trees		
Juniperus scopulorum Sarg.	leafy twigs, seeds	+++
Pinus ponderosa var. *scopulorum* Engelm	leaves, cone-scales	++
Shrubs		
Rhus trilobata Nutt.	seeds, leaves	++
Artemisia cana Pursh	leaves	+
Atriplex canescens (Pursh) Nutt.	winged fruits	+
Cercocarpus montanus Raf.	leaves, fruits	+
Chrysothamnus viscidiflorus (Hook.) Nutt.	involucres, achenes	+
Eurotia lanata (Pursh) Moq.	leaves, fruits	+
Prunus besseyi Bailey	endocarps	+
Agavaceae, Cactaceae		
Yucca glauca Nutt.	capsules, seeds, leaves	+++
Opuntia polyacantha Haw.	areoles, seeds	++
Grasses and forbs		
Oryzopsis hymenoides (R. & S.) Ricker	fruits	++
Stipa comata Trin. & Rupr.	fruits	+
Lappula fremontii (Torr.) Greene	nutlets	+
Lithospermum incisum Lehm.	nutlets	+

Symbols for relative abundance: +, low; ++, intermediate; +++, high.
* Species no longer growing in vicinity of deposit.
[1] The other middens show a similar composition, with the notable exception of content of *Pinus ponderosa*, which appears in only one other dated deposit (radiocarbon age, 1860±80 years).

tered over an area of several square kilometers in the vicinity of the ancient *Neotoma* sites. This is the first instance of a coincidental occurrence of other macrofossil evidence complementary to that of *Neotoma* middens. Despite the fact that the exposed wood ranges in radiocarbon age from about 200 to 1700 years, the logs are remarkably well preserved, and even the strong fragrance of terpenes characteristic of red cedar wood is retained. Weathering is limited to complete removal of the thin shreddy bark and to graying of the intact outermost sapwood; charring was observed on only one log.

Some of the logs are of gigantic dimensions by present standards, the largest having a massive trunk more than four feet in diameter; persistent stubs of branches and roots indicate little or no transport (Fig. 2). A ring count of a 20-inch section at the apical end of the main axis showed about 450 growth rings, implying a life span of as much as 1000 years for the tree. The radiocarbon age of the outermost sapwood records the approximate time of death of the tree, in this instance 940±105 years ago (GX-1407). Hence, the largest juniper log should yield a dendrochronology covering the greater part of the second millennium BP—that is, about the first millennium A.D.

Radiocarbon ages of the outermost sapwood of 10 juniper logs at as many different sites scattered on the surface of the eolian sand deposits range from a remarkably recent 205±95 to 1735 ±80 years. Thus the massive logs elegantly corroborate the more detailed macrofossil evidence in the *Neotoma* middens as to former abundance and extent of juniper woodland on the floor of the basin during the overlap-

Fig. 2. Log of western red cedar (*Juniperus scopulorum*) with a trunk more than four feet in diameter, exposed at the surface of eolian sand deposits in the arid southwestern corner of the Laramie Basin, Wyoming. Radiocarbon age of the outermost sapwood is 940±105 years, indicating the approximate time of death. A ring count on a half-meter cross-section at the apical end showed about 450 growth rings, suggesting that favorable climatic conditions prevailed on the floor of the basin during much of the first millennium A.D. Existing semidesert vegetation is dominated by the silver sagebrush (*Artemisia cana*).

ping time interval from about 1100 to 1800 years ago (Table 3). A uniquely sheltered juniper log, lodged in a rock crevice and partly buried by a more recent *Neotoma* deposit, has a radiocarbon age of 5625±140 years, making it the oldest paleoecological record so far obtained from the area. It is apparent that rock-sheltered material, whether it be a *Neotoma* midden or a log, affords a greater reach backward in time than material exposed at the surface. This is true even in the most arid deserts of southwestern North America, where the most delicate structures from 78 woodland or semidesert species are perfectly preserved in a mummified condition in numerous rock-sheltered *Neotoma* middens with radiocarbon ages ranging from about 8000 to greater than 40,000 years (Wells and Jorgensen, 1964; Wells, 1966; Wells and Berger, 1967). However, because of weathering and sheet-flood erosion for at least the last 8000 years, surface occurrences of logs or other macrofossils dating from the pluvial woodland period are extremely scarce.

Whereas the dated *Neotoma* middens record the presence of *Juniperus scopulorum*, *Pinus ponderosa*, and 13 other species at the time of deposition, the radiocarbon ages of the outermost sapwood of the logs pinpoint incidents of mortality in the juniper population. The available chronology of woodland conifers on the floor of the Laramie

TABLE 3. Radiocarbon chronology of woodland conifers on the floor of the Laramie Basin at sites where no conifers exist today*

Material Dated	Radiocarbon Age, Years	Sample Number	Paleoecological Significance
Log of *Juniperus scopulorum* Sarg.	205±95	GX-1406	death of juniper tree
" " " " "	385±85	GX-1400	" " " "
" " " " "	580±105	GX-1405	" " " "
" " " " "	735±95	GX-1404	" " " "
" " " " "	790±95	GX-1408	" " " "
" " " " "	940±105	GX-1407	" " " "
" " " " "	1060±75	GX-1401	" " " "
Neotoma midden, abundant juniper	1100±80	UCLA-1406	presence of *J. scopulorum*
Log of *J. scopulorum*	1145±80	GX-1402	death of juniper tree
" " " "	1445±95	GX-1403	" " " "
Neotoma midden, abundant juniper	1600±80	UCLA-1404	presence of *J. scopulorum*
Log of *J. scopulorum*	1735±80	UCLA-1098E	death of juniper tree
Neotoma midden, abundant juniper	1860±80	UCLA-1098A	presence of *Pinus ponderosa* var. *scopulorum* and juniper
" " " "	2020±80	UCLA-1405	presence of *J. scopulorum*
" " " "	2320±80	UCLA-1098D	" " " "
" " " "	2900±80	UCLA-1098C	" " " "
" " " "	4060±80	UCLA-1098B	presence of *Pinus ponderosa* var. *scopulorum* and juniper
Log of *J. scopulorum*	5625±140	GX-1426	death of juniper tree

* Dating of juniper logs based on outermost sapwood; dating of *Neotoma* middens based chiefly on the abundant coniferous plant material. Radiocarbon ages determined at Institute of Geophysics, University of California, Los Angeles (UCLA), courtesy of Drs. Rainer Berger and W. F. Libby; and at Geochron Laboratories, Inc., Cambridge, Massachusetts (GX).

Basin, based on 18 radiocarbon dates scattered over a span of 5600 years, is summarized in Table 3. The well-documented record of *Juniperus* over the past 2300 years is based on 15 evenly dispersed radiocarbon dates, and hence there is no indication of any particularly catastrophic mortality of juniper during this time interval. Some clumping of juniper-sapwood dates would be expected if such were the case. As it is, the life span of the larger junipers, indicated by ring counts on the logs, is much greater than the average interval between the radiocarbon dates, which is only about 150 years (Table 3). This implies a continuous existence of junipers on the floor of the basin during the last 2300 years, except for the past two centuries. Nevertheless, the juniper population has indeed declined to the point of imminent extinction here at present.

A probable explanation is that the reproduction rate of the junipers gradually declined in response to a secular trend toward increasingly arid climate during the past several centuries, while individuals with well-established root systems eked out their different life spans before succumbing at staggered intervals. The desiccated condition of the three surviving individuals of *Juniperus scopulorum*, their lack of reproduction, and their restriction to one site uniquely favored by runoff indicate that the trend to greater aridity has continued into the present century. The imminent demise of these individuals will carry to completion a slow process of elimination of juniper from the floor of the southwestern corner of the Laramie Basin, an end-point that evidently was not reached at any time during the past 2000 years or more.

A POST-HYPSITHERMAL MAXIMUM OF ARIDITY IN THE LARAMIE BASIN

The subfossil remains of *Juniperus scopulorum* and *Pinus ponderosa* from an arid part of the Laramie Basin document the occurrence of coniferous woodland at 16 different sites where none exists today. Since the record thus far obtained extends back some 5600 years, it follows that at least in the latter part of the Hypsithermal interval of maximum post-glacial warmth it was significantly less arid here than it is today. According to Deevey and Flint (1957), who proposed the term, the Hypsithermal interval extended from about 9000 to about 2500 years ago. It is broader in scope than the Altithermal interval of Antevs (1955), which extended from 7500 to 4000 years ago. Relative to the European Blytt-Sernander sequence, the Hypsithermal corresponds in time to the Boreal, Atlantic, and Sub-Boreal zones, whereas the Altithermal is essentially equivalent in time with the Atlantic zone. The Sub-Boreal zone, now known to date from about 4500 to 2500 years ago, was thought to have been a time of maximal post-glacial dryness, a European Xerothermic (Sernander, 1910). The radiocarbon-dated record of woodland on the floor of the Laramie Basin during the last 5600 years spans the latter part of the Atlantic and all of the subsequent Sub-Boreal and Sub-Atlantic zones.

The common use in America of the term Xerothermic for a warm post-glacial interval (Sears, 1942) implied that it was a time of greater aridity than today. This interpretation is probably appropriate where precipitation is concentrated in the cool winter months and severe drought regularly coincides with the time of maximum evapotranspiration stress in summer, as in regions influenced by the climatic rhythm of the Mediterranean type along the Pacific coast, and inland in the Mohave and Great Basin deserts. However, in regions further east, as in the Rocky Mountains and Great Plains, there is a profoundly different, monsoonal type of climatic regime with a strong summer maximum of precipitation from incursions of humid gulf air, that ameliorates the peak in evapotranspiration stress on vegetation corresponding to the summer maximum in temperature. The ecological differences between the two contrasting climatic rhythms should be enhanced by a general increase in summer temperatures. Hence, the time interval of maximum post-glacial warmth (the co-inciding peaks of the "Hypsithermal," "Altithermal," or "Atlantic" intervals) should be expected to show different effects on climate and vegetation in regions that are fundamentally different today. Surprisingly, this basic point seems to have escaped proper emphasis. On this view, there is no conflict in the divergent indications of a rather moist Hypsithermal in the summer-rain area of southeastern Arizona (Martin, 1963) and a dry Hypsithermal in the summer-drought area of the Mohave Desert (Wells and Jorgensen, 1964). Similarly, in the Great Plains and the adjacent Laramie Basin a Hypsithermal climate less arid than that of today may not be anomalous.

In summary, the macrofossil record of coniferous woodland on the floor of the Laramie Basin extends over the past 5600 years, on the basis of 18 radiocarbon dates for only part of the abundant material. The record indicates a climate significantly less arid than today's during a time span that encompasses the latter half of the Hypsithermal interval (or nearly the same fraction of the Altithermal) and most of the post-Hypsithermal (or Medithermal of Antevs). There is especially abundant subfossil material of *Juniperus scopulorum* that speaks for a continuous existence of juniper on the floor of the basin during at least the last 2000 years, and *Pinus ponder-*

osa is represented in the *Neotoma* middens as recently as 1860±80 years ago. A series of dated logs extends the juniper chronology to 205±95 years ago, and the last three junipers, barely surviving at a uniquely favorable bedrock site, establish a tenuous continuity with the present. Because the ancient juniper logs and *Neotoma* deposits are at 16 different sites scattered over an area of several square kilometers, it is certain that at some time in the past, reproduction of juniper declined to zero, as at present.

The available evidence also points to a distinct possibility that the climate in the southwestern corner of the Laramie Basin may be more arid at present than at any time during the Hypsithermal interval of maximum post-glacial warmth. In other words, the drastic shrinkage of woodland over the past several centuries, now culminating in a semidesert steppe on the floor of the basin, may constitute the first climatically induced episode of treelessness at this locality in post-Wisconsin time.

RELEVANCE TO THE ECOLOGICAL HISTORY OF THE GREAT PLAINS

The indication of a climate significantly less arid than at present on the floor of the Laramie Basin, during much of the latter half of post-glacial time, has major implications for the history of climate and potential vegetation on the Great Plains, only 30 miles to the east. The edge of the plains lies immediately to the east of the Front (Laramie) Range of the Rocky Mountains, the physiographic boundary on the east side of the Laramie Basin. Precipitation increases progressively toward the east on the plains, despite a gradient of declining elevation in that direction. At the same latitude, the Great Plains province is decidedly less arid than parts of the Laramie Basin, where the existing vegetation bears a desertic stamp, as described above for the subfossil woodland locality in the southwestern corner. Correspondingly, a significant feature of the existing vegetation of the Great Plains is the wide distribution of upland forests or woodlands in the vicinity of escarpments and other major topographic breaks. Grasslands occupied most of the smooth topography, the flat or rolling plains and gentle slopes. Under the widest range of climate, nonriparian woodlands or forests coexisted with grassland on the uplands, but the forests were more or less restricted to rough, dissected topography associated with the bolder scarps, or wind-swept mesas and buttes isolated by extensive grassy plains (Wells, 1965). Trees also occurred in relatively level upland situations on the leeward sides of lakes and rivers that afforded shelter from the wind-driven prairie fires (Gleason, 1913; Daubenmire, 1936), and a similar explanation is applicable to the scarp woodlands.

Distribution of existing scarp woodlands in the plains region.—The zone of segregated coexistence of nonriparian, scarp-restricted woodland, and grassland, is not limited to a narrow ecotonal transition between regional forest and regional grassland. It extends across the entire width of the Central Plains in the latitude of the Prairie Peninsula, from Indiana through Illinois, Iowa, and Nebraska to Wyoming. In accordance with the east-west climatic gradient of decreasing precipitation and humidity, there is a shift from broad-leaf deciduous forest to open, xerophilous, coniferous woodland dominated by *Pinus ponderosa* Laws. and *Juniperus scopulorum* Sarg. along the scarps in central Nebraska. Similarly, the grasses on the adjacent plains decrease in stature and density westward, as dominance shifts from the tall Andropogoneae in the east to the dwarf Chlorideae in the west. Nevertheless, the abrupt topographic segregation of woodland and grassland prevails throughout the gradient in vegetation and climate.

Even in the relatively dry western sector of the plains, numerous wooded scarps interrupt the flat monotony of the short-grass steppe from eastern Montana south to New Mexico and Texas. For example, coniferous forests or open woodlands dominated by pines or junipers, with a lower synusia of grasses, occur in the midst of immense grasslands of arid aspect (because of their treelessness) at such topographic irregularities as Piney Buttes in eastern Montana; Pine Ridge escarpment in South Dakota, Nebraska, and Wyoming; Scotts Bluff and Wildcat Hills in western Nebraska; Pawnee Buttes, Cedar Point, and Two Buttes in eastern Colorado; Black Mesa in western Oklahoma; and the Llano Estacado in New Mexico and Texas (Fig. 1).

Well out on the Great Plains, along the westward-facing escarpment of the High Plains (Ogallala formation) near Pine Bluffs, in the southeastern corner of Wyoming, there are extensive woodlands dominated by *Pinus ponderosa* var. *scopulorum* and more remarkably by limber pine, *P. flexilis* James, with western red cedar, *Juniperus scopulorum,* in exactly the same latitude as the subfossil woodland locality in the Laramie Basin, but 80 miles east of it. The popular conception that the High Plains physiographic province is too arid for the growth of trees, except along streams, colors these upland occurrences of coniferous woodland with an aura of anomaly. The idea of climatically induced treelessness is extrapolated from the more arid regions to the west of the High Plains, where the prevalent vegetation is a desertic shrub-steppe dominated by *Artemisia,* and trees are indeed restricted to watercourses, being obligately riparian, broad-leaf deciduous species of *Populus* and *Salix.*

In the desert, coniferous woodland is now lacking at the lower elevations, even along streams or on the boldest escarpments, as in the southwest corner of the Laramie Basin, where the manifest trend toward more arid climate is now eliminating conifers from the floor of the basin altogether. In fact, the ecological contrasts between the scarp woodlands and grasslands (or grass-steppe) of the Great Plains, and the treeless shrub-steppe vegetation of the semideserts or deserts to the west, are readily explained by the long-term climatic records. Meteorological data show that the wooded Pine Bluffs locality on the grassy plains has a distinctly wetter climate than the floor of the Laramie Basin, which lies just to the west of the Front Range of the Rocky Mountains and in its rain-shadow with respect to moist Gulf airmasses. The mean annual precipitation at Pine Bluffs is 16 inches, compared to 12 inches at Laramie, while the temperature regimes are quite similar (U.S. Weather Bureau, 1954, 1960).

The present wide distribution of scarp woodlands in the Great Plains, their recent increase (Phillips and Shantz, 1963; see references in Wells, 1965 and this volume), and a number of ancillary facts, such as the success of experimental tree plantations from North Dakota south to Texas (Munns and Stoeckler, 1946; Bates and Pierce, 1913), suggest an existing potential for a natural upland growth of various xerophytic tree species throughout much of the grassland province of North America. The potential must have been greater at times when the now much more arid southwestern corner of the Laramie Basin was wooded. The well-documented macrofossil record of the woodland conifers, *Juniperus scopulorum* and *Pinus ponderosa,* on the floor of the Laramie Basin at various times during the last 5600 years speaks for significantly more humid conditions, which should have had a counterpart on the less arid Great Plains, a short distance to the east.

A Neotoma record of woodland on the Great Plains about 1500 years ago. —A rock-sheltered *Neotoma* midden, containing a woodland record with a radiocarbon age of 1530±85 years (GX-1427), has been discovered in the Great Plains sector of southeastern Wyoming, about 30 miles east of the Front (Laramie) Range, southeast of Guernsey (Fig. 1). The deposit consists of abundant remains of woodland conifers, and it documents the presence of both *Juniperus scopulorum* and *Pinus ponderosa* on the plains at a time synchronous with the evidence of juniper woodland from the more arid Laramie Basin, about 95 miles to the southwest. On the other hand, it does not prove that grassland was lacking on the plains at that time. The Guernsey *Neotoma* site is in the midst of a scarp woodland of *Juniperus scopulorum* and *Pinus ponderosa* today, and the scarp bounds an extensive grassy plain. The record does contribute to the history of the scarp woodland, showing that the existing conifers were in the Great Plains about 1500 years ago, but it offers no information on treeless grassland. However, then as now, the potential for spread of xerophytic tree species, well adapted to upland growth on the plains, may have been realized only close to seed-sources that had been sheltered from prairie fires, perhaps for time periods of thousands of years, by abrupt breaks in topography.

The interaction of topography and fire in the origin and maintenance of grasslands.—For the well-entrenched point of view that the distribution of treeless grasslands is governed principally by climate (Clements, 1916; Livingstone and Shreve, 1921; Thornthwaite, 1931; Borchert, 1950), the wide distribution of scarp woodlands throughout the climatically diverse grassland province poses some difficult questions. Regardless of local or regional variations in climate and the species composition of both woodland and grassland in the plains region, the rougher and the more dissected the topography, the greater the former extent and current spread of woody vegetation at the expense of grassland (Wells, 1965). Over and above the droughty climate, which undoubtedly has been a contributing factor, it is the vast flat or rolling smoothness and continuity of surface of the relatively undissected sedimentary mantle that appears to have played a powerful role in the development of the great expanses of treeless grassland on the plains. The topographic control of vegetation pattern implied by the distribution of extensive grassy plains and scarp-restricted woodlands in the Central Plains region can be explained as a resultant of many interacting factors. However, wind-driven grass fires, whether ignited by lightning or by man, must be accorded a key position in the hierarchy.

The seasonally dry perennial grasses serve as a highly flammable fuel that may burn with sufficient heat to injure or kill woody plants; and yet a new crop of grass is regenerated from protected buds after each fire. The establishment of seedlings of xerophytic tree species in grassland is hazardous under the droughty climate of the Central Plains, even with protection from fire, and successful reproduction is usually limited to the relatively rare sequences of wetter years. But the pattern of recurrent fire prevalent in "natural" grasslands is disastrous to seedlings or saplings of trees during the many precarious years of vegetative growth required to attain seed-bearing maturity. Fire is particularly disastrous to conifers that lack the capacity to sprout from older wood or stumps following destruction of the resinous tops. With frequently recurrent fire, even the most xerophytic tree species, such as the native junipers of the plains, are unable to reproduce successfully in grasslands. This is true not only under semiarid conditions, but even in the humid eastern prairies, as on Long Island, New York (Blizzard,

1931; Taylor, 1923; Wells, this volume).

The immense continuity of the fire-conducting matrix of tinder-dry grass on the smooth topography of the plains formerly favored the spread of vast prairie fires. The regional extent of these now long-extinguished conflagrations is well documented historically, beginning with Cabeza de Vaca's 1528 account of fires set by Indians on the plains of Texas (see original narrative in Hodge, 1907). For a review of literature on prairie fires see Gleason (1913), Sauer (1920, 1927, 1944, 1950, 1952, 1956), Stewart (1951, 1953, 1956), Roe (1951), Humphrey (1958), and Curtis (1959). Lewis and Clark made a number of observations on burning of grasslands by Indians in the course of their traverse of the Northern Great Plains in 1804 and 1805, and they attributed the scarp-restriction of upland woody vegetation to this cause (see original diaries, Thwaites, 1904).

George Catlin, the artist who lived among the plains Indians and recorded their way of life prior to the onslaught of the mid-nineteenth century, was much impressed by the extent and frequency of burning that was practiced throughout the grassland province. According to Catlin (1842): "Every acre of these vast prairies (being covered for hundreds and hundreds of miles with a crop of grass, which dies and dries in the fall) burns over during the fall or early in the spring, leaving the ground of a black and doleful color. [Is the characteristic blackness of the prairie soils and chernozems due in part to the long history of prairie fires, with attendant accumulation of carbonized residues of burned plant materials in the soil profile?] There are many modes by which the fire is communicated to them, both by white men and by Indians—par accident; and yet many more where it is voluntarily done for the purpose of getting a fresh crop of grass, for the grazing of their horses, and also for easier travelling during the next summer, when there will be no old grass to lie upon the prairies, entangling the feet of man and horse as they are passing over them."

Prairie fires were known to have burned with considerable speed before the wind, moving great distances at one sweep. As recently as 1910, a prairie fire in central Nebraska burned over a linear distance of 125 miles in a single day (Nebraska National Forest, 1963). The wavelike motion of a windswept grass fire across a flat or rolling plain would continue indefinitely until quenched by rain or checked by an abrupt break in topography. A bold, rocky, sparsely-grassed escarpment would serve as a natural firebreak, harboring fire-sensitive trees in safe sites, as on rocky promontories, and in the rincons or reentrants. From these seed-sources, a slow lateral encroachment on the adjacent grassy plains could proceed, as at present (see references in Wells, 1965 and this volume). The scarp woodlands of the plains may be viewed as insular relicts of woody vegetation in a formerly fire-swept sea of grass.

These relations would have held with equal or greater force for the Late Wisconsin boreal spruce flats of the Northern Plains, when they began to suffer from the effects of post-glacial warming of climate. A *coup de grâce* from forest fire is a likely fate for non-sprouting boreal conifers stranded under an unfavorable climate. An elegant analogue is the presence of charcoal horizons, with radiocarbon ages ranging from 900 to 3500 years, over forest podzols in the Keewatin tundra, as much as 175 miles north of the present tree-line (Bryson et al., 1965). Advances of the boreal forest into the tundra during periods of milder climate (the older dates correspond to the latter part of the Hypsithermal interval) were followed by catastrophic forest fires and a failure of forest regeneration under periods of adverse climate (the latest being the last 900 years). Although the modern vegetation pattern on the plains has had all of post-glacial time to develop (potentially, 8000 to 10,000 years), an early elimination of boreal conifers and scarp-restriction of temperate tree species would be expected, as indeed the pollen record from the Northern Plains indicates (McAndrews, 1966; Watts and Bright, 1968; Horr, 1955). A catastrophic elimination of the highly combustible and ill-adapted spruce forests would have opened a vast expanse of Northern Plains to the rapid spread of mobile, opportunistic plants, such as herbaceous Compositae and Gramineae and the wind-dispersed aspen, *Populus tremuloides* Michx., which sprouts after fire and proliferates laterally from the roots. The sheer extent of deforested plains or aspen parklands (to the north) would insure a long hiatus of open vegetation, because migration of potentially adapted, xerophytic, temperate tree species from distant refugia into the central sectors of the plains must have required a great deal of time, even under the most favorable climatic conditions. The migration of trees is slowed by the relatively long interval of maturation, often as much as 10 to 20 or more years between dispersals. Mass migration of tree species, and hence of forest, would proceed at a much slower rate than isolated colonizations due to occasional long dispersals. But recurrent fire on the wind-swept plains would be sufficient to delay indefinitely the centripetal migration of trees from peripheral, sheltered sites. Nevertheless, the pollen record also suggests that there were times during the post-glacial when upland trees were probably less scarp-restricted than they are today, as in the Sandhills of Nebraska about 5000 years ago (Sears, 1961).

A rational explanation of these phenomena is that scarp-restriction of forest or woodland vegetation throughout the region of grassy plains in the central part of the continent originated as a consequence of regional forest and prairie fires. Certainly, the vast conflagrations helped to maintain a treeless condition by sweeping the seasonally dry grasslands on the smooth surface of the plains for great distances, until stopped by an abrupt topographic break. When all the facts are admitted, the generalization that the treeless grasslands of the Great Plains are climatically determined becomes a superficial one. Obviously, climate is a factor, as is always true in plant geography, but together with the regional flatness and continuity of the physiography, the smooth, relatively undissected mantle of unconsolidated Pleistocene sediments, the fuel-producing, annual die-back of grasses characteristic of the herbaceous way of life, and, *sine qua non,* fire.

ACKNOWLEDGMENTS

The research on which this paper is based was supported by National Science Foundation grant GB-5002. I thank Dr. Rainer Berger for radiocarbon determinations and Dr. Paul S. Martin for information.

LITERATURE CITED

Antevs, E.
  1955. Geologic-climatic dating in the West. Amer. Antiquity, 20:317-335.
Bates, C. G., and R. G. Pierce
  1913. Forestation of the Sandhills of Nebraska and Kansas. U.S.D.A. Forest Serv. Bull., 121:1-49.
Blizzard, A. W.
  1931. Plant sociology and vegetational change on High Hill, Long Island, New York. Ecology, 12:208-231.
Borchert, J. R.
  1950. The climate of the central North American grassland. Ann. Assoc. Amer. Geog., 40:1-39.
Breitung, A. J.
  1954. A botanical survey of the Cypress Hills. Canadian Field-Nat., 68:55-92.

Bryson, R. A., W. N. Irving, and J. A. Larsen
1965. Radiocarbon and soil evidence of former forest in the southern Canadian tundra. Science, 147:46-48.

Catlin, G.
1842. Letters and notes of the manners, customs and conditions of the North American Indians. London, 2 vols.

Chaney, R. W., and M. K. Elias
1938. Late Tertiary floras from the High Plair Publ. Carnegie Inst. Washington, 476:1-46.

Clements, F. E.
1916. Plant succession: an analysis of the development of vegetation. Publ. Carnegie Inst. Washington, 242:1-512.
1920. Plant indicators: the relation of plant communities to processes and practice. Publ. Carnegie Inst. Washington, 290:1-388.

Conard, H. S.
1938. The fir forests of Iowa. Proc. Iowa Acad. Sci., 45:69-72.

Curtis, J. T.
1959. The vegetation of Wisconsin. Univ. Wisconsin Press, Madison, 657 pp.

Daubenmire, R. F.
1936. The "Big Woods" of Minnesota: its structure, and relation to climate, fire and soils. Ecol. Monogr., 6:233-268.

Deevey, E. S., and R. F. Flint
1957. Postglacial Hypsithermal interval. Science, 125:182-184.

Elias, M. K.
1942. Tertiary prairie grasses and other herbs from the High Plains. Geol. Soc. Amer. Spec. Paper, 41:1-176.

Frye, J. C., A. B. Leonard, and A. Swineford
1956. Stratigraphy of the Ogallala formation (Neogene) of northern Kansas. Bull. Kansas Geol. Surv., 118:1-92.

Gleason, H. A.
1913. The relation of forest distribution and prairie fires in the Middle West. Torreya, 13:173-181.

Hafsten, U.
1961. Pleistocene development of vegetation and climate in the Southern High Plains as evidenced by pollen analysis. Pp. 59-91, in Paleoecology of the Llano Estacado (F. Wendorf, ed.), Mus. New Mexico Press, Santa Fe, 144 pp.

Hodge, F. W. (ed.)
1907. The narrative of Alvar Nuñez Cabeza de Vaca. Pp. 12-126, in Spanish explorers in the southern United States 1528-1543, New York.

Horr, W. H.
1955. A pollen profile study of the Muscotah Marsh. Univ. Kansas Sci. Bull., 37:143-149.

Humphrey, R. R.
1958. The desert grassland—a history of vegetational change and an analysis of causes. Bot. Rev., 24:193-252.

Kapp, R. O.
1965. Illinoian and Sangamon vegetation in southwestern Kansas and adjacent Oklahoma. Contrib. Mus. Paleontol., Univ. Michigan, 19:167-255.

Little, E. L.
1939. The vegetation of the Caddo County canyons, Oklahoma. Ecology, 20:1-10.

Livingstone, B. E., and F. Shreve
1921. The distribution of vegetation in the United States, as related to climatic conditions. Publ. Carnegie Inst. Washington, 284:1-590.

MacGinitie, H. D.
1962. The Kilgore flora. Univ. California Publ. Geol. Sci., 35:67-158.

Martin, P. S.
1963. The last 10,000 years. A fossil pollen record of the American Southwest. Univ. Arizona Press, Tucson, 87 pp.

Maze, J. R.
1968. Past hybridization between *Quercus macrocarpa* and *Quercus gambelii*. Brittonia, 20:321-333.

McAndrews, J. H.
1966. Postglacial history of prairie, savanna and forest in northwestern Minnesota. Mem. Torrey Bot. Club, 22 (2):1-72.

McGregor, R. L.
1968. A C-14 date for the Muscotah Marsh. Trans. Kansas Acad. Sci., 71:85-86.

McIntosh, A. C.
1931. A botanical survey of the Black Hills of South Dakota. Black Hills Eng., 19:159-276.

Mehringer, P. J., C. E. Schweger, W. R. Wood, and R. B. McMillan
1968. Late Pleistocene boreal forest in the western Ozark highlands? Ecology, 49:567-568.

Moir, D. R.
1958. Occurrence and radiocarbon date of coniferous wood in Kidder County, North Dakota. North Dakota Geol. Surv. Misc. Ser., 10:108-114.

Munns, E. N., and J. H. Stoeckler
1946. How are the Great Plains shelterbelts? Jour. Forestry, 44:237-257.

Nebraska National Forest
1963. Self-guided tour, Bessey nursery and plantation. Govt. Print. Office, Washington, D.C.

Phillips, W. S., and H. L. Shantz
1963. Vegetational changes in Northern Great Plains. Rep. Univ. Arizona Agric. Expt. Sta., 214:1-185.

Pool, R. J.
1914. A study of the vegetation of the Sandhills of Nebraska. Minn. Bot. Studies, 4:189-312.

Ritchie, J. C., and B. deVries
1964. Contributions to the holocene paleoecology of west central Canada: a late glacial deposit from the Missouri Coteau. Canadian Jour. Bot., 42:677-692.

Roe, F. G.
- 1951. The North American buffalo. Univ. Toronto Press, Toronto, Ontario, viii+957 pp.

Sauer, C. O.
- 1920. Geography of the Ozark Highland of Missouri. Univ. Chicago Press, 245 pp.
- 1927. Geography of the Pennyroyal. Kentucky Geol. Surv., 25:1-299.
- 1944. A geographic sketch of early man in America. Geogr. Rev., 34:529-573.
- 1950. Grassland climax, fire, and man. Jour. Range Manag., 3:16-21.
- 1952. Agricultural origins and dispersals. Amer. Geogr. Soc., New York, 110 pp.
- 1956. The agency of man on the earth. Pp. 49-69, in Man's role in changing the face of the earth (W. L. Thomas, ed.), Univ. Chicago Press.

Sears, P. B.
- 1942. Xerothermic theory. Bot. Rev., 8:708-736.
- 1961. A pollen profile from the grassland province. Science, 134:2038-2040.

Sernander, R.
- 1910. Die schwedischen Torfmoore als Zeugen postglacialer Klimaschwankungen. Pp. 195-246, in Die Veränderung des Klimas seit dem Maximum der letzten Eiszeit, XI Congr. Géol. Internat., pp. 195-246.

Stebbins, G. L.
- 1950. Variation and evolution in plants. Columbia Univ. Press, New York, 643 pp.

Stewart, O. C.
- 1951. Burning and natural vegetation in the United States. Geogr. Rev., 41:317-320.
- 1953. Why the Great Plains are treeless. Colorado Quart., 2:40-50.
- 1956. Fire as the first great force employed by man. Pp. 115-133, in Man's role in changing the face of the earth (W. L. Thomas, ed.), Univ. Chicago Press.

Taylor, N.
- 1923. The vegetation of Long Island I. The vegetation of Montauk: a study of grassland and forest. Mem. Brooklyn Bot. Garden, 2:7-107.

Thornthwaite, C. W.
- 1931. The climates of North America. Geogr. Rev., 21:633-654.

Thwaites, R. G. (ed.)
- 1904. Original journals of the Lewis and Clark expedition. Dodd, Mead & Co., New York, 7 vols.

Tucker, J. M., and J. R. Maze
- 1966. Bur oak *(Quercus macrocarpa)* in New Mexico? Southwestern Nat., 11:402-405.

U.S. Weather Bureau
- 1954. Climatic summary of the United States—supplement for 1931 through 1952. Section 42, Wyoming, Govt. Print. Office, Washington, D.C., 44 pp.
- 1960. Climatic summary of the United States. Section 14, southeastern Wyoming, Govt. Print. Office, Washington, D.C., 18 pp.

Watts, W. A., and R. C. Bright
- 1968. Pollen, seed and mollusk analysis of a sediment core from Pickerel Lake, northeastern South Dakota. Bull. Geol. Soc. Amer., 79:855-876.

Watts, W. A., and H. E. Wright
- 1966. Late Wisconsin pollen and seed analysis from the Nebraska Sandhills. Ecology, 47:202-210.

Wells, P. V.
- 1965. Scarp woodlands, transported grassland soils, and concept of grassland climate in the Great Plains region. Science, 148:246-249.
- 1966. Late Pleistocene vegetation and degree of pluvial climatic change in the Chihuahuan Desert. Science, 153:970-975.

Wells, P. V., and R. Berger
- 1967. Late Pleistocene history of coniferous woodland in the Mohave Desert. Science, 155:1640-1647.

Wells, P. V., and C. D. Jorgensen
- 1964. Pleistocene wood rat middens and climatic change in Mohave Desert: a record of juniper woodlands. Science, 143:1171-1174.

A slightly modified version of this paper, entitled "Postglacial vegetational history of the Great Plains," was published in Science, vol. 167, pp. 1574-1582, March 20, 1970. Copyright © 1970 by the American Association for the Advancement of Science.

Effects of Historical Droughts on Grassland Vegetation in the Central Great Plains

G. W. Tomanek and G. K. Hulett

ABSTRACT

Variation in the weather of the Central Great Plains has been well documented during the past century. During the past 100 years, 48 years have been below average precipitation, 44 years above, and eight years near average. Several drought periods have been recorded, but the two most recent extended droughts, 1933–1939 and 1952–1956, can be used to illustrate drought effects on vegetation. Vegetation records have been continuously kept of the area near Hays, Kansas, for the past 36 years. The reaction of the cover, composition, and production of the vegetation to drought is reviewed.

Basal cover of native vegetation of short-grass prairies decreased from 85 percent before the drought of the 1930's to less than 20 percent at the close of the drought. In an extensive survey of damage caused by the 1952–1956 drought, Albertson *et al.* (1957) found losses of more than 90 percent of the original basal cover on heavily grazed areas.

Production of grasslands near Hays, Kansas, has varied from less than 900 pounds of grass per acre to more than 4100 pounds during a 27-year period. Low production years coincided with, or followed, years of low precipitation.

The composition of three major communities in grassland near Hays was greatly altered by the two drought periods. Dominant plants often were completely replaced by other plants, and often the changes still are apparent 10 years after the last drought.

Although detailed studies on animal responses to drought are not available for the area, observations indicate that certain species such as the prairie vole *(Microtus ochrogaster)* and the dickcissel *(Spiza americana)* almost disappear during drought periods. However, some species such as the white-footed mouse *(Peromyscus maniculatus)* seemed unaffected by drought, whereas others such as the black-tailed jackrabbit *(Lepus californicus melanotis)* increased three-fold during drought years.

Extended drought periods have resulted in changes in the nature of our native grasslands and the animals they support.

Variations in the weather of the Great Plains has been well documented during the past century. In order to illustrate some of these variations and their effect on our native vegetation, data collected at Hays, Kansas, will be used. Hays is a good location for at least two reasons: (1) it is centrally located in the Great Plains, and (2) weather records have been kept for 100 years and vegetation records for the past 36 years.

Precipitation records collected at the Fort Hays Branch Experiment Station show considerable variation (Fig. 1). During the last 100 years, 1868 through 1967, there have been 48 years below average in precipitation, 44 years above average, and eight years near average. The annual precipitation has ranged from a low of 9.21 inches, in 1956, to a high of 43.34 inches just five years earlier, in 1951. Drought periods of more than three years duration occurred during the intervals of 1868–1873, 1879–1884, 1910–1914, 1933–1939, and 1952–1956. Other short-period droughts have oc-

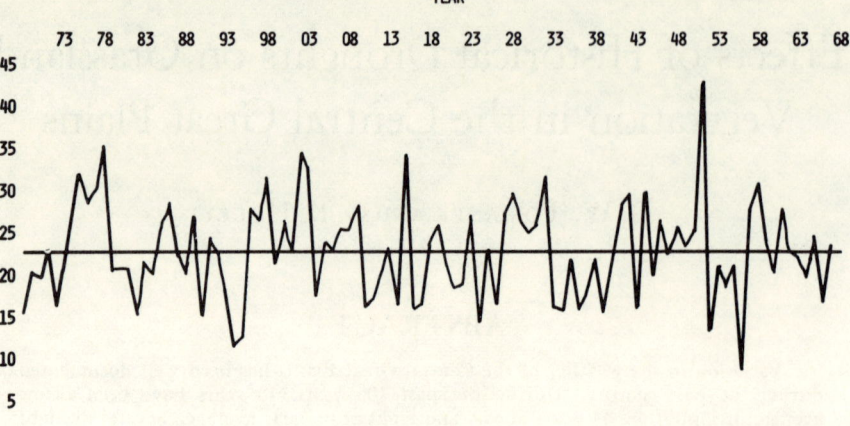

Fig. 1. Annual precipitation at Hays, Kansas, 1868–1967.

curred, and their effects may have been as severe as some of the longer dry periods. The drought periods with which we will be primarily concerned are 1933–1939 and 1952–1956, because they occurred during the period for which we have vegetation records (Fig. 2). Two other dry years occurring during the past 36 years were 1943 and 1946. Actually 1946 had above-average

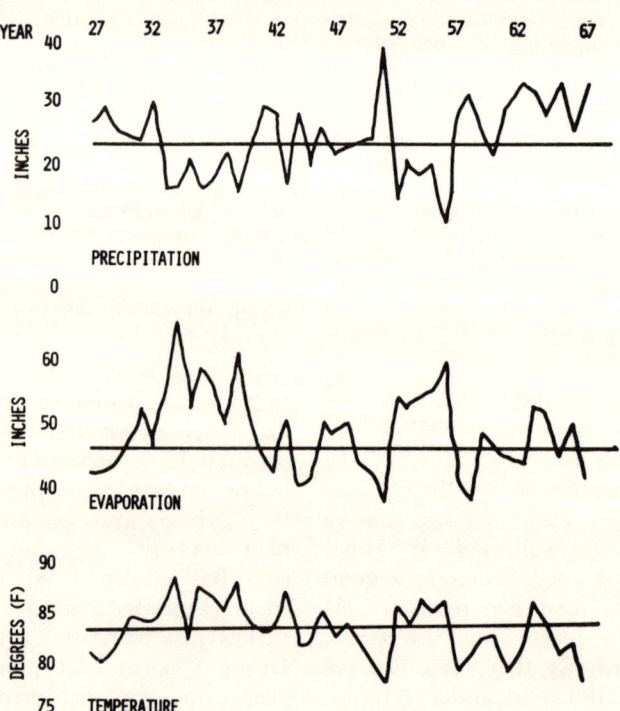

Fig. 2. Annual precipitation and evaporation and average daily maximum temperatures for growing seasons, 1927–1967, Hays, Kansas.

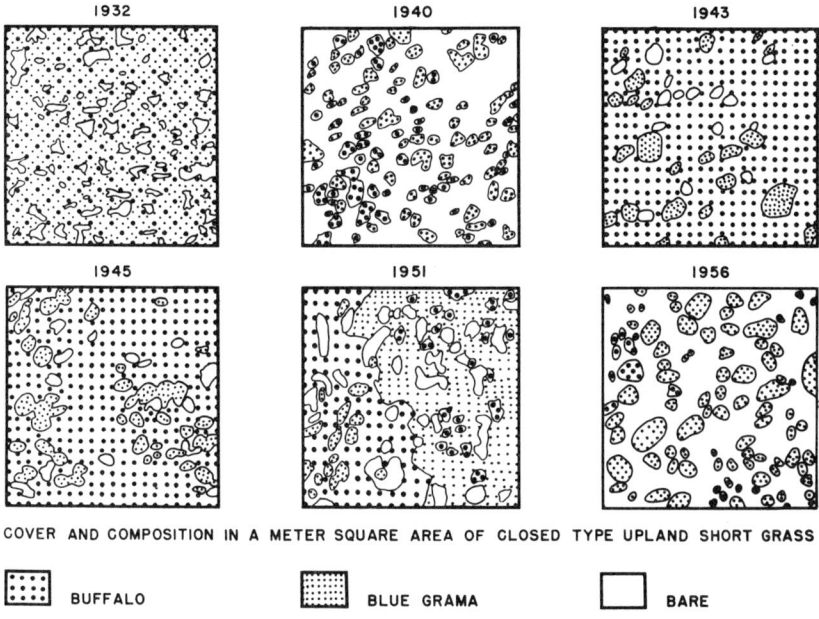

FIG. 3. Changes in cover of short grass on meter quadrat from 1932 to 1956. Large dots—buffalo grass *(Buchloe dactyloides)*, small dots—blue grama *(Bouteloua gracilis)*.

precipitation, but the distribution of it was so poor that it was essentially a drought year for vegetation.

Evaporation, which averages about 48 inches from a free-water surface during the growing season at Hays, was considerably above average during the droughts of the 1930's and 1950's. Average evaporation during the 12 years of drought was 56.2 inches, or more than eight inches above the overall average. Average maximum daily temperatures also were considerably above the overall average during the drought periods. In fact, the drought years had an average of 34.5 days with temperatures above 100°F, whereas the remaining years of a 30-year period (1932–1961) averaged only 12.7 days (Albertson and Tomanek, 1965). Average wind velocities also were higher during drought years.

The vegetation showed a tremendous reaction to these two drought periods in terms of cover, composition, and production. Changes in cover of the two dominant species on one quadrat from 1932 to 1956 are shown diagrammatically in Figure 3. Cover exceeded 85 percent in 1932, but was reduced to less than 20 percent in 1940, immediately following the drought. When the rains came, cover in this short-grass habitat was rapidly reestablished, due primarily to the growth of buffalo grass *(Buchloe dactyloides)*. Buffalo grass and blue grama *(Bouteloua gracilis)* each furnished about 50 percent of the cover in 1932, but in 1940 there was more blue grama left than buffalo grass. Just three years later, however, buffalo grass had recovered so rapidly that it furnished the major portion of the cover. During that period, buffalo-grass stolons were observed growing as much as an inch a day (Tomanek, 1948). By 1951, after more than 10 years of good rainfall, each of the two species again furnished about 50 percent of the cover. However, by 1956, after five years of drought, the cover was again drastically reduced. Studies conducted by Weaver and Albertson during and af-

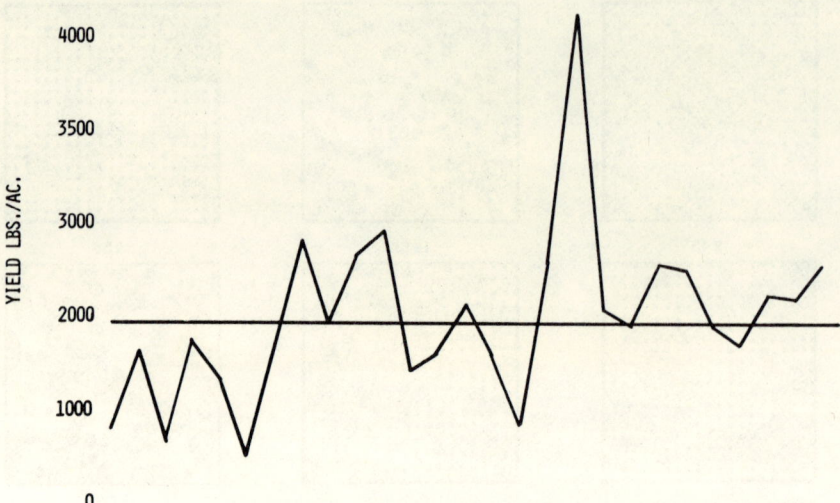

Fig. 4. Production of grass on moderately grazed short-grass pasture from 1941 to 1967.

ter the drought of the 1930's have documented many examples of losses in cover in the vegetation of western Kansas (Albertson and Weaver, 1942, 1944, 1946; Weaver and Albertson, 1939, 1940, 1943, 1944). Albertson et al. (1957) made an extensive survey throughout the Central Great Plains at the close of the drought of the 1950's and found losses of more than 90 percent of the cover, especially in areas that had previous histories of heavy grazing. Albertson and Weaver (1944) also found drought effects on vegetation in the 1930's intensified by heavy grazing.

Production of the native vegetation in the short-grass prairies of western Kansas has been measured from 1941 to the present. At the end of each

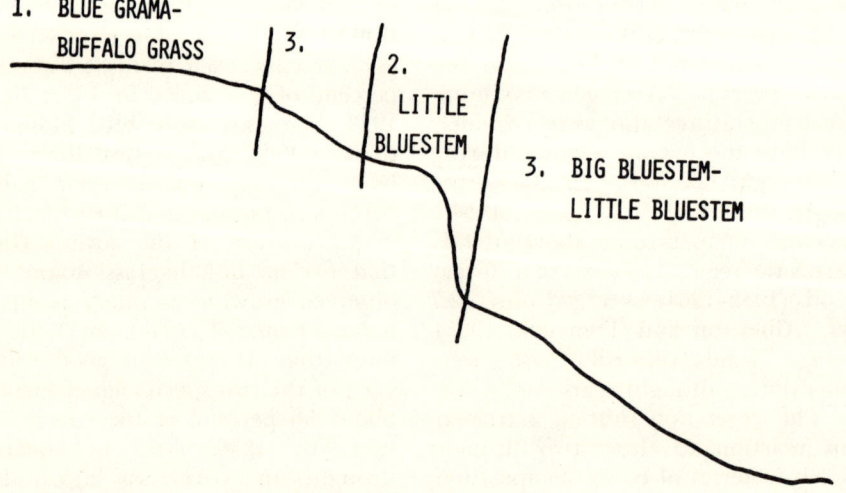

Fig. 5. Topographical location of three major plant communities near Hays, Kansas.

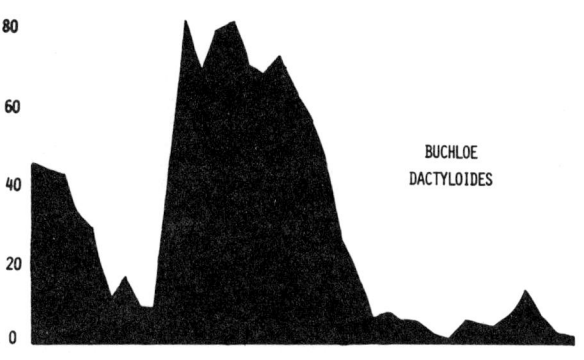

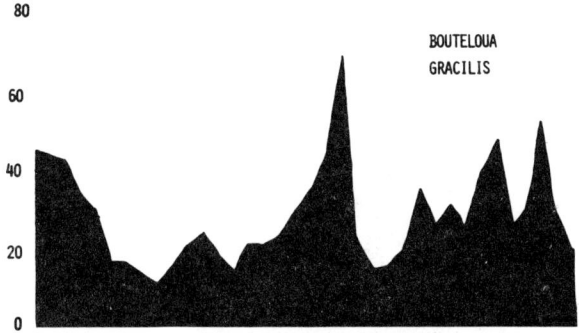

FIG. 6. Changes in cover of two dominants on short-grass prairie from 1932 to 1967, near Hays, Kansas.

season, herbage from clip quadrats was collected and air-dried. Annual production of grass has varied from less than 900 pounds to more than 4100 pounds per acre during this 27-year period (Fig. 4). The lowest yield was encountered in 1941, immediately following the drought of the 1930's. Yields stayed below average from 1941 to 1947, partially because of the after-effect of the drought, but also because two years during that period, 1943 and 1946, were poor moisture years. Production again fell below average from 1952 through 1956 except for 1954. Twelve years of this 27-year period were above the nearly 2000-pound-per-acre average and 12 were below; three years were nearly average. Other factors do affect production in the grasslands, but a definite relation does exist between precipitation and yield.

In the mixed prairie area near Hays, Kansas, there are three major grassland communities, and the change in the composition of these communities as a result of drought is an interesting phenomenon. The three communities are characterized by differences in the relative abundance of dominant grasses as primarily influenced by soils and topography (Fig. 5). The blue grama–buffalo grass community, often called the short-grass community, is found on the deep, mature soils of the uplands. The soils were developed from loessial material deposited on top of the Fort Hays Limestone. The little-bluestem community was developed on the shallow soil derived from the Fort Hays Limestone. This community is found primarily on the rocky breaks and is dominated by widely-spaced bunch

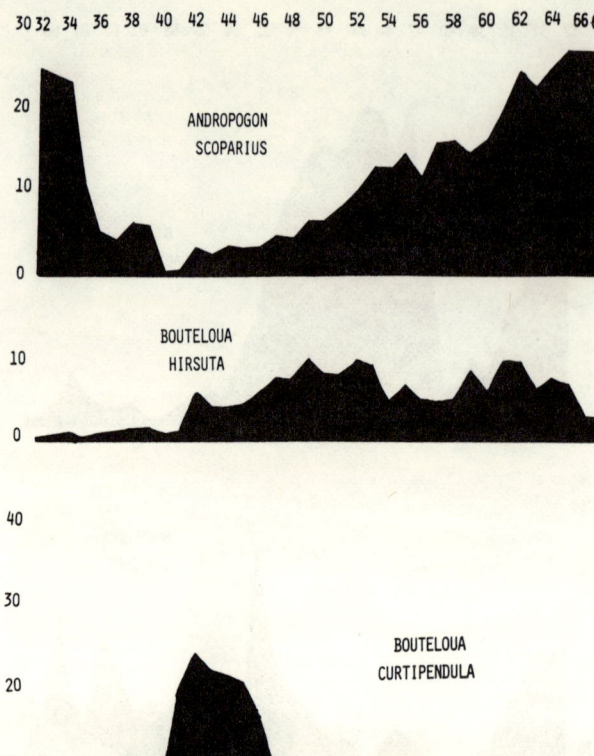

Fig. 7. Changes in cover of three grasses on the little-bluestem *(Andropogon scoparius)* habitat from 1932 to 1967, near Hays, Kansas.

grasses, principally little bluestem. The third community is found on immature soil that developed over Fort Hays Limestone but is usually more than 20 inches deep. It is found immediately above and below the rocky breaks or outcrops and is dominated by big and little bluestem, although side-oats grama and blue grama often become quite important.

We have charted the vegetation on a number of meter quadrats in each of these communities each year since 1932. We have averaged the results of the quadrats of each community each year to show changes in the relative abundance or cover furnished by dominant species.

In the short-grass community the two dominant species each furnished about 44 percent of the cover in 1932; but after the beginning of the drought in 1933, the cover of both species decreased until 1940, when buffalo grass constituted only eight percent and blue grama nine percent of the cover (Fig. 6). Total cover had been reduced from 88 percent to 17 percent. In only two years (1941–1942), cover of buffalo grass increased from eight to 80 percent and stayed relatively high until the drought of the 1950's, when it was reduced again to less than eight percent and has remained low to the present time. Blue grama, on the other hand, recovered more slowly following the drought of the 1930's and did not reach its peak until 1952, after which it decreased rapidly due to the drought. After the drought of the 1950's it

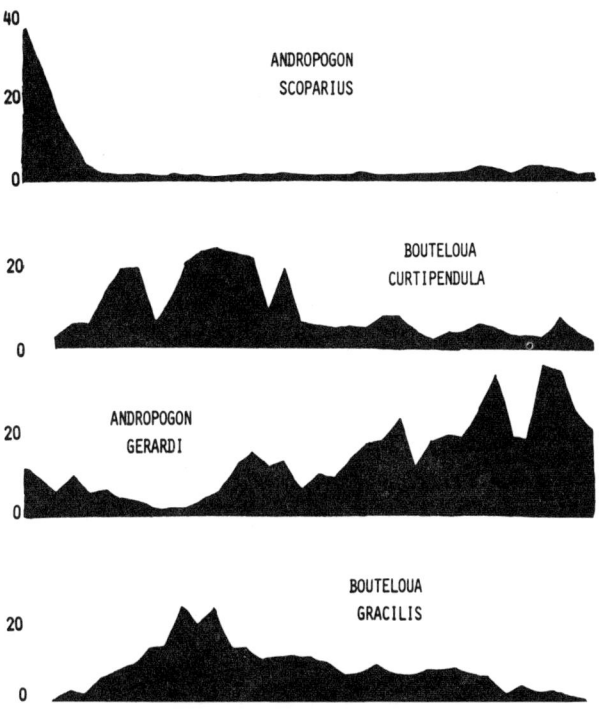

FIG. 8. Changes in cover of four grasses on the big bluestem (*Andropogon gerardi*)–little bluestem (*A. scoparius*) habitat from 1932 to 1967, near Hays, Kansas.

recovered rapidly and has provided a cover of 20 to 50 percent for the last 10 years. This community illustrates the rapid and extensive changes that can take place in the relative abundance of the dominant plants. For example, in 1932 it could aptly be called the blue grama–buffalo grass community, whereas in 1942 through 1948 it might have been called simply a buffalo-grass community; in the past 10 or 12 years it has been dominated solely by blue grama.

In the little-bluestem community there was only one dominant species in 1932, little bluestem (Fig. 7). However, the drought greatly reduced this species from a cover of 25 percent in 1932 to less than 0.5 percent in 1940. As precipitation and other environmental conditions improved, little bluestem gradually increased its cover, until in 1967 it was back up to 27 percent. Side-oats grama (*Bouteloua curtipendula*) and hairy grama (*B. hirsuta*) were rare in 1932, but came in during the drought years to fill the void left by the temporary disappearance of little bluestem. As the good years returned, side-oats grama diminished in importance, but hairy grama remained more important than it was in 1932. In the early 1940's the community might have been called a side-oats grama–hairy grama, rather than a little-bluestem, type.

The third community was dominated by big and little bluestem in 1932, but with the onslaught of drought their place was taken by side-oats grama and blue grama (Fig. 8).

Little bluestem did not start recovery until the early 1960's, after the second drought, but big bluestem made a gradual recovery, and in 1967 it was the dominant grass of the community. As big bluestem increased its cover, the two gramas decreased. Thus, over a period of 36 years, we have one community that has had three different combinations of dominant plants: big bluestem and little bluestem; side-oats grama and blue grama; and big bluestem alone.

Animal populations near Hays, Kansas, have not been analyzed in detail over a long period, as have the plants. However, Wooster (1935, 1939) made some observations that indicate trends in animal populations in relation to drought. He found that the prairie vole *(Microtus ochrogaster)* decreased in number to practically zero during the first two years of drought in the 1930's and did not come back until several years after the drought. On the other hand, the white-footed mouse *(Peromyscus maniculatus)* seemed to retain its abundance; and trapping records also indicated that other rodent species did not decrease in abundance to any marked degree. Gier (1967) conducted a statewide small-mammal census from 1951 to 1963. He found that the numbers of prairie voles and cotton rats *(Sigmodon hispidus)* decreased significantly during the drought of the 1950's, but that the number of white-footed mice seemed to be unaffected. Most small mammals were reduced in number by cold weather in January and by heavy snow in February or March. The black-tailed jackrabbit *(Lepus californicus melanotis)* increased from 158 per square mile in March, 1933, to 484 per square mile in March, 1935 (Wooster, 1935). Some birds, such as the dickcissel *(Spiza americana)*, disappeared from the Hays area during the drought. Although no measurements are recorded, the fish and amphibian populations must have been greatly reduced by the drying of ponds and streams.

We can summarize by pointing out that the environment, particularly the weather and, more specifically, drought conditions, greatly affect plant and animal populations in the Central Great Plains. The nature of the native grasslands of the area and the animals they support do vary with variable weather.

LITERATURE CITED

Albertson, F. W., G. W. Tomanek, and A. Riegel
  1957. Ecology of drought cycles and grazing intensity on grasslands of Central Great Plains. Ecol. Monogr., 27:27-44.
Albertson, F. W., and G. W. Tomanek
  1965. Vegetation changes during a 30-year period on grassland communities near Hays, Kansas. Ecology, 46:714-720.
Albertson, F. W., and J. E. Weaver
  1942. History of the native vegetation of western Kansas during seven years of continuous drought. Ecol. Monogr., 12:23-51.
  1944. Effects of drought, dust, and intensity of grazing on cover and yield of short-grass pastures. Ecol. Monogr., 14:1-29.
  1946. Reduction of ungrazed mixed prairie to short grass as a result of drought and dust. Ecol. Monogr., 16:449-463.
Gier, H. T.
  1967. Vertebrates of the Flint Hills. Trans. Kansas Acad. Sci., 70:51-59.
Tomanek, G. W.
  1948. Pasture types of western Kansas in relation to the intensity of utilization in past years. Fort Hays Kansas State Coll. Studies, Sci. Ser., no. 3, General Ser., no. 13:171-196.
Weaver, J. E., and F. W. Albertson
  1939. Major changes in grassland as a result of continued drought. Bot. Gaz., 100:576-591.
  1940. Deterioration of midwestern ranges. Ecology, 21:216-236.
  1943. Resurvey of grasses, forbs, and underground plant parts at the end of the great drought. Ecol. Monogr., 13:63-117.
  1944. Nature and degree of recovery of grassland from the great drought of 1933 to 1940. Ecol. Monogr., 14:393-479.
Wooster, L. D.
  1935. Notes on the effects of drought on animal population in western Kansas. Trans. Kansas Acad. Sci., 38:351-352.
  1939. The effects of drouth on rodent population. Turtox News, 17:26-27.

Historical Factors Controlling Vegetation Patterns and Floristic Distributions in the Central Plains Region of North America

PHILIP V. WELLS

ABSTRACT

Grasslands, even when narrowly defined as both treeless and shrubless, have a ubiquitous geography in North America. In addition to the well-known region of the Great Plains and the eastward salient of the Prairie Peninsula, which extended to Indiana with outliers in Kentucky, Ohio, and Michigan, there were extensive grasslands in regions of much higher precipitation and greater humidity, as along the Atlantic seaboard from Long Island south to Florida. The eastern grasslands are floristically, as well as physiognomically, similar to the tall-grass prairie of the Middle West, and are dominated by races of the same species of grasses. Throughout its wide range in the eastern half of North America, the tall-grass prairie has been an unstable vegetation since settlement put a stop to prairie fires. Unless burned or mowed, the grasses decline in vigor, and formerly treeless grasslands have been rapidly invaded or replaced by trees and shrubs. In like manner, the short-grass steppe vegetation of the western Great Plains occurs at a wide range of elevations in the Rocky Mountains as more or less extensive treeless "parks" scattered through the coniferous forest and woodland zones. Grasslands of similar physiognomy and species composition also occur in extensive, disjunct areas beyond the continental divide on the high plains of northern Arizona. Even under the more arid climates of the western plains, the short-grass prairie is also unstable in contact with woody vegetation; where seed sources are available, it is freely invaded by xerophytic junipers, mesquite, or sagebrush.

It is a notable fact that the dominant grasses, and most other species in the so-called grassland flora of the Central Plains region, have a major part of their diverse ranges as synusial components of vegetation dominated by woody plants. In brief, much of the flora occurring in the tall-grass prairie also occurs in open woodland habitats throughout the Middle West and eastward for various distances to the Atlantic seaboard; also, the flora of the short-grass plains ranges widely in open woodland throughout much of the southern Rockies and the Southwest. Hence, the flora of the Central Plains region may be regarded as a derivative one, recruited from herbaceous components of diverse surrounding vegetations dominated by woody plants. A good measure of the recency of derivation is the paucity of species endemic to the treeless grasslands of the plains. Virtually none of the endemics are grasses. This stands in striking contrast to the richly endemic floras of adjacent vegetations dominated by woody plants, which have many hundreds of endemic species of trees, shrubs, and herbs, including numerous endemic grasses.

Another significant attribute of the grasslands of North America is the abrupt nature of their boundaries with widely different woody vegetations. Topographic control is primary: under the widest range of climate, grasslands occupy the smooth topography, the flat plains, and gentle slopes; whereas woodlands or forests are usually restricted to the rough, dissected topography, the scarps, the abrupt breaks. The zone of coexistence of scarp-restricted woodland and grassland extends across the entire width of the Central Plains in the latitude of the Prairie Peninsula, from Indiana through Illinois, Iowa, and Nebraska to Wyoming, and across the Southern Plains in the latitude of the Edwards Plateau and the Llano Estacado. A rational explanation of these phenomena, which fits the known facts, is that scarp-restriction of woodland vegetation in the plains region ultimately originated as a consequence of the regional prairie fires. These vast conflagrations, now largely forgotten but well documented historically, formerly swept the tinder-dry grasslands on the smooth surface of the plains for great distances, until stopped by an abrupt topographic break. In view of these facts, the traditional assumption of climatic control of distribution for an extensive, physiognomically uniform formation (Clements) is basically unsound when applied to vegetation with a fire history.

INTRODUCTION

Before attempting to understand the history of the vegetation and climate of the Central Plains region it is essential to evaluate our dogmatic assumptions about the relation of existing vegetation to climate. The central dogma is the climatic determinism of treeless grasslands. Grasslands may be defined as herbaceous vegetation dominated by grasses. Presence of a tree synusia over the grass layer defines a savanna or open woodland. Codominance of woody shrubs and grasses may constitute a shrub-steppe, as in the various combinations involving species of *Artemisia*. The distributions of mixed vegetations dominated by woody plants often have been mapped as grassland on the strength of unproven assumptions that the woody component would be eliminated in competition with grasses, if the latter were not overgrazed. However, long-term exclosure studies in this type of "grassland" indicate that xerophytic woody plants are not eliminated, but tend to increase at the expense of the grasses, even in the absence of grazing, if fire is excluded.

DISTRIBUTION OF GRASSLANDS IN NORTH AMERICA

Grasslands, even when narrowly defined as both treeless and shrubless, have a ubiquitous geography in North America. In addition to the well-known region of the Great Plains and the eastward salient of the Prairie Peninsula, which extended to Indiana with outliers in Kentucky, Ohio, and Michigan (Transeau, 1935), there were extensive grasslands in regions of much higher precipitation and greater humidity. For example, consider the grasslands along the Gulf coast of Louisiana and Texas, the Jackson Prairie of Mississippi, parts of the Black Belt of Mississippi and Alabama, the Kissimmee Prairie of Florida, the Hempstead Plains, and Shinnecock and Montauk grasslands of Long Island. The eastern grasslands are floristically, as well as physiognomically, similar to the tall-grass prairie of the Middle West, and are dominated by races of the same species of grasses. For example, the grasslands of Long Island, New York, which have received much attention (Harper, 1911; Taylor, 1923; Blizzard, 1931; Conard, 1935; Cain et al., 1937), are dominated by the little bluestem, *Andropogon scoparius* Michx.; and other characteristic tall-grass species are abundant, including the big bluestem, *A. gerardi* Vitman, Indian grass, *Sorghastrum nutans* (L.) Nash, and switch grass, *Panicum virgatum* L.

The extensive prairie on the rolling moraine at Montauk, Long Island, is known from historical records to have been in existence at the time of settlement by the English during the 17th century. A pact for the grazing of livestock on the Montauk grassland was negotiated in 1658 by Lion Gardiner, acting for the Town of Easthampton, and Wyandanch, chief of the Montauk Indians. It is significant that a written agreement made in October, 1665, specified: "Indians not to set fire to the grass before the month of March, without consent of the town" (Taylor, 1923). Burning of the dry grass in the spring was still practiced at Montauk, and the tall-grass prairie, monographed by Taylor (1923), was still extensive and essentially shrub-free as recently as the 1940's, when I first became familiar with the vegetation of the area. However, postwar development caused a cessation of burning, and by 1968 most of the renowned Montauk grassland had already been invaded and suppressed by dense thickets of bayberry (*Myrica pensylvanica* Loisel.) and winged sumac (*Rhus copallina* L.). Only patches of prairie, small and fragmented, persist to mark out the former extent of grassland, and the shrubs are encroaching upon even these pitiful remnants. Meanwhile, invasive arboreal species such as black cherry (*Prunus serotina* Ehrh.), which were formerly confined to the edge of the fire-restricted forest on the crest of the moraine and in the kettles, are now colonizing the newly established shrublands.

Throughout its wide range in the eastern half of North America, the tall-grass prairie has been an unstable vegetation since settlement put a stop to prairie fires, even where it has been spared the destructive effects of severe overgrazing. Unless burned or mowed, the grasses decline in vigor, and the formerly treeless grasslands have been more or less rapidly invaded or replaced by trees and shrubs, as on Long Island (Blizzard, 1931), in Kentucky (Shull, 1921), Wisconsin (Curtis, 1959), Iowa (McComb and Loomis, 1944), Missouri (Beilmann and Brenner, 1951), central Texas (Bray, 1906; Foster, 1917), and eastern Kansas (Wells, unpublished data).

In like manner, the short-grass steppe vegetation of the western Great Plains occurs at a wide range of elevations in the Rocky Mountains as more or less extensive treeless "parks" scattered through the coniferous forest and woodland zones. Grasslands of similar physiognomy and species composition also occur in extensive, disjunct areas beyond the continental divide on the high plains of northern Arizona (Nichol, 1952). Even under the more arid climates of the western plains, the short-grass prairie is also unstable in contact with woody vegetation; where seed sources are available, it is freely invaded by xerophytic junipers, mesquite, or sagebrush (Glendening, 1952; Humphrey, 1958; Johnsen, 1962). The grasses are usually not eliminated (unless severely overgrazed), but coexist as a lower synusia under or between the woody plants in a stable relationship that may be closer to climax.

Also, the Palouse bunch-grass prairie dominated by *Agropyron spicatum* (Pursh) Scribn. and *Festuca idahoensis* Elmer, which extends from the loess-mantled eastern sector of the Columbia Plateau eastward into the Snake Plains and across the Northern Rockies to Montana, enters into a stable synusial relationship with the ubiquitous sagebrush (chiefly *Artemisia tridentata* Nutt.). The *Stipa*-dominated bunch-grass prairies, or oak and bunch-grass savannas, of California are similar to the Palouse Prairie in growing under a winter-rain and summer-drought climatic regime radically different from that of the grasslands of the Central Plains region of North America. The anomalous rain-drenched prairies of the Northwest coast, some existing under an annual precipitation of more than 80 inches, are surrounded by magnificent coniferous rain forest.

High-elevation, montane grasslands or "parks" in the coniferous forests of the Cordilleras, often dominated by bunch-grasses of the genus *Festuca*, have a counterpart in the Northern Great Plains, where bunch-grass prairie dominated by *Festuca scabrella* Torr. occupies the relatively cool climatic zone of the aspen parkland bordering the boreal forest in the Prairie Provinces of Canada.

Inasmuch as these heterogeneous assortments of herbaceous vegetation are usually dominated throughout by grasses, it is natural to assemble them under the physiognomic rubric "grassland." But what is meant by the concept grassland climate? No other physiognomic class of vegetation covers such extensive areas under such a wide array of climatic regimes as did the grasslands, with single-family dominance. In fact, the only category of vegetation with a comparably wide distribution is the heterogeneous one of fresh-water marshes, dominated by sedges, rushes, and grasses. Of course, one does not speak of a "marsh climate" even for formations of regional extent, as for example, the Everglades of Florida, where fire probably also played a role.

DERIVATION OF THE FLORA OF THE CENTRAL PLAINS

It is a notable fact that the dominant grasses, and many of the other species in the so-called grassland flora of the Central Plains region, have a major part of their diverse ranges as synusial components of vegetation dominated by woody plants. For example, the flora occurring in the tall-grass prairie ranges through the eastern deciduous forest for various distances toward the Atlantic coast, not only as outliers of treeless grassland, but also as part of the herbaceous understory of open woodland, as in oak, cedar, or pine barrens. The open habitat may be a transient seral stage generated by recurrent fire, or more permanent, as on bedrock outcrops or cliffs where trees do not form a closed canopy.

Within the relatively small and mostly forested area of North and South Carolina, a large number of typical "prairie" species are native. In fact, most of the principal dominants of the tall-grass prairie of the eastern half of the plains are present in open habitats in the Carolinas. A comparison of species diversity in genera common to the Carolinas and the grasslands of the plains shows a consistent pattern of relatively few species per genus in the vastly more extensive Central Plains region (Table 1). The exceptions are mainly in genera with major development west or south of the plains, as in *Artemisia, Haplopappus, Hymenopappus,* and *Petalostemon*. This pattern of low species diversity in the plains is especially anomalous in the grasses, the dominant life-form.

The genera that include most of the characteristic species of the tall-grass prairie have major concentrations of species not in the treeless prairie of the plains, but in the wooded regions of the Middle West and Southeast. This is true of *Andropogon, Panicum, Sorghastrum, Sporobolus, Sphenopholis, Aster, Echinacea, Eryngium, Helianthus, Hypoxis, Liatris, Silphium,* and *Solidago*. The grass genera *Andropogon* and *Panicum* are especially rich in species in the Southeast, from North Carolina south to Florida. Three of the species occurring in the East, *Andropogon scoparius, A. gerardi,* and *Panicum virgatum,* range westward to the Rocky Mountains, contributing heavily to the treeless grassland vegetation of the eastern plains. The important point is that throughout much of the predominantly wooded region of eastern North America there is not only sufficient open habitat to accommodate many dominant and minor species of the "grassland" flora of the plains, but the same heliophile genera show a greater diversity of species in the grassy "fields," "meadows," savannas, or woodland openings of the East than in the "prairies" of the plains. Furthermore, the flora of the grasslands, savannas, or openings along the Atlantic seaboard, and westward around the Gulf to eastern Texas, has a large endemic component that is not shared with the grasslands of the plains at the generic level. There is a clear implication that the direction of migration of the "grassland" flora of the eastern plains, since the end of the Wisconsin glacial about

10,000 years ago, has been mainly northward and westward from ancestral centers in southeastern North America.

Similarly, the flora of the short-grass steppe ranges widely throughout much of the southern Rockies and the Southwest as a synusial component of open ponderosa pine forest, pinyon-juniper or oak woodland, and desert shrubland. The rich herbaceous flora of the Southern Plains has a center of diversity not in the treeless grasslands, but in the open oak or thorn woodland and shrubland extending from the Edwards Plateau south through Tamaulipas and Nueva Leon and westward into the Chihuahuan Desert. The two large genera of grasses that have dominant species in the short-grass steppe, namely *Bouteloua* and *Hilaria,* have a notable concentration of species in the Chihuahuan Desert and in the shrubby desert grassland. Only a few species extend northward on the steppes of the Great Plains, for example the blue grama, *Bouteloua gracilis* (HBK) Lag.

Hence, the flora of the Central Plains region may be regarded as a derivative one, recruited from herbaceous components of diverse surrounding vegetations dominated by woody plants. A good measure of the recency of derivation is the relative paucity of species endemic to the treeless grasslands of the plains. It is most anomalous that virtually none of the endemics are grasses, even though 41 genera and 112 species of this dominant life-form occur natively in the grasslands (Rydberg, 1932). By way of contrast, there are 62 genera and 244 species of grasses indigenous to the relatively small area of North and South Carolina (Radford *et al.,* 1968), and some of the genera and a great many of the species are endemic to the Southeast.

The flora of the grasslands in the Central Plains region is actually much smaller than is indicated by the 177 families, 1066 genera, and 3988 species described in the classic *Flora of the Prairies and Plains of Central North America* (Rydberg, 1932). Since Rydberg treated an area delineated by political boundaries from North Dakota south to Kansas, and parts of adjacent states, a number of natural vegetational boundaries were transgressed. As a consequence, he included generous samples of the floras restricted to the eastern deciduous forest, and the boreal and Rocky Mountain coniferous forests, as Fernald (1932) was quick to notice. Despite the fact that Fernald failed to establish the floristic composition of the physiognomic classes of vegetation within the range of Gray's Manual, he criticized Rydberg for failing to define explicitly the fraction of his "grassland" flora actually occurring in grassland vegetation. If the many taxa restricted to forest, aquatic, or other nongrassland communities are disregarded, and the finely segregated taxonomy at the family and generic level is brought into alignment with current treatment, the size of the "grassland" flora is drastically reduced.

A careful compilation, which accepts most of Rydberg's species and distributional data at face value, yields a total of only 57 families, 296 genera, and 954 species occurring naturally in grassland vegetation on the plains. None of the families, few, if any, genera, and remarkably few species (about five percent) are endemic to the grasslands of the Central Plains. Considering the great geographic extent of the grasslands treated by Rydberg, the relatively small size of the flora and the weak development of both wide and narrow endemism are most anomalous. There is a striking contrast in the large and richly endemic floras of adjacent vegetational provinces dominated by woody plants, which have some endemic families, numerous endemic genera, and hundreds of widely or narrowly endemic species of trees, shrubs, and herbs, including many endemic grasses. For example, the local flora indigenous to the Great Smoky Mountains comprises 117 families, 438 genera, and 1111 species, many of which are endemic to the Southeast (Hoffman, 1964).

There are comparatively few genera with large representations of species in the plains region, and among the grasses, only *Aristida* has more than 10 species in the plains (Table 2). The largest genus, *Astragalus*, has been thoroughly revised by Barneby (1964),

and it is instructive to examine its distribution in detail. *Astragalus* (Leguminosae) is one of the largest genera of vascular plants in North America. However, of the 368 species, only 31, or less than 10 percent, are known from the Great Plains and vicinity. There are 25 species that range more or less widely in the plains, but 22 of

TABLE 1. Regional Differences in Species Diversity in Genera Ranging from the Carolinas to the Grasslands of the Central Plains*

Genus	'Prairie' Species Native in Carolinas	Species per Genus in Carolinas	Species per Genus in Grasslands of Central Plains	Genus	'Prairie' Species Native in Carolinas	Species per Genus in Carolinas	Species per Genus in Grasslands of Central Plains
GRASSES				**COMPOSITAE**			
Andropogon	A. scoparius Michx.	7	4	Parthenium	P. integrifolium L.	1	2
	A. gerardi Vitm.			Echinacea	E. purpurea (L.) Moench	3	3
Sorghastrum	S. nutans (L.) Nash	3	1		E. pallida Nutt.		
Panicum	P. virgatum L.	62	6				
Sphenopholis	S. obtusata (Michx.) Scribn.	5	1	Coreopsis	C. lanceolata L.	11	5
				Hymenopappus	H. scabiosaeus L'Her.	1	5
Sporobolus	S. heterolepis Gray	7	6	Gaillardia		1	5
Aristida	A. oligantha Michx.	12	13	Artemisia	A. caudata Michx.	1	15
	A. curtissii (Gray) Nash			Cacalia		4	1
	A. purpurascens Poir.				**OTHER HERBS**		
Elymus	E. canadensis L.	4	5	Tradescantia	T. ohiensis Raf.	6	2
	E. virginicus L.			Hypoxis	H. hirsuta (L.) Cov.	5	1
Calamovilfa		1	2	Anemone	A. caroliniana Walt.	4	5
Spartina	S. pectinata Link	4	2	Crotalaria	C. sagittalis L.	6	1
Tripsacum	T. dactyloides L.	1	1	Psoralea		5	14
COMPOSITAE				Petalostemon		1	10
Liatris	L. squarrosa (L.) Michx.	14	7	Lespedeza	L. capitata Michx.	10	3
	L. aspera Michx.			Euphorbia	E. corollata L.	12	18
Kuhnia	K. eupatorioides L.	1	1		E. dentata Michx.		
Eupatorium	E. altissimum L.	24	2	Viola	V. pedata L.	26	4
Solidago	S. rigida L.	39	15	Eryngium	E. yuccifolium Michx.	6	2
	S. altissima L.			Zizia	Z. aptera (Gray) Fern.	3	2
	S. nemoralis Ait.				Z. aurea (L.) Koch		
	S. graminifolia (L.) Salisb.			Sabatia	S. campestris Nutt.	11	1
Haplopappus	H. divaricatus (Nutt.) Gray	1	10	Gentiana	G. quinquefolia L.	9	3
Aster	A. oblongifolius Nutt.	40	14	Asclepias	A. viridiflora Raf.	19	19
	A. patens Ait.				A. tuberosa L.		
Erigeron	E. strigosus Muhl.	6	7		A. verticillata L.		
Silphium	S. terebinthinaceum Jacq.	7	5	Phlox	P. pilosa L.	13	7
	S. perfoliatum L.			Lithospermum	L. canescens (Michx.) Lehm.	4	6
Helianthus		16	9	Salvia	S. azurea Lam.	4	2
Berlandiera		1	2	Monarda	M. fistulosa L.	5	5
					M. citriodora Cerv.		
				Pycnanthemum	P. virginianum (L.) D. & J.	8	2

* Data from Radford *et al.* (1968) and Rydberg (1932).

TABLE 2. Extralimital Versus Restricted Distributions in Genera Represented in the Grasslands of the Central Plains of North America by 10 or More Species*

Genus	Species Ranging Widely in and Beyond the Plains	Species Restricted to the Plains or Vicinity
Astragalus	22	9
Oenothera (s.l.)	18	9
Asclepias	17	2
Euphorbia	13	5
Aster	12	4
Artemisia	12	3
Solidago	11	4
Psoralea	7	7
Aristida	13	0
Carex (dry-land)	11	2
Eriogonum	11	2
Potentilla	10	3
Oxytropis	10	1
Gilia	10	0
Haplopappus	8	2
Petalostemon	5	5

* Based mainly on the taxonomic and geographic data in Rydberg (1932), except for *Astragalus* (Barneby, 1964). Rydberg's taxonomy requires extensive revision; most pertinently, many of the species with restricted ranges are weakly defined.

these species also range far beyond its borders in nongrassland vegetation, some as far as California and Alaska. Of the nine species with ranges restricted to the vicinity of the plains, only *A. plattensis* Nutt. has a range more or less congruent with the Great Plains region, and can be regarded as a wide endemic of the grassland province. The remaining eight species are either narrowly endemic in the foothills of the Rocky Mountains or on the adjacent edge of the plains, or range widely in both (Table 3).

A compilation of geographic distributions of a major part of the flora of Kansas, representative of the Central Plains region (Bare, 1968), uncovered no endemics, and indeed, there appear to be no vascular plants endemic to the state. A total of four species of flowering herbs occurring in Kansas are endemic to the Southern Plains, mainly as edaphic specialists on calcareous substrata. Edaphic diversity plays a leading role as an isolating mechanism in the evolution or subsequent accommodation of narrowly endemic species of plants (Mason, 1964). But contrary to popular opinion, the plains are by no means lacking in edaphic diversity. A mosaic of heterogeneous substrata exists; for example, sand deposits, loess, shale, gypsum, limestone, sandstone, basalt, granite, and saline or alkaline, wet or dry soils (Wells, 1965). Furthermore, the stereotyped conception of physiographic uniformity breaks down, because the continuity of the plains is interrupted at long intervals by numerous escarpments, badlands, sand hills, and playas or "buffalo wallows" that offer beautifully isolated, diversified habitats. Nevertheless, the considerable potential for the evolution of narrow endemics has not been realized in the existing flora of the Central Plains region.

TABLE 3. Distribution Patterns of Species of *Astragalus* in Grasslands of the Central Plains of North America*

RANGING WIDELY BEYOND THE PLAINS REGION	
Reaching Western Plains Only	Ranging Widely Across Plains
A. kentrophyta Gray	A. canadensis L.
A. nuttallianus DC.	A. adsurgens Pall.
A. ceramicus Sheld.	A. lotiflorus Hook.
A. spatulatus Sheld.	A. missouriensis Nutt.
A. drummondii Hook.	A. tenellus Pursh
A. aboriginum Richards	A. bisulcatus (Hook.) Gray
A. vexilliflexus Sheld.	A. flexuosus (Hook.) Don
A. bodini Sheld.	
A. mollissimus Torr.	A. agrestis Dougl.
A. purshii Dougl.	A. racemosus Pursh
A. praelongus Sheld.	A. crassicarpus Nutt.
A. gilviflorus Sheld.	

ENDEMIC TO EAST SLOPE OF THE ROCKY MOUNTAINS OR THE PLAINS	
Endemic to Foothills of Rocky Mountains	Wide in Foothills and Western Plains
A. shortianus Nutt.	A. gracilis Nutt.
A. puniceus Osterh.	A. pectinatus Dougl.
A. tridactylicus Gray	
Endemic to West Edge of Plains	Wide Endemic of Plains
A. barrii Barneby	A. plattensis Nutt.
A. sericoleucus Gray	
A. hyalinus Jones	

* Data from Barneby (1964).

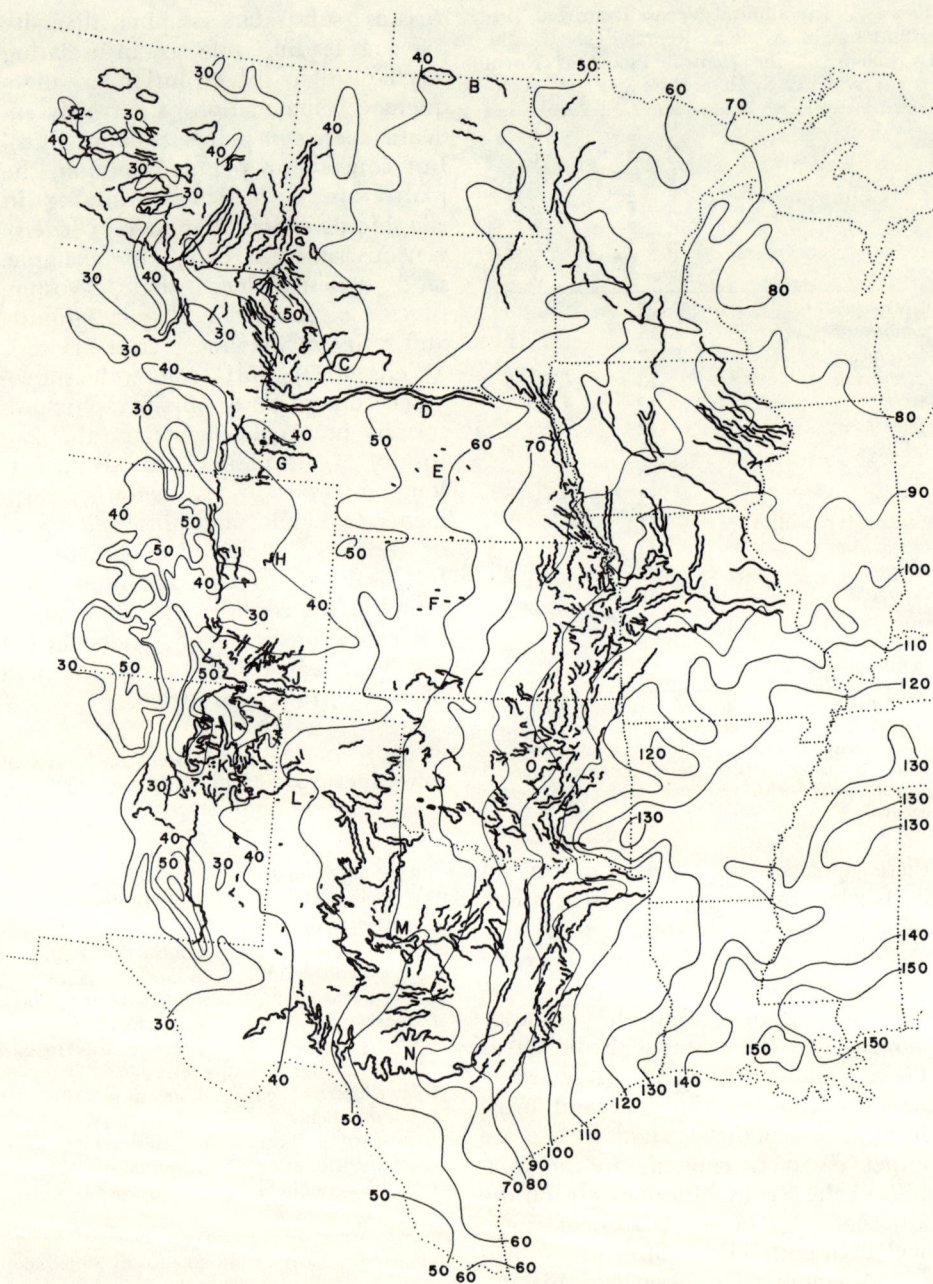

FIG. 1. Provisional mapping of some nonriparian woodlands in the grassland province of the Central Plains region of the United States. Heavy lines indicate location of scarps or other rough, broken, or steep topography with indigenous woodland vegetation. Thin, continuous lines are isohyets at 10-centimeter intervals, ranging from 30 to 150 centimeters of mean annual precipitation. Dotted lines indicate state or other boundaries. Some dominant species of nonriparian vegetation from selected areas indicated by letters from A to O on map are as follows:

   A. Montana, eastern Wyoming, and western Dakotas. Woodland: *Pinus ponderosa* Laws., *Juniperus scopulorum* Sarg. Grassland: *Bouteloua gracilis* HBK, *Agropyron smithii* Rydb., *Stipa comata* Trin. and Rupr.

In essence, the grasslands of the plains are defined by the overwhelming dominance of several widespread species of grasses, in particular, members of the relatively subtropical tribes Chlorideae and Andropogoneae. It has been shown that some of the widest-ranging species in the plains, *Bouteloua curtipendula* (Michx.) Torr. and *Andropogon scoparius* Michx., vary geographically in genetically determined physiological requirements, indicating a clinal differentiation into ecotypic races (Olmsted, 1944; McMillan, 1956). However, evolution on this scale is probably rapid; the idea of "catastrophic" evolution in plants described by Lewis (1962) seems applicable (Wells, 1969). The deficiency of endemism in the "grassland" flora in general, and among the grasses in particular, does not suggest a long evolutionary history in the Central Plains region for most of the existing genera. Although some groups of herbaceous plants could have evolved in the Great Plains, as Elias (1942) has supposed for certain borages and Stipeae, it is not necessary to infer that regional treelessness was required for their evolution.

The relative shortage of endemics in the grassland vegetation is a significant fact, inasmuch as it bears on the recency of tenure of grasslands in the Central Plains region. It is also true that the present extensive distribution of desert vegetation in North America is a relatively recent development. Tertiary leaf-impression floras indicate that the present distribution of deserts is younger than Upper Pliocene, probably Pleistocene (Axelrod, 1950, 1958). Nevertheless, there is a rich development of endemism in the North American desert floras; some of the families, many of the genera, and hundreds of species are endemic to the deserts. For example, the endemism rate in the Sonoran Desert is 30 percent. This implies a long history of evolution and extinction, extending back to the less arid regional Tertiary climates, per-

---

B. Pembina escarpment and Turtle Hills, North Dakota; Erskine and Big Stone moraines, Minnesota. Woodland: *Populus tremuloides* Michx., *Quercus macrocarpa* Michx. Grassland: *Andropogon scoparius* Michx., *A. gerardi* Vitman, *Agropyron trachycaulum* (Link) Malte.

C. Pine Ridge Escarpment, South Dakota, Nebraska, Wyoming. Vegetation similar to *A*.

D. Niobrara escarpments, Nebraska. Woodland: *P. ponderosa, J. scopulorum, J. virginiana* L., *Q. macrocarpa.* Grassland: *B. gracilis, S. comata, A. scoparius.*

E. Sandhills area, Nebraska. Woodland: isolated stands of *P. ponderosa, J. virginiana, Celtis occidentalis* L. Grassland: *Andropogon hallii* Hack., *A. scoparius, Calamovilfa longifolia* (Hook.) Scribn.

F. Western Kansas. Woodland: isolated stands of *J. virginiana*. Grassland: *A. scoparius, B. gracilis, Buchloe dactyloides* (Nutt.) Engelm.

G. Wildcat Hills, Nebraska. Vegetation similar to *A*.

H. Cedar Point, Colorado. Woodland: *P. ponderosa, J. scopulorum*. Grassland: *B. gracilis, Buchloe dactyloides.*

J. Black Mesa-Mesa de Maya area, Oklahoma, New Mexico, Colorado. Woodland: *Pinus edulis* Engelm., *P. ponderosa, Juniperus monosperma* (Engelm.) Sarg., *J. scopulorum, Quercus undulata* Torr. Grassland: *B. gracilis, Buchloe dactyloides.*

K. Canadian Escarpment, New Mexico. Vegetation similar to *J*.

L. Llano Estacado, New Mexico, Texas. Woodland: on northwest, *P. edulis, J. monosperma, Q. undulata;* on east (Break of the Plains), *Juniperus pinchotii* Sudw., *Quercus mohriana* Buckl. Grassland: *B. gracilis, Buchloe dactyloides.*

M. Callahan Divide, Texas. Woodland: *Quercus virginiana* Mill., *Q. shumardii* Buckl., *Q. mohriana, Juniperus ashei* Buchholz, *J. pinchotii.* Grassland: *B. gracilis, Buchloe dactyloides.*

N. Edwards Plateau, Texas. Woodland: Similar to *M*. Grassland: *A. scoparius, Bouteloua curtipendula* (Michx.) Torr.

O. Oklahoma, eastern Kansas. Woodland: *Quercus stellata* Wang., *Q. marilandica* Muenchh., *Q. muehlenbergii* Engelm., *Q. shumardii, J. virginiana*. Grassland: *A. gerardi, A. scoparius, Sorghastrum nutans* (L.) Nash, *Panicum virgatum* L.

haps in localized pockets of aridity in the rain shadow of high mountains. If a regional expansion of the richly endemic desert floras is a recent development, what can be said about the history of the almost entirely derivative flora of the grasslands of the Central Plains region?

TOPOGRAPHIC CONTROL OF VEGETATION PATTERNS IN THE PLAINS

Another significant attribute of the grasslands of North America is the abrupt nature of their boundaries with widely different woody vegetations. Because the main features of this phenomenon have been outlined elsewhere (Wells, 1965 and this volume), I will summarize by saying that topographic control is primary. Under the widest range of climate, grasslands occupy the smooth topography, the flat plains and gentle slopes. Woodlands or forests coexisting on the uplands with grassland are usually restricted to the vicinity of rough, dissected topography, the bolder scarps, the abrupt breaks (Fig. 1).

The zone of segregated coexistence of nonriparian, scarp-restricted woodland and grassland is not limited to a narrow ecotonal transition between regional forest and regional grassland. It extends across the entire width of the Central Plains in the latitude of the Prairie Peninsula, from Indiana through Illinois, Iowa, and Nebraska to Wyoming. In accordance with the east-west climatic gradient of decreasing precipitation and humidity, there is a shift from broad-leaf deciduous forest to open, xerophilous, coniferous woodland dominated by *Pinus ponderosa* Laws. and *Juniperus scopulorum* Sarg. along the scarps in central Nebraska. Similarly, the grasses on the adjacent plains decrease in stature and density westward, as dominance shifts from the tall Andropogoneae in the east to the dwarf Chlorideae in the west. Nevertheless, the abrupt topographic segregation of woodland and grassland prevails throughout the gradient in vegetation and climate.

Numerous scarp woodlands interrupt the flat monotony of the arid western part of the plains, from Montana and North Dakota south to the Llano Estacado of western Texas, and from there eastward across the Southern Plains via the Callahan Divide to the Edwards Plateau (Fig. 1). These remarkable distributions of woodland throughout the plains region create internal ecotonal boundaries between upland growths of woodland and grassland even in the most arid sectors of the plains. Not only treeless grasslands but also nonriparian woodlands grow under a wide range of precipitation and temperature regimes throughout the extensive plains region, and some species of juniper—*Juniperus scopulorum* Sarg., *J. monosperma* (Engelm.) Sarg., *J. pinchotii* Sudw.—inhabiting the scarp woodlands of the plains also penetrate the margins of the winter-cold sagebrush deserts or the hot creosote-bush deserts, under climates more arid and fluctuating than in any part of the plains. It is indeed doubtful, therefore, that the distribution of treeless grasslands is guided solely by "the master hand of climate," as assumed from circumstantial evidence by Borchert (1950). Circumstantial evidence also suggests that "the master hand" of physiography, as well as other factors, including fire and climate, are involved.

ACKNOWLEDGMENT

The research on which this paper is based was supported in part by a grant from the National Science Foundation.

LITERATURE CITED

Axelrod, D. I.
1950. Evolution of desert vegetation in western North America. Publ. Carnegie Inst. Washington, 590:215-306.
1958. Evolution of the Madro-Tertiary geoflora. Bot. Rev., 24:431-509.

Bare, J.
1968. An introduction to the phytogeography of Kansas. Unpubl. Ph.D. thesis, Univ. Kansas.

Barneby, R. C.
1964. Atlas of North American Astragalus. Mem. New York Bot. Garden, 13:1-1188.

Beilmann, A. P., and L. G. Brenner
1951. The recent intrusion of forests in the Ozarks. Ann. Missouri Bot. Gardens, 38: 261-282.

Blizzard, A. W.
1931. Plant sociology and vegetational change on High Hill, Long Island, New York. Ecology, 12:208-231.

Borchert, J. R.
1950. The climate of the central North American grassland. Ann. Assoc. Amer. Geogr., 40:1-39.

Bray, W. L.
1906. Distribution and adaptation of the vegetation of Texas. Bull. Univ. Texas, 82, Sci. Ser. 10, Austin, 108 pp.

Cain, S. A., M. Nelson, and W. McLean
1937. Adropogonetum Hempsteadi: a Long Island grassland vegetation type. Amer. Midland Nat., 18:334-350.

Conard, H. S.
1935. The plant associations of central Long Island. Amer. Midland Nat., 16:433-516.

Curtis, J. T.
1959. The vegetation of Wisconsin. Univ. Wisconsin Press, Madison, 657 pp.

Elias, M. K.
1942. Tertiary prairie grasses and other herbs from the High Plains. Geol. Soc. Amer. Spec. Paper, 41:1-176.

Fernald, M. L.
1932. Rydberg's *Flora of the Prairies and Plains*. Rhodora, 34:243-247.

Foster, J. H.
1917. The spread of timbered areas in central Texas. Jour. Forestry, 15:442-445.

Glendening, G. E.
1952. Some quantitative data on the increase of mesquite and cactus on a desert grassland range in southern Arizona. Ecology, 33:319-328.

Harper, R. M.
1911. The Hempstead Plains: a natural prairie on Long Island. Bull. Amer. Geogr. Soc., 43:351-360.

Hoffman, H. L.
1964. Check list of vascular plants of the Great Smoky Mountains. Castanea, 29: 1-45.

Humphrey, R. R.
1958. The desert grassland: a history of vegetational change and an analysis of causes. Bot. Rev., 24:193-252.

Johnsen, T. N.
1962. One-seed juniper invasion of northern Arizona grasslands. Ecol. Monogr., 32: 187-207.

Lewis, H.
1962. Catastrophic selection as a factor in speciation. Evolution, 16:257-271.

Mason, H. L.
1946. The edaphic factor in narrow endemism. Madrono, 8:209-226, 241-257.

McComb, A. L., and W. E. Loomis
1944. Subclimax prairie. Bull. Torrey Bot. Club, 71:46-76.

McMillan, C.
1956. Nature of the plant community II. Variation in flowering behavior within populations of Andropogon scoparius. Amer. Jour. Bot., 43:429-436.

Nichol, A. A.
1952. The natural vegetation of Arizona. Univ. Arizona Agric. Expt. Sta. Tech. Bull., 127:187-230.

Olmsted, C. E.
1944. Growth and development in range grasses IV. Photoperiodic responses in twelve geographic strains of side-oats grama. Bot. Gaz., 106:46-74.

Radford, A. E., H. E. Ahles, and C. R. Bell
1968. Manual of the vascular flora of the Carolinas. Univ. North Carolina, Chapel Hill, 1181 pp.

Rydberg, P. A.
1932. Flora of the prairies and plains of central North America. New York Bot. Garden, New York, 969 pp.

Shull, C. A.
1921. Some changes in the vegetation of western Kentucky. Ecology, 2:120-124.

Taylor, N.
1923. The vegetation of Long Island I. The vegetation of Montauk: a study of grassland and forest. Mem. Brooklyn Bot. Garden, 2:7-107.

Thwaites, R. G. (ed.)
1905. Original journals of the Lewis and Clark expedition. Dodd, Mead & Co., New York, 7 vols.

Transeau, E. N.
1935. The Prairie Peninsula. Ecology, 16: 423-437.

Wells, P. V.
1965. Scarp woodlands, transported grassland soils, and concept of grassland climate in the Great Plains region. Science, 148:246-249.
1969. Discussion. Ecological aspects of the systematics of plants. Pp. 206-211, *in* Systematic biology, Proc. Int. Conf. Syst. Biol., Nat. Acad. Sci., Washington, D.C.

ZOOLOGY

The Ecological History of the Great Plains: Evidence from Grassland Insects

HERBERT H. ROSS

ABSTRACT

The total distribution of dominant grasses of the Great Plains prairie biome suggests that the various distinctive types of the biome are arrested subclimax communities similar to those occurring in adjacent forest biomes. Evidence from two groups of insects closely associated with grasses, the grasshoppers and the lataline leafhoppers, give additional support to this concept.

Of 108 grasshopper species occurring extensively on the prairie, 105 occur widely also to the north, south, east, or west. Only three are prairie endemics. These distributions match those of the grasses, and demonstrate the ecological affinity of the prairie grasshoppers with surrounding biomes. These numerical data do not indicate, however, the direction of ecological evolution, that is, whether the species originated in the prairie biome or the subclimax grass communities.

The lataline leafhoppers offer two types of evidence bearing on the origin of the prairie fauna. Subclimax grass communities in forest biomes have a greater lataline fauna than comparable and neighboring grass communities in the prairie biome, supporting the "arrested subclimax" theory. As is the case among the grasshoppers, few lataline leafhoppers are prairie endemics.

Each of the lataline genera *Diplocolenus* and *Rosenus* has one species occurring in the prairie and also in neighboring subclimax communities, and each has several other species occurring only in subclimax communities to the north and west of the prairie. The phylogeny of the two genera indicates that they evolved in subclimax grass communities of coniferous forests, and that subsequently one member of each became able to live also in the prairie.

Fossil evidence indicates that during the Pleistocene maxima the prairie was reduced to a remarkably small area and may have been essentially a savanna having isolated areas of grass. The grasshopper and leafhopper species discussed would have been able to persist in these areas as they do now in similar subclimax habitats.

INTRODUCTION

To the traveler and the ecologist, the prairie biome, which constitutes the primary vegetation clothing the Great Central Plains of North America, is a unique and remarkable phenomenon. As an ecological unit of existing vegetation it is naturally compared with the steppes of Eurasia and the pampas of South America. A closer scrutiny shows that the North American grassland biome of the Great Plains comprises three main types (Fig. 1): (1) a tall-grass eastern strip, characterized by tall dominant grasses of which *Andropogon* is the most conspicuous; (2) a central and western area variously called the short- or mixed-grass prairie, characterized by a mixture of the moderately tall genera *Stipa* and *Agropyron* and short grasses, chiefly *Bouteloua* and *Buchloe;* and (3) a northern strip of bunch-grass prairie, characterized by *Festuca* and *Agropyron* (Evers, 1955; Coupland, 1950; Moss and Campbell, 1947). These three types or associations of grass communities intergrade and interfinger in such a way that one ap-

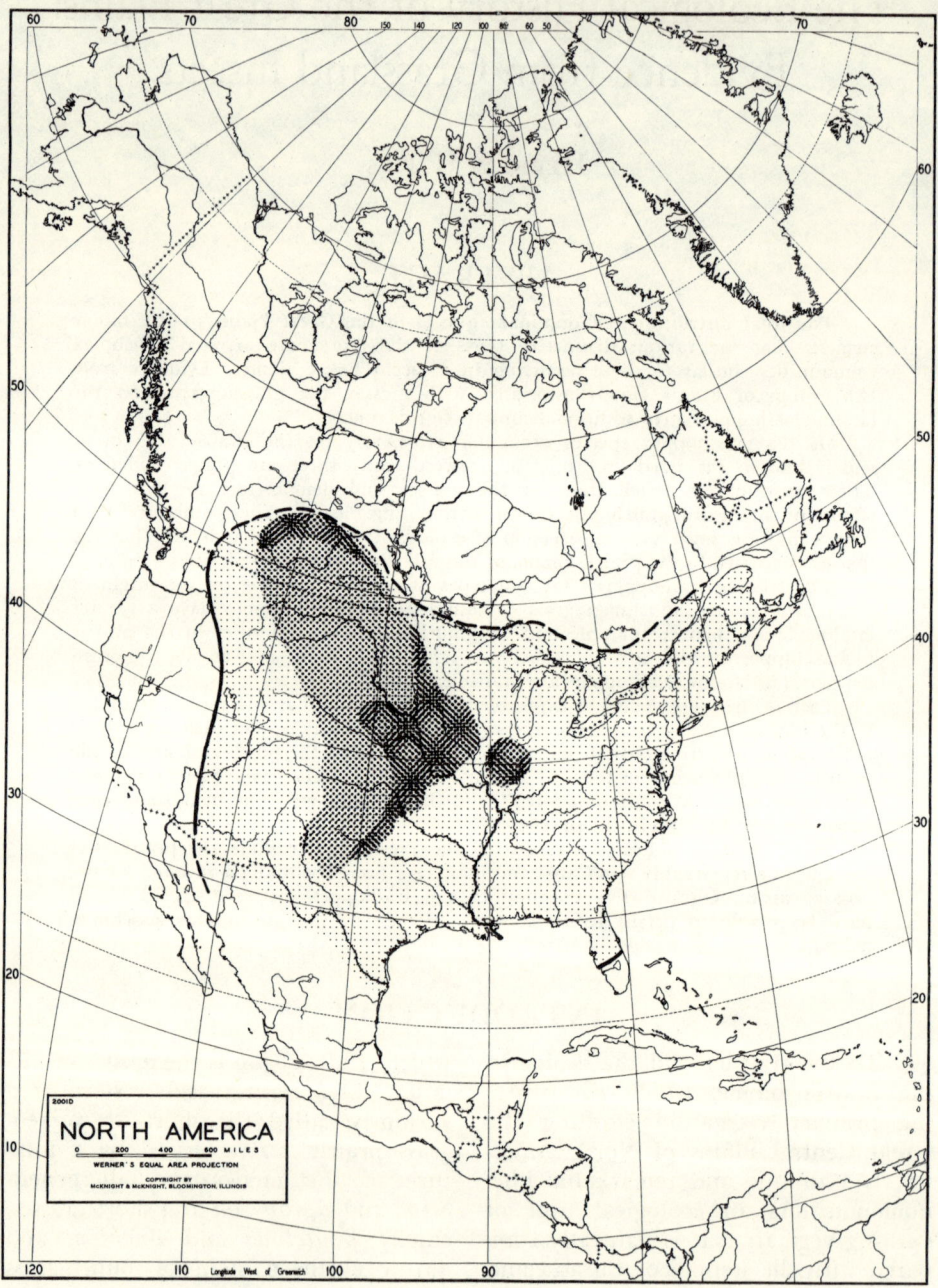

Fig. 1. Map of prairie biome of North America. The two heavier patterns of stippling indicate the prairie biome; the lighter one, the mixed-grass prairie; the northern heavier one, the northern fescue prairie; and the eastern heavier one, the tall-grass prairie. The lightest stipple shows the range of *Andropogon gerardii* and *A. scoparius*.

pears to merge imperceptibly with another. Although much of the Central Plains area is now under cultivation, enough remnants of the original prairies remain so that one can visualize what a unique feature this thousand-mile expanse of grasses must originally have been.

If one looks at the total range of each dominant grass in this prairie biome, the uniqueness of the prairie as a vegetative unit begins to disappear. This is especially striking in the case of the more eastern tall-grass prairie. Two common dominants in this prairie are *Andropogon gerardii* and *A. scoparius*. Figure 1 illustrates the approximate areas of the prairie biome in which both are important dominants and also the remarkably large total known area in which both species occur, extending far beyond the prairie biome as a whole (Hitchcock and Chase, 1950). Under what circumstances do these dominants grow outside the tall-grass prairie? To the west of the area they are associated with seeps and gullies, local spots or slopes enjoying more moisture than is typical for the area as a whole. To the east of the prairie area, these species are the dominant grasses in certain seral stages that are part of the normal succession leading from a denuded area to a climax forest. Here, then, we have two species of grasses that are dominants both in the climax community of a prairie biome and also in a subclimax community of a forest biome (Ross, 1962, p. 266). Although the seral stages of the semiscrub desert of the Southwest, the various montane forests to the west of the Great Plains, and the boreal forests in the north and northwest of the Great Plains are poorly documented, there is considerable evidence that the dominant grasses of the short or mixed prairie and the northern fescue prairie extend into these surrounding desert or forest biomes and are important elements of seral stages occurring in these habitats. This circumstance leads to the concept that the North American prairie biome is a collection of arrested subclimax stages that have become the actual climax communities of the area because climatic conditions would not permit the existence of the next stages in the succession.

In a situation of this sort there is always the question as to whether common elements have moved from the forests into the grassland area or from the grassland area into the forests. If the latter were the case, then members of the grassland areas would be spreading into successional stages of the forest and competing with and displacing former seral-stage dominants. Intrigued with these possibilities, I have tried to find indications among the insects that might offer some clues as to the probability of one explanation rather than the other being the correct one. Although grassland insects have seldom been studied as a faunal unit, a considerable amount of material is accumulating regarding two groups that are common and possibly dominant grassland animals. One group is the range grasshoppers, the other the grass-feeding leafhoppers.

Recently an extremely useful summary of the distribution of range grasshoppers (Newton and Gurney, 1956-1957; Gurney, personal communication) has made possible the characterization of the faunal affinities of the grasshopper species occurring in the Great Plains. These are indicated in Figure 2. Of these prairie grasshoppers, 108 species have ranges that embrace more than a small peripheral portion of the prairie biome. Of these, 17 occur also in the eastern temperate deciduous forest, 31 occur also to the west and northwest of the prairie, 34 occur also to the south and southwest of the prairie, and 23 are transcontinental. Only three can be classified as endemic prairie species—*Hesperotettix speciosus, Hypochlora alba,* and *Pardalophora haldemani*. These distributions are consistent with the idea that the grasshopper fauna of the Great

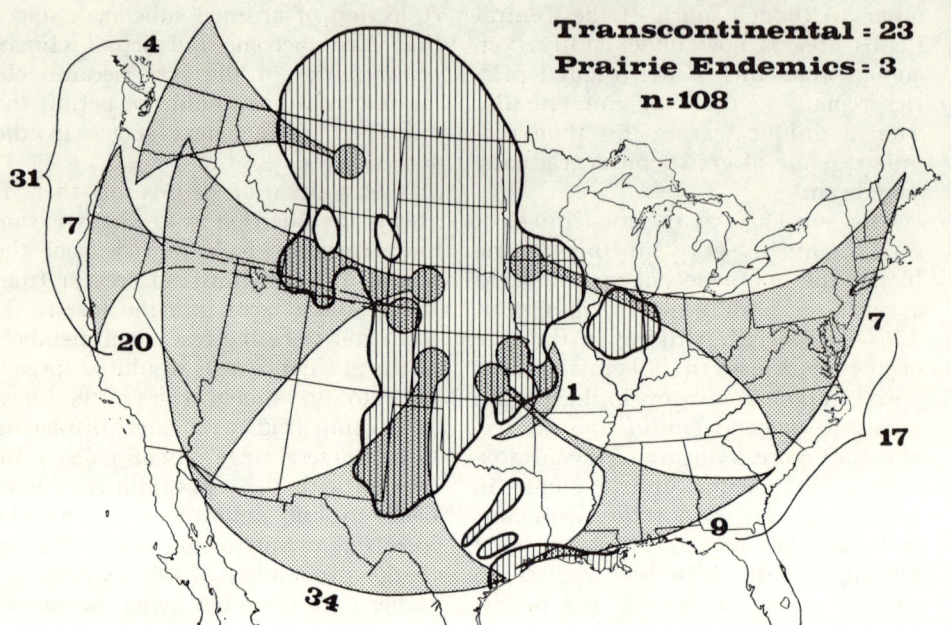

Fig. 2. Faunal affinities of grasshoppers occurring in the Great Plains grassland of North America. Explanation in the text.

Plains is not a distinctive ecological unit but has strong ecological affinities with all of the surrounding biomes. This is emphasized not only by the small number of Great Plains endemics, but also by the fact that each of the three endemic species of grasshoppers belongs to a different genus.

Data from the grassland leafhoppers are not so extensive concerning details of distribution for each species but do help by adding two other ways of looking at the problem. First, we have comparable collections of the entire leafhopper fauna from specific prairie communities and from visually similar grass-dominated seral stages in forest biomes. Second, we have some phylogenetic information concerning a few of the grassland leafhoppers, with assurances that characters are available to provide eventually a reliable network of grassland leafhopper relationships. How useful these avenues of inference will prove is not now known. It must be emphasized that the remarks that follow are a progress report rather than a presentation of highly deliberative results, and are drawn from the leafhopper tribe Deltocephalini, popularly known as "delts."

COMPARATIVE COMMUNITY COMPOSITION

Two Deltocephalini examples will be discussed, dealing with the tall-grass and mixed-grass prairies, respectively. A comparison of the grass-feeding Deltocephalini leafhopper fauna occurring in certain tall-grass *Andropogon* communities in the Great Plains with comparable leafhopper groups collected from *Andropogon* seral stages situated in the temperate deciduous forest shows that the leafhopper species in the latter situations are much more diverse than those in the Great Plains communities. Several

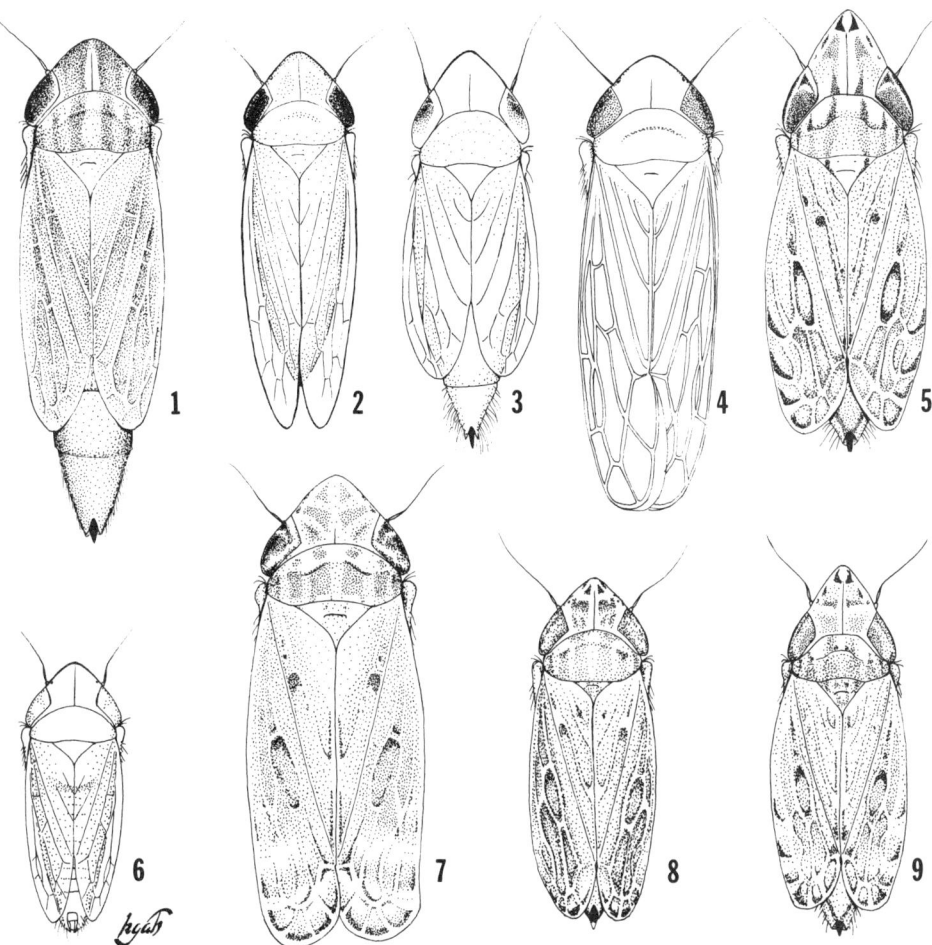

FIG. 3. Nine grass-feeding leafhopper subdominants of the mixed prairie region of North America: 1, *Mocuellus caprillus*; 2, *Auridius ordinatus*; 3, *Auridius flavidus*; 4, *Auridius auratus*; 5, *Flexamia flexulosa*; 6, *Orocastus perpusillus*; 7, *Diplocolenus configuratus*; 8, *Rosenus cruciatus*; 9, *Flexamia abbreviata*. (Drawings by K. G. A. Hamilton.)

leafhopper genera that feed on *Andropogon*, such as *Amblysellus* and *Stirellus*, are abundant in the forest communities but absent in the prairies. In the leafhopper genera *Deltocephalus* and *Graminella*, at least 30 species feed on subclimax grasses in the deciduous forest, but only three of these have been found extending into prairie communities. Many of the grass-feeding leafhoppers restricted to the temperate deciduous forest area feed on species of grasses that likewise are restricted to the deciduous forest area.

The leafhopper genus *Flexamia* is the only example among the Deltocephalini in which the tall-grass prairie communities have a faunule as numerous and diverse as that of the deciduous forest seral stages.

The Deltocephalini leafhopper faunule of the mixed-grass or short-grass prairie of the mid-northern Great Plains exhibits a comparable relationship with that of the grass communities occurring in the western montane and northwestern coniferous forests. Even though only limited pertinent data are

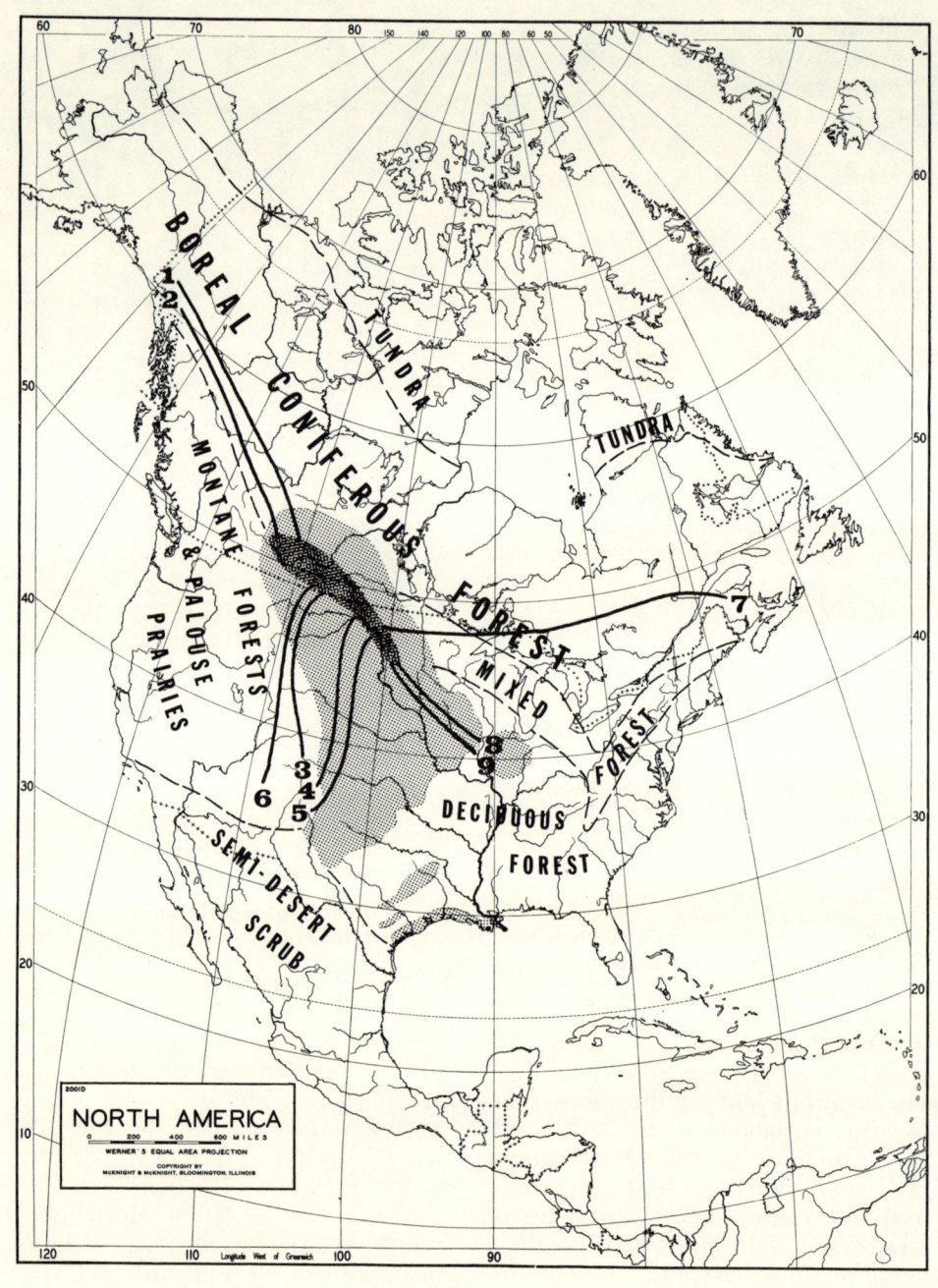

FIG. 4. Greatest distance that the nine subdominant leafhoppers shown in Figure 3 have been found away from the north-central portion of the mixed prairie. Numbers refer to the same species as in Figure 3. Darker area is the area of mixed prairie sampled; lighter area is the extent of the entire prairie biome of the Great Plains.

available in these forest biomes, it is evident that the presumably successional grassy communities in them contain three times as many species of Deltocephalini as does the northern mixed-grass prairie. Looking at this in the other direction, from the prairie outward, we see that almost all the nine abundant leafhopper species in the climax community of the northern mixed-grass prairie occur to the west or northwest of the mixed-grass association; one occurs far to the east. Two come close to being prairie endemics, but both of these occur in savanna-like subclimax grass communities southeastward into Illinois (Figs. 3 and 4).

Both the tall-grass prairie and the mid-northern mixed-prairie examples give numerical support for the idea that the grassland fauna of the Great Plains came from the surrounding biomes according to the model I suggested several years ago for the colonization of abiotic terrestrial areas (Ross, 1962, pp. 287-289). In this model it was suggested that the earliest seral stage would need to colonize an abiotic area before the next seral stage would be able to do so, and that colonization would proceed stepwise by successive seral stages. It was further postulated that not all members of a seral stage would necessarily move into the newly colonized area synchronously, but that certain species that by chance developed the ability to live in an environment previously inimical to the species would be the first colonizing organisms. If this model were correct, one would expect to find fewer species in the colonizing community than in the parental community. As regards the Deltocephalini leafhoppers and the grass genera on which they feed, there is a much greater species diversity in certain seral stages of the deciduous forest biome than in comparable grass communities of the tall-grass prairie. These data are consistent with the model.

The attempted direct application of this model to Great Plains prairie biome ecology has one peculiar drawback. If the climatic conditions prevailing in the Great Plains region during pregrassland periods were as described by MacGinitie (1953) and Dorf (1960), then in the Great Plains area the above ecological model operated in reverse. Until Early Miocene, the Great Plains area seems to have been covered by temperate and tropical forests. After that, the central part of the Great Plains showed evidence of true prairie or prairie-savanna conditions (Elias, 1942), presumably associated with a climate that became more typically continental, hotter in summer and colder in winter. The vegetational changes of the Great Plains heartland would have been, therefore, a die-back or disappearance of those biotic elements unable to withstand increasing continentality of climate to areas having a climate in which they could persist. Presumably spruce and aspen, able to withstand extremes of cold, but not heat, became restricted to the north of the incipient prairie because of inability to withstand summer extremes; many desert and semidesert species would have become restricted to the south because of inability to withstand winter extremes; montane and temperate deciduous forests would have become restricted to the west and east, respectively, because of decreasing moisture in the Great Plains region. In these biotic changes it is obvious that if the dominant plant species disappeared from an area, the entire community of which these plants were dominants would also disappear. As continentality of climate increased from Miocene to Late Pleistocene there was probably a stepwise disappearance of the original communities living in the Great Plains backwards through their normal course of succession.

If this sequence of events is correct, then the grasshoppers and leafhoppers I have been discussing would represent either species of successional stages that have been able to maintain viable population levels in the face of changing climate, or species that have evolved from such parental stocks. The fact that there are so few prairie endemics in these two groups might seem to indicate that little evolution of this latter sort had taken place. It is still possible, however, that grasshopper and leafhopper species now common to forest and prairie biomes might have evolved in the prairie itself since its formation in the Miocene and might subsequently have invaded adjacent biomes having subclimax stages essentially like those of the prairie. No matter what the numerical figures suggest, without phylogenetic evidence they do not furnish a basis for determining direction of ecological movement.

DIRECTION OF ECOLOGICAL EVOLUTION

In an attempt to obtain information concerning this dilemma, W. J. Knight of the British Museum and I have been attempting to work out probable phylogenies for many of the Deltocephalini having species occurring in grassland communities in various biomes. It is in one way fortunate and in another way unfortunate that we have run into the two problems inherent in such world studies of insect groups. First, there is the ever present problem of having to describe and name the new species that are encountered, a time-consuming process, but one essential to providing a nomenclature or language to use in recording data. Second, there is the problem of locating type material of older species that have been poorly diagnosed, in order that the records and morphological information they hold can be incorporated into more inclusive concepts. Although most of this material is still in the process of analysis, two deciphered case histories indicate the promise of this type of investigation.

One typical mixed-prairie Deltocephalini is *Diplocolenus configuratus*. The genus to which *D. configuratus* belongs is a Holarctic one, the six species of which are found chiefly in subclimax grass communities of northern deciduous or northern coniferous forests, or in arctic tundra (Ross and Hamilton, 1970). Based on morphological characters, the relationships of the species are expressed in Figure 5. The genus apparently evolved in boreal areas of North America, and subsequently one branch dispersed to Eurasia, where it has evolved into three northern or montane species. The North American lineages also evolved into three species, two of them forming a closely related pair of sister species. All three American species occur in subclimax communities of the boreal coniferous forest; one of them, *D. configuratus*, is abundant also in the northern portions of the prairie biome. Ecologically, the simplest explanation is that the evolution of all six known species of the genus occurred on grasses in subclimax communities of the northern coniferous forest or in northern tundra areas and that subsequent to its evolution as a distinctive species, *D. configuratus* has evolved the ability to live also in the Great Plains grasslands. Here there is excellent evidence that *D. configuratus* moved from the forest to the prairie biome.

A similar example involves the prairie species *Rosenus cruciatus*, to date known only from the Central Plains prairie with outlying populations in the sand prairies of western Illinois oak savanna forests. This genus also is Holarctic, and a ring of its species inhabits the subclimax grass areas west, northwest, and northeast of the Great Plains prairie. *R. cruciatus* is most closely related to a tundra species, *R. abiskoensis*, known from Alaska to northern Sweden. As is the case with *Diplocolenus configuratus*, *R. cruciatus* also would appear to be a prairie "offshoot" of a genus occupying tundra or seral grass communities in boreal or sub-boreal forests. The slight differences observable between *R. cruciatus* and its tundra ally *R. abiskoensis* would also indicate a relatively recent origin of these two species from their common ancestor. It is highly likely that (1) the distribution of this ancestor was broken up during one of the later glacial maxima into a population in the unglaciated part of Alaska and one in the central part of North America south of the glacial lobes, (2) these two elements evolved into the two present species, and (3) the ranges of the two have never become approximate.

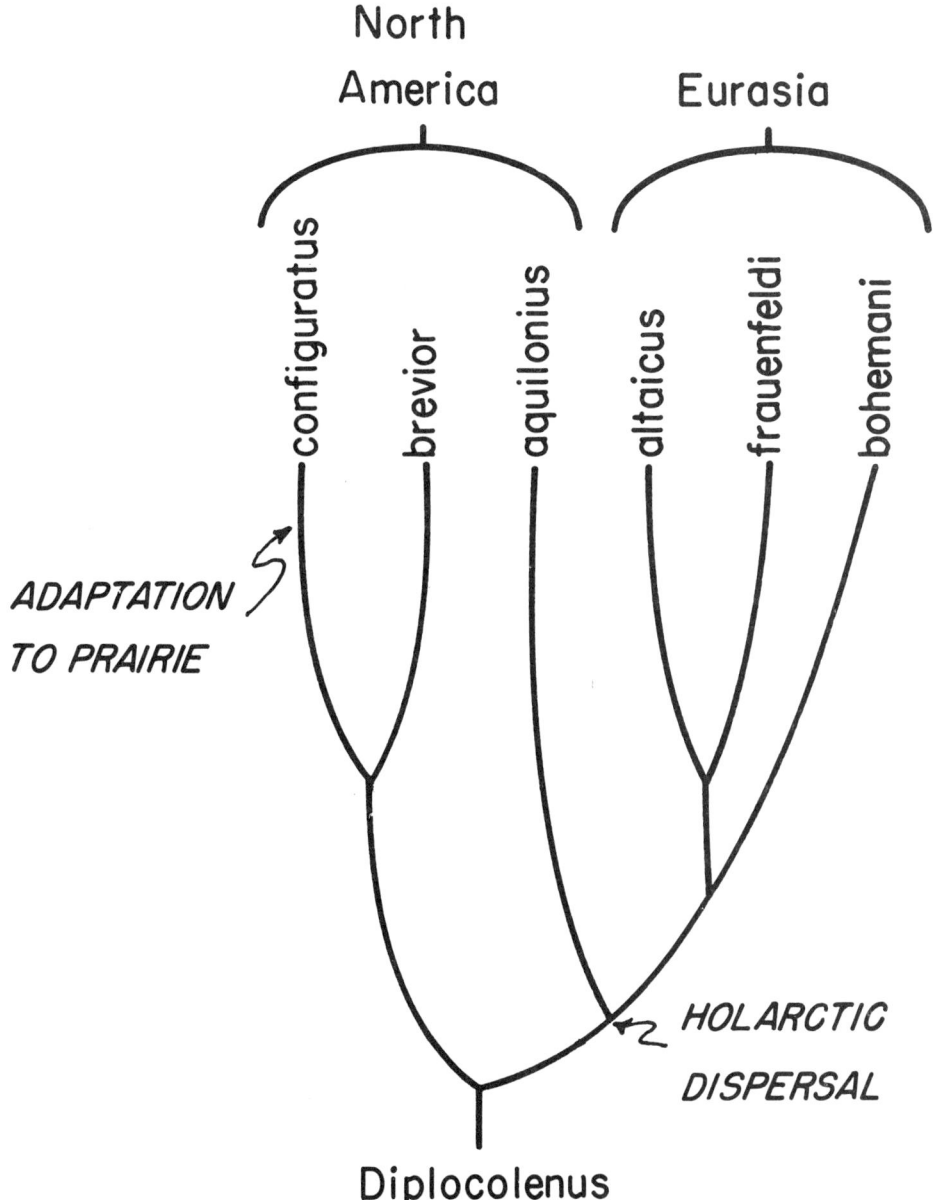

Fig. 5. Family tree of known world species of *Diplocolenus*. (From Ross and Hamilton, 1970.)

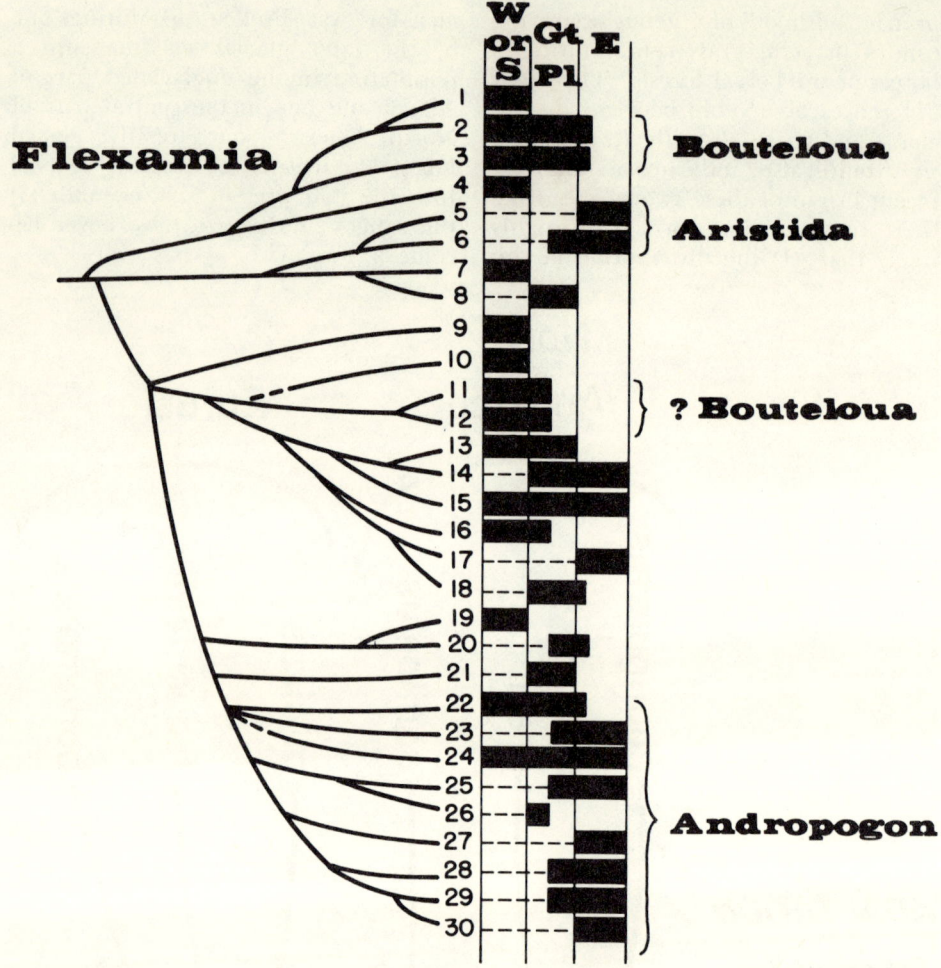

Fig. 6. Family tree of known world species of *Flexamia*. Known host associations are indicated by the grass genera at top. Distribution is indicated by bars on the central columns. *W or S*, western or southern; *Gt.Pl.*, Great Plains; *E*, eastern. (From Ross and Cooley, 1970.)

A more puzzling situation is found in the wholly North American genus *Flexamia* (Ross and Cooley, 1970). The 30 known species comprising this genus (Fig. 6) exhibit a variety of geographic ranges. Six occur only west or south of the Great Plains, four occur only east of the Great Plains, and five appear to be either Great Plains endemics, or nearly so. The remaining 15 species occur commonly either in all or part of the Great Plains area and also to the west, south, and east of it. The family tree gives some suggestion that the more primitive species (Fig. 6, nos. 1-22) arose from primarily western and southern ancestors and that the more specialized members (nos. 23-30) arose from *Andropogon*-feeding species in eastern North America. The picture is not clear. Here again we suffer from a lack of detailed distributional information concerning both geography and the grass-host preference of the species. Perhaps future studies will clarify some of these points and give us a better understanding of what has happened in the ecological evolution of *Flexamia* and its prairie associates.

The three phylogenetic examples discussed point up another interesting feature concerning the dispersal of

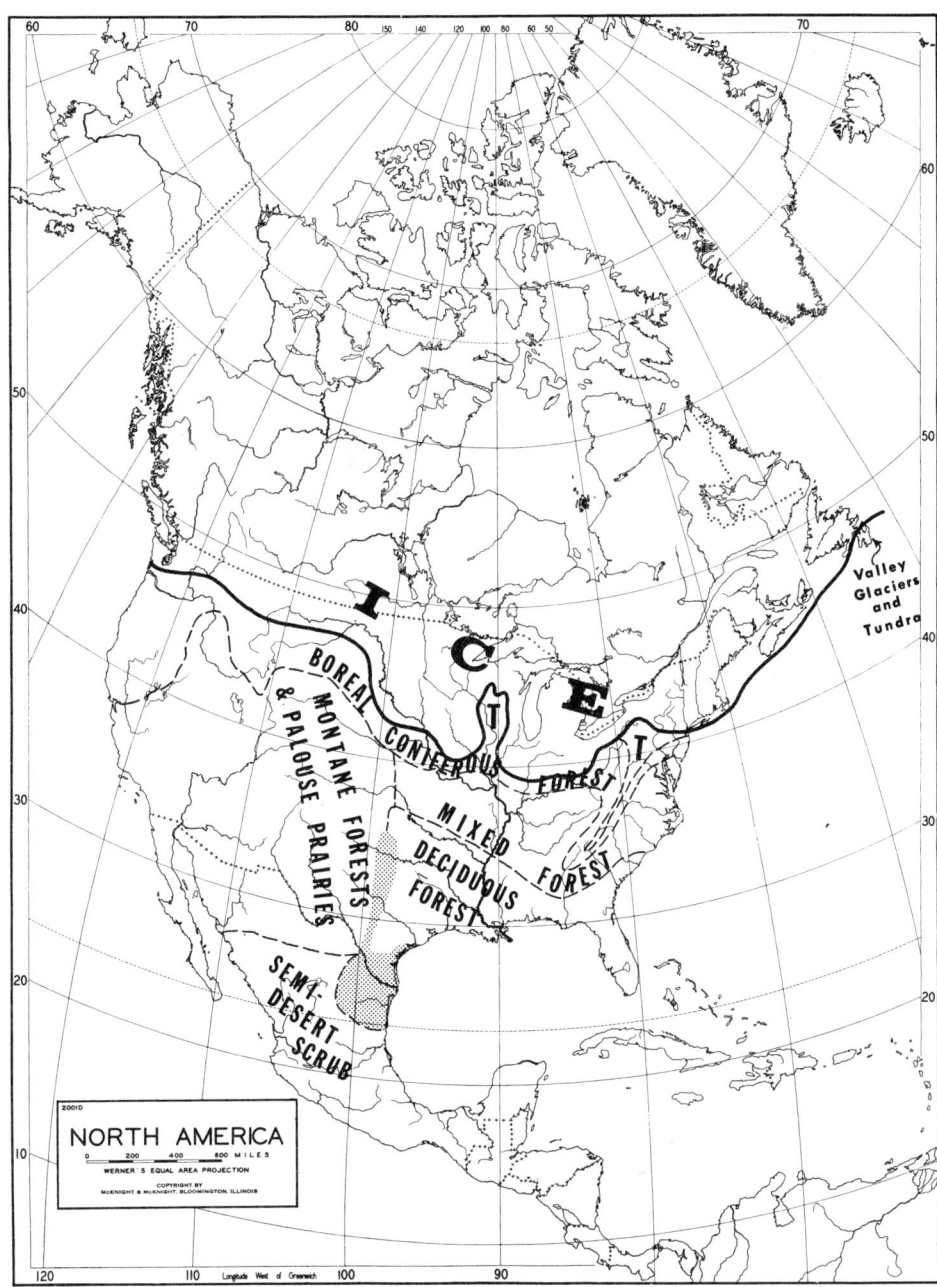

Fig. 7. Probable distribution of principal biomes in North America during glacial optimum of the Wisconsinan. Stippled area is the prairie biome.

Fig. 8. A typical climax mixed-grass prairie. View taken on the Matador tract southwest of Beechy, Saskatchewan.

some of these insect inhabitants of the Great Plains prairies. Some of the species, exemplified by members of *Diplocolenus* and *Rosenus,* represent a large group of more northern species resulting from a complex intercontinental exchange between North America and Eurasia. Another large set, exemplified by *Flexamia* and *Auridius,* represents a large faunal unit that (1) is restricted to North America and (2) has a pattern of evolution and speciation that has occurred south of the boreal coniferous forest. It is obvious that events of the Pleistocene must have exerted a great influence on the speciation pattern of the more northern species. It is not yet clear how great has been this Pleistocene influence on the endemic North American grassland leafhoppers occurring more to the south.

SURVIVAL DURING THE PLEISTOCENE

Considering that the Great Plains prairie biome or something like it has probably been in existence since Late Miocene or Early Pliocene, I find it puzzling that there appear to be so relatively few prairie biome endemics in the Deltocephalini fauna of this area. It is possible that true prairie endemics did not come under the same factors of species fission as did the inhabitants of subclimax grass communities in forested areas. As more information has accumulated concerning Pleistocene climates, especially that of the Wisconsinan maxima, a second possible explanation emerges. If one combines the pollen analyses for western and southwestern areas (Martin and Mehringer, 1965) with those from eastern and central North America (Cushing, 1965), plus indications taken from the fossil molluscan faunas of present Great Plains areas (Leonard, 1952; Frye and Leonard, 1952; Taylor, 1965) and indications from the distribution of aquatic insects (Ricker and

Fig. 9. South-facing bluffs along the north bank of the Yukon River at Taylor, British Columbia. The lower half of the apparently bare areas is clothed with a dense *Agropyron*-dominated grass community.

Fig. 10. Natural grass opening two miles north of Haines Junction, Yukon Territory. This short-grass prairie is dominated by *Agropyron*.

Fig. 11. Grass and sagebrush community on steep south-facing slope near Richardson, Alaska. The dominant grass is *Agropyron*.

Ross, 1969), it becomes apparent that during the Wisconsinan maxima the grassland biome could have been at most as large as the area shown in Figure 7. There is a possibility that it extended southward a considerable distance into Mexico, although this is doubtful. Extensive leafhopper collecting in that area has not shown evidence of relict populations that such a grassland extension should have produced. Concerning the narrow finger of prairie biome indicated in Figure 7, it is possible that there existed no truly large tracts of grassland such as that of the Saskatchewan Matador prairie (Fig. 8), rather the grass communities were represented by small patches occupying south- or west-facing slopes as part of a savanna-type, prairie-forest mixture.

If this were the case, where would the grassland leafhoppers have persisted? A glance at Figure 4 shows that this is no problem. This map shows the distance from the northern mixed prairie at which each of its nine most abundant leafhopper species are found. Two of these leafhoppers occur into northern British Columbia and the Yukon, where they are extremely abundant on hillside grass patches (Fig. 9) or natural open meadows (Fig. 10); they will probably be found ultimately as far north as central Alaska on the similar wheatgrass–silver sagebrush prairie areas covering the steep, south- or west-facing eroded banks abundant in that area (Fig. 11). Four others occur in alpine meadows in the Rocky Mountains from central Alberta south to at least southern Colorado. One occurs in grass patches of the northern coniferous forest as far east as Labrador. Two occur in seral grass patches associated with oak forests occurring on sandy areas in Illinois. During the Wisconsinan maxima at least, small grass patches of these types must have occurred in favorable ecological situations south of the ice sheets and would have provided ample opportunities for these prairie leafhopper species to persist, just as they are doing now in small areas outside the limits of the prairie biome.

This raises the possibility that during the Wisconsinan maxima only those species could persist that were adapted to living in such small scattered grass patches. If grassland leafhopper endemics requiring other situations had been in existence before this period, they might have become extinct. Perhaps this possible set of circumstances is an explanation for the small observed number of endemic prairie insect species. The northern prairie endemic *Flexamia dakota* (Fig. 6, no. 26), could be explained on the basis of the novel postulate of Wright (this volume) concerning the existence of Wisconsinan open dune areas in the Sandhills of Nebraska. Such patches of open sand might well have supported at least small areas of early successional stages dominated by grasses and might have provided areas of persistence for certain grass-feeding leafhopper endemics.

Assuming that climatic conditions in Sangamon time were as dry as, or drier than, those of the present, one would infer that since then there had occurred a drastic contraction, then an equally great expansion of the prairie biome in the Great Plains area. Such an idea leads to the thought that contractions and expansions of the prairie might have occurred in association with each major glaciation. If such vegetational changes had been similar geographically, they would not have reflected similar climates, because evidence is mounting that different glacial maxima were accompanied by markedly different climates (Hibbard, this volume). This all adds up to the proposition that during the Pleistocene, patterns of prairie leafhopper evolution, dispersal, and extinction have been much more complex than the relatively simple patterns outlined in this article.

ACKNOWLEDGMENTS

During the course of this work I have received valuable aid from Dr. Ashley B. Gurney, U.S. Department of Agriculture; Dr. Cedric Milner, University of Saskatchewan; Dr. George E. Ball, University of Alberta; Dr. Max E. Britton and Dr. Max Brewer, Department of the Navy; and Mr. Andrew Hamilton and Mr. Tim A. Cooley, Illinois Natural History Survey. To these colleagues I tender my sincere thanks.

This project was supported by a research grant from the National Science Foundation.

LITERATURE CITED

Coupland, R. T.
  1950. Ecology of mixed prairie in Canada. Ecol. Monogr., 20:271-315.
Cushing, E. J.
  1965. Problems in the Quaternary phytogeography of the Great Lakes region. Pp. 403-416, in The Quaternary of the United States (H. E. Wright, Jr., and D. G. Frey, eds.), Princeton Univ. Press, Princeton, New Jersey, x+922 pp.
Dorf, E.
  1960. Climatic changes of the past and present. Amer. Scientist, 48:341-364.
Elias, M. K.
  1942. Tertiary prairie grasses and other herbs from the High Plains. Geol. Soc. Amer. Spec. Paper, 41:1-176.
Evers, R. A.
  1955. Hill prairies of Illinois. Bull. Illinois Nat. Hist. Surv., 26:367-446.
Frye, J. D., and A. B. Leonard
  1952. Pleistocene geology of Kansas. Bull. Kansas Geol. Surv., 99:1-230.
Hitchcock, A. S., and A. Chase
  1950. Manual of the grasses of the United States. U.S.D.A., Misc. Publ., 200:1-1051.
Leonard, A. B.
  1952. Illinoian and Wisconsinan molluscan faunas in Kansas. Univ. Kansas Paleontol. Contrib., 9:1-38.
MacGinitie, H. D.
  1953. Fossil plants of the Florissant beds, Colorado. Publ. Carnegie Inst. Washington, 599:1-198.
Martin, P. S., and P. J. Mehringer, Jr.
  1965. Pleistocene pollen analysis and biogeography of the Southwest. Pp. 433-451, in The Quaternary of the United States (H. E. Wright, Jr., and D. G. Frey, eds.), Princeton Univ. Press, Princeton, New Jersey, x+922 pp.
Moss, E. H., and J. A. Campbell
  1947. The fescue grassland of Alberta. Canadian Jour. Res., 25:209-227.
Newton, R. C., and A. B. Gurney
  1956-1957. Distribution maps of range grasshoppers in the United States. Cooperative Economic Insect Report, 6:597-600, 710-712, 743-744, 775-776, 838-840, 883-884, 920, 938-940, 972, 987-988, 1002-1004, 1019-1020, 1036, 1050-1052, 1090-1092, 1122-1124, 1149-1150; 7:50-52, 71-72, 89-90, 109-110, 129-130, 150, 188, 208, 225-226, 247-248, 263-264, 315-316, 368, 388, 409, 432, 455-456, 479-480.
Ricker, W. E., and H. H. Ross
  1969. The genus *Zealeuctra* and its position in the family Leuctridae (Plecoptera, Insecta). Canadian Jour. Zool., 47:1113-1127.
Ross, H. H.
  1962. A synthesis of evolutionary theory. Prentice-Hall, Inc., Englewood Cliffs, New Jersey, 387 pp.
Ross, H. H., and T. A. Cooley
  1970. A phylogenetic arrangement of the grass-feeding leafhopper genus *Flexamia*. Ohio Jour. Sci., in press.
Ross, H. H., and K. G. A. Hamilton
  1970. Philogeny and dispersal of the grassland leafhopper genus *Diplocolenus*. Ann. Entomol. Soc. Amer., 63:328-331.
Taylor, D. W.
  1965. The study of pleistocene nonmarine mollusks in North America. Pp. 597-611, in The Quaternary of the United States (H. E. Wright, Jr., and D. G. Frey, eds.), Princeton Univ. Press, Princeton, New Jersey, x+922 pp.

Fishes as Indicators of Pleistocene and Recent Environments in the Central Plains

FRANK B. CROSS

ABSTRACT

The diversity of the plains fish fauna diminishes mainly from east to west, but also from south to north; more than 150 native fresh-water species are known from Missouri and Illinois, fewer than 50 from New Mexico and Montana. Most plains fishes occur widely in the Mississippi River system, and many inhabit other Gulf, Atlantic, or Arctic drainages as well. The range-limits of most species, and of recognizable faunal assemblages, conform less well with drainage divides than with environmental differences that transect drainage basins.

All Pleistocene fishes that have been determined specifically from plains sites seem identical with (or at least closely resemble) modern species—implying conservative evolutionary rates and lengthy opportunity for dispersal to suitable natural waters. Changes in drainage patterns that are thought to have occurred since the Pliocene may explain the wide dispersal accomplished by many species and the preponderant influence of local habitat conditions on the present composition of local faunas. Therefore, fishes are useful indicators of post-Pliocene environmental change. Hibbard and his associates have equated fish faunas and climates in Illinoian time in southwestern Kansas and northwestern Oklahoma with present fish faunas and climates in southern Wisconsin. Equivalence with eastern Missouri is also plausible, as shown herein.

Supplementing fossil evidence, the presently discontinuous ranges of several species and records of recent distributional change in other species provide evidence of change in plains environments from Pleistocene to present time. Most adjustments of range imply reduction in the amount of surface water and diversity of aquatic habitats in the Central and Southwestern Plains, with warming ambient (air) temperature a likely corollary.

Historically, man's alteration of the environment has modified the distribution and abundance of fishes. With reference to "northern" or "cold-water" species, both decimation of the southernmost populations and establishment of new populations southward from the original ranges have occurred. Some obligatory stream- or marsh-inhabiting species have receded northward, whereas lake inhabitants have been introduced successfully in southern impoundments. These and other facts suggest caution in using distributional data on fishes for purposes of climatic interpretation. The composition of a local fish fauna reveals much about its aquatic environment, but inferences extended to include climatic variables such as air temperature cannot be precise if based on fishes alone.

INTRODUCTION

Fishes, like other organisms, have limited tolerance of environmental variables, so information about fishes that lived at a particular time and place should provide insight into the nature of the environment at that time and place. In simplest form, the occurrence of fossilized parts of fishes in, say, Pliocene time in western Kansas suggests the occurrence of more or less "permanent" (year-round) surface water then and there. That assumption could be challenged but is not likely to be debated seriously. If the particular kinds of fishes found seem to be morphologically like living species, then we may also assume that the Pliocene environment matched that now occupied by these species. This assumption is more likely to be challenged than the former one, especially if the environmental similarity is taken to include climatic equivalence. Temporal change in the sorts of habitats used by some of the species seems possible; precise knowledge of the factors that restrict them to their present areas of occurrence seems impossible; and the suitability of aquatic organisms as a basis for climatic inferences seems questionable. The acceptability of the first assumption (that water accompanied the fish) and its extensions differ considerably in degree, but their premises are parallel. Problems inherent in the sequential inferences have been apparent to most workers, and the interpretive rationale has been considered by Hibbard (1960), Deevey (1965), Wilson (1968), and others.

Several authors have used the ichthyological record in central North America to characterize prehistoric environments, based partly on their fish faunas (C. L. Smith, 1954, 1958, 1962; Hibbard and Taylor, 1960; Uyeno and Miller, 1962; G. R. Smith, 1963; Schultz, 1965; Getz and Hibbard, 1965; Miller, 1965; Taylor, 1965; Semken, 1966; Wilson, 1968), or to relate recent environmental change in specified areas to changes in composition of their fish faunas within historic time (for example, Ellis, 1914; Starrett, 1951; Carlander, 1954; Bailey and Allum, 1962; Larimore and Smith, 1963; Mills et al., 1966; Metcalf, 1966; extralimitally, see especially Trautman, 1957, and Miller, 1961).

In this paper, my objective will be reasonably critical appraisal of the utility of distributional data on fishes for deriving inferences about unknown environments, within the plains region since Pliocene time, using the following lines of evidence: (1) the distributional patterns of species that now occupy the region, and the apparent extent to which their distributions have been determined by ecologic factors other than purely mechanical barriers to their dispersal; (2) the kinds of fishes reported from certain Pleistocene localities, primarily of Illinoian age, and the habitat conditions they seem to signify; (3) species significant by reason of discontinuities in their geographic ranges; and (4) changes in the ranges occupied by certain species in historic time.

DISTRIBUTIONAL PATTERNS OF RECENT FISHES

About 259 species of fishes are indigenous to the region from the continental divide eastward through the states of Minnesota, Illinois, Missouri, Oklahoma, and Texas. That number (259) excludes euryhaline fishes, species confined to lakes Superior and Michigan, subterranean species (amblyopsids plus two kinds of catfishes), and the eel (Anguilla); none of these seems appropriate to the zoogeographic purposes at hand. The list is not limited to species belonging to primary groups of fresh-water fishes, because I think it more appropriate to include than to exclude those vicarious species, and members of secondary groups, which are essentially a part of the inland fauna of the region during the period considered.

These 259 species represent 27 families, as follows: six lampreys, three sturgeons, one paddlefish, four gars, one bowfin, five herrings, five trouts (including whitefishes), two mooneyes, one mudminnow, four pikes, one tetra, 86 minnows, 24 suckers, 16 catfishes, 11 killifishes, seven livebearers, one cod, two sticklebacks, one troutperch, one pirate perch, two sea basses, 20 sunfishes, 48 perches and darters, one drum, one cichlid, three sculpins, and two silversides.

Twenty-five of these species are

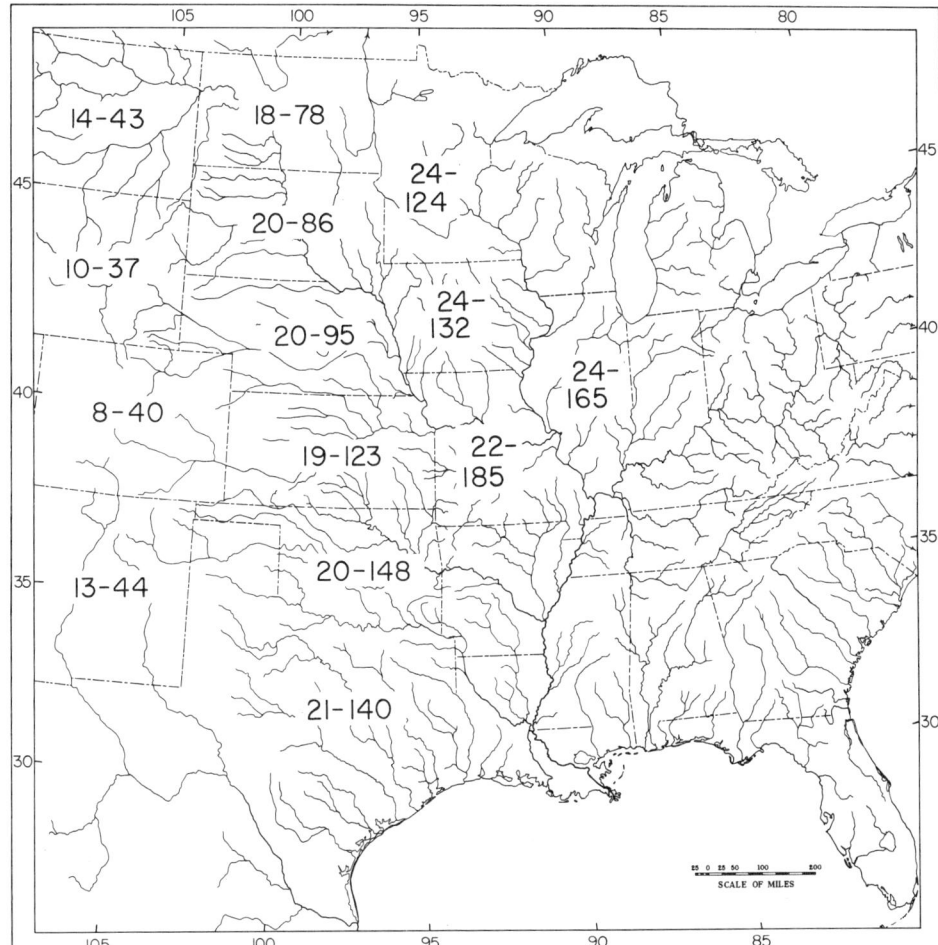

FIG. 1. Numbers of families and species of fresh-water fishes indigenous to the central states and useful for zoogeographic interpretations.

confined to coastal streams of Texas, or belong to groups (tetras, cichlids) that are indigenous northward only to the Rio Grande basin. Another 15 species occur only in streams of the Ozark and Ouachita uplands in Missouri and Oklahoma. Southern Texas and the Ozark-Ouachita highlands are the only areas having pronounced endemism, although a few additional species in the region are confined to single drainageways (for example, the pallid sturgeon and the sicklefin chub in the Missouri-Mississippi mainstream, and the related shiners *Notropis girardi,* *N. bairdi,* and *N. buccula* in the Arkansas, Red, and Brazos river systems, respectively). The remaining species—approximately 200—range through various parts of the Mississippi system, and many of them occur in Arctic, Atlantic, or other Gulf coastal drainages as well. They represent almost half of the total fresh-water fauna (restricted as stated above) of eastern North America.

Figure 1 illustrates the amount of diversity in the fish fauna as it varies in the region; numerals in each state indicate the number of families and species that are thought to be indigenous to that state within historic time.

The diversity diminishes mainly from east to west, along major drainageways rather than across drainage divides. Missouri has a fauna 50 percent larger than that of Kansas and more than three times larger than that of Colorado, although the three states share the same latitude and drainage basins (Missouri and Arkansas systems). Most species found in the western areas occur also in eastern areas. The western fauna consists basically of that fraction of the eastern fauna capable of survival in the more rigorous prairie environments. Moore (1950) has discussed specializations shown by several species that are common in the western streams, particularly in sensory structures that enhance their tolerance of turbidity. Diversity in the regional fauna diminishes also from south to north, but that trend is subordinate to the east-west trend. It is noteworthy that about twice as many kinds of fishes now inhabit Minnesota as are found in New Mexico and eastern Colorado, despite reestablishment of a Minnesota fauna only in post-Wisconsinan time.

Regional changes in species composition, not apparent in the numerical trends as I have presented them, are greater from north to south than from east to west. Northern and southern species associations exist, but the two groups are recognizable mainly at the latitudinal extremes. Range limits of their members do not coincide well in the mid-latitudes; and some northern species extend southward at high altitudes, with the result that their ranges cut across one or more drainage divides. Less than 20 percent of the fauna of the Missouri River system is absent from the Arkansas River system. Approximately 25 percent of the species in the Arkansas system are lacking in the Red River basin. The Arkansas system has the most diverse fauna in any basin west of the Mississippi—a richness relating to its mountainous headwaters, its receipt of drainage from most of the Ozark upland, and penetration of the Gulf coastal plain in its lower course. The large Ozark fauna accounts for most species lacking in the Red River basin.

The degree of similarity in species composition among states of the region is generally high. Approximately half of the fishes native to the east slope in Montana occur also in Texas, and about three-fourths of them inhabit Illinois. Most Montana species absent from Texas or Illinois are montane kinds that bridge the continental divide, or are species that extend nearly across the continent at high latitudes.

Locally within the region, species distributions seem to be determined by environmental variables. Hubbs (1957) concluded that the distributional patterns of fishes in Texas conform better to "biotic districts," predetermined from terrestrial organisms, than to drainage basins. That result might not be expected in Texas, where the principal rivers discharge separately into the Gulf of Mexico. Blair (1959) found that the distributions of darters in northeastern Oklahoma are best explained in terms of biotic districts. In that area, the southward-flowing Grand (Neosho) River roughly separates the Ozark and Cherokee Prairie biotic districts (of W. F. Blair and Hubbell, as cited by A. P. Blair, 1959, p. 4). Hall (1954) reported a total of 52 species of fishes from eastern and western tributaries of the Grand River, but found only 12 (23 percent) of these in creeks on both sides; thus, proportionately, differences between the faunas on opposing slopes of that valley exceed differences between the general faunas of any two states in the plains region. The boundary of the Ozark upland in western Missouri similarly defines the ranges of many species, largely independently of directional trends in the principal drainageways (W. L. Pflieger, A distributional study of the fishes of Missouri, doctoral dissertation, University of Kansas, 1969).

In the preceding discussion of distributional patterns I probably overemphasize the degree to which they are dependent on environmental conditions alone. Surely limitations of access have influenced the composition of local faunas, and I do not mean to imply that each species now occupies all places suitable for it. Too many recent introductions of species outside their natural ranges have succeeded, at least temporarily, for that thought to be tenable. Most of the successful introductions, however, have been made in environments greatly modified by man, involving impoundments especially.

A high level of environmental control of the species composition in streams of the region is not surprising. Most of the region now drains into one river system, affording continuous routes of dispersal. Aided by many scattered introductions, the carp became nearly ubiquitous in the Mississippi basin in less than 100 years.

While records of that exotic species demonstrate a potentiality, species indigenous to the region are not likely to have capabilities for dispersal equal to those of a large, vagile, introduced fish like the carp. Complex factors probably inhibit extensions of range by native fishes, preventing their rapid dispersal to the extremities of the system while the environment retains general stability.

More direct routes of access to various parts of the plains have been open to fishes in the past. Pliocene drainage patterns differed considerably from those of today; Pleistocene events grossly altered the older stream systems and conditions of life within them. Resultant displacement of fishes may have amalgamated subfaunas that once were relatively distinct in various parts of the region. Metcalf (1966) has given one interpretation of the way faunal mixing came about in plains streams.

PALEONTOLOGICAL EVIDENCE AND ENVIRONMENTAL INTERPRETATIONS

Although few living species are known as fossils, there is reason to suppose that much of the present fauna has existed long enough to undergo shuffling of species distributions in several stages of the Pleistocene. C. L. Smith (1962) made specific determinations for 15 kinds of Pliocene fishes from Kansas, Oklahoma, and Nebraska, eight of which seem inseparable from Recent species; none is generically distinct from modern kinds. Wilson (1968) reported another Pliocene fauna in Kansas, mostly without specific determinations, but none of his material seems to represent species clearly distinct from Recent kinds. Miller (1965, p. 571) has stated that "most if not all of the Pleistocene fishes recorded thus far . . . are osteologically indistinguishable from living species . . . ." Therefore, both the age of the fauna and the environmental significance of its composition seem adequate to justify inferences about aquatic habitats where local faunas are known.

Fossil fishes have been recorded from several sites of varying age in Texas: late Kansan (Getz and Hibbard, 1965), Sangamonian (Uyeno and Miller, 1962; Uyeno, 1963; Lundberg, 1967), and Wisconsinan (Swift, 1968). All species comprising these faunas occur at present near the places where they were found as fossils. Therefore little departure from Recent environmental conditions is needed to explain their Pleistocene occurrence, although some of the species reported usually inhabit larger streams than now exist at the fossil sites. [Other evidence, mainly the molluscan fauna, led Getz and Hibbard (1965) to conclude that Kansan climates in north-central Texas differed considerably from climatic conditions in that area today, perhaps approximating present conditions in eastern South Dakota and southern Minnesota.]

TABLE 1. Species of Fishes Reported from Four Illinoian Sites in Northwestern Oklahoma and Southwestern Kansas, and the Present Distributional Relationships of Those Species

	In same area today	In Arkansas Basin today	Not in Arkansas Basin today
Lepisosteus (platostomus)		X	
Lepisosteus osseus		X	
Esox masquinongy			X
Notemigonus crysoleucas		X	
Semotilus atromaculatus		X	
Hybopsis gracilis	X		
Dionda nubila		X	
Hybognathus hankinsoni			X
Pimephales promelas	X		
Campostoma anomalum	X		
Ictiobus sp.		X	
Moxostoma duquesnei		X	
Catostomus commersoni		X	
Ictalurus melas	X		
Ictalurus punctatus	X		
Micropterus salmoides	X		
Lepomis cyanellus	X		
Lepomis humilis	X		
Perca flavescens			X
19 species, all Recent	8	8	3

Hibbard and his associates have intensively studied fossil faunas of Illinoian deposits in Meade County, Kansas, and adjacent Beaver County, Oklahoma, contributing an important series of publications on fishes from that limited area and time (C. L. Smith, 1954, 1958; G. R. Smith, 1963; Schultz, 1965, 1967; Hibbard and Taylor, 1960; Taylor, 1965). Cumulatively, 19 species have been determined from these sites (Table 1), the largest number of fossil fishes yet known from one locality in North America. Semken (1966) recorded four species from other Illinoian deposits not far away, in central Kansas, including one species (the walleye, *Stizostedion vitreum*) not found in Meade County.

For purposes of climatic inference, most authors cited above correlated the fossil faunas with present areas of occurrence of the same species, and concluded that the Illinoian climate in southwestern Kansas was much cooler in summer and also wetter than today, approximating the present climate of southern Wisconsin. Their conclusion is supported by the facts that most of the species in the Illinoian fauna have not been found recently in streams of Meade County, and a maximal proportion of those species now occurs in the general area of southern Wisconsin. Species whose present ranges are most indicative of environmental change are the brassy minnow, the yellow perch, and the muskellunge. Figure 2 (slightly modified from G. R. Smith, 1963) outlines the native ranges of those three species, in relation to the Illinoian fossil locality. All three inhabit the region about southern Wisconsin, with the minimum distance being established by the muskellunge.

While I am inclined to accept the climatic analogy with southern Wisconsin as most reasonable, an alternative interpretation is possible and ought to be recorded. Of the 19 species represented in the Meade County Illinoian fauna, eight still occur in streams within 20 miles of the fossil site and eight others are now found elsewhere in the Arkansas River basin in Kansas. The brassy minnow persists about 130 miles northwest in the Smoky Hill River, and yellow perch now inhabit impoundments in Kansas and much farther south (though the perch probably is not native to the state in the historic sense). Since the Recent range of the muskellunge extends southward through the Ohio and Tennessee river systems, all species in the Illinoian fauna seem to be able to exist at a latitude approximating that of Meade County. The fossil fauna is a composite of habitat groups: some are big-river fishes (gars, and to a lesser extent buffalofishes and the flathead chub); some are characteristic of clear, vegetated pools or lakes (muskellunge, yellow perch, golden shiner); some are now characteristic of clear, cool, permanent streams (ozark minnow, black

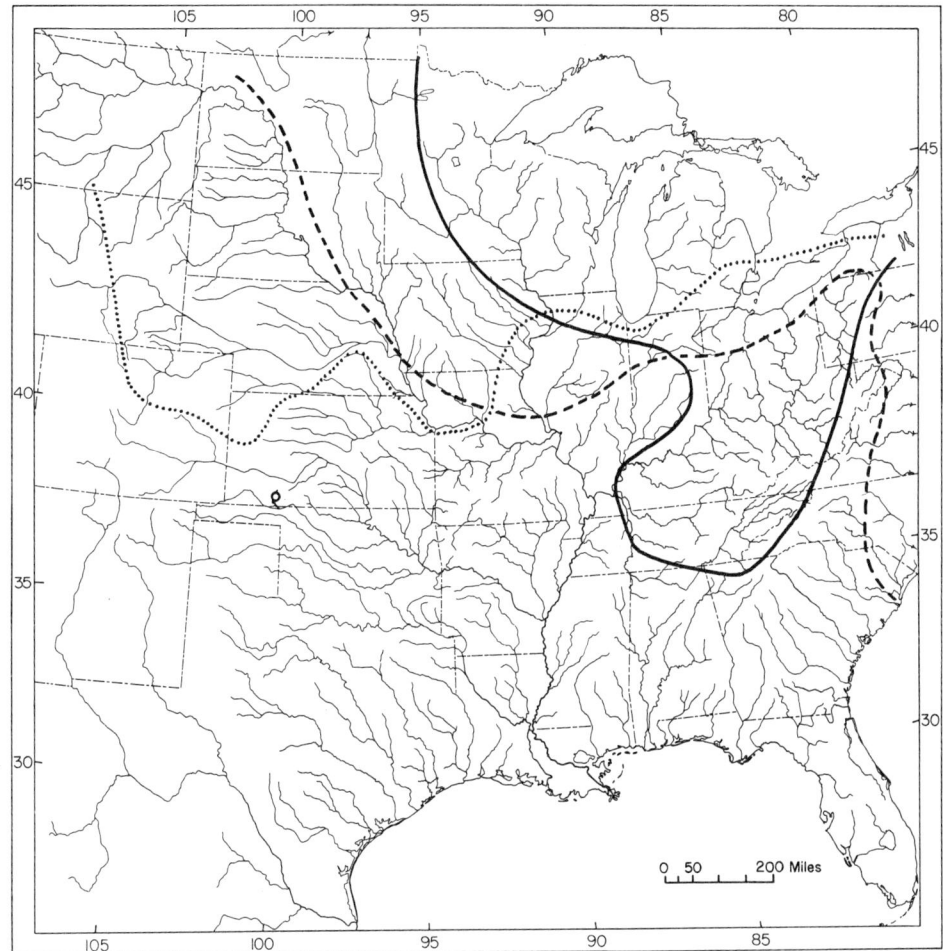

Fig. 2. Recent distributional limits in natural waters of the brassy minnow (dotted line), the yellow perch (dash line), and the muskellunge (solid line), in relation to the site in southwestern Kansas where those species occurred in Illinoian time (circle).

redhorse, common sucker, brassy minnow); and some are now found most abundantly in intermittent, commonly turbid creeks (black bullhead, green sunfish, fathead minnow).

Much more water than is now present in Meade County would seem to be the environmental change most needed to accommodate its Illinoian fauna. Prior to Yarmouthian time, water from external sources might have provided the necessary habitats, because the combined flows of the Arkansas and Cimarron rivers passed through the Meade County area (Frye and Leonard, 1952, p. 194). By Illinoian time, stream-flow probably had been reduced by separate discharge of the Arkansas River along a more northerly course; hence, increased moisture of local origin is indicated as a climatic feature of the area in which the Illinoian fish fauna existed.

The water must have been cooler, as well as more plentiful, than the scanty surface supply now present near the fossil sites; but the requisite atmospheric temperature is difficult to estimate on the basis of the fauna alone. The temperature of surface water is dependent on its volume, depth, rate of flow, distance from springs, and extent of shading by aquatic vegetation, none of which is known accurately. Ground water in Meade County is cool enough now to support any of the species found as fossils if enough of it emerged in springs. The fossil fauna is not composed of species that require exceptionally cold water; apart from the muskellunge and the yellow perch, few species that are now essentially northern in distribution are represented.

The same number of Meade County Illinoian fishes that now occur in southern Wisconsin also occur in streams along the lower Missouri River and adjacent parts of the Mississippi system. Missouri is more nearly central to the present ranges of most of the fishes involved. Both states have the abundant surface water and diverse aquatic habitats that are implied by the Meade County Illinoian fauna. Each now lacks one species represented in the fauna: the flathead chub, *Hybopsis gracilis*, is not known from Wisconsin, the muskellunge is not known from Missouri. The habitats of these two species differ radically. The flathead chub is found mainly in fluctuating sandy rivers that are devoid of vegetation, from the Mackenzie basin to the Rio Grande, eastward to the Missouri-Mississippi mainstream (but not the upper Mississippi River). The muskellunge inhabits lakes and clear, low-gradient streams, usually in association with aquatic vegetation and relatively shallow water. Of the two species, the muskellunge seems the more significant for purposes of climatic inference. The habitat of the muskellunge—definable as relatively clear, cool, shallow water—seems more subject to the influence of local climate than does the less precisely definable habitat of the flathead chub. Nevertheless, the relationship of muskellunge habitat to a specific atmospheric temperature regimen is not so direct as to permit precise determination of one from the other. The Recent range of the muskellunge spanned several isothermal zones, and has receded greatly within the last century.

DISJUNCT RANGES OF RECENT SPECIES ON THE CENTRAL PLAINS

The ozark minnow, *Dionda nubila*, signifies nearly as much about environmental conditions at the Meade County Illinoian site as do the species whose ranges are shown in Figure 2. The recent range of the ozark minnow has further interest because of its discontinuity (Fig. 3). *Dionda* now occurs only in clear, permanent streams. Its long intestinal tract and ventral mouth indicate that it feeds on benthic microorganisms, including algae, and that it requires the firm substrates and clear water that support such organisms. *Dionda* is not essentially a rifflefish, however. It inhabits all major streams in the Ozarks, but drops out of their fauna when they enter lowlands and become fluctuating, silt-laden, and warm. Within historic time, Graham (1885) reported the ozark minnow from the Neosho River in Kansas, where it has not since been found; the exact locality was not designated, but it would require a slight westward extension of the range that is shown in Figure 3. The ozark minnow seems to avoid areas that have been glaciated, and it may have been eliminated from an extensive preglacial range by erosion of till and loess into streams that it formerly occupied. Soil-type as influenced by glaciation affects the occurrence of other fishes. For example, 50 species are known from tributaries that enter the lower Kansas River from the south, whereas only 30 species have been recorded from tributaries that drain glaciated areas north of that river.

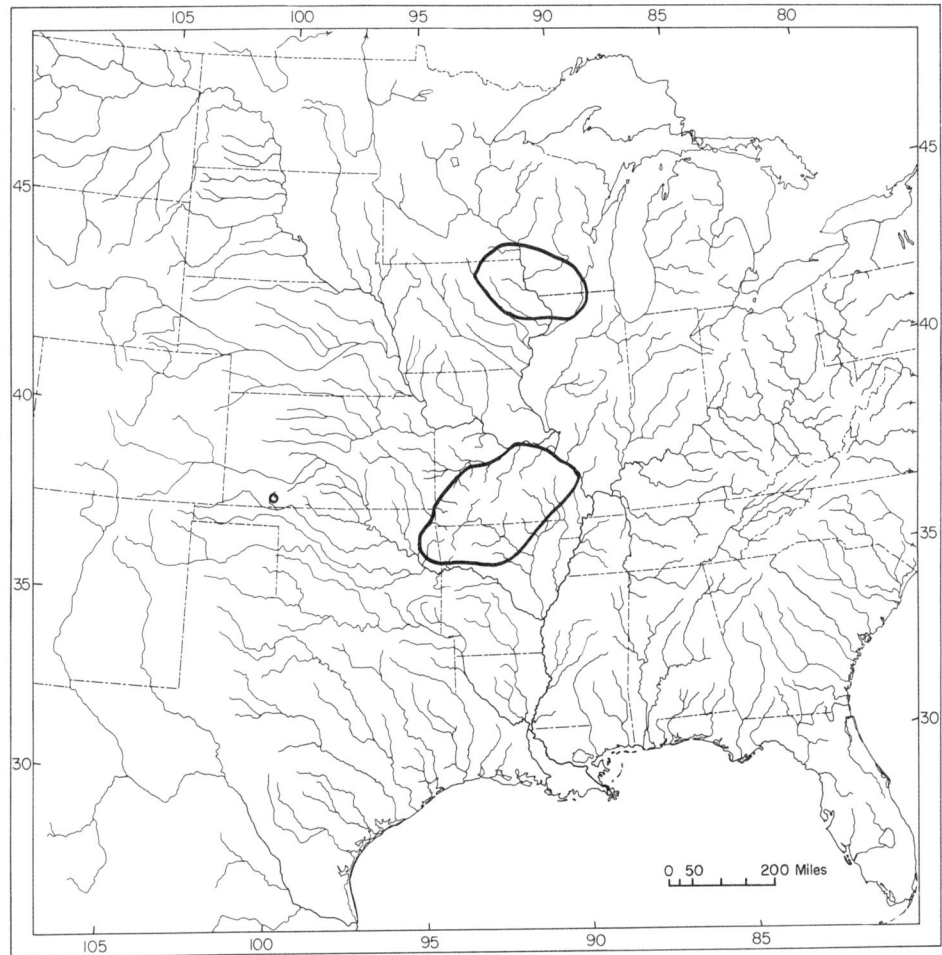

FIG. 3. Recent range of the Ozark minnow, *Dionda nubila* (outlined), in relation to the Illinoian fossil site in Meade County, Kansas.

A species not excluded from sandy streams, but having significant discontinuity in its range, is the plains topminnow, *Fundulus sciadicus* (Fig. 4). That species is found along weedy shorelines of small streams, near their headwaters, in the Ozark segment of its range; in the Platte River system *F. sciadicus* inhabits shallow, sandy spring-runs that have much vegetation, commonly filamentous algae. The only locality where the plains topminnow has been taken in the Republican (Kansas) River basin is below the stilling basin at Enders Dam (southwestern Nebraska, isolated dot in Fig. 4). That population may result from introduction, because in the years prior to our collection most bait sold at Enders Reservoir was seined from the Platte. The population may be sustained by cool water discharged from the reservoir and by seepage during times of low flow.

The disjunct range of the southern redbellied dace, *Chrosomus erythrogaster*, implies a formerly extensive southwestern distribution (Fig. 5). Two isolated 19th-century records from western Kansas are not shown. The species has declined recently in Missouri, although it persists in all major drainage systems in the Ozark region. This dace inhabits small

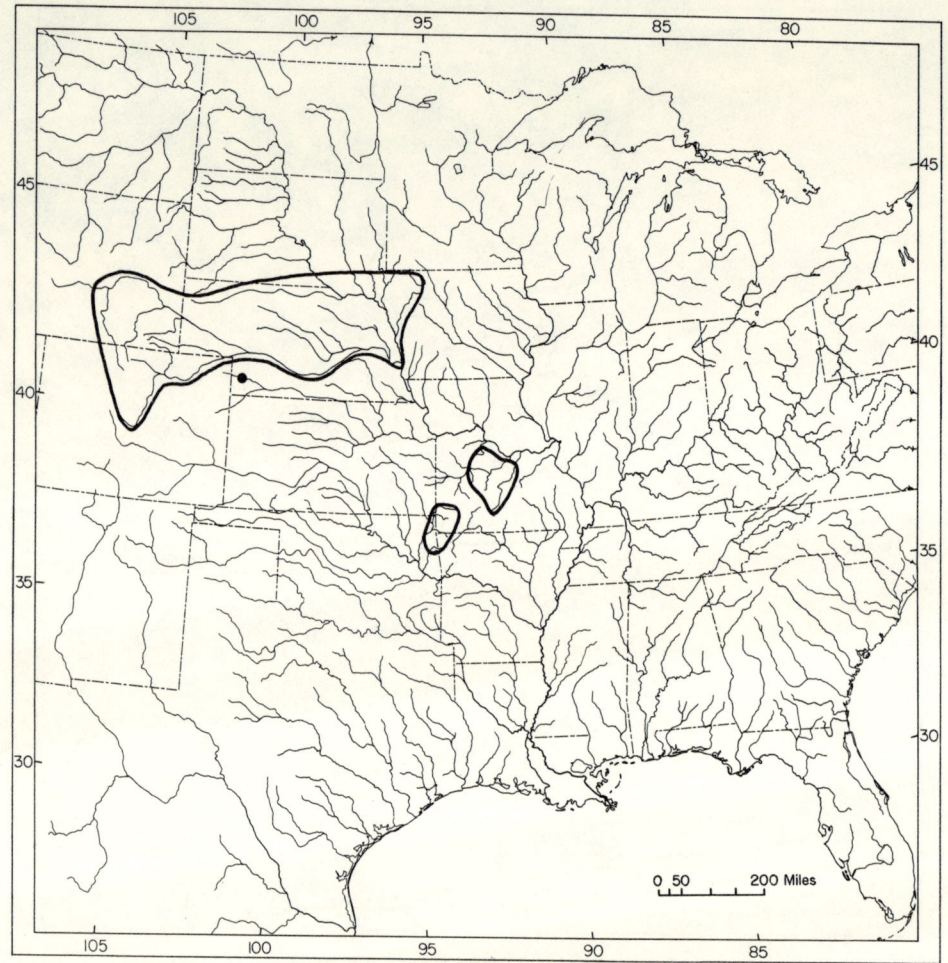

Fig. 4. Recent range of the plains topminnow, *Fundulus sciadicus* (outlined areas).

streams where springs or seeps emerge from the banks, especially if aquatic vegetation grows and organic sediments accumulate, but the water remains clear. It is adapted for an herbivorous diet: "Its food is evidently obtained by nibbling or sucking the surface slime from stones and other objects on the bottom. It consists, in all cases examined by us, mainly of mud containing algae with an occasional trace of *Entromostraca*" (Forbes and Richardson, 1909, p. 113). *Chrosomus* almost certainly represents an old group, though known only from Wisconsin time (Miller, 1965, p. 574). The ranges of its other plains species are similarly fragmented, in states from Colorado and Nebraska northward into Canada (see Bailey and Allum, 1962, pp. 40-42).

The three species discussed because of their disjunct ranges are alike in inhabiting small, clear streams. Cool temperatures seem more important to them than current, and two of them—the killifish and the dace—prefer streams with enough stability to support aquatic vegetation. Because the shallow habitats of these species are exposed to the atmosphere, their distributional histories may reasonably be interpreted relative to changing atmospheric temperatures as follows: (1)

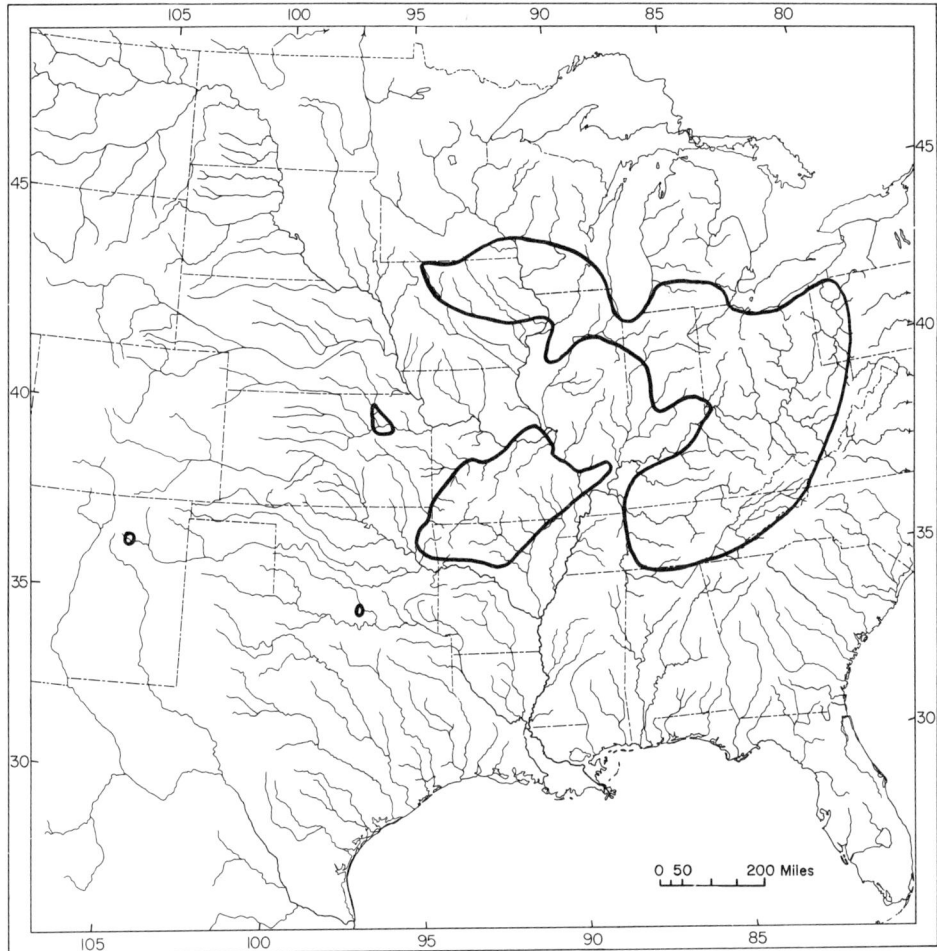

Fig. 5. Recent range of the southern redbellied dace, *Chrosomus erythrogaster* (outlined areas).

southward dispersal during glacial advances, due to atmospheric cooling and consequent cooling of streams, probably supplemented by increased moisture and more consistent flow; (2) subsequent extirpation of most southern populations as atmospheric temperatures became warmer, heating the shallower streams and causing many of them to dry, but leaving remnant populations of these species near springs. In this sense, the Ozark region represents a refugium for many additional kinds of fishes, because of its abundant springs (see Thornbury, 1965, p. 269, fig. 14-4). Isolated populations of some northern fishes (*Semotilus margarita*, *Chrosomus eos*, and *C. neogaeus*) in the Sandhills of Nebraska may be explained by similar reasoning.

HISTORIC RESTRICTIONS OF RANGE

The range of the blacknose shiner, *Notropis heterolepis*, has receded farther within historic time than that of any other species in the plains region. The southwesternmost records, obtained between 1880 and 1900 in Wallace and Sedgwick counties in Kansas, seem questionable to me. But more

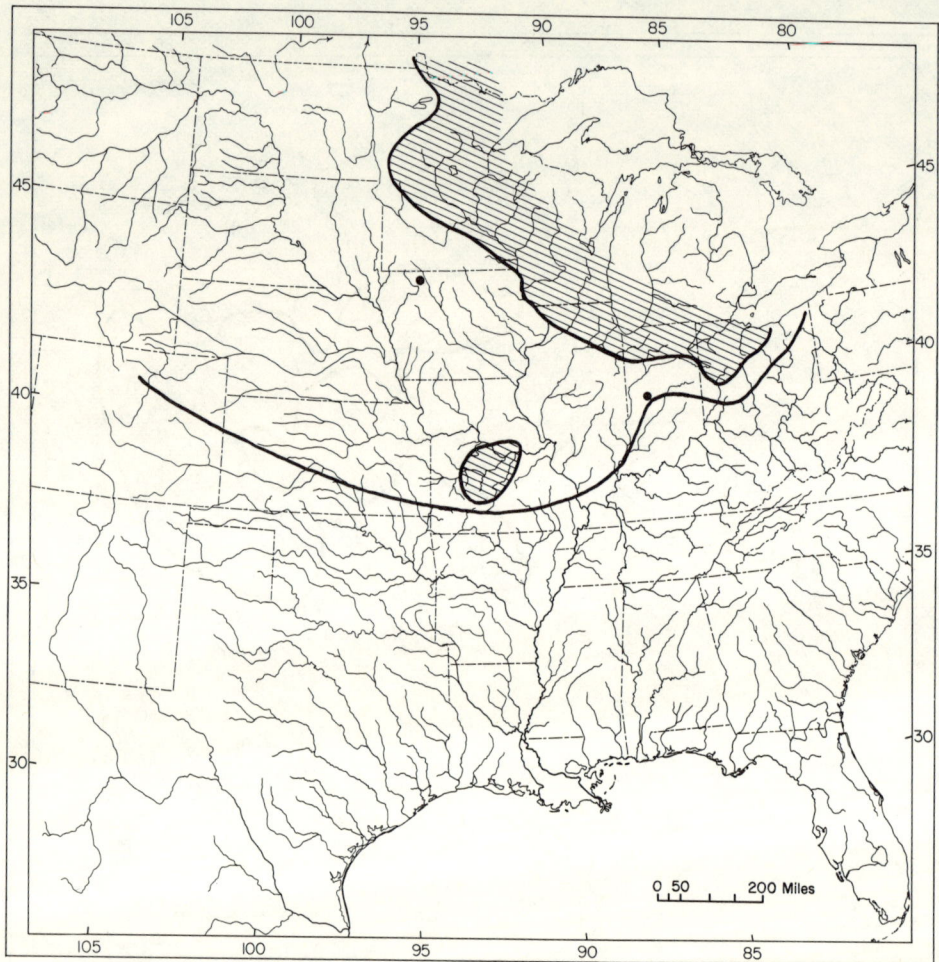

FIG. 6. Southern limits of the range of the blacknose shiner, *Notropis heterolepis,* as known prior to 1900 (single dark line) and presently (shaded areas). Dots indicate other localities of recent record.

than 50 localities of record that cannot now be duplicated—in Nebraska, the Dakotas, Iowa, and Illinois—lie between the two lines in Figure 6 that delimit its 19th-century and present ranges. The species has become scarce in Indiana and Ohio. The enclosed area in Missouri, where the blacknose shiner persists, designates records obtained recently from four tributaries of the Missouri River plus approximately 20 localities where the species may have been extirpated. It inhabits clear, weedy pools having organic sediment.

In Kansas the hornyhead chub, *Nocomis biguttatus,* and the common shiner, *Notropis cornutus,* have disappeared from some areas they occupied before 1900 (Figs. 7 and 8). Circles on the figures represent records in the late 1800's; dots locate records in the last two or three decades. The common shiner occurred in nearly all collections recorded from western Kansas before 1900, and the hornyhead was found at nearly half the localities seined by early investigators. Both species frequent small, permanent streams. Patches of silt-free bottom are required for their reproduction.

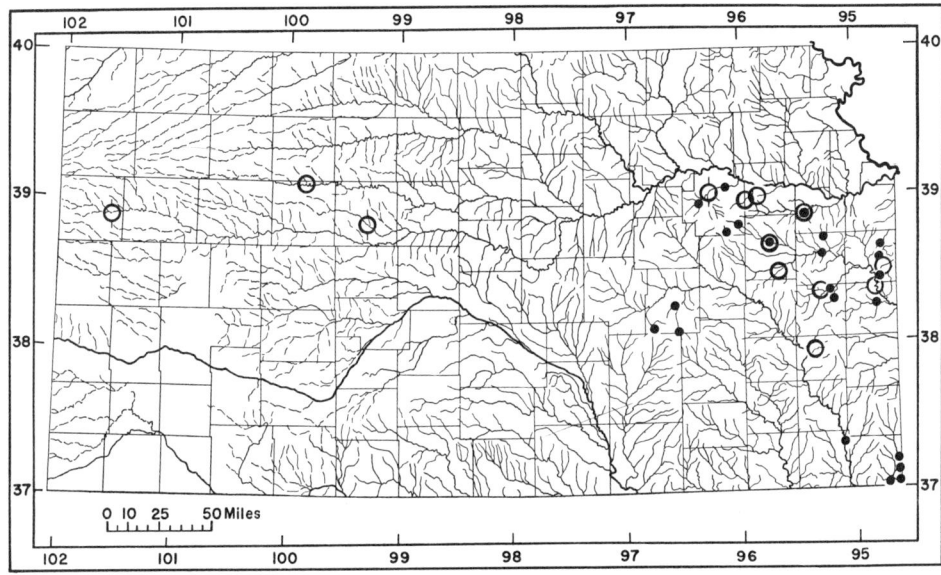

Fig. 7. Distribution of the hornyhead chub, *Nocomis biguttatus*, in Kansas. Circles indicate records prior to 1915, mostly prior to 1900; dots represent more recent records.

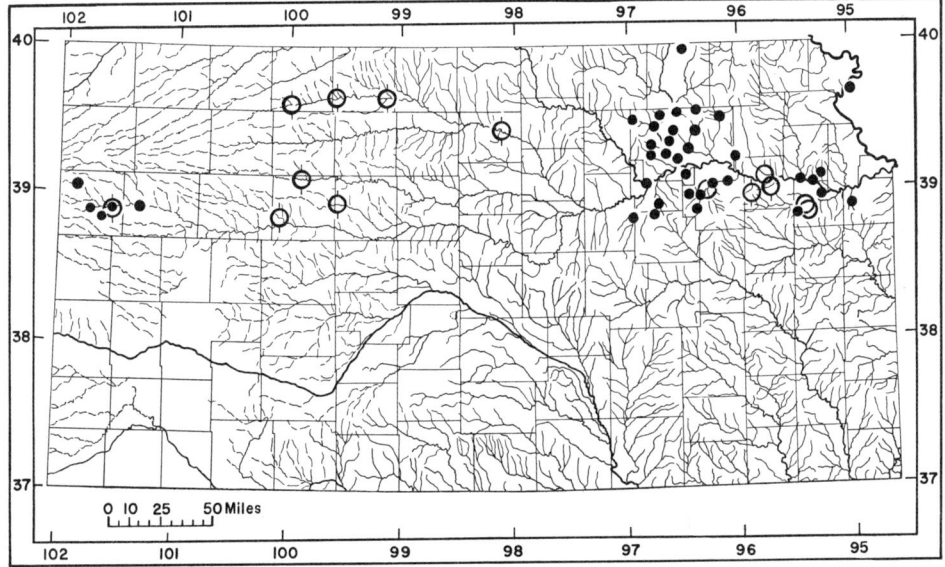

Fig. 8. Distribution of the common shiner, *Notropis cornutus*, in Kansas. Circles indicate records prior to 1915, mostly prior to 1900; dots represent later records.

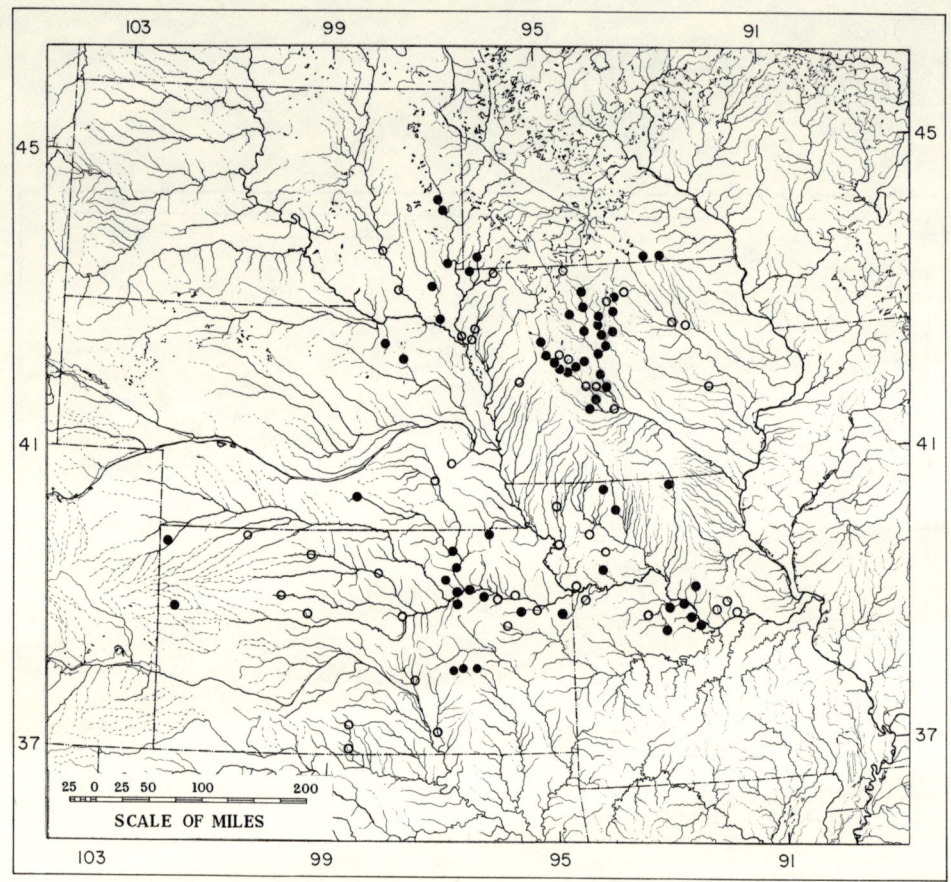

FIG. 9. Distribution of the Topeka shiner, *Notropis topeka*. Circles indicate localities of record where the species has not been reported for 30 or more years. (Modified from Bailey and Allum, 1962, fig. 6).

The distribution of the common shiner (Fig. 8) is noteworthy also because it indicates probable restriction of range by the barrier of a drainage divide. Occurrence of the species throughout the Kansas River system (Missouri River basin), coupled with an absence of verifiable records from the Arkansas River system, implies such late (Wisconsin) entry of this fish into the fauna that it was unable to utilize connectives with the Arkansas River that existed in Illinoian time (Smith and Fisher, this volume; Cross, 1967, p. 114). It is disconcerting, however, that the same assumption might be made from the present range of the brassy minnow, were it not for the single fossil record of that species in the Meade County Illinoian fauna.

Occurrences of the Topeka shiner seem now to be dwindling in parts of its range (Fig. 9). Some areas of depletion coincide with those in which the hornyhead and common shiner have disappeared, areas having poor aquifers. The Topeka shiner occupies pools of relatively small, clear streams.

The four species discussed above, like those for which discontinuities suggest earlier restriction of range, seem dependent on small, cool streams as habitats. Some kinds of fishes that require large amounts of water have also declined recently in rivers of the central United States (lake sturgeon, paddlefish, blue sucker, river redhorse, and skipjack). These adjustments of range seem attributable to modern use of land and water by man rather than to climatic changes that have occurred in historic time.

DISCUSSION

The relationship of fish distributions to environmental change in the Central Plains of North America can be interpreted from the following evidence: (1) a Recent fauna in which the number of species diminishes westward as the amount and diversity of aquatic habitat diminishes, but that shows less correlation with latitude and drainage divides; (2) a Pleistocene (Illinoian) fauna that implies more water and more diverse fish habitats on the Western Plains than exist there now; (3) discontinuities in the ranges of some species, indicating that they were formerly more widespread; and (4) historic disappearance of several species from areas of known occurrence.

With reference to the first point above, westward diminution of the fauna coincides with several environmental changes of which temperature is one. Streams on the Western Plains are not only smaller and more subject to intermittency than streams farther east, but also are shallower and more exposed to intense radiation. Temperatures in their broad, generally barren, sandy channels sometimes exceed 90°F on hot summer days. Segments of these stream systems that extend into the Rocky Mountains—and there regain diversified habitat including steep gradient, rocky substrate, pools, and cool temperatures—also regain some species of fishes prevalent on the eastern fringe of the prairie region but absent from intervening areas. The influence of temperature alone is difficult to dissociate from the potential influence of other environmental factors that change simultaneously with temperature.

It is likewise difficult to discern the importance of temperature in explaining the combination of species recorded from Illinoian sites in Meade County, Kansas, and Beaver County, Oklahoma. Each of the following hypotheses could be advanced. (1) Mean seasonal air temperatures were not much different from those today, because the species represented in the Illinoian fauna still exist somewhere else at similar latitude and altitude. Changes in moisture alone may have increased Illinoian supplies of surface water enough to sustain the species that have since been extirpated. Increased humidity probably moderated air temperatures, so that maxima were less extreme than today. Nearby, surface flows were augmented by the flow of the Smoky Hill River and possibly other streams from the north, facilitating access of big-river fishes to the Meade County area. (2) Atmospheric cooling in Illinoian time caused an increase in precipitation and supplies of surface water, and enabled southwestward dispersal of some northern fishes. The cooling was not extreme, because few exclusively "northern" species are known to have reached the Meade County site. Climatic warming subsequently reduced moisture, raised water temperatures, and forced northward retreat by those species that require cool temperatures. (3) Atmospheric temperatures in Illinoian time were a great deal cooler than now, but only a few species from the "northern" fauna became established in the Meade County area, because a tenacious resident fauna resisted displacement there. Precipitation and surface water probably were more generous, but need not have been greatly increased, because the amount of water needed to sustain the northern species is inversely related to its temperature.

Most authors who have drawn climatic inferences from the Illinoian fish fauna have adopted a version of the second hypothesis above: cooler weather, accompanied during at least part of the Illinoian period by more evenly distributed rainfall than occurs now in Meade County. That interpretation finds some support in the kinds of fishes that have discontinuous distributions on the plains, because many of these species inhabit small streams subject to influence by atmospheric temperatures. But the fact that still other species have undergone strikingly similar adjustments of range within the last few decades suggests caution in assuming a direct relationship between fish distributions and air temperatures. Recent distributional changes in walleye populations provide a useful example. In the late 19th century, walleyes occurred southward at least to Nebraska and Arkansas, and probably to Kansas and Oklahoma although natural occurrence there is uncertain. As settlement progressed, the walleye ceased to be reported from those states. Following introductions into impoundments, however, the walleye became established in all states of the region with the possible exception of Texas. Yellow perch now inhabit all plains states, and northern pike have been transplanted successfully in many states south of their native range. Few transplants of muskellunge have been attempted. The artificiality of the stocks and habitats makes these extensions of range less significant than natural adjustments of range for judging natural environmental change. Nevertheless, the transplants do suggest that present climatic conditions, per se, do not preclude the existence of any plains fishes in any areas from which they ever have been recorded.

The artificiality of the reservoir habitats relates to the amount of water stored, which controls its temperature. Thus, any condition that altered supplies of surface water might account for the changing distributional patterns of fishes. The central question is whether surface water could have varied independently of mean temperatures over the plains. Unless this question can be answered with assurance, data on fishes alone—or other aquatic organisms alone —cannot yield precise information about specific climatic factors.

ACKNOWLEDGMENT

I am grateful to Gerald R. Smith for calling my attention to some of the references cited and for valuable suggestions concerning the manuscript.

LITERATURE CITED

Bailey, R. M., and M. O. Allum
  1962. Fishes of South Dakota. Misc. Publ. Mus. Zool., Univ. Michigan, 119:1-132.

Blair, A. P.
  1959. Distribution of the darters (Percidae, Etheostomatinae) of northeastern Oklahoma. Southwestern Nat., 4:1-13.

Carlander, H. B.
  1954. History of fish and fishing in the upper Mississippi River. Publ. Upper Mississippi R. Conserv. Comm., 96 pp.

Cross, F. B.
  1967. Handbook of fishes of Kansas. Misc. Publ. Mus. Nat. Hist., Univ. Kansas, 45: 1-357.

Deevey, E. S., Jr.
  1965. Pleistocene nonmarine environments. Pp. 643-652, in The Quaternary of the United States (H. E. Wright, Jr., and D. G. Frey, eds.), Princeton Univ. Press, Princeton, New Jersey, x+922 pp.

Ellis, M. M.
  1914. Fishes of Colorado. Univ. Colorado Studies, 11:1-136.

Forbes, S. A., and R. E. Richardson
  1909. The fishes of Illinois. Illinois Nat. Hist. Surv., Urbana, cxxxi+357 pp.

Frye, J. C., and A. B. Leonard
  1952. Pleistocene geology of Kansas. Bull. Kansas Geol. Surv., 99:1-230.

Getz, L. L., and C. W. Hibbard
  1965. A molluscan faunule from the Seymour Formation of Baylor and Knox

counties, Texas. Papers Michigan Acad. Sci., Arts, Letters, 50:275-297.

Graham, I. D.
1885. Preliminary list of Kansas fishes. Trans. Kansas Acad. Sci., 9:69-78.

Hall, G. E.
1954. Observations on the fishes of the Fort Gibson and Tenkiller reservoir areas, 1952. Proc. Oklahoma Acad. Sci., 33:55-63.

Hibbard, C. W.
1960. An interpretation of Pliocene and Pleistocene climates in North America. Ann. Rep. Michigan Acad. Sci., 62:5-30.

Hibbard, C. W., and D. W. Taylor
1960. Two Late Pleistocene faunas from southwestern Kansas. Contrib. Mus. Paleontol., Univ. Michigan, 16:1-223.

Hubbs, C.
1957. Distributional patterns of Texas freshwater fishes. Southwestern Nat., 2:89-104.

Larimore, R. W., and P. W. Smith
1963. The fishes of Champaign County, Illinois, as affected by 60 years of stream changes. Bull. Illinois Nat. Hist. Surv., 28:295-382.

Lundberg, J. G.
1967. Pleistocene fishes of the Good Creek formation, Texas. Copeia, 1967:453-455.

Metcalf, A. L.
1966. Fishes of the Kansas River system in relation to zoogeography of the Great Plains. Univ. Kansas Publ., Mus. Nat. Hist., 17:23-189.

Miller, R. R.
1961. Man and the changing fish fauna of the American Southwest. Papers Michigan Acad. Sci., Arts, Letters, 46:365-404.
1965. Quaternary freshwater fishes of North America. Pp. 569-581, in The Quaternary of the United States (H. E. Wright, Jr., and D. G. Frey, eds.), Princeton Univ. Press, Princeton, New Jersey, x+922 pp.

Mills, H. B., W. C. Starrett, and F. C. Bellrose
1966. Man's effect on the fish and wildlife of the Illinois River. Illinois Nat. Hist. Surv., Biol. Notes, 57:1-24.

Moore, G. A.
1950. The cutaneous sense organs of barbeled minnows adapted to life in muddy waters of the Great Plains region. Trans. Amer. Microscop. Soc., 69:69-95.

Schultz, G. E.
1965. Pleistocene vertebrates from the Butler Spring local fauna, Meade County, Kansas. Papers Michigan Acad. Sci., Arts, Letters, 50:235-265.
1967. Four superimposed Late-Pleistocene vertebrate faunas from southwest Kansas. Pp. 321-336, in Pleistocene extinctions: the search for a cause (P. S. Martin and H. E. Wright, Jr., eds.), Yale Univ. Press, New Haven, Connecticut, x+453 pp.

Semken, H. A.
1966. Stratigraphy and paleontology of the McPherson Equus Beds (Sandahl local fauna), McPherson County, Kansas. Contrib. Mus. Paleontol., Univ. Michigan, 20:121-178.

Smith, C. L.
1954. Pleistocene fishes from the Berends fauna of Beaver County, Oklahoma. Copeia, 1954:282-289.
1958. Additional Pleistocene fishes from Kansas and Oklahoma. Copeia, 1958:176-180.
1962. Some Pliocene fishes from Kansas, Oklahoma, and Nebraska. Copeia, 1962: 505-520.

Smith, G. R.
1963. A Late Illinoian fish fauna from southwestern Kansas and its climatic significance. Copeia, 1963:278-285.

Starrett, W. C.
1951. Some factors affecting the abundance of minnows in the Des Moines River, Iowa. Ecology, 32:13-27.

Swift, C.
1968. Pleistocene freshwater fishes from Ingleside Pit, San Patricio County, Texas. Copeia, 1968:63-69.

Taylor, D. W.
1965. The study of Pleistocene nonmarine mollusks in North America. Pp. 597-611, in The Quaternary of the United States (H. E. Wright, Jr., and D. G. Frey, eds.), Princeton Univ. Press, Princeton, New Jersey, x+922 pp.

Thornbury, W. D.
1965. Regional geomorphology of the United States. John Wiley and Sons, New York, 609 pp.

Trautman, M. B.
1957. The fishes of Ohio. Ohio State Univ. Press, Columbus, xvii+683 pp.

Uyeno, T.
1963. Late Pleistocene fishes of the Clear Creek and Ben Franklin local faunas of Texas. Jour. Grad. Res. Center, Univ. Michigan, 31:168-173.

Uyeno, T., and R. R. Miller
1962. Late Pleistocene fishes from a Trinity River terrace, Texas. Copeia, 1962:338-345.

Wilson, R. L.
1968. Systematics and faunal analysis of a Lower Pliocene vertebrate assemblage from Trego County, Kansas. Contrib. Mus. Paleontol., Univ. Michigan, 22:75-126.

Factor Analysis of Distribution Patterns of Kansas Fishes

GERALD R. SMITH AND DAVID R. FISHER

ABSTRACT

The distribution patterns of fishes within Kansas were analyzed to determine (1) a small number of generalized patterns based on similarities of distributions and intended to summarize the available distributional data, and (2) correlations between distributions and environmental variables. Ninety-six drainage units in Kansas were assigned a value for each of 105 species of fishes. A 132 by 132 matrix of correlation coefficients between distributions of species and environmental variables was calculated; eight factors were extracted by the complete centroid method. Factor scores for each drainage unit were computed and plotted as trend-surface maps for each factor. Seven of the factors enable summarization of seven classes of fish distribution patterns and environmental patterns.

The generalized patterns are: (I) 21 species of large-river fishes; (II) 13 species of cool-water prairie and plains fishes; (III) 19 widespread prairie and plains species; (IV) 21 warm-water species with high correlations with environmental variables; (V) 10 species associated with the Osage River drainage; (VI) fishes with distributions centered in the Neosho River drainage—four species plus five with higher correlations with factor VII; (VII) fishes with Kansas distributions centered in the Spring River drainage in the southeast corner of the state—14 species plus seven species with higher correlations with other factors. Nineteen species independent of the factors had relatively unique distribution patterns.

INTRODUCTION

The pattern of distribution of a species will reflect limitation by existing environmental features as well as effects of past environmental features on the distribution and dispersal of ancestral populations. In this paper we describe and summarize the patterns of distribution of fishes in Kansas and attempt to analyze the environmental and historical influences on them. Our method is to arrange the distributional data objectively into a few generalized patterns and to look for correlations between patterns of environmental variables and fish distributions. Our working hypothesis is that one or more of the tested environmental variables will correlate with, and help explain, each distribution pattern in the study area. Two alternative possibilities are that nontested environmental variables or historical explanations are necessary to account for species patterns.

The usual approach to such a zoogeographical study involves the subjective assignment of species patterns into groups or units which presuppose an explanatory hypothesis. For example, a conceptual unit, "plains forms of southern origin," is assigned species patterns which, in the judgment of the investigator, fit such a category better than other categories. A disadvantage is imposed by the practice of assuming and including the explanatory hypothesis, for example, "plains forms of southern origin," as a descriptor of the group before assigning species to the group. Subsequent use of the phrase "plains forms of southern origin" as an explanatory conclusion might be subject to the criticism of circularity.

The method outlined below attempts to avoid the problems of subjectivity and circularity and to increase the repeatability of zoogeographic analysis (Fisher, 1968; Orloci, 1967).

The generalized pattern groups are formed strictly on the basis of similarity of distribution in Kansas. The descriptors and summary statements for each group are suggested by the form of the generalized patterns and the known ecological requirements of the included species.

TREATMENT OF DATA

The basic data for this analysis consist of spot distribution maps of Kansas fishes compiled by Cross (1967). The study is restricted geographically because Kansas is one of few areas for which such thorough fish distributional maps are published. We subdivided a hydrographic map of the state into 96 drainage units of relatively consistent size—the largest being only several times the size of the smallest. The shape of the units was variable. Additional drainage units of similar size were drawn around the perimeter of the state and used to mollify the edge-effect of the trend-surface mapping program, but this is not an essential part of the analysis. A matrix of fish distributions by drainage units was composed by scoring each drainage unit for the presence or absence of each of 105 species of fishes known from more than one drainage unit in the state.

ENVIRONMENTAL VARIABLES

Data for 27 environmental variables were included in the above matrix by scoring each drainage unit for a value taken from trend-surface maps copied from sources as follows.

Geological substrate and elevation.—The 1:500,000 scale Kansas Geologic Map prepared by the State Geological Survey of Kansas (1937) was used as the basis for scoring each drainage unit for the presence or absence of nine variables. The strata used as variables are: (1) Carboniferous and Permian limestones, shales, etc.; (2) Permian Cimarron series (sandstone, dolomite, etc.); (3) Cretaceous Dakota group (sandstone, etc.); (4) Cretaceous Carlile shale; (5) Cretaceous Niobrara chalk; (6) Cretaceous Pierre shale; (7) Tertiary Ogallala group (sands, gravels, silts, etc.); (8) Pleistocene alluvium and terrace deposits. Elevation was recorded from the United States Relief Map (U.S. Geological Survey, 1929).

Temperature variables.—Trend-surface maps for the following data in Kansas were taken from Flora (1948): (1) length of growing season (number of frost-free days); (2) number of days in which the maximum temperature exceeds $90°F$; (3) number of days in which the minimum temperature is less than $32°F$; (4) number of days in which the maximum temperature is less than $32°F$.

Precipitation and water variables.—Trend-surface maps for the following were taken from Flora (1948) and Miller *et al.* (1963): (1) average annual precipitation; (2) average annual runoff; (3) number of April–September dry periods (number of times during 45-year interval that 30 consecutive days passed with not more than 0.25 inch of precipitation on any day; (4) mean evapotranspiration; (5) mean lake evaporation; (6) hardness of surface waters (as ppm $CaCO_3$); (7) salinity of surface waters (measured as areas where surface waters may contain more than 1000 ppm of dissolved solids).

Drainage basin.—Correlations with individual major drainage basins (degree of endemism) were checked by including the following as variables in the matrix: (1) Kansas River drainage; (2) Osage River drainage; (3) Neosho River drainage; (4) Verdigris River drainage; (5) Arkansas River drainage; (6) Cimarron River drainage; (7) drainage of Missouri River and immediate tributaries.

FACTOR ANALYSIS

The basic data matrix consists of values for 132 fish distributions and environmental variables scored for each of 96 drainage units. A 132 by 132 correlation matrix was computed, giving the correlation between each of the distributional and environmental variables. Factor analysis—in this case, R-type factor analysis—is a method of reducing a large data matrix (here the 132 by 132 correlation matrix) to a smaller number of summarizing reference axes. The complete centroid method of Thurston (1945) was used for factor calculation because of the speed of the method and because of limitation of computation facilities when the study was begun. The number of factors to be extracted, eight, was determined by the method of Rohlf (1962). Extractions of six and 17 factors were also calculated, but six summarized the data too severely, with too much loss of information, and 17 factors failed to summarize concisely enough for our purposes. In these extraction exercises the same kinds of patterns—in some cases exactly the same patterns—emerged; they simply were divided up and expressed in more or less concise ways, with more or less information reduction. The eight factors were rotated to simple structure, using the method of oblique rotation (MTAM) developed by Sokal (1958). These factors are no longer uncorrelated. The method of Wilkinson and Householder (Ralston, 1965) was also used to extract eight factors. The differences provide an interesting comparison of the methods and a test of the repeatability of this kind of analysis.

Factor scores for each drainage unit were computed and plotted as sixth-degree trend-surface maps for each factor, using a program made available by John C. Davis of the Kansas Geological Survey. Lists of the individual distribution patterns correlated with each of the factors were compiled, using a correlation coefficient of .300 as the lower limit for inclusion of a species or environmental variable in a factor-group. Some species had higher correlation coefficients than .300 with more than one factor; 19 species were not correlated as high as .300 with any factor. Choice of .300 as the cut-off point is one of the arbitrary aspects of the analysis.

The factor-groups are summarizations based on similarity of distribution patterns and warrant explanatory generalizations. The 19 independent patterns require more specialized explanations. Tables 1-6, showing the degree of correlation between individual distribution patterns and some of the dominant environmental variables, enable analysis of the role of those variables in limiting the distribution patterns in each factor group.

GENERAL DISTRIBUTION PATTERNS AND ENVIRONMENTAL CORRELATIONS

The eight factor-groups extracted by Thurstone's complete centroid method summarize seven kinds of fish distributions and one pattern for the environmental variables. Trend-surface maps of the geographical patterns of factor loadings for each factor are shown in Figures 1 and 2. Two of the factor-groups, V and VI, are drainage-centered and suggest historical corridors of dispersal or origin for some of the species; group I reflects a unique habitat restriction; groups III and IV reflect environmentally determined distributions; groups II and VII reflect special historical-environmental interactions. Maps of the fish distributions referred to in the following discussion appear in Cross's (1967) *Handbook of fishes of Kansas.*

Factor I: Large river habitat.—Twenty-one species of Kansas fishes (Table 1, Figs. 1, 4) occur principally in large rivers in the state. The Missouri, Kansas, and Arkansas rivers provide the primary habitat for these fishes. The distribution patterns reflect the occurrence of this restricted, but relatively stable, habitat and are not usually correlated with climatic variables. It is interesting to note that most of the fishes belonging to relatively ancient families belong to this distributional group.

Factor II: Cool-water prairie and plains species.—Thirteen species are correlated with a pattern that centers primarily in northeastern Kansas, with a second cluster in the springs area at the contact between the Ogallala formation and underlying Cretaceous formations in northwestern Kansas. Twelve of the species are also correlated with one or more other factors, especially III (Table 2, Figs. 1, 4). The six primary members of this group are basically northern or cool-water species with inferred wider distribution in Kansas during times with more widespread cool-water habitat, but which are now common to areas fed by springs. They are isolated peripheral populations, the species-ranges of which presently are centered farther

TABLE 1. Species associated with factor I*

Species[1]	Correlation with factor I	Correlation with other factors	Elevation	Number frost-free days	Number days reaching 90°F	Number days falling to 32°F	Annual Precipitation	Annual runoff	Lake evaporation	Pliocene Ogallala Formation
Ichthyomyzon castaneus	+	(VII:+)						+	+	−
Acipenser fulvescens	++									
Scaphirhynchus platorynchus	++									
S. albus	+++									
Polyodon spathula	+	(V:+)								
Lepisosteus platostomus	+	(V:+)		−		−		+	+	−
Hiodon alosoides	+									
Hybopsis gracilis	++									
H. storeriana	+									
H. meeki	+++									
H. aestivalis	+									
H. gelida	++									−
Notropis shumardi	+++									
N. blennius	+									
Hybognathus nuchalis	+++									
Cycleptus elongatus	++									
Ictiobus niger	+			−		−	−		+	+	−	−
Ictalurus furcatus	+	(VI:+)								
Lota lota	+++									
Morone chrysops	+	(VII:+)								
Stizostedion canadense	++									−

* Distribution in large rivers. The − or + indicates negative or positive correlation coefficient between 0.300 and 0.499; double symbols indicate values between 0.500 and 0.699; triple symbols indicate correlations higher than 0.700. Symbols in parentheses indicate existence of a higher correlation with another factor.

[1] Four additional species, *Lepisosteus osseus, Amia calva, Anguilla rostrata,* and *Ictiobus cyprinellus,* were added to this group when the analysis was run using the Wilkinson-Householder factor-extraction method. This method also showed this group to be correlated with the Missouri River drainage.

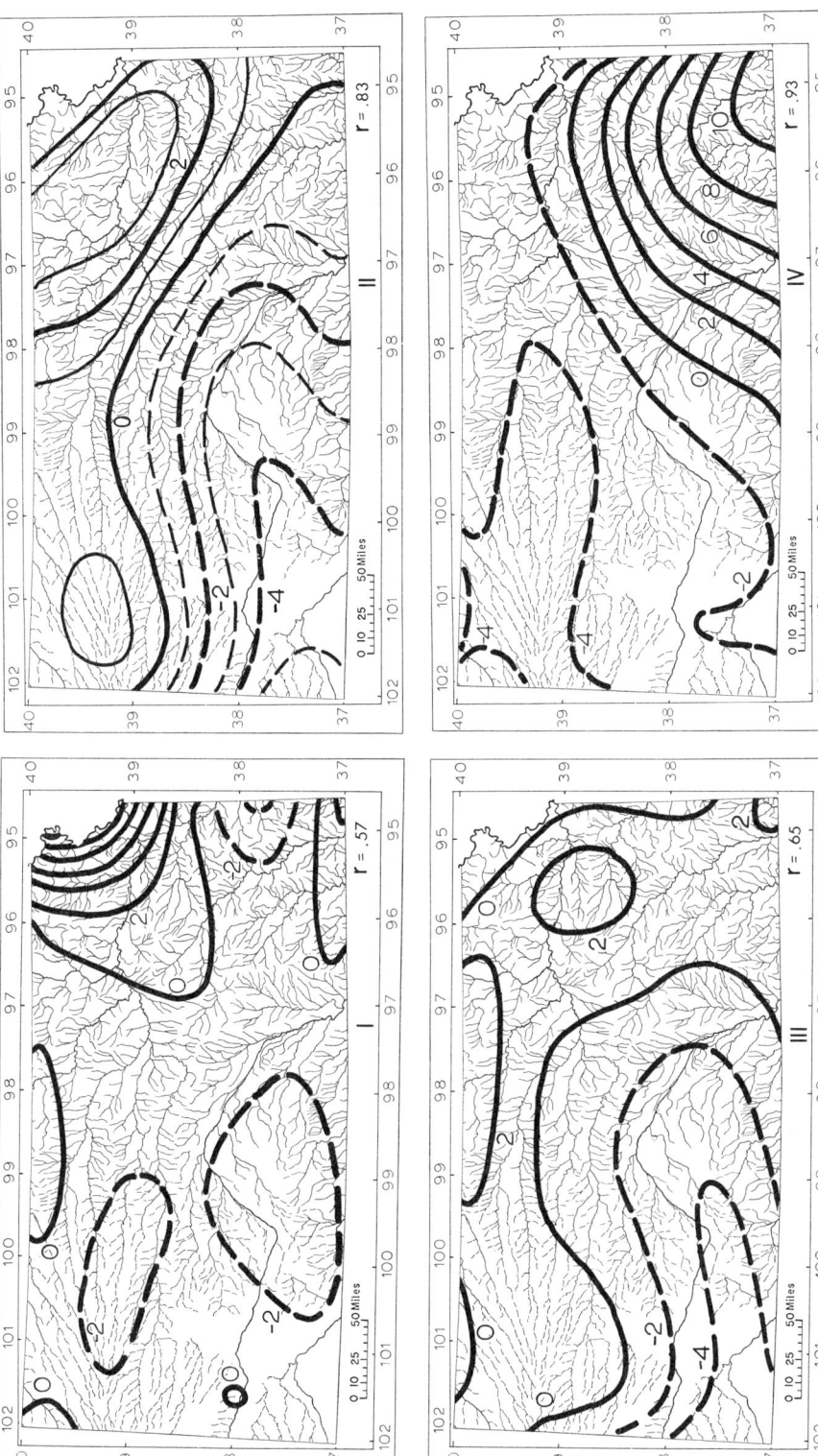

FIG. 1. Trend-surface maps of factor loadings for factors summarizing distribution patterns of fishes in Kansas. Factor I, large river species; factor II, cool-water prairie and plains species; factor III, widespread prairie and plains species; factor IV, warm-water species. The cophenetic correlation coefficient, r, reflects the degree to which the factor pattern is depicted by the trend surface.

north. There is a tendency for these species to be negatively correlated with the frequency of days exceeding 90°F and high lake evaporation (Table 2), further suggesting limitation of these species by historical reduction of the availability of permanent, cool-water habitat. The correlation between groups II and III ($r=.8$) suggests that II could be regarded as a subgroup of III.

Factor III: Prairie and plains species.—Nineteen species (Table 2, Figs. 1, 4) correlate with this factor; 11 of these are also correlated with factor II, and two of these are also correlated with factor IV. The pattern of factor III is interesting in that it is lacking in areas of high concentration. The species are inhabitants of prairie and plains streams and are widespread in Kansas. We suspect that the trend-surface pattern also reflects areas of collecting concentration. Some of the distributions, for example that of *Pimephales notatus*, are correlated with environmental gradients. *Fundulus kansae* is unique in that the signs of its correlations are the reverse of the usual pattern—its distribution appears to be affected in the opposite way by most environmental variables. The introduced carp, *Cyprinus carpio*, fell into this factor-group.

Factor IV: Warm-water species.—Twenty-one species (Table 3, Figs. 1, 4) are correlated with factor IV. Two of these are also correlated with factor VII. The species are southern in Kansas and show consistent high correlation with environmental variables, indicating that their present ranges are strongly affected by climate and water conditions. A favorable environment

TABLE 2. Species associated with factors II (northern) and III (prairie and plains streams)*

Species	Correlation with factor II[1]	Correlation with factor III	Correlation with other factors	Elevation	Number frost-free days	Number days reaching 90°F	Number days reaching 32°F	Annual precipitation	Annual runoff	Lake evaporation	Pliocene Ogallala Formation
Cyprinus carpio		++									
Semotilus atromaculatus	(+)	+	(IV:+)		— —						
Chrosomus erythrogaster	+		(VII:+)								
Phenacobius mirabilis	(+)	++		—					+	+	—
Notropis cornutus	+	(+)				—				—	
N. lutrensis	(+)	++		—					+	+	—
N. topeka	++	(+)									
N. stramineus		++		—							
Hybognathus placitus		+									
Pimephales promelas	(+)	++									
P. notatus		+		— —	+			—	++	+	—
Campostoma anomalum	(+)	++		—					+	+	—
Carpiodes carpio		+									
Catostomus commersoni	++	(+)	(IV:+)						—		
Ictalurus melas	(+)	++									
I. punctatus		+									
Fundulus kansae		+		+		+	+	—		+	—
Lepomis cyanellus		+			—			—	+	+	—
L. humilis	(+)	+						—		+	—
Etheostoma nigrum	+				—						
E. spectabile	+	(+)		—	+	—			++	—	—

* Symbols as in Table 1.
[1] The species of groups II and III, excepting *Chrosomus erythrogaster* and *Catostomus commersoni*, were combined as one group when the factor-extraction was done with the method of Wilkinson and Householder.

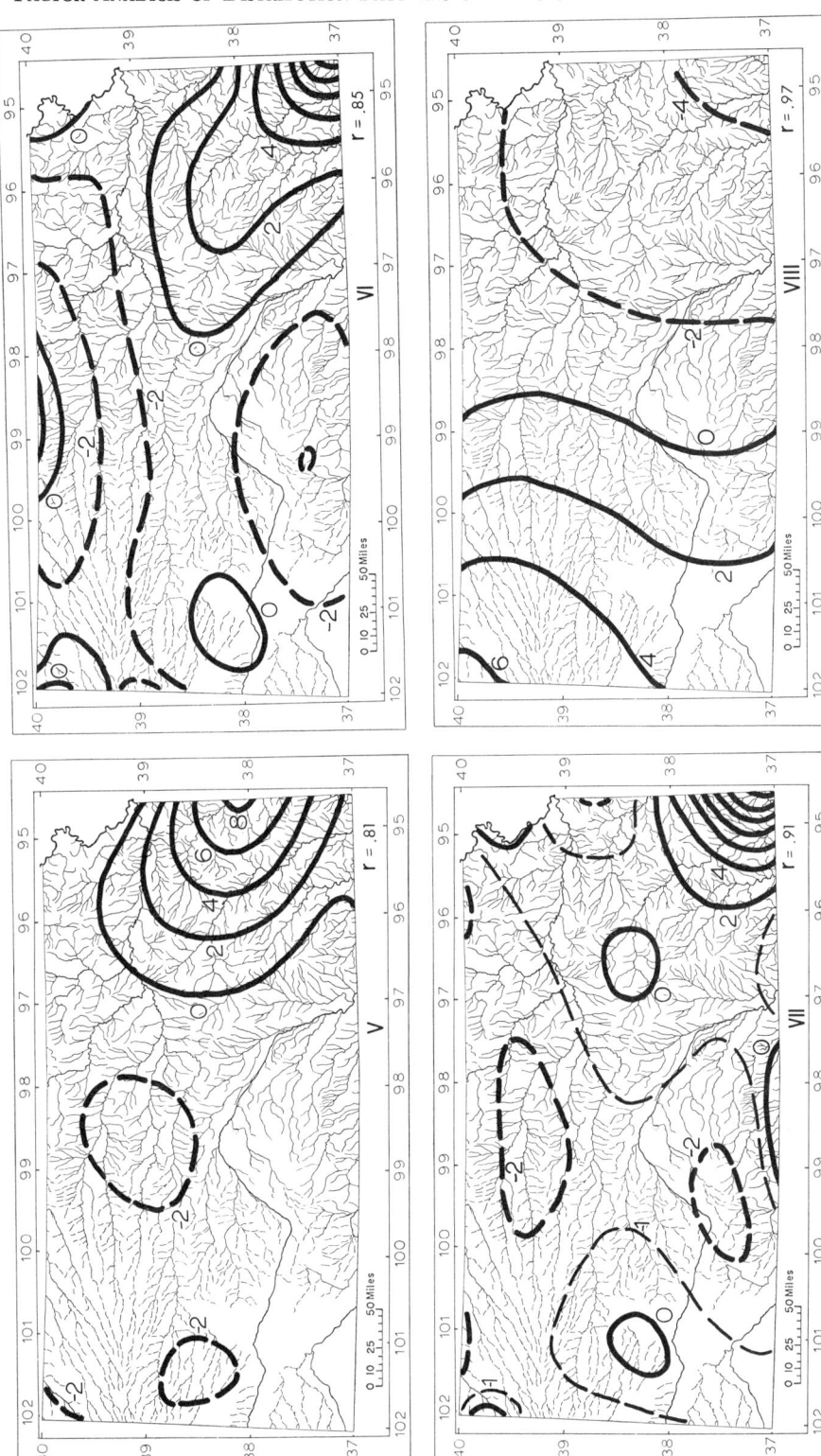

Fig. 2. Trend-surface maps of factor loadings for factor V, species centered in the Osage River drainage; factor VI, species centered in the Neosho River drainage; factor VII, species centered in the Spring River drainage; factor VIII, environmental variables.

TABLE 3. Species associated with factor IV—southern distribution patterns*

Species[1]	Correlation with factor IV	Correlation with other factors	Elevation	Number frost-free days	Number days reaching 90°F	Number days reaching 32°F	Annual precipitation	Annual runoff	Lake evaporation	Pliocene Ogallala Formation
Dorsoma cepedianum	+		−	+	−		+	+	−	−
Notropis umbratilis	+		− −	++	−	− −	++	+++		
N. boops	+	(VII:+)		+	−		+	+		
N. camurus	+++		−	++			+	+		
N. volucellus	++	(VII:+)	−	+		−	+	+		
N. buchanani	+		− −	++	−	− −	++	+++	− −	−
Pimephales vigilax	++		−	++		−	++	+		
P. tenellus	++		−	++			+	++		
Ictiobus bubalus	+		−	+	−		++	++	−	
Minytrema melanops	+		−	++		−	++	++	−	
Moxostoma erythrurum	++		− −	++		− −	++	++	−	−
Fundulus notatus	++		−	++		− −	++	++		
Gambusia affinis[2]	++			++	−			+		
Labidesthes sicculus	+++		−	++		− −	++	++		+
Micropterus punctulatus	++		−	++		− −	+	+		+
Lepomis macrochirus	+		−	+	−	− −	++	++		
L. megalotis	++		−	++		− −	++	++	−	
Percina phoxocephala	++		−	++		− −	++	++	−	
P. caprodes	+		− −	++		− −	++	++	−	
P. copelandi	++		−	+	−		+	++		
Etheostoma whipplei	+		−	+		−	+	++		

* Symbols as in Table 1.
[1] Factor-extraction by the method of Wilkinson and Householder included two additional species, *Noturus nocturnus* and *Etheostoma spectabile*, and two environmental variables, number of frost-free days (+) and number of days with the maximum temperature less than 32°F (−−) with this group. *Opsopoeodus emiliae*, known in Kansas from one locality in the Verdigris drainage, was not included in the factor analysis, but belongs with this group.
[2] Range artificially extended northward by recent introductions.

TABLE 4. Species associated with Factor V—the Osage River distributions*

Species	Correlation with factor V	Correlation with other factors	Elevation	Number frost-free days	Number days reaching 90°F	Number days reaching 32°F	Annual precipitation	Annual runoff	Lake evaporation	Pliocene Ogallala Formation
Polyodon spathula[1]	(+)	I:+								
Lepisosteus platostomus[1]	(+)	I:+	−		−		+	+	−	
Notemigonus crysoleucas	+		−	+		−	+	++	−	
Hybopsis biguttata	+				−		+	+	−	
Carpiodes velifer	+				−		+	+		
Hypentelium nigricans	(+)	VII:++					+	+		
Ictalurus furcatus	(+)	I:+							−	
Noturus gyrinus	+++									
N. nocturnus[1]	+			+			−	+	+	
N. exilis	++	(VII:+)	−			−	+	++	−	

* Symbols as in Table 1.
[1] Factor-extraction by the method of Wilkinson and Householder dropped these three species from this group.

TABLE 5. Species associated with factors VI (Neosho River) and VII (Spring River in Southeastern Kansas)[1]

Species	Correlation with factor VI	Correlation with factor VII	Correlation with other factors	Elevation	Number frost-free days	Number days reaching 90°F	Number days reaching 32°F	Annual precipitation	Annual runoff	Lake evaporation	Pliocene Ogallala Formation
Ichthyomyzon castaneus		(+)	I:+					+	+	−	
Chrosomus erythrogaster		(+)	II:+								
Hybopsis x-punctata	+++	(+)			−	+	−	+	+		
Notropis pilsbryi	++	(+)							+		
N. boops		(+)	IV:+	+		−		+			
N. spilopterus		++							+		
Hypentelium nigricans		++	(III:+)					+	+		
Noturus exilis		(+)	V:++	−		−	−	+	++	−	
N. placidus	++	+	(II:+)	−		+	−	+	+		
N. miurus	(+)³	++			+				+		
Morone chrysops		(+)	I:+								
Micropterus dolomieui	(+)	+			+		−	+	+		
Chaenobryttus gulosus		+						+	+	−	
Ambloplites rupestris	(+)³	+++							+		
Pomoxis nigromaculatus		+									
Etheostoma chlorosomum	(+)³	+++						+	+		
E. zonale		++							+		
E. blennioides	(+)³	+++						+	+		
E. cragini		+					+				
E. flabellare	+	(+)			+		−	+	+		
E. gracile		++			+			+	+		

[1] Symbols as in Table 1. Eleven additional species were known from Kansas from one or two localities only and were not included in the factor analysis. *Dionda nubila*, *Moxostoma duquesnei*, *Fundulus sciadicus*, *Cottus carolinae*, *Etheostoma punctulatum*, and *E. microperca* are known in Kansas only from the Spring River and its tributaries, and belong to group VII. *Notropis chrysocephalus* (see Cross, this volume), *Percina shumardi*, *Etheostoma stigmaeum* are rare in the Neosho and Spring River drainages in Kansas, and can be regarded as members of group VII. *Lepisosteus occulatus* and *Hybopsis amblops* are known from single localities in the Neosho drainage (group VI).

[2,3] *Moxostoma carinatum* was added to group VII, and the species indicated [3] were dropped from group VI in the results of the factor-extraction using the method of Wilkinson-Householder.

for these species would appear to involve relatively low elevations, a long, warm growing season, and abundant permanent water. The broad pattern of these species suggests expanding range, as opposed to the contracting, relict ranges of factor-group II. Factor IV is relatively highly correlated ($r = .7$) with factor VI.

Factor V: Eastern species associated with the Osage River drainage.—Ten species (Table 4, Figs. 2, 5) are correlated with factor VI, but only six of these have their highest correlation with this factor. Six of the species are also correlated with either I or VII. The group includes a species restricted in Kansas to the Osage drainage, *Noturus gyrinus*, several species common in the Osage and adjacent tributaries to the Kansas River, and several large-river species for which the Osage drainage provides favorable habitat. A number of the species show high correlations with the environmental variables associated with abundant permanent water. It is possible that association with the Osage drainage is historically important to the dispersal of some of the species in this factor-group, but habitat appears to be the primary influence on these patterns.

Factor VI: Species associated with the Neosho River drainage.—Nine species (Table 5, Figs. 2, 5) are correlated with this factor; all of these are also

TABLE 6. Distribution patterns independent of factors I-VII*

Species	Factor with highest correlation	Correlations with environmental variables							
		Elevation	Number frost-free days	Number days reaching 90°F	Number days reaching 32°F	Annual precipitation	Annual runoff	Lake evaporation	Pliocene Ogallala Formation
Lepisosteus osseus	IV	−	+			+	++	−	−
Amia calva	I								
Anguilla rostrata	I								
Notropis atherinoides	I								
N. dorsalis[1]	I								
N. rubellus[2]	V	−	+	−	− −	++	++		
N. girardi	IV			+				−	−
N. heterolepis[2]	IV						+		
Hybognathus hankinsoni[1,2]	I								
Ictiobus cyprinellus	I	−			−	+	+	−	−
Carpiodes cyprinus	III								
Moxostoma carinatum	VI		+		−	+	++		
M. macrolepidotum[2]	IV	− −	+		−	++	++	−	
Ictalurus natalis[2]	III	−			−	+	+		
Pylodictis olivaris[2]	V								
Noturus flavus[2]	III	−			−	+	++	−	−
Micropterus salmoides[2]	V		−			+	+	−	
Stizostedion vitrium[2]	V								
Pomoxis annularis	III	−	+		−	++	++	−	−
Percina maculata[2]	II								
Aplodinotus grunniens[2]	V	−				+	+	−	−

* Symbols as in Table 1.

[1] Actually appeared correlated with factor I as a result of having been taken in small streams adjacent to the Missouri and Kansas rivers in Kansas.

[2] Species having no factor-group correlations higher than $r=.28$ in results of the Wilkinson-Householder method of factor-extraction.

correlated with factor VII. The distributions of these species in Kansas tend to be centered in the Neosho River drainage, suggesting a possible historical relationship to this drainage. The species are also consistently correlated with high annual runoff, indicating restriction to permanent streams. The correlation between factors VI and VII is high ($r=.8$).

Factor VII: Species associated with Spring River.—Twenty-one species (Table 5, Figs. 2, 5) are correlated with this factor; all but six of these are also correlated with some other factor or factors (an additional six species are limited in Kansas to Spring River and belong to this group, although they were not included in the computer analysis because they are known from one locality only). Spring River, in extreme southeastern Kansas, is a stream situated at the edge of the Ozark Highlands. Because of this, Spring River is the focal point in the distribution of Ozarkian fishes in Kansas. It is also in the vicinity with the highest climatic equability in Kansas (Fig. 3). These fishes show high correlations with environmental variables associated with permanent streams. Although the distribution of these fishes in Kansas is distinctly southern, they do not show the high degree of positive correlation with length of growing season that is characteristic of the members of group IV.

Factor VIII: Environmental variables.—The design of the analysis was intended to provide maximum opportunity for correlation between environmental variables and factor-groups.

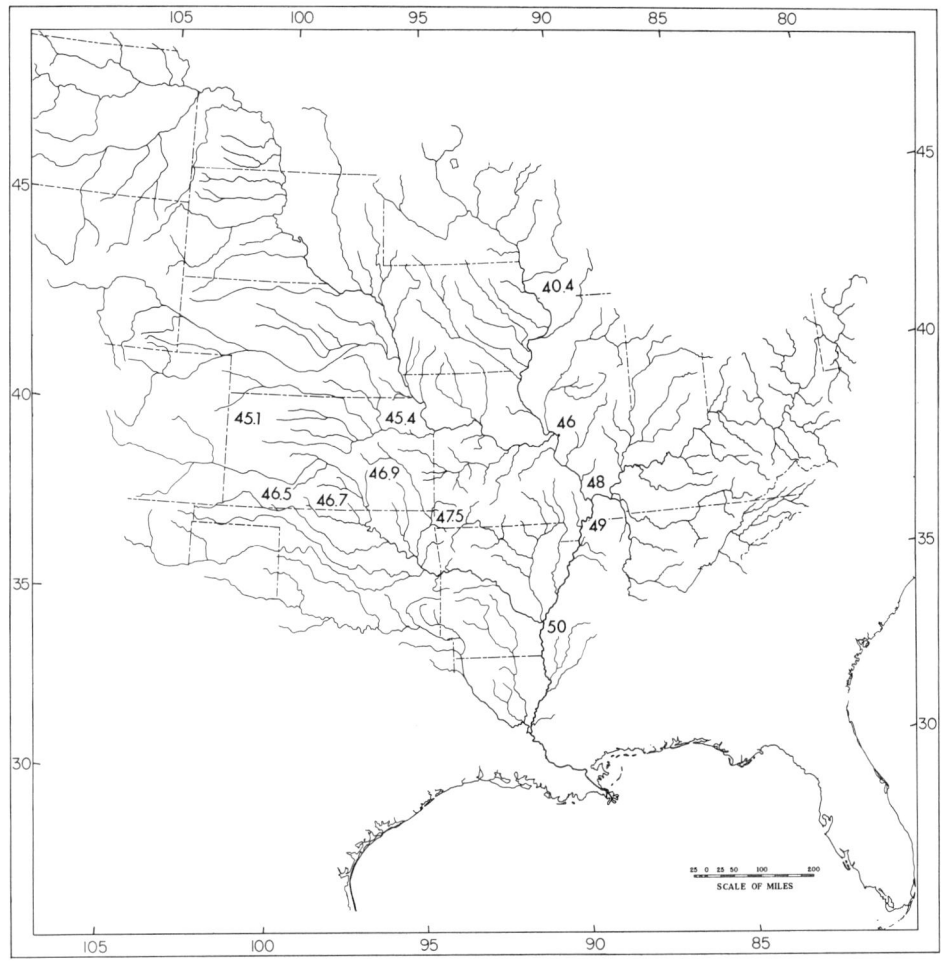

FIG. 3. Kansas is drained by the Kansas-Missouri system, which joins the Mississippi River at St. Louis, Missouri, and the Arkansas, which joins the Mississippi River in southeastern Arkansas. The figures give estimates of climatic equability, calculated according to Axelrod and Bailey, 1968.

However, the distribution patterns of environmental variables were more similar to each other (Fig. 2) than any was to any of the generalized factor distributions. The high correlations among environmental variables made it possible to select a smaller number as examples for expression in Tables 1-6.

The inclusion of major drainage patterns among the variables showed little about endemism or centers of dispersal relative to the factor-groups. The Neosho and Osage drainages as variables were highly correlated with factors VI and V, as expected. The Verdigris drainage is correlated ($r = .66$) with the warm-water factor-group (V). The prairie factor-group (III) is negatively correlated ($r = -.36$) with the Arkansas River drainage west of the Verdigris River, although all but one of the 19 species correlated with factor III occur in the Arkansas as well as the Kansas drainage.

Most of the fishes whose distributions appear highly correlated with climatic variables also show high positive correlations with the Carboniferous and Permian limestones and shales of eastern Kansas. Many of these species are negatively correlated with the Tertiary Ogallala group (sands, gravels, silts, etc.) of western Kansas. Limestone may contribute to the favorable habitat for these fishes through influence on bottom type, control of stream flow, and contribution to nutrient richness in the aquatic environments. The contact between the water-bearing Ogallala formation and the underlying aquicludes, the Niobrara chalk and the Pierre shale, provides spring-fed, permanent streams important to the distribution of plains fishes (Metcalf, 1966), including some of those in factor-group II and *Hybognathus hankinsoni*, which has a positive correlation ($r=.38$) with the Pierre shale.

Among the climatic variables tested, length of the growing season and annual runoff showed the highest, most consistent correlations with the largest number of fish distributions, and are interpreted to be the most influential limiting factors examined. Elevation and evaporation are negative limiting factors associated with the above.

Independent distributions.—Nineteen fishes have sufficiently unique distribution patterns that they were not highly correlated with the eight factors extracted by the complete centroid method (Table 6). However, six of these were allocated to groups by the method of Wilkinson and Householder (see below), and can be regarded as relatively nonconformable members of those groups. In addition, *Notropis atherinoides, N. girardi,* and *Carpiodes cyprinus* bear special relationship to a group associated with the Arkansas River drainage. *Notropis rebellus,* an upland stream species, is approximately equally correlated with factors IV, V, and VI ($r=.23$) and is highly correlated with environmental variables. *Ictalurus natalis* and *Noturus flavus* are clear-stream fishes of eastern Kansas with distributions otherwise correlated with factor III. *Moxostoma macrolepidotum* is a large-stream inhabitant of eastern Kansas, and its distribution is correlated with environmental variables. *Notropis heterolepis* has undergone recent range restriction (see Cross, this volume). The distributions of *Pylodictis olivaris, Micropterus salmoides, Stizostedion vitreum, Aplodinotus grunniens,* and perhaps *Ictaluris natalis* have been influenced by their occurrence in impoundments or their transport for fisheries in the state, which may account for their failure to sort out with the factor-groups. *Percina maculata,* limited (in Kansas) to clear streams in Wabaunsee and Riley counties, is clearly associated with factor-group II ($r=.25$). Three species on the "independent" list, *Notropis atherinoides, Notropis rubellus,* and *Moxostoma macrolepidotum,* are taxonomically complex and probably include more than one adaptive response to environmental limitation. The same may be true of others, including *Hybopsis aestivalis. Hybognathus hankinsoni* and *Notropis dorsalis* are limited to a few cool, sandy-bottomed streams in Kansas. Their highest correlations in this study were with factor I, but this is an artifact of their occurrence adjacent to the Missouri and Kansas rivers in northeastern Kansas. *Hybognathus hankinsoni* also shows relictual distribution in spring-fed headwaters of the Smoky Hill and Republican rivers in northwestern Kansas.

Competitive exclusion.—Correlations between species were searched for high negative values as indicators of possible competitive exclusion. The method is generally nonproductive because two of three possible expressions of such an interaction would not be recognized by correlations based on occurrence or absence in sample quadrats of the size used in this analysis. One possible type would involve species geographically sympatric but exclusive at the microhabitat level. A second would involve species with complementary distributions, but whose mu-

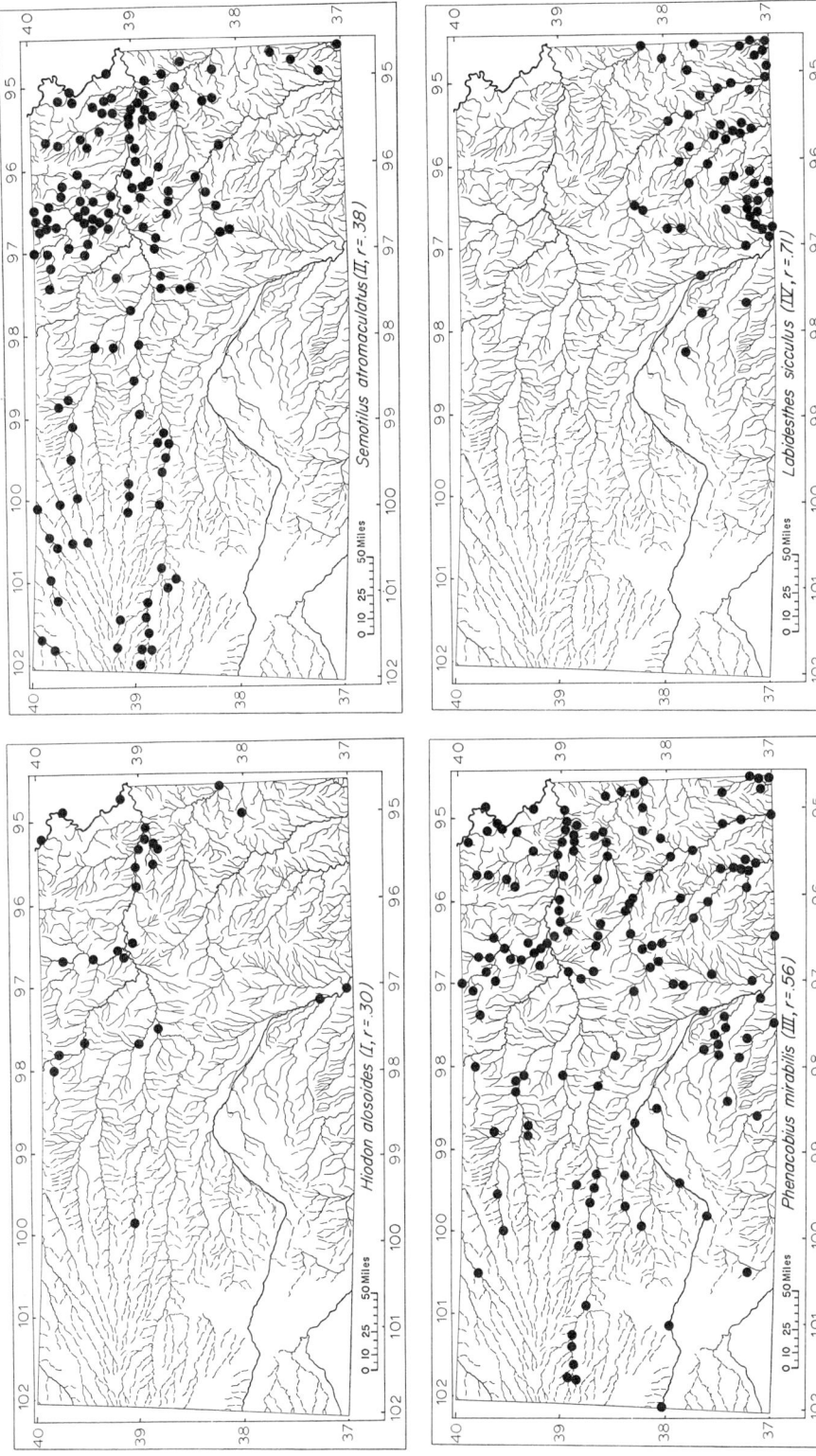

FIG. 4. Examples of species distributions representing factor-groups, I—*Hiodon alosoides*, II—*Semotilus atromaculatus*, III—*Phenacobius mirabilis*, and IV—*Labidesthes sicculus*. The r value gives the correlation between the species distribution pattern and the factor.

tual absence from a large part of the study area would depress the correlation coefficient. The noticeable form would involve two widespread species such as *Fundulus kansae* and *Fundulus notatus* (Fig. 5) whose ranges are largely complementary. The correlation coefficient for the distributions of these two species is —.24, and competitive exclusion might be involved.

TEST OF REPEATABILITY

The factor extraction was repeated with the method of Wilkinson and Householder (Ralston, 1965) to test the objectivity and repeatability of this analysis. Eight factor-groups appeared that corresponded to the eight factor-groups extracted by the complete centroid method, except that group II was included in group III and a new group was comprised of fish distributions in southern Kansas. The other seven groups corresponded well with those outlined in Tables 1 to 6, except as noted in the footnotes on those tables. The new group consists of forms found in the Arkansas River drainage in Kansas, including *Hybopsis aestivalis, Notropis atherinoides, Notropis girardi, Hybognathus placitus, Carpiodes cyprinus, Gambusia affinis, Etheostoma cragini, Fundulus kansae,* and the widespread species *Carpiodes carpio* and *Ictalurus punctatus*. High correlations suggest that the annual number of days over 90°F, salinity, absence of extreme cold (as measured by number of days with maximum less than 32°F), and some Arkansas River endemicity are causally related to this general pattern. (The extraction of six factors by the complete centroid method also deviated only in the recognition of this group, but without the two widespread species mentioned above or *Fundulus kansae*.) The environmental variables had relatively higher correlations among the eight fish factor-groups. The "independent" group was reduced from 19 to 10 distributions.

The method of Wilkinson and Householder is the more powerful of the two techniques used for factor extraction in this study. It is regarded as one of the "exact" methods, as opposed to the much faster "approximate" methods such as centroid. The results are comparable, however, and the centroid technique can be regarded as satisfactory for cases in which computation facilities or time are limiting. We conclude that factor analysis provides a direct and powerful tool for determining groups based on similarity of distribution and that the structure of these groups is largely independent of the method of factor extraction.

HISTORICAL ASPECTS

About 38 percent of the rich, fresh-water fish fauna of the Mississippi Basin is native to at least some Kansas waters. The major river systems draining Kansas—the Missouri and the Arkansas—join the Mississippi River several hundred miles to the east and southeast (Fig. 3). All but three of the 121 species native to Kansas have their geographical affinities to the east, northeast, or southeast, most of them in or near the center of the Mississippi Basin, a lesser number in the Ozark highlands, and a few each in the Missouri and Arkansas river basins. Forty-four species are restricted to either the Arkansas or Missouri systems in Kansas, but only 19 of these are so restricted when their entire range is considered. Five of these species are large-river inhabitants (group I) absent from the Arkansas River. Of the remainder, three are in the Kansas drainage, 11 are in the Arkansas drainage.

The general picture is one of distribution through the central confluence area of the Mississippi Basin (the dispersal "crossroads," as well as the

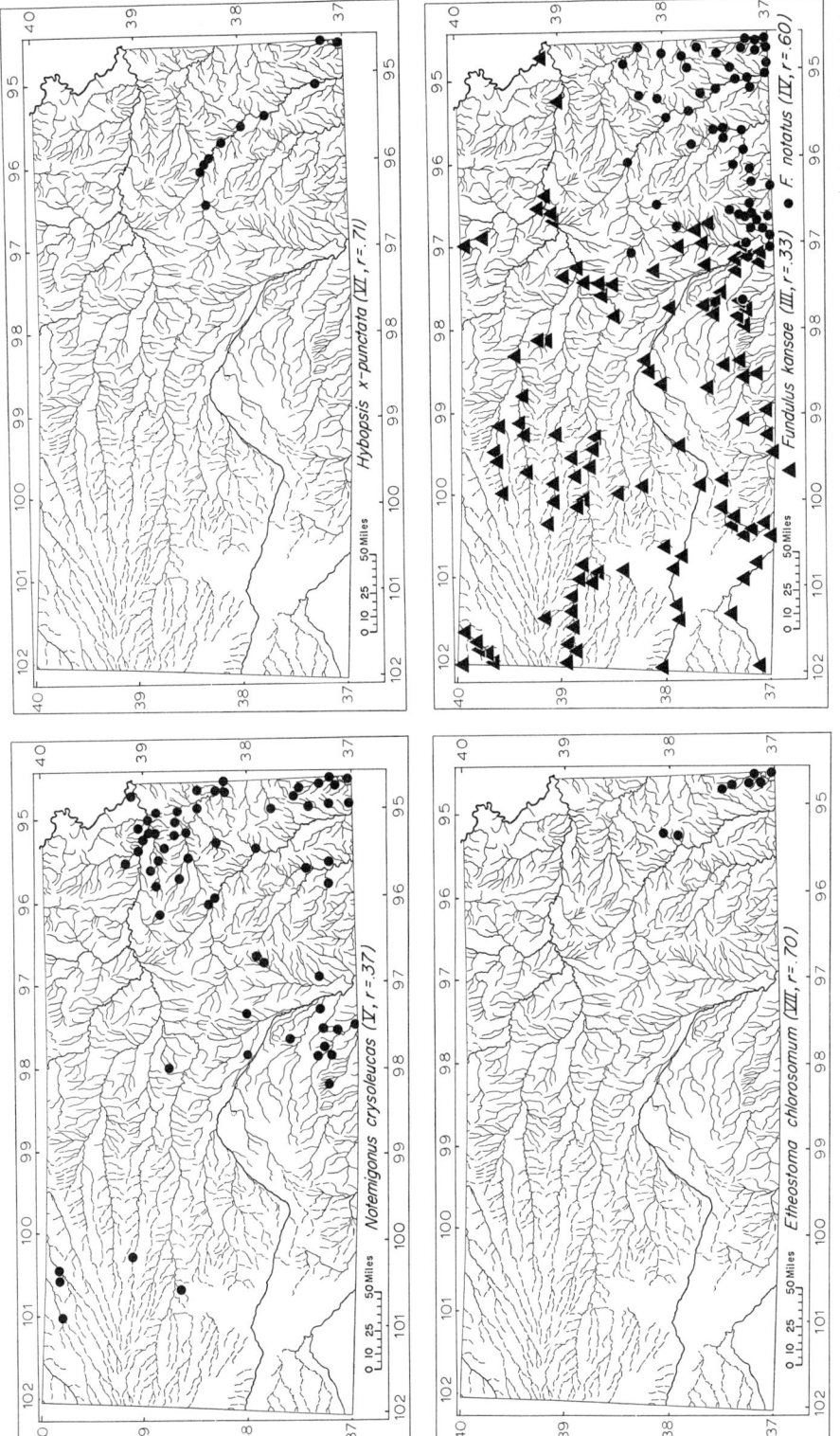

Fig. 5. Examples of species distributions representing factor-groups, V–*Notemigonus crysoleucas*, VI–*Hybopsis x-punctata*, VII–*Etheostoma chlorosomum*, V–*Note* and two complementary distributions, *Fundulus kansae* and *Fundulus notatus*.

area with the highest climatic equability, Fig. 3), with western range limitation primarily by elevation, climate, or habitat restriction, rather than by lack of access across divides. There is little evidence of differentiation of plains stream fishes in isolation by drainage divide barriers in western Kansas, and few instances in eastern Kansas. This indicates a history of significant faunal transfers accompanying stream captures in the evolution of the drainage basins, especially in the western half of the state. However, the distributions of several restricted species argue against the postulation of frequent drainage transfers. *Notropis cornutus,* in the east and west ends of the Kansas River drainage, and the southeastern species *Notropis camurus, Pimephales vigilax, P. tenellus, Micropterus punctulatus,* and *Percina phoxocephala* in the Arkansas drainage, have been found close enough to the drainage divide to be potential transfers, but remain restricted. The distribution of *Notropis cornutus* (group II) clearly argues against southward stream captures in the vicinity of its range during its present occupation (see Cross, this volume). However, considered in conjunction with the distribution of the many species on both sides of the divide, the pattern suggests relatively recent occupancy of the Kansas River drainage by *Notropis cornutus.* The type II distribution pattern of this species indicates dispersal into the present range during colder times. The period of access can be estimated as Wisconsin on the basis of the argument for recency plus the consideration that the upper Smoky Hill River, which it occupies, has been tributary to the Kansas River only since early Illinoian time (Bayne and Fent, 1963). The five southeastern species mentioned above all belong to factor-group IV, indicating primary limitation of their distribution in Kansas by length of growing season and permanence of stream flow. The distribution of *Percina phoxocephala* supports arguments against transfers from the Neosho and Marais des Cygnes (Osage) drainages into the Kansas, and the other three species weigh against transfers from the Neosho drainage to the Marais des Cygnes or Kansas drainages.

It is assumed in these arguments that if any of these six species gained access to the waters on the other side of the divides, their successful dispersal would not be limited by competition with species presently occupying the area. This is supported by the distributions of ecologically associated species that do occupy all major drainages in the vicinity. It is the distribution of these species, more than 36 of which have patterns that freely span the drainage divides in question, that suggests that the divides have not been significant barriers to the majority of the fauna. The well-documented transfer of the upper Saline and Smoky Hill rivers from the Arkansas to the Kansas drainage in early Illinoian time (Bayne and Fent, 1963) effected a significant northward faunal transfusion and probably allowed limited southward exchange for plains headwater species. Two endemics of the Arkansas drainage, *Notropis girardi* and *Etheostoma cragini,* are not found north of the Arkansas River on the plains and were probably unaffected by the shift of the divide.

In the above discussion, the terms divide and barrier have been used in the sense of the perimeter of the basin, which is uncrossed by aquatic habitat and fishes except by stream piracy. However, the first- and second-order streams near the perimeter of a basin are firstly, the streams most frequently involved in piracy and secondly, the environment of a distinctive headwater habitat characterized by higher elevation and smaller stream dimensions. In eastern Kansas, because of the relief and hard-rock substrate, the headwater habitat is also characterized by more rapid current and cooler waters; the reverse may hold in many environments of western Kansas. In either case it is clear that the headwater habitat provides an ecological barrier to some fishes (for example, group I in this analysis) and a habitable or optimum site for others (groups II and III), and thus directly determines their potential access to transfer by stream capture.

DISCUSSION

The numerous high correlations between individual distribution patterns and environmental variables further suggest the predominant potential of ecological explanations to account for fish distributions. Most systematically oriented investigations have traditionally sought historical explanations involving hypothetical past zoogeographic relations. The few patterns that are logically approachable owing to uniquely restrictive circumstances are highly interesting, but the vast majority of distributions lack logically useful historical controls and remain unsatisfactorily explained. Sources of ecological information remain generally unused, possibly owing to the lack of methods for detailed analysis and objective summarization. Environmental data were somewhat naively included in the present study to assay for patterns of correlation. The results suggest that a more detailed inclusion of more and better ecological data would enable more powerful explanations of distribution limits and patterns. Such general data as climatic equability and effective temperature (Axelrod and Bailey, 1968), soil types, nutrient availability, incident radiant energy, and the like, as well as more specific limnological data such as stream flow means and extremes, proportion of riffles and pools, daily temperature fluctuations, substrate, current, and the availability of benthic macroinvertebrates would be worth testing. Inclusion of specific environmental data would require concomitant collection of fish samples and would warrant inclusion of abundance estimates as well as data on trophic position and age structure. Past systematic sampling of localities has usually stressed effort to obtain the most complete listing of all, including rare, members of a taxonomic group to the exclusion of data on abundance and population structure. It might become necessary for collections also to include samples designed to reflect abundance and population structure, in addition to environmental data, for future zoogeographic studies. As regards paleozoogeography, if inferences to past climates are to use data on distribution and ecology of Recent faunas, then more detailed information on ecological limitation on Recent species is necessary.

CONCLUSIONS

Fishes occurring in Kansas can be divided into seven or eight groups based on similarity of geographical distribution in the state. The generalized patterns of these groups and the correlations between species ranges and environmental variables suggest possible environmental and historical explanations of fish distributions.

Group I includes 21 species, 17 percent of the native fauna, common to the distinctive habitat provided by large rivers. Five of these species do not occur in the Arkansas River, but the remainder are in both major drainages within the state. These species appear to be minimally influenced by the climatic variables tested, possibly because of the relatively stable and unique large-river habitat.

Group II is associated with group III, prairie and plains species, but emphasizes fishes requiring cooler waters. The most highly correlated members of this group are cool-water relicts whose distribution on the plains was more widespread in cooler, wetter times.

Group III encompasses prairie and plains species, most of which are widespread in Kansas. Few of the distributions are highly correlated with environmental variables. Most of the species occupy small streams and occur on the fringes of drainage basins, where they have probably been subject to transfer by capture of small streams during the evolution of the basins. A restricted exception is *Notropis cornutus*, limited to the Kansas drainage, which it probably invaded during Wisconsin time.

Group IV contains 21 species that have a generally southeastern distribution in Kansas but often more northerly limits in the center of the Mississippi Basin. High correlations suggest that these species are responsive to lower elevations, long and warm growing seasons, and abundance and permanence of water. Twelve of the species are excluded from the Kansas Basin, but only four of these penetrate the headwater habitat far enough to be available for transfer by small stream capture, and these four apparently ascend first-order streams but rarely.

Group V consists of a small group of species found, or formerly found, mainly in the Marais des Cygnes (Osage) drainage in Kansas. The Osage Basin has probably been especially important to the plains dispersal of some of these forms, but the historical and ecological interplay is unclear.

Group VI is made up of several species found in Kansas only, or primarily, in the Neosho drainage. These species are found in the larger streams in the basin and are probably limited by permanence of stream flow.

Group VII comprises a total of 28 species (or 23 percent of the native fauna) associated with Spring River, in the extreme southeastern corner of the state on the edge of the Ozark highlands. This area enjoys the highest climatic equability and supports the richest fauna in the state. At least 74 species of fishes have been collected in the Spring River drainage in Kansas. The 28 species of factor-group VII are Ozarkian species with Kansas distributions radiating out from a focal point in the Spring River corner. These species are not as dependent on growing season as are those of group IV, but are sensitive to stream permanence.

Several analyses of our data suggested the existence of another group of species—10 plains taxa with distributions widespread in the Arkansas drainage. Nineteen species have distinctive patterns not highly correlated with the factors. These species will require more specialized explanations, although nine of them were associated with factors when a more powerful analytic technique was employed. Several methods of factor analysis were tried and produced basically the same kinds of groupings, usually with essentially the same composition.

The Kansas fish fauna consists of Mississippi Basin fishes that range as far west as Kansas because they are adapted to either large rivers, the Ozark streams, or the prairie and plains streams, via the Arkansas or Missouri river drainages. The majority of the species occur in or near the center of the Mississippi Basin and occur in both the Kansas and Arkansas drainages. Drainage transfers have been effective in distributing at least those members of the fauna that inhabit streams near the edges of the basins.

Distributional evidence bearing on the centers of origin of the species of this fauna is virtually erased by the occupation of the central, climatically equable part of the Mississippi Basin by many species that might have arisen elsewhere and by the dominance of ecological factors in the determination of species limits. Nineteen species are restricted to the Arkansas or Kansas basins; however, some unknown but significant number of these were not so restricted in earlier times in the Pleistocene (see Cross, this volume). Only one species in Kansas is distinctly western. There is little direct evidence in the fish distribution patterns to suggest that other members of the fauna had western origins separate from the Mississippi Basin (but see Metcalf, 1966).

Considering that the habitation of Kansas waters by fresh-water fishes is a matter of individual degrees of adaptation to a continuum of environmental conditions, it is perhaps surprising that the distribution patterns do indeed fall into groups based on similarity. The nature of the drainage is such that fishes have had dispersal access to Kansas waters by groups; but in fact, the generalized patterns reflect ecology more than they reflect dispersal routes. It appears that solid data bearing on distribution are to be found in fossil occurrences and ecology; speculations on centers of origin and dispersal remain difficult.

ACKNOWLEDGMENTS

We are indebted to F. B. Cross for collecting and compiling the distributional records on which this study is based and to R. R. Sokal and F. J. Rohlf for introducing us to the multivariate techniques used in the analysis. These colleagues and teachers at the University of Kansas have also provided patient help and numerous suggestions regarding fish distributions and numerical analysis. The trend-surface program was provided by J. C. Davis of the Kansas Geological Survey. The research of D. R. Fisher was supported by grant GM 11935 from the National Institute of General Medical Sciences to Robert R. Sokal. The computations were carried out at the University of Kansas Computation Center using the NT-SYS system of multivariate computer programs developed by F. J. Rohlf, J. R. L. Kishpaugh, and R. L. Bartcher.

LITERATURE CITED

Axelrod, D. I., and H. P. Bailey
1968. Cretaceous dinosaur extinction. Evolution, 22:595-611.

Bayne, C. K., and O. S. Fent
1963. The drainage history of the upper Kansas River Basin. Trans. Kansas Acad. Sci., 66:363-377.

Cross, F. B.
1967. Handbook of fishes of Kansas. Misc. Publ. Mus. Nat. Hist., Univ. Kansas, 45: 1-357.

Fisher, D. R.
1968. A study of faunal resemblance using numerical taxonomy and factor analysis. Syst. Zool., 17:48-63.

Flora, S. D.
1948. The climate of Kansas. Rep. Kansas State Bd. Agric., 67:1-297.

Metcalf, A. L.
1966. Fishes of the Kansas River system in relation to zoogeography of the Great Plains. Univ. Kansas Publ., Mus. Nat. Hist., 17:23-189.

Miller, D. W., J. J. Geraghty, and R. S. Collins
1963. Water atlas of the United States. II. Water Inf. Center, Port Washington, New York, 40 pl.

Orloci, L.
1967. An agglomerative method for classification of plant communities. Jour. Ecol., 55:193-205.

Ralston, A.
1965. A first course in numerical analysis. McGraw-Hill, 587 pp.

Rohlf, F. J.
1962. A numerical taxonomic study of the genus *Aedes* (Diptera, Culicidae) with emphasis on the congruence of larval and adult classifications. Unpubl. Ph.D. Dissertation, Univ. Kansas, Lawrence, 98 pp.

Sokal, R. R.
1958. Thurstone's analytical method for simple structure and a mass modification thereof. Psychometrika, 23:237-257.

Thurstone, L. L.
1945. Multiple factor analysis. Univ. Chicago Press, 535 pp.

The North American Central Plains as an Isolating Agent in Bird Speciation

ROBERT M. MENGEL

ABSTRACT

The Central Plains and the environments immediately surrounding them are considered, with special reference to their respective roles in the speciation and evolutionary radiation of birds.

The temperate grasslands and cool sagebrush deserts of the Central Plains support a small avifauna in which some 37 essentially grassland-adapted species are important; only 12 are endemic. Relations with Old World steppe are slight and mostly at family and generic levels. Only three species (one passerine) are shared. (The Horned Lark—the only tundra-adapted and only New World lark— may have invaded the New World by way of alpine tundra, which breaches the world-wide band of taiga that otherwise separates the steppes and arctic tundra.) Relations with South American savannah are little closer, five species being shared (three passerines, again including the Horned Lark). Only four genera *(Tympanuchus, Sturnella, Ammodramus, Spizella)* presently contain as many as two important grassland species.

Thus, despite considerable antiquity and regional variations in plant composition, the grasslands and Central Plains are poor in avian niches, perhaps chiefly for want of vertical stratification. The scarcity of closely related congeners—not to mention species groups—suggests there has been little past fractionation of the Central Plains into isolated environments. Thus the Central Plains seem to have played an unimportant role as a stage for bird speciation.

The more complex environments peripheral to the Central Plains—for example, various scrubs and, especially, forest and woodland formations—support larger and more complicated sets of bird species. Here the number of species per genus is presently near 3.0—compared to about 1.1 in the Central Plains—and many genera possess several species (14 have five or more, and six possess eight to 21 species!). Thus there are not only many cases of closely related congeners in these varied formations, but also many superspecies and complex species groups occur.

Attention is directed to such complexes, starting with species and semispecies pairs and proceeding to species groups. Some 100 taxa are considered, arranged in seven groups containing 34 different complexes whose ranges are mapped.

Finally considered—as the last of several bases for discussions to follow—are some 69 species important in the taiga (whether or not previously discussed), divided into truly boreal (cold-adapted) and "summer boreal" (warmth-adapted) groups.

Clearly, speciation around the Central Plains has been much more extensive than in them, so that the most important role of the grasslands has been that of an isolating agent. Evidence for the effectiveness of this isolation is considered to be good. The analogy is then developed of the Central Plains as a changeable but persistent "sea" separating the eastern deciduous forest "continent" from western montane forest and woodland "islands." The northerly taiga serves as a "land bridge" that periodically connects the two. This bridge is relatively broad during interglacials, when the grasslands are large; and it is narrow and frequently broken during glacial periods, when the grasslands are small and invaded by savannah and woodland. In full-glacial times, taiga species would tend to be isolated in eastern and western refugia and could begin to differentiate. During interglacial times, western differentiates would be still more effectively isolated as islands of montane forest retreated higher in the mountains and shrank in the surrounding grasslands sea. New evidence for full-glacial fractionation of taiga is considered, and evidence concerning the pattern of Wisconsin glacial retreat is examined, the latter showing why eastern species have reached the northwest in so many more instances than have western species.

A model is developed showing what would theoretically happen if an eastern species invaded the taiga after the first major glaciation, was disjoined by a second, and differentiated into two species. Then the eastern species again expanded upon the second glaciation, and so on, the entire process being repeated four times. The result, if no steps are omitted, is a "continental" eastern descendant species, presently with a large and expanding taiga distribution, and three more or less isolated western montane "insular" forms. The most recently formed of the latter should tend to be most like the eastern taxon and should tend to be allopatric *inter se* (as in superspecies) but sympatric with the earlier-formed western montane isolates (which should be less like the eastern species). Further, the more newly formed differentiates in the western group should tend to displace their predecessors southward or into marginal environments, or both (being continental forms competing with theoretically less fit and aggressive insular forms).

This model, developed in an earlier paper to account for the speciation of certain groups of wood warblers, is examined for its further relevance to these and other groups. Besides the Black-throated Green Warbler and Nashville Warbler groups of five taxa each, which fit the model remarkably well, a very good fit is found in the Least Flycatcher group of five species, and some fit is found in the northern juncos (which, however, are thought to have a somewhat different history). Various others of the species groups considered earlier are shown to have conformations compatible with the model but variously incomplete (because of shorter histories or extinctions). Among these are the Mourning Warbler and Acadian Flycatcher groups of three species each, and various species pairs. It is noted that many taiga species today are theoretically vulnerable to isolation in case of another glaciation.

Probable exceptions to, and variations upon, the theme of the model are considered, among them Pliocene isolation of stocks (some of which may then have followed the model course), simultaneous differentiation of more than one western isolate, and the possibility of some radiations occurring within the Wisconsin alone (considered unlikely). It is shown that western differentiates should be less likely to reinvade the taiga than eastern ones, and that this seems in fact to have happened but rarely.

Many groups, of course, must have histories entirely different from that of the model, as is indicated by consideration of the birds of the taiga from another standpoint. This shows that these birds fall into two distinct and quite different groups. These are the true boreal, cold-adapted species, which are sedentary or only moderately migratory, and the summer-boreal, warmth-adapted species, which are usually long-distance migrants. The first belong chiefly to groups of Old World origin, the second to groups derived in warm-temperate or tropical America. They have on the whole quite different distributional patterns. The boreal taiga species often occur widely in western montane forest but rarely invade eastern deciduous forest; the summer boreals display exactly the opposite pattern. Although quite a few of them show evidence of disjunction and speciation on opposite sides of Bering Strait, the boreals, curiously, while they greatly exceed the summer boreals in infraspecific geographic variation, have undergone comparatively little radiation of the type postulated by the model and have produced few eastern differentiates (the Carolina Chickadee is a rare exception). The matter is puzzling. Possibly the ability of boreals to occupy truly cold breeding ranges decreases their isolation by glaciation; some montane derivatives of the past may have perished in warm interglacials.

The summer boreals, in any case, show little infraspecific geographic variation but have undergone marked radiations, developing more than four times as many members of superspecies and species groups as have the true boreals—these being those considered in relation to the model. Of 31 summer-boreal species, 13 (42 percent) are members of nine superspecies or species groups, which contain 27 species, for an average of three per group; 26 true boreals have only two (8 percent) members belonging to such groups, which contain six species in all. The summer boreals, as we have seen, further include in their various species groups numerous western montane species not found in the taiga. The boreal groups contain virtually no species of this type (the Mexican Chickadee is a notable exception). Inability to breed in truly cold environments is suggested as a possibility favoring their isolation by glaciation. A possible tendency to establish marked allohiemy because of their long-distance migrations is examined as facilitating rapid speciation of western differentiates. Extensive panmixia due to migration, however, may at the same time inhibit infraspecific geographic variation.

Evidence for bridges south of the taiga is briefly considered. A rather large number of species pairs suggests the past existence of woodland and scrub connections between eastern forest and Mexican forests and woodlands of different kinds. Some of these pairs may also provide evidence for Pleistocene refugia in Florida and Mexico. The evidence, however, is limited and requires caution in its interpretation. Finally, the full-glacial occupancy of the Central Plains by woodland and savannahs (shrinkage of the sea) is seen as having afforded intermittent, broad bridges that prevented the full speciation of various woodland and forest-edge species—for example, Yellow-shafted and Red-shafted flickers, Baltimore and Bullock's orioles, Rose-breasted and Black-headed grosbeaks, Indigo and Lazuli buntings, which are more or less fully isolated under benign interglacial climates—while being inadequate to break down the isolation (and prevent the full speciation) of many truly forest-adapted taxa.

INTRODUCTION

Because this symposium is devoted to the Central Plains, I think it appropriate to begin with a comparatively brief survey of the grassland avifauna most characteristic of that region. Because there seems to have been little speciation within the grassland formation, this effort will necessarily tend toward the descriptive rather than the analytical. I shall define a Central Plains avifauna and then examine its diversity, distinctness, and relationships with other steppe avifaunas.

Having done so, I shall move on to surrounding areas that are more inviting to the study of speciation and biogeographical analysis. What we shall be studying are various aspects of island effects. As MacArthur and Wilson (1967, p. 3) have properly pointed out, far from being strictly oceanic features, islands are "a universal feature of biogeography." There is no reason, therefore, that radiations cannot also be studied on continents, and inferences made about their histories.

It is by movement from one island to another, followed by periods of isolation, that populations evolve into new species, and species increase in number. When one stock does much of this in a comparatively short time, radiation results, and we are familiar with the classical examples afforded by animals on oceanic island groups, for instance Darwin's finches in the Galápagos (Lack, 1947; Bowman, 1961) and the Hawaiian honeycreepers (Amadon, 1950; Baldwin, 1953). In the instance of radiations on continents, to be revealing, it is necessary only that they be of sufficient recency so that the resultant taxa have not had time to establish the extensive sympatry that ultimately obscures origins. Hence the great basic importance of superspecies (Mayr, 1963, pp. 499 ff.; Amadon, 1966, p. 245; Moreau, 1966b, p. 9), as defined below.

Thus the second, and by far the larger, part of this paper is devoted to the analysis of critical superspecies and species groups that occur chiefly in the wooded and forested areas of the continent. These areas form a complex of ecological islands of greatly varied sizes. I justify this emphasis in a symposium on the Central Plains, because the grassland and other open environments that occupy the plains have provided the sea in which these islands are set.

Before proceeding further, I should like to emphasize two points. First, the notion is common that birds disperse at random because they fly, and therefore little biogeographical sense can be made of their distribution. This is probably less than half true. The subject has been well discussed by Darlington (1957, pp. 239-242). Sedentary birds of weak flight are obviously poor dispersers, but in many cases even strongly migratory birds are persistently loyal to their breeding sites, wintering grounds (Nickell, 1968), and migratory routes, hence notably conservative in distribution. Also, my colleague Robert S. Hoffmann (personal communication, manuscript in press) has shown that the exchange of arctic birds between North America and Eurasia has been little greater than that of arctic mammals.

Second, while some paleontologists hold that biogeographical analyses approach futility in the absence of a fossil record (and with birds this record is usually absent), I think this position is extreme. Strictly biogeographical phenomena demand (and deserve) our best efforts to understand them as surely as mountains demand to be climbed. It is true that one could suffer a painful fall from some hypothetical peak upon the eventual appearance of the appropriate fossils, but this emergence of truth should compensate for the anguish. Moreover, it is doubtful that a fossil record of the usual kind would help much (if at all) to interpret the history of the superspecies and species groups, usually passerine, which are most interesting in connection with avian species formation (see Moreau, 1954, p. 420). Finally, fossils also present problems, in that only their geographic localities are fixed: their geological ages, taxonomic allocation, and ecological affinities are all variously subject to interpretation. As facts, they are usually more equivocal than the phenetic distances between, and distributions of, the living members of a fauna. Thus the existence of relevant fossils, while eminently desirable, does not remove the necessarily hypothetical aspect from biogeographical conclusions.

Critical to the comprehension of much that follows is the correct understanding of several much-used terms as employed in this paper. They may be ranged in the sequence subspecies-semispecies-superspecies-species group, representing the hypothetical course of evolution of a group of diverging natural entities. "Subspecies" and "superspecies," like the basic category "species," are here used as defined by Mayr in a series of works (see glossary in Mayr, 1969, pp. 397 ff.). "Semispecies" is here limited, as urged by Amadon (1966, p. 245), to part of Mayr's definition (see also Short, 1969, p. 89). Thus: *subspecies,* a geographically defined aggregate of local populations that differ taxonomically from other such divisions of the species; *semispecies,* populations that have acquired some, but not yet all, attributes of species rank, or borderline cases between species and subspecies (*not* the component species of superspecies, as this term is also used by Mayr); *superspecies,* a monophyletic group of entirely or largely allopatric species (the component species of superspecies, after Amadon, are here considered *allospecies*); *species group,* for the purposes of this paper, a monophyletic group further differentiated than in a superspecies, with consequent establishment of extensive sympatry among at least some of the members. The term *species,* of course, is used in Mayr's sense: groups of actually (or potentially) interbreeding natural populations that are reproductively isolated from other such groups.

PART I: BIRDS OF THE CENTRAL PLAINS

While the Central Plains as defined for the over-all purposes of this symposium are essentially coextensive with the somewhat restricted Great Plains Province of Fenneman (1931, map in pocket), additional areas are importantly related to these plains. These are the Grassland (most, but not all, of which lies within the Central Plains), the Sagebrush, and the Desert Scrub formations of Weaver and Clements (1938, frontispiece). Alternatively, they are the following, somewhat more finely divided biomes of Shelford

(1963, figs. 1-9, p. 14): Temperate Grassland, Cold Desert, Hot Desert, Chaparral, and Scrub. With some exceptions in the scrub group, these areas are open habitats with vegetation so low as to permit a standing man to see to the horizon.

It is this essential lack of a vertical component that sharply distinguishes most of these environments from the more or less closed environments—woodlands and, especially, forest. It would doubtless be rewarding to consider the birds of all of these open formations, but neither time nor space permits this, and, fortunately, it has already been done to a considerable degree by Udvardy (1958; 1963). Therefore, I shall here limit myself to the grasslands formation, in which, for present purposes, I shall include also the Cold Desert biome of Shelford (1963), which is equivalent to the Sagebrush formation of Weaver and Clements (1938). This I do because of the many large, complex areas where the grasslands and Cold Desert intergrade and because, while a few bird species of a larger Great Basin complex (Miller, 1951, pp. 591-595) are intimately associated with the sagebrush (*Artemisia*) component of Cold Desert and do not invade true grassland, various of the more widespread grasslands species are important in the Cold Desert environment.

I shall not try to enumerate, much less analyze, all of the species normally occurring or breeding within this great area, which inevitably includes many special habitats. Rather, as is discussed more fully beyond, I shall concentrate on species intimately adapted to and dependent upon open, plains environments.

PRIMARY AND SECONDARY GRASSLAND SPECIES

As was noted by Udvardy (1958, p. 62; 1963, pp. 1155-1156), perhaps the most remarkable thing about the endemic avifauna of the present plains is its meagerness. Aside from the few endemics, there are various species that are more or less widespread in, and typical of, the central grasslands, but that also occur beyond the limits of this area—usually in comparable grassy environments—or occupy specialized kinds of habitats (such as marsh) whether or not within the central grasslands. These species make the number of species of the grasslands avifauna, while still not especially large, considerably more difficult to establish. The difficulty, really, lies in defining grassland species, which results from the fact that grassland itself is not easy to define precisely. How small may a prairie be before it is a mere opening? Where does grassland stop and very open woodland begin? Are ground-nesting birds in the latter habitat grassland birds, woodland birds, or forest-edge birds? How much sagebrush is required before grassland becomes some form of desert scrub? Such questions can be compounded at length.

The results of my efforts to define the grasslands avifauna are shown in Table 1. (Although both scientific and vernacular names appear in this table, and scientific names are occasionally used elsewhere when convenient, the scientific names of all species of birds mentioned will be found in Appendix A.) In Table 1 I have placed endemics (primarily grassland species) in Group I and secondary species, that is, birds having strong affinities with the grasslands, although not restricted to them, in Group II. Constituting the second group involved much reading and thought, and I doubt that any two workers would make it quite the same (see Kendeigh, 1961, p. 326; and Shelford, 1963, chapter 13, pp. 328 ff.).

The endemics of Table 1, perhaps because of long specialization and consequently narrow environmental tolerances, are all somewhat restricted in range—few if any occur throughout the entire grasslands formation (see Fig. 1). The species of Group II seem to

TABLE 1.—The North American Grasslands Avifauna (monotypic genera in capitals)[1]

Non-passerines	Passerines
I. Primary Species (Endemic)	
1. *Buteo regalis* Ferruginous Hawk[ws]	1. *Anthus spraguei* Sprague's Pipit[ne]
2. *EUPODA* (cf. *Charadrius*) *montana* Mountain Plover[c]	2. *CALAMOSPIZA melanocorys* Lark Bunting[n]
3. *Numenius americanus* Long-billed Curlew[ws]	3. *Ammodramus bairdii* Baird's Sparrow[ne]
4. *Limosa fedoa* Marbled Godwit[n]	4. *Aimophila cassinii* Cassin's Sparrow[se]
5. *STEGANOPUS* (cf. *Phalaropus*) *tricolor* Wilson's Phalarope[n]	5. *RHYNCHOPHANES* (cf. *Calcarius*) *mccownii* McCown's Longspur[nc]
6. *Larus pipixcan* Franklin's Gull[n]	6. *Calcarius ornatus* Chestnut-collared Longspur[n]
II. Secondary Species (More Widespread)	
1. *Ictinia misisippiensis* Mississippi Kite	1. *Eremophila alpestris* Horned Lark
2. *Buteo swainsoni* Swainson's Hawk	2. *OREOSCOPTES montanus* Sage Thrasher
3. *Circus cyaneus* Marsh Hawk	3. *Sturnella magna* Eastern Meadowlark
4. *Falco mexicanus* Prairie Falcon	4. *Sturnella neglecta* Western Meadowlark
5. *Tympanuchus cupido* Greater Prairie Chicken	5. *SPIZA americana* Dickcissel
6. *Tympanuchus pallidicinctus* Lesser Prairie Chicken	6. *CHLORURA* (cf. *Pipilo*) *chlorura* Green-tailed Towhee
7. *PEDIOECETES phasianellus* Sharp-tailed Grouse	7. *Passerculus sandwichensis* Savannah Sparrow
8. *CENTROCERCUS urophasianus* Sage Grouse	8. *Ammodramus savannarum* Grasshopper Sparrow
9. *BARTRAMIA longicauda* Upland Plover	9. *Passerherbulus henslowii* Henslow's Sparrow
10. *SPEOTYTO* (cf. *Athene*) *cunicularia* Burrowing Owl	10. *POOECETES gramineus* Vesper Sparrow
11. *Asio flammeus* Short-eared Owl	11. *CHONDESTES grammacus* Lark Sparrow
	12. *Amphispiza belli* Sage Sparrow
	13. *Spizella breweri* Brewer's Sparrow
	14. *Spizella pallida* Clay-colored Sparrow

Key: ws, widespread; c, central; w, western; e, eastern; n, northern; s, southern.

[1] Classification of A.O.U. (1957). Possible combinations with these monotypic genera are indicated in parentheses, for example, *EUPODA* (cf. *Charadrius*).

have much broader ranges of adaptation. Some occur not only throughout the grasslands, or nearly so, but also have wide distributions in comparable habitats elsewhere. A few are widespread on more than one continent (for example, Marsh Hawk, Short-eared Owl, Burrowing Owl, Horned Lark).

Special problems.—Faunal and ecological analyses of aquatic species and of other birds of specialized ecological requirements (often large non-passerines—for example, the Prairie Falcon, which requires cliffs for breeding) pose special difficulties (Miller, 1951, p. 605). One wonders whether their presence in an area depends solely on the appropriate aquatic (or other) environmental element, or whether their requirements also include some features of the surrounding terrain or some geographically varying complex of conditions.

Here, it seems to me, there is no doubt of the correctness of including the Mountain Plover and the Long-billed Curlew in the primary grassland group, because each is totally independent of bodies of water and entirely typical of upland plains. The similarly water-independent Upland Plover, however, is not endemic to the grasslands. As for the rest of the endemics, which are more or less dependent upon bodies of water at nesting time, I have included in Group I all aquatic species whose known ranges in historic times fall wholly within the central grasslands area. They are found nowhere else, and in the absence of contrary evidence we may suppose there are

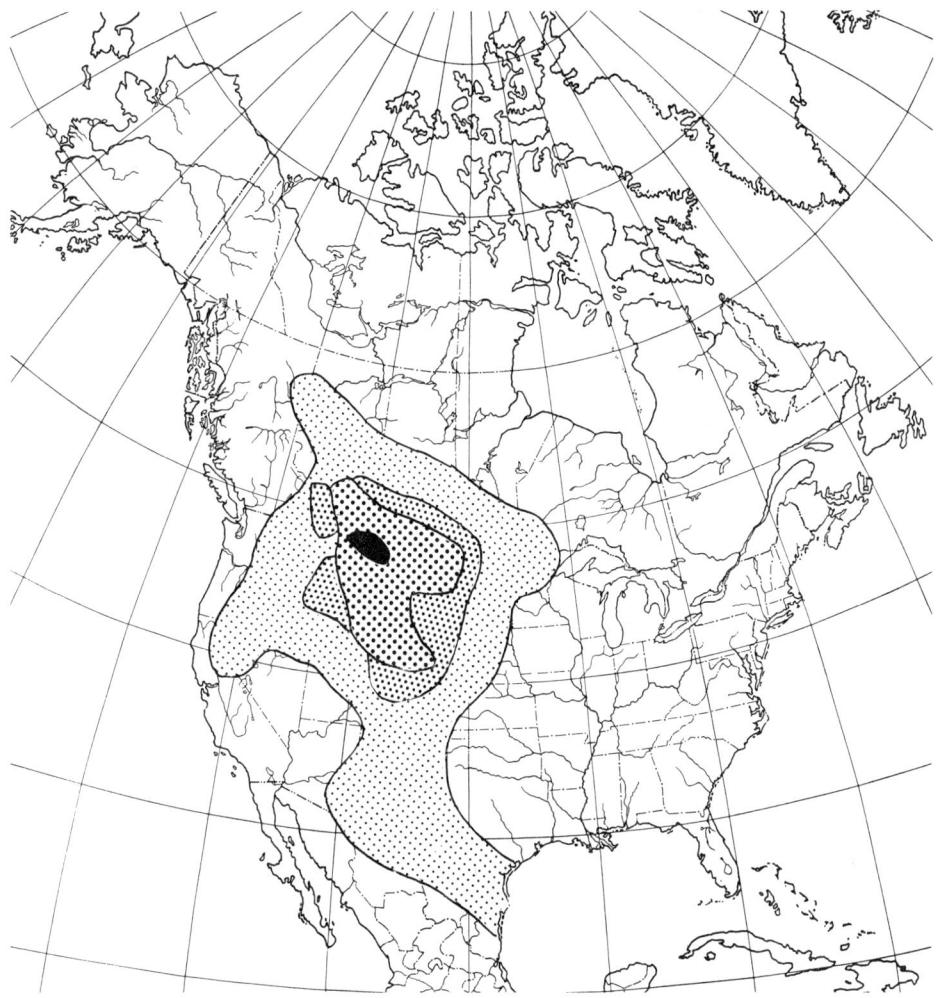

Fig. 1. Over-all distribution of endemic grasslands species (Group I of Table 1). Species densities from the palest peripheral pattern to the black center are: 1–3; 4–6; 7–9; and 10–12.

adaptive reasons for this that are somehow related to grassland conditions.

In making up Group II, I found that the problem of aquatic, especially marsh, species was more acute. As Miller (1951, p. 559) stated: "The highest relationship [of grasslands] is with fresh water marsh, mainly because of the . . . frequent proximity and resemblance in life-form of the plants involved in the two." Several marsh-inhabiting passerine species (some with secondary preferences for moist or even dry grassland) that are more or less widespread in the central grasslands were excluded from Group II. These include at least the Long-billed and Short-billed marsh wrens, the Yellowthroat, the Red-winged Blackbird, the major inland race (*Ammospiza caudacuta nelsoni*) of the Sharp-tailed Sparrow, and the Le Conte's Sparrow. All but the last two are widespread, racially complex, and found in marshes in numerous ecogeographic areas. This is also true of the Sharp-tailed Sparrow when all races are considered. The inland Sharp-tailed Sparrows, as is the Le Conte's Sparrow (Murray, 1969, fig. 4), are found to a considerable extent in marshes within the general range of the taiga as well as within the grasslands (following Hoffman, 1958, I regard "taiga" as synonymous with northern coniferous forest).

Among non-passerines, the American Avocet is certainly a conspicuous and typical water bird of the central grasslands, but it also originally bred eastward to New Jersey (A.O.U., 1957, p. 209). Also, there are various gulls and terns, and numerous ducks, associated with prairie marshes and sloughs, all of which occur also in other formations; I cannot help but think that their primary affiliations are with some feature of their specific habitat and not with the grasslands. Hopefully, I have solved the dilemma by eliminating all of these widespread marsh species from the table.

Group II, while consisting of birds more or less adapted to open, grassland environments, predictably contains birds with a wide variety of relationships to these environments. Some, like the Mississippi Kite and Prairie Falcon, evidently require extensive grassy plains for feeding at certain times of year. The falcon is, over large areas, highly dependent upon two grassland species for food: Horned Larks in winter and Richardson's ground squirrels *(Spermophilus richardsonii)* in summer (Behle, 1942, p. 210; Enderson, 1964, pp. 340, 346). They have also, however, extrinsic (that is, non-grassland) requirements for nesting sites (riparian woodland or shelter belts for the kite, cliffs for the falcon). The Sharp-tailed Grouse has races of differing ecological adaptations and on the whole is at least as much a species of scrub and broadleaf second growth in taiga as an inhabitant of brushy subclimaxes in true grasslands (Aldrich, 1963, pp. 538-539). The Sage Grouse, Sage Thrasher, Green-tailed Towhee, Sage Sparrow, and Brewer's Sparrow depend upon intrusions of sagebrush into the grasslands. They represent an element of an essentially Great Basin avifauna (Miller, 1951, p. 591). Nearly all the remaining species range out of the Central Plains into secondary and often transient grassy habitats in other ecogeographic regions, and several (Marsh Hawk, Burrowing Owl, Short-eared Owl, Dickcissel, Savannah Sparrow, Henslow's Sparrow, and, most notably, Horned Lark) have ranges outside the Central Plains that exceed those within them. I could enlarge Group II still further by including other widespread birds that are abundant in or over, or are scattered through, the grasslands and probably adapted to them locally, as for example the Red-tailed Hawk, White-necked Raven, and Mourning Dove; but all of these are wholly or to some extent tree nesters and hence seem less intimately associated than ground nesters with the grassland environment. There are, finally, Central Plains subspecies of some widespread species, for example the plains race of the Common Nighthawk *(Chordeiles minor sennetti)* and the western, disjunct race of the Willet *(Catoptrophorus semipalmatus inornatus)*. Unlike some grasslands subspecies of the poorly defined continental variety, the last is rather distinct and is disjunct in distribution.

I should explain the absence from Table 1 of the Bobolink, a species attributed to the grasslands by various authors. It seems in early historic times to have been an eastern meadow species and only to have invaded the Central Plains with the intrusion of agriculture (see sources cited in Bent, 1958, pp. 29-30). The Brown-headed Cowbird, although it may have originated on the plains, is rare or absent in open, treeless grassland (Mayfield, 1965, p. 15), and so has been excluded.

Finally, there are to be considered a few highly restricted peripheral species, some in isolated environments and all more or less adapted to grassy areas. They may possibly be anciently derived from widespread, typically grassland ancestors. The list is incomplete, I expect, but one may mention the Striped Sparrow *(Oriturus superciliosus)* of high Mexican pine-forest openings, the little-known Sierra Madre Sparrow *(Xenospiza baileyi)* of grassy habitats in the central Mexican highlands, and the Rufous-winged Sparrow *(Aimophila carpalis)* of southern Arizona and northern Mexico. Additional members of the genus *Aimophila* probably qualify as grassland birds in some areas.

RELATIONS TO THE OLD WORLD STEPPE

The distinctness of the American grasslands avifauna from its Eurasian steppe counterpart is striking. All but four of the 37 species of Table 1 belong to families that were considered by Mayr (1946) to be either unanalyzable or of North American origin. One of these four, the Horned Lark, is a Holarctic species, being widespread in both the Old and New worlds, and is the only Nearctic representative of its large family. Another is Sprague's Pipit of the north-central plains, one of but four of its family regularly breeding in the northern Nearctic and the only one endemic there (two others are the Eurasian Yellow and White wagtails, which breed in the New World only in arctic Alaska and migrate back across Bering Strait). Its congener, the Holarctic Water Pipit, while an essentially arctic species, breeds in alpine tundras as far south as north-central New Mexico. Seemingly, however, like the more or less sympatric rosy finches *(Leucosticte),* it has never broken away from the alpine breeding environment. Sprague's Pipit, interestingly, more nearly resembles certain Eurasian species, for example, the Meadow Pipit, than it does the Water Pipit of nearby mountain tops, and indeed may be more nearly related to some of them (see also, however, below).

The last of the four members of Old World families in Table I are the two owls. The Burrowing Owl, as noted by Voous (1960, pp. 159-160) is probably congeneric with the Little Owl of the Old World and presumably descends from a common ancestor. This ancestor, however, must have come to the New World long enough ago to permit development of the marked differences that separate the Burrowing and Little owls. The Short-eared Owl, by contrast, is a Holarctic species today that is not even subspecifically differentiated in its northern range, being widespread and a notable disperser.

On the Eurasian steppes and in Old World grasslands generally, the complex of niches occupied in the New World by one lark and one pipit and by various blackbirds (Icteridae) and finches (Emberizinae) seems to be filled by another complex of this same subfamily—chiefly species of *Emberiza*—and by a multitude of larks, pipits, and, at least locally, some Old World warblers (Sylviidae). The shore-bird and predator niches, on the other hand, are occupied, except for the nearly cosmopolitan Short-eared Owl and the Holarctic Marsh Hawk, by the same families and genera but by different species. The role of the plains grouse in North America is, in the Old World, evidently played (Udvardy, 1958, p. 63) by an assortment of bustards (Otididae), sand grouse (Pteroclidae), coursers (Glareolidae), and (although Udvardy does not mention them) various pheasants and partridges (Phasianidae). Contrasted with the great similarity in species and groups of vertebrates shared by the circumpolar tundras (Rausch, 1963; Hoffmann and Taber, 1967) and with the considerable similarity of the New World and Old World taiga (Udvardy, 1958; Stegmann, 1963), this difference in the steppe avifaunas is striking.

The reason, of course, seems fairly obvious. Although tundra and steppe provide what, for many birds, are probably quite similar habitats—except for considerations of cold-adaptation—the tundras of the Old and New worlds are more or less continuous, and have been intermittently connected during the Pleistocene, whereas the steppes of the two northern land masses are widely separated and, more important, isolated from the tundras by a broad, world-wide band of taiga.

Since the Eurasian mountain chains tend to be both southerly and oriented east-west, there is only one tenuous breach in this broad barrier of taiga, namely, the somewhat broken north-south arc provided by the alpine tundras of the East Siberian mountains and, especially, the American cordillera.

The highly adaptable, subspecifically plastic Horned Lark, the only alaudid with extensive tundra adaptations, probably used this avenue to invade the American grasslands. Various populations today breed in alpine tundra of both the Cascades and the Rocky Mountains (Behle, 1942, p. 208; Verbeek, 1967; R. S. Hoffmann, personal communication), and at least one race is confined to alpine tundra (A.O.U., 1957, p. 354). Long ago such mountain populations evidently made the physiologically considerable, but geographically short, jump to the warmer grassland habitats nearby. Then they spread widely and as far as northern South America, albeit with a broad present disjunction between the Isthmus of Tehuantepec and Colombia. The Water Pipit, on the other hand, likewise an inhabitant of both arctic and alpine tundra, seems never to have made a jump from alpine tundra to nearby prairie, although it seems possible that the ancestor of Sprague's Pipit did so.

There are, certainly, other ways of dispersal from Eurasian to American steppe, or vice versa, and such notable travelers as the Marsh Hawk and Short-eared Owl may have moved much more directly from one to the other. The shore-birds may descend from remote, widespread ancestors as probably the Prairie Falcon of the American steppe and its Eurasian counterparts the Saker and the Altai Falcon do also. It is conceivable that the Ferruginous Hawk and the Old World Long-legged Buzzard and Upland Buzzard—the last two being closely related, if not conspecific —have a similar, though more remote, history (see Voous, 1960, p. 55).

RELATIONS TO SOUTH AMERICAN GRASSLAND AND SAVANNAH

Three passerine species of the North American Central Plains (Table 1) presently range to northern South America, and each has subspecifically distinct populations scattered southward through Mexico and Middle America. These are the Horned Lark, Eastern Meadowlark, and Grasshopper Sparrow. While these distributions might tempt one to postulate either southward displacements of the grasslands, or a formerly greater extent of this formation, such hypotheses are probably unnecessary.

All three species are able to occupy the savannah type of habitat, and examination of figures 1 and 2 of Haffer (1967, pp. 316-317, see also text) shows what an impressive amount of this type of habitat occurs today throughout much of Central America and northern South America. With moderate changes in climate it is easy to imagine these habitats increasing to the southward; it would require only moderate shifting and recombinations of habitat for the three species in question to get to South America.

That grassland birds have reached South America by this or some other means at least once, much earlier than the three species above-mentioned, is shown by the presence in southern and central South America of five endemic pipits of the genus *Anthus*. Although the ancestors of some or all South American pipits may have arrived there by subantarctic island-hopping, it is also quite possible that Sprague's Pipit is a relict of an invasion via Beringia, from which stock all or some of them ultimately stemmed. In any case, Hall (1961, pp. 283-284) has suggested that the South American Short-clawed Pipit (*Anthus furcatus*) and Sprague's Pipit are sufficiently alike, at least morphologically, to be considered conspecific. She considered them most like the Old World complex *Anthus campestris–A. godlewskii–A. berthelotii*. Also, an earlier invasion of South America by a Grasshopper Sparrow-like ancestor is shown by the presence there of two polytypic species of the genus *Myospiza*.

Perhaps it should be added in conclusion that the meadowlarks as a group (*Leistes, Pezites, Sturnella*; united in *Sturnella* by Short, 1968) show every evidence of South American origin, with six species occurring there today. The Eastern and Western meadowlarks, however, seem to me to be probably of North American (grasslands or savannah) origin, with the former reinvading South America.

Finally, the Short-eared Owl and the Burrowing Owl, both notable dispersers, are also shared with various South American savannahs.

DISTINCTNESS AND SIZE OF THE GRASSLANDS AVIFAUNA

The small size of the grasslands avifauna was emphasized above. Its distinctness is somewhat more equivocal. As classified by the rather conservative A.O.U. (1957) Check-list Committee, four of the 12 endemic species of Table 1 (33.3 percent) and nine of the 25 secondary species (36 percent) are members of monotypic genera (or 13 of 37 in all, for 35.1 percent). However, as K. C. Parkes has correctly emphasized (letter, August 1, 1969), generic combinations already proposed by various authorities —as indicated in parentheses in Table 1—could reduce the monotypic genera to one in Group I (8 percent) and six in Group II (24 percent) for a total of seven in 37 (or 19 percent). Conversely, various specific combinations could increase the percentage of monotypic genera—for example, *Tympanuchus cupido* with *T. pallidicinctus* and *Passerculus sandwichensis* with *P. princeps* (the "Ipswich Sparrow"). Still other changes are possible, but enough has been said already to indicate the essential futility of computing hard percentages from such soft bases, much less drawing therefrom very firm conclusions. It is doubtless enough in this regard to say that even should widespread lumping prove to be the best taxonomic course, a modicum of distinctness, albeit faint, is evidenced for the grasslands avifauna by the original recognition of approximately one-third of its genera as monotypic, some four of these endemic.

Distinctness at the specific level, relative to other formations, is at least considerably lower than the upper limits of distinctness at the generic level. Still, of the 37 species here admitted to the avifauna, a respectable 12 (32.4 percent) are endemic.

More interesting is the small number of species per genus (1.1—a value not likely to change much with forseeable taxonomic rearrangements) in the grasslands avifauna. This argues strongly for a rarity of fragmentation of this environment for periods long enough to permit the specific differentiation of isolated parental stocks. In fact, there are only four cases in which closely related congeners are important in the Central Plains. The Eastern and Western meadowlarks are sibling species and are now known to hybridize, albeit with doubtful hybrid fertility (Lanyon, 1966, p. 23). They occur sympatrically over a fairly extensive range (Fig. 5, F) without blurring of the characters that distinguish them morphologically (Lanyon, 1957, pp. 53-58). Although Short (1968, p. 27) considered them to form a superspecies, they are close to the limits of allopatry that I should think permissible in this category. The prairie chickens are barely distinct at the species level, in my opinion, and have been united by some (Short, 1967, p. 26, for example). More distinct are the endemic Baird's Sparrow and the Grasshopper Sparrow, but these, too, may well be the result of Pleistocene differentiation, as, I think, is probable in the first two cases. The same may be said of the Clay-colored and Brewer's sparrows.

Why so few species have entered the grasslands at all is a different matter, because the formation does vary regionally. According to Weaver and Clements (1938, p. 516), "the grassland . . . is the most extensive and most varied of all the climaxes of the North American continent." They go on (pp. 516-528) to recognize six climax associations and one essentially permanent disclimax based on various combinations of dominant and subdominant plants. Although composition of plant species varies widely, what is far more important to birds (Pitelka, 1941; Miller, 1951, pp. 540-541) is the life form or essential aspect of the vegetation in providing places for birds to perch, roost, feed, nest, and so on. The comparatively uniform aspect of the grasslands, together with their intrinsic lack of vertical stratification, sharply limits the number of niches available (Cody, 1968, p. 145). Neither, apparently, do the grasslands differ enough from place to place to provide much pressure for infraspecific geographic variation, Weaver and Clements (as quoted just above) notwithstanding. Only one geographical variant was nomenclaturally recognized by the A.O.U. (1957) among the species of Group I (Table 1), while the same source recognized subspecific variation in just over half (13) of the 25 species of Group II. Further, subspeciation within the central grasslands, as in the more homogenous parts of continents generally, is mosaic and clinal and poorly described by trinomial designation.

ANTIQUITY OF THE GRASSLANDS AVIFAUNA

The American grasslands and scrubs, separately or together, have fewer species than do similar areas in the Palearctic and probably have a considerably shorter and less complex history (Udvardy, 1958, p. 63). They may also have been subject to much greater trauma from the Pleistocene, with resulting extinctions (John P. Hubbard, letter). Even the American grasslands, however, appear to be of respectable antiquity. They evidently had their beginnings in the Miocene and attained considerable extent and importance by mid-Pliocene as the Rocky Mountains approached impressive heights and the climate became drier (Axelrod, 1950, 1952; MacGinitie, 1953, p. 59, and 1958, p. 69). While this period of 8–10 million years is not long geologically, it is much greater than the hypothetical average life-expectancy of 500,000 years ingeniously if precariously calculated for bird species by Brodkorb (1960, p. 41—note objections of Moreau, 1966a). The fact that these grasslands have such a small and relatively undifferentiated, yet fairly distinct, avifauna is, therefore, somewhat paradoxical. The controversial fate of the grasslands during the Pleistocene (of which more below) is obscure, yet the adaptations and distinctness of the present avifauna strongly suggest chronological continuity of at least some grassland—as opposed to geographical stability—through the glacial periods.

PART II: SPECIATION IN ENVIRONMENTS PERIPHERAL TO THE PLAINS

In the beginning I remarked that the grasslands have played a limited role as a stage for speciation, but a more important one as an isolating agent in species formation. Evidence for this statement must be sought in the avifaunas surrounding the grasslands as here defined.

It should be mentioned at this point that the usual requirements for species formation, that is, the differentiation of a single stock into two or more species, are here taken to include most importantly physical isolation of the differentiating stocks (whether geographical or ecological so long as effectively complete) for a period of sufficient duration to allow the cumulation of genetic changes adequate to ensure the requisite genetic isolation. This assumption seems to require no particular defense at present (Mayr, 1963, pp. 480, 512).

COMPARATIVE PROLIFERATION OF AVIFAUNAS PERIPHERAL TO THE GRASSLANDS

We have just seen that the small grasslands avifauna shows little evidence of recent speciation, possessing but 1.1 species per genus. A limited survey of the birds of the various formations surrounding the plains produced results, which, although unrefined, seem adequate for rough comparison (Table 2). The procedure I employed, for maximum simplicity, was to take as a basis the 1957 A.O.U. *Check-list of North American Birds* (which covers Baja California and continental North America north of the Mexican border), although I accepted changes adopted in the "Peters" *Check-list of Birds of the World* (various dates). I excluded all tundra species (because the tundra does not contact the grasslands), casual wanderers or "accidentals," difficult aquatic species (see above), species of genera barely established north of the Rio Grande (*Platypsaris*, for example), and all the species of Table 1 (that is, those of the grasslands). For the sake of simplicity, and because all that was required was that the samples be comparable, the analysis was restricted to passerines (in-

TABLE 2.—Generic Size and Other Qualities of Grassland and Non-grassland Avifaunas

	Grassland	Non-grassland
Number of genera	33	78[1]
Number of species	37	227
Species per genus	1.1	2.9
Number and percent of monotypic genera[2]	13 (39%)	21 (27%)
Species per genus, excluding monotypic genera	1.3	3.6

[1] Qualifications given in text. See also Appendix B.
[2] Note warning in text.

formal examination of the data suggests that inclusion of non-passerines would not change the results appreciably). This left 78 genera (see Appendix B) with the characteristics noted in Table 2.

It should immediately be stressed that little should be made of the precise values indicated in Table 2, because they can be no firmer than the classification upon which they rest. Since this classification is that of the 1957 A.O.U. check-list (which tended toward generic splitting), altered by incorporation of changes adopted in the completed volumes of the "Peters" check-list (which generally tends to generic lumping), not only is the coverage somewhat uneven, but still further changes may also be expected. However, it is important to note that the changes still to come will almost surely involve more lumping than splitting, so that the number of species per genus—however subjective a category the latter inevitably is—may be expected to rise. The interesting thing about Table 2 is that there are at least two and one-half to three times as many species per genus, depending on whether or not monotypic genera are excluded, in the complex circum-grassland avifauna than in the grasslands. (There may be something innately conservative about the species in monotypic genera, wherever found, but clearly the plastic circum-grasslands groups have proliferated more than have plastic grasslands groups.) Fur-

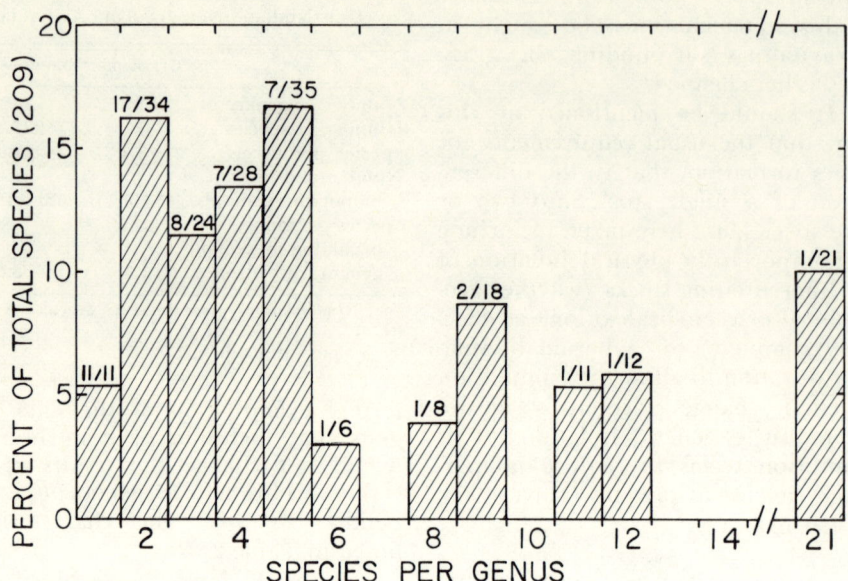

FIG. 2. Species per genus, exclusive of monotypic genera, of the composite passerine avifauna of all formations contiguous with the central grasslands. The actual numbers of genera possessing each number of species (and the resultant number of species) are given above the appropriate bars of the histogram.

ther, this factor would have been more than three times as large had the tabulation included various Mexican circum-grassland species. Thus, members of this complex avifauna seem to have encountered far more opportunities for geographical and ecological isolation than have the grasslands taxa.

This is to be expected. The circum-grasslands "avifauna" with which I have been concerned is not a homogeneous or single avifauna in the sense often implied by the word. Rather, it is a complex amalgam of forest, woodland, and assorted scrub avifaunas arranged in a crudely circular way around the great grassland barrier. Not only does each of the formations provide more niches than the grasslands, permitting occupancy by more species, more or less in direct proportion to the magnitude of the vertical component of the habitat (see MacArthur and MacArthur, 1961), but also each provides a distinctive set of environments suitable for the differentiation of uniquely adapted species. Each such species, furthermore, is variously capable of competitively excluding species adapted to other formations. And finally, various of the formations involved, although containing similar niches, are widely isolated from one another by the central grasslands.

Another interesting feature of the avifaunas peripheral to the plains is revealed by Figure 2. Of the 57 nongrassland, polytypic genera of Table 2 (see Appendix B), with their 208 component species, one-third (33.7 percent) of all the species (or 70) are comprised in only one-tenth (10.5 percent) of the genera. Conversely, nine-tenths of all the genera (those with six or fewer species) provide only about two-thirds of all the species. There are six notably large genera, namely *Toxostoma* (eight species), *Empidonax* and *Vermivora* (nine each), *Parus* (11), *Vireo* (12), and the remarkable *Dendroica* (21). To this list (which would feature still more species per genus if a larger area were covered) could also be added *Junco* and perhaps other genera if the survey extended to Central America.

Clearly, the large genera are those that have had the most abundant opportunities for species formation. This means, simply, being at the right places at the right times to become isolated by some geological or ecological circumstances. In the following consideration of speciation patterns found in the formations surrounding the central grasslands, I shall repeatedly use examples from these large genera as well as some from somewhat smaller groups.

I have arranged the distributional and speciation patterns considered into several groups, which are discussed roughly in order of increasing complexity. The seven groups include more than 70 taxa, most of which are distinct at the species level. When these patterns have been considered, the historical and evolutionary implications will be examined.

SIMPLE DISTRIBUTIONAL COMPLEXES

Most of these complexes involve two or three taxa, and in no case does the number exceed four. Several more or less distinct patterns may be discerned, but of course some complexes are difficult to place in a particular group.

Distributions (Figs. 3-8) indicated as solid in any part of the Central Plains are often far from it; in this environment all but truly grassland species are restricted to widely scattered riparian trees and shrubs, shelterbelts, plantings about farms and towns, and so forth.

It should be noted that maps of the type used here are impossible to construct with much accuracy, and indeed I designed them crudely precisely because the present interest is in the relations of distributions rather than their minutiae. Therefore they should not be trusted for details.

Group I (wholly or nearly disjunct east-west species pairs).—The six pairs (Fig. 3) of this group are arranged roughly in decreasing order of the distance between their ranges. All of the pairs have the allopatric or essentially allopatric distribution of superspecies. The *Cyanocitta* jays and *Strix* owls, however, seem distinct beyond the level usually associated with the superspecies. This is not the case with the Scarlet and Western tanagers, which apparently hybridize on occasion (Tordoff, 1950; Mengel, 1963), presumably as a result of individuals straying across the narrow gap in Saskatchewan between the regular ranges of the species. The swifts also are closely related, and it has even been suggested that they are conspecific (see Lack, 1956, p. 10).

The Tufted and "Black-crested" titmice of the 1957 A.O.U. Check-list (*Parus bicolor* and *P. atricristatus*) hybridize freely in a few small zones of contact in Texas and probably are not distinct at the species level (Dixon, 1955, p. 190). I consider them to be a single eastern complex separate from the similar, but quite distinct and allopatric, Plain Titmouse. The Plain and Tufted groups together perhaps constitute allospecies of a superspecies. The Tufted and Black-crested populations are interesting in themselves, however, in showing evidence of a former disjunction in Texas, which is displayed in several of the following groups.

The Eastern Bluebird and the similar Western Bluebird differ from the other pairs in this group in having established limited sympatry in southeastern Arizona and northern Mexico, where they demonstrate their distinctness by failure to interbreed despite an apparent lack of ecological separation (Marshall, 1957, p. 103).

Group II (semispecies and subspecies with zones of contact where they hybridize).—The distributions of the four complexes in this group are shown in Figure 4. Hybridization between the buntings has been studied by Sibley and Short (1959); between the gros-

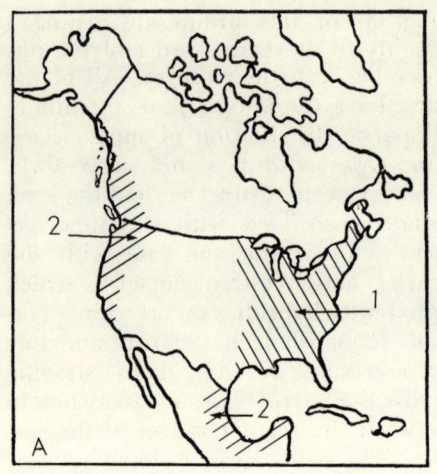

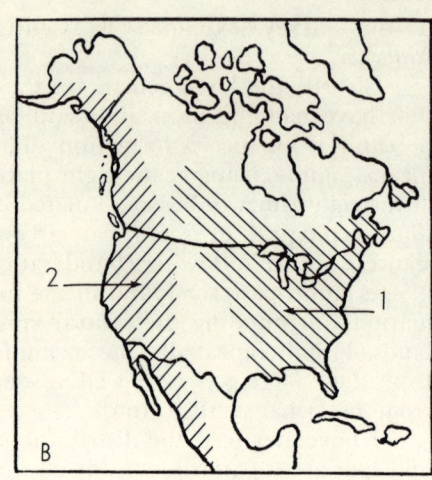

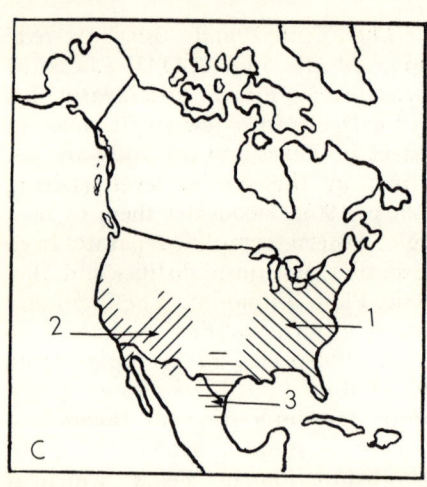

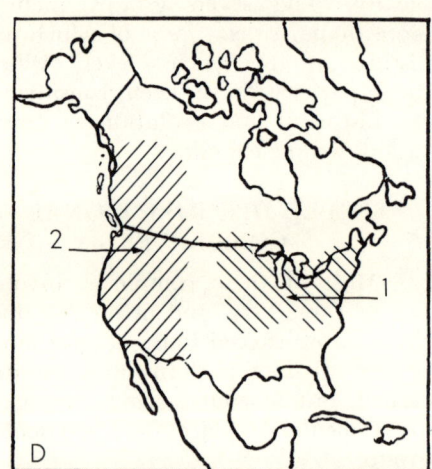

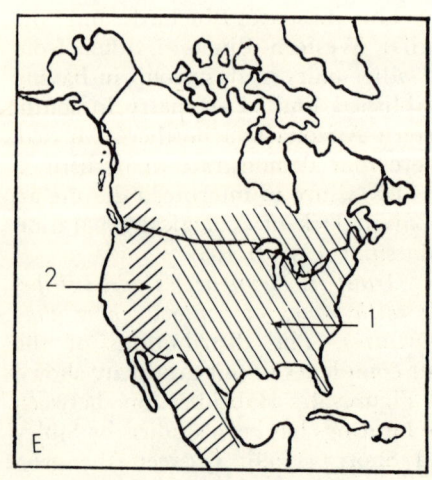

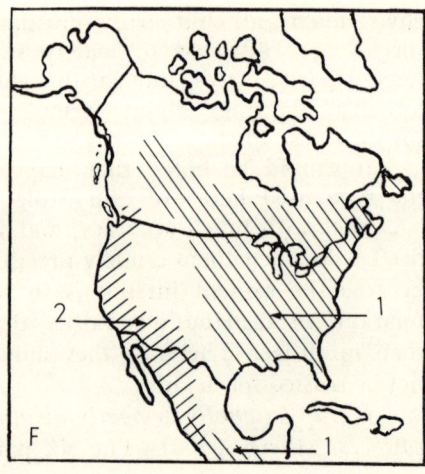

Fig. 4. The distributions of the species of Group II. A: 1, Indigo Bunting; 2, Lazuli Bunting; 3, Varied Bunting. B: 1, Rose-breasted Grosbeak; 2, Black-headed Grosbeak. C: 1, Baltimore Oriole; 2, Bullock's Oriole. D: 1, Yellow-shafted Flicker; 2, Red-shafted Flicker; 3, Gilded Flicker; 4 and 5, respectively, the Cuban subspecies *(Colaptes auratus chrysocaulus)* and the Grand Cayman subspecies *(C. a. gundlachi)* of the Yellow-shafted Flicker. The last two, with the Central American *C. a. mexicanoides* (not shown) are mentioned in passing because of the similarity of their West Indian distributions to those of certain other groups discussed, notably the Yellow-throated Warbler and Ivory-billed Woodpecker groups (Fig. 5, B, C).

beaks, by West (1962); between the orioles, by Sutton (1938), Sibley and Short (1964), and James D. Rising (Ph.D. thesis, The University of Kansas, 1968); and between the flickers, by Short (1965). In all cases hybrids seem to be fertile and without known reduction in viability, hybrid zones seem to be relatively stable and for the most part comparatively narrow (long-range introgression is limited), and establishment of many of the hybrid zones ap-

---

Fig. 3. The distributions of the species of Group I. A: 1, Chimney Swift; 2, Vaux's Swift (note peculiar distribution of disjunct Mexican races). B: 1, Blue Jay; 2, Steller's Jay. C: 1, Tufted Titmouse; 2, Plain Titmouse; 3, Black-crested Titmouse. D: 1, Scarlet Tanager; 2, Western Tanager. E: 1, Eastern Bluebird; 2, Western Bluebird. F: 1, Barred Owl; 2, Spotted Owl.

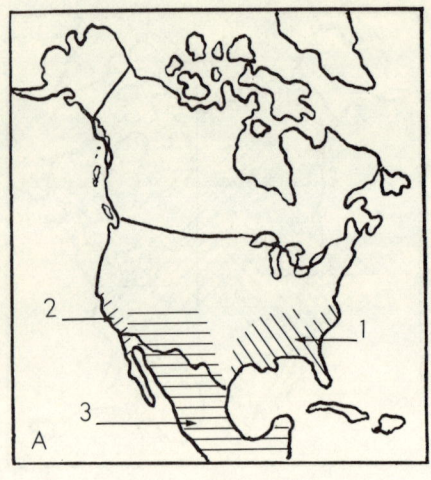
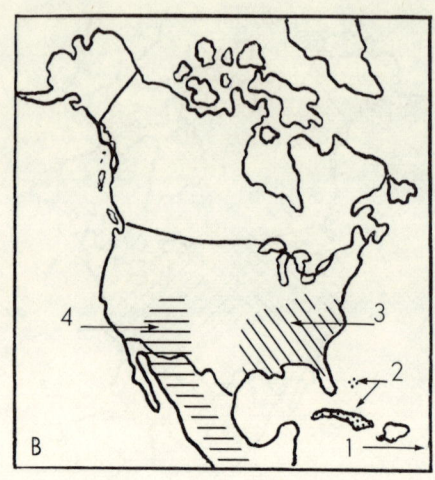

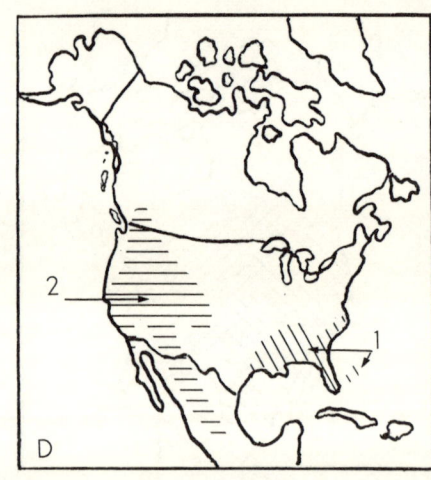
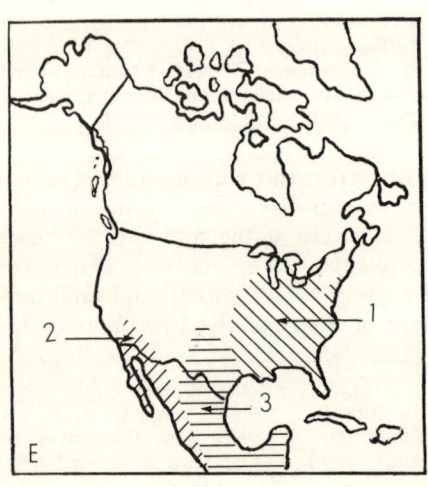
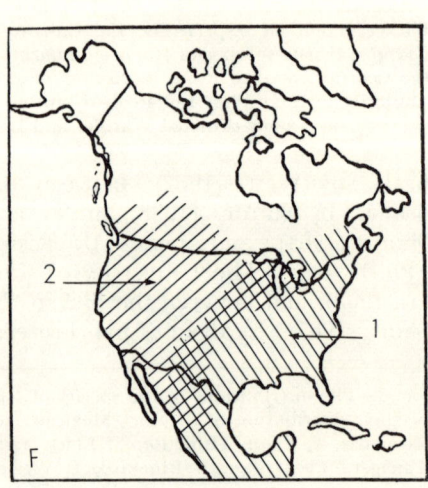

pears to have been facilitated by man-induced changes in the environment.

These essentially static hybrid zones seem to serve here as effectively as does competitive exclusion between more fully differentiated taxa to prevent the establishment of sympatry (see commentary of Bigelow, 1965).

Among the buntings, the Indigo and Lazuli seem to be semispecies. The Varied Bunting seems to form a superspecies with that pair, and it forms a species group with these and others of the genus (it has hybridized with the Painted Bunting—Storer, 1961). It is another species with a southern disjunct range, which suggests the long-standing separation of a jointly ancestral form in Texas or Mexico.

However, as in Group I, the main lines of disjunction in Group II are east-west. Both complexes vary considerably in the latitudes of distributions. The rather wide taiga distribution of the Rose-breasted Grosbeak and the Yellow-shafted Flicker in the present group is characteristic of many additional eastern-northern species soon to be discussed (Figs. 8-12).

Hybridization between the pairs of orioles and of flickers seems to be more extensive than between the pairs of grosbeaks and of buntings. In the first two cases, over appreciable "hybrid zones" there are no clearly parental types of either form. The hybrid zone of the orioles is apparently somewhat narrower than that of the flickers. The breadths of both hybrid zones, however, and the distance over which some introgression seemingly may be detected, suggests that these birds are best regarded as well-marked subspecies at a secondary zone of contact than as semispecies. Still, it is far from certain that one will ever swamp the other. It is even less likely that the Gilded Flicker—which also hybridizes with the Red-shafted Flicker in limited areas—will ever be swamped. Indeed, while Short (1965) concluded that all the flickers should be thought of as a single species, Johnson (1969, p. 227) held, in a thorough review of Short's work, that the Gilded Flicker is best considered a full species. Like the ivory-billed woodpeckers and the Yellow-throated Warbler group of Group III, the flickers also have disjunct representatives in the mountains of southern Mexico and Middle America and in the West Indies.

Certain towhees form a complex that might also have been considered here, except that the northern forms, at least, seem not to have reached the level of semispecies. These towhees are the eastern black-backed and western spotted-backed races of the Rufous-sided Towhee (studied by Sibley and West, 1959). In the Mexican highlands the spotted-backed population hybridizes with the green-backed form *(Pipilo ocai)* of the Mexican volcanic belt. While possibly semispecies, these two have perhaps attained specific distinctness (Sibley, 1954; Sibley and Sibley, 1964).

Group III (some representative superspecies).—Here are grouped six complexes (Fig. 5) with allopatric distributions. Most of them do not meet at all, but where overlap does occur, hybridization is limited or nonexistent. In the first four complexes all of the species are more or less adapted to pine or pine-oak woodland, except for the Ladder-backed Woodpecker, the east-

---

Fig. 5. The distributions of the species in Group III. A: 1, Red-cockaded Woodpecker; 2, Nuttall's Woodpecker; 3, Ladder-backed Woodpecker. B: 1, Adelaide's Warbler; 2, Olive-capped Warbler; 3, Yellow-throated Warbler; 4, Grace's Warbler (it is impossible to show at this scale that a resident race of the Yellow-throated Warbler is sympatric in the Bahamas with the Olive-capped Warbler). C: 1, Cuban Ivory-billed Woodpecker, a subspecies *(Campephilus principalis bairdi)* of the mainland form; 2, mainland Ivory-billed Woodpecker; 3, Imperial Woodpecker. D: 1, Brown-headed Nuthatch; 2, Pygmy Nuthatch. E: 1, Red-bellied Woodpecker; 2, Gila Woodpecker; 3, Golden-fronted Woodpecker. F: 1, Eastern Meadowlark; 2, Western Meadowlark.

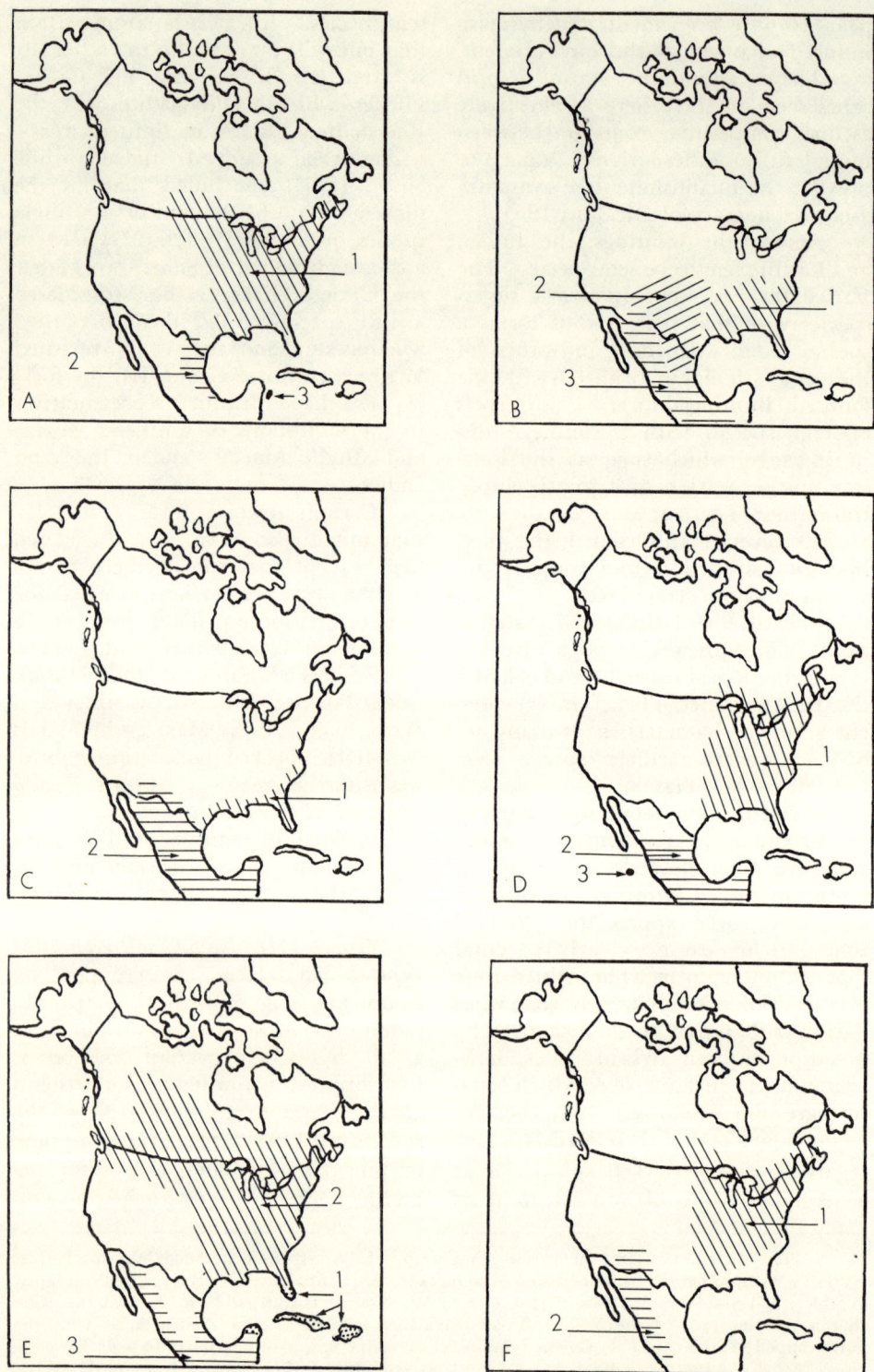

Fig. 6. Distributions of the species of Group IV. A: 1, Brown Thrasher; 2, Long-billed Thrasher; 3, Cozumel Thrasher. B: 1, Bachman's Sparrow; 2, Cassin's Sparrow; 3, Botteri's

ern Ivory-billed Woodpecker, and Adelaide's Warbler. The Ladder-backed Woodpecker group has been discussed by Voous (1947, pp. 91-92), who regarded the differentiation of the Red-cockaded Woodpecker from its western relatives as predating the Pleistocene (that is, in the Pliocene). I have previously commented (Mengel, 1964, pp. 29-30, 36) upon the distributions of the Yellow-throated Warbler and Ivory-billed Woodpecker groups, noting the possession by each of one or more West Indian taxa. I suggested that these differentiations also may date from the Pliocene.

The nuthatches have been studied extensively by Norris (1958), who concluded (p. 292) that they were distinct at the species level. It is perhaps worth noting that the Cuban Ivory-billed Woodpecker, which until recent years was regarded as specifically distinct, inhabits upland pine woods, as does the Imperial Woodpecker. For that matter, the eastern Ivory-bills sometimes foraged in pines (Allen, 1939, p. 2).

The Red-bellied Woodpecker group has no pine-woodland representative; otherwise it fits well enough with the preceding ones. The group has been studied by Selander and Giller (1963), who concluded (p. 267) that the three taxa mapped are fully distinct at the specific level because they fail to interbreed extensively in Texan and Mexican zones of sympatry.

Note the more or less broad gap, centered in Texas, between the distributions of one pair of each of these complexes.

The meadowlarks, finally, really do not belong in this section, having been treated under secondarily grassland species above; they are included, however, because of their east-west pattern, which is typical of so many other North American birds.

Group IV (additional superspecies, with distributions in eastern deciduous forest and in Mexican woodland or scrub).—Figure 6 shows the distributions of six pairs of decidedly similar species or—in two or three cases—perhaps of rather distinct subspecies or semispecies. Blake (1968, p. 123) and Sutton (1951, p. 242) considered the Red-eyed and Yellow-green vireos to be conspecific, whereas Hamilton (1962, p. 45) considered them to form a superspecies with the addition of the Black-whiskered Vireo (including the Yucatan Vireo, *Vireo magister*). The warblers were considered to be possibly conspecific by Lowery and Monroe (1968, p. 11), as were the grackles by most authors until Selander and Giller's work (1961) seemed to demonstrate their distinctness at the species level. The widely disjunct Whip-poor-wills seem in all probability to be subspecies, although this cannot be certain.

Storer (1955, p. 201), in a brief survey, regarded the species of the *Aimophila* complex as a natural group. Superficially, at least, Botteri's and Bachman's sparrows seem to me to be a little more similar to each other than either is to Cassin's Sparrow. Cassin's and Bachman's, taken together, would fit well in Group I; Bachman's and Botteri's, taken together, fit well in the present group. The Brown and the Long-billed thrashers closely resemble each other, and were once regarded as races of one species (Bent, 1948, p. 375). Paynter (1955, pp. 223-224) lumped the Cozumel Thrasher with the Long-billed, but K. C. Parkes (personal communication) thinks the three are best regarded as a superspecies.

Sparrow (note that Cassin's Sparrow is a grassland species and the other two may be essentially savannah forms). C: 1, Great-tailed Grackle; 2, Boat-tailed Grackle. D: 1, Parula Warbler; 2, Olive-backed Warbler; 3, Socorro Warbler. E: 1, Black-whiskered Vireo; 2, Red-eyed Vireo; 3, Yellow-green Vireo. F: 1, eastern race *(Caprimulgus vociferus vociferus)* of the Whip-poor-will; 2, western race of the same species *(C. v. arizonae)*, a well-marked subspecies. Other Mexican races are not mapped.

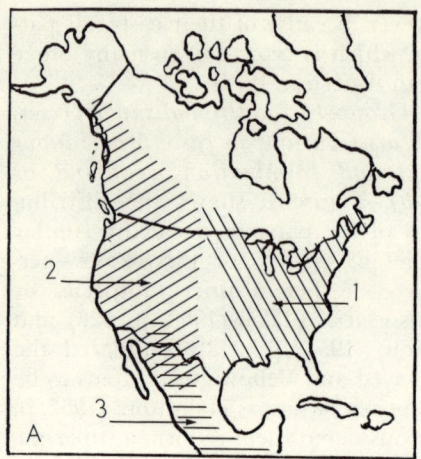 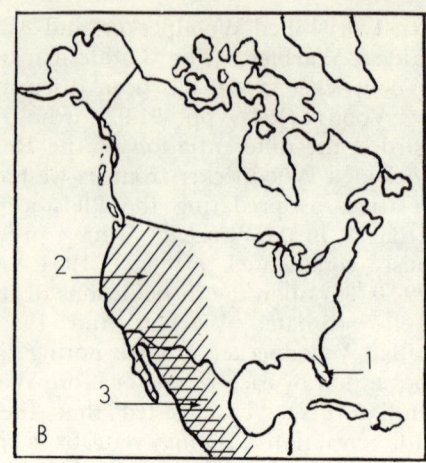

Fig. 7. Distributions of the species in Group V. A: 1, Eastern Wood Pewee; 2, Western Wood Pewee; 3, Coues' Flycatcher. B: 1, Florida race (*Aphelocoma coerulescens coerulescens*) of the Scrub Jay; 2, western races of the Scrub Jay; 3, Mexican Jay.

A common characteristic of the distributional patterns of Group IV is the indication in all of them of past or present disjunction in the vicinity of Texas (see above). The most different, probably the thrashers, sparrows, warblers, and (perhaps) the grackles, may derive from Pliocene separations; some or all of the others may date from Pleistocene disjunctions. It is also possible, of course, that some have been separated as long ago as the Pliocene but have intermittently reestablished contact without loss of conspecificity.

Group V (two fairly simple species groups).—We come now to some distributions (Fig. 7) featuring extensive sympatry among some members of the complexes. The members of the present group are on the whole more southerly in distribution than those of the following group.

The scrub jays have been studied by Pitelka (1951), who concluded that the separations leading to the differentiation of the Scrub Jay and the Mexican Jay, and of the now widely disjunct Florida subspecies of the Scrub Jay (*Aphelocoma c. coerulescens*) were both Pliocene. In view of recent findings and theories concerning speciation rates in passerines (see Johnston and Selander, 1964; Mengel, 1964, p. 39; Moreau, 1966a, pp. 408-410), I wonder if the latter disjunction is not somewhat more recent.

The wood pewees could descend from an ancient stock common with the ancestor of Coues' Flycatcher (and, more remotely still, with that of the Olive-sided Flycatcher). The two pewees themselves, however, with distributions suggesting those of groups I and II, apparently form a superspecies and seem likely to be of Pleistocene differentiation. Closeness of relationship between the Western Wood Pewee and Coues' Flycatcher, despite their considerable differences and broad sympatry, is suggested by the recent report (Phillips and Short, 1968) of a hybrid between them. The Tropical Pewee, ranging north only to southern Mexico, may be more nearly related to the northern wood pewees than to Coues' Flycatcher (K. C. Parkes, personal communication).

Group VI (patterns involving montane forest, eastern deciduous forest, and especially taiga).—Whereas a few of the species have been northerly, most of the patterns considered earlier have involved southerly distributions. If the taiga has been occupied at all, it has, with rare exceptions, been mainly in its southern ecotones or in its seral stages; it has been occupied all the way across the continent only by the Yel-

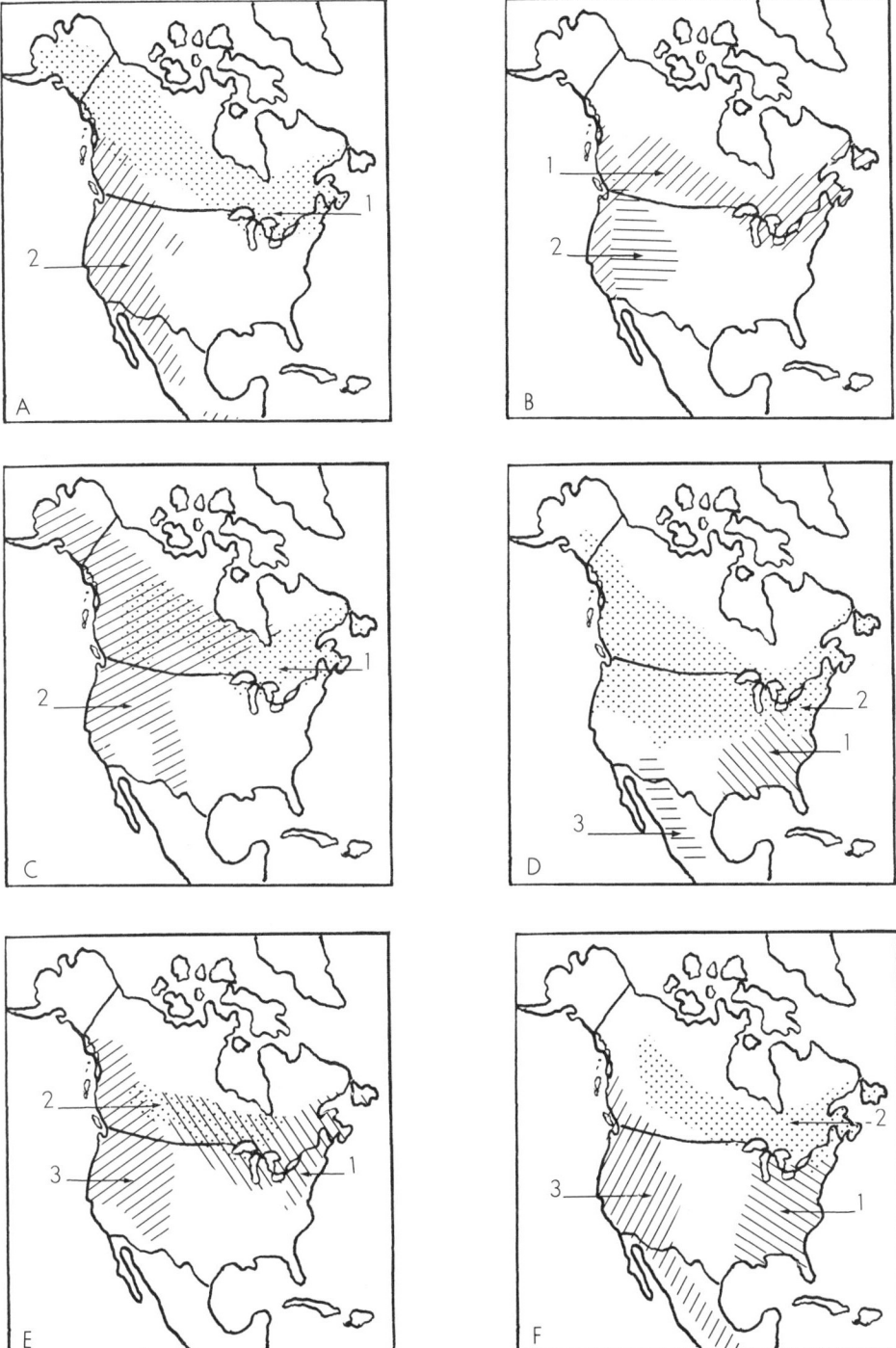

Fig. 8. Distribution of the species of Group VI. A: 1, Myrtle Warbler; 2, Audubon's Warbler. B: 1, Purple Finch; 2, Cassin's Finch. C: 1, Tennessee Warbler; 2, Orange-crowned Warbler. D: 1, Carolina Chickadee; 2, Black-capped Chickadee; 3, Mexican Chickadee. E: 1, Mourning Warbler; 2, Connecticut Warbler; 3, MacGillivray's Warbler. F: 1, Acadian Flycatcher; 2, Yellow-bellied Flycatcher; 3, Western Flycatcher.

low-shafted Flicker. We come now (Fig. 8) to distributional patterns that display considerable use by their owners of this huge and important environment. Such use involves seral stages as well as the climax. The species we are here concerned with possess complex interrelationships with species of other major environments, notably the deciduous and montane forests of the eastern and western parts of the continent, respectively.

In the present group the Myrtle and Audubon's warblers are probably well-marked subspecies. They hybridize freely in a narrow area of contact in Alberta and British Columbia (Hubbard, 1969, p. 395). MacGillivray's and Mourning warblers—both entirely distinct from the Connecticut Warbler—and Black-capped and Carolina chickadees both appear to be semispecies pairs constituting superspecies. Sympatry of the first pair is uncertain; the chickadees hybridize in some areas, apparently with some loss of fertility (Brewer, 1963; Rising, 1968), but do not seem to hybridize in other areas (Tanner, 1952). The remaining taxa of this group all appear to be distinct at the species level and to form superspecies or species groups.

Full occupancy of the vast transcontinental taiga greatly increases the probability that a species possessing such a range will be fragmented into isolated refugia in the event of any major changes in the environment, such as glaciation. Earlier, I (Mengel, 1964) enlarged upon this theme with respect to speciation in wood warblers and, in so doing, considered among various others the three parulid groups shown here. Further consideration of the wood warblers and other groups in this section will be found in the discussions below.

COMPLICATED DISTRIBUTIONAL PATTERNS (GROUP VII)

Since publication of my 1964 paper, I have searched about for distributions of avian complexes that parallel those within the wood warblers—not only the simpler patterns but also the complex and strikingly similar patterns of the Black-throated Green and Nashville warbler groups. Various similarities of distributions in pairs or trios of taxa of other families are indicated in the maps of groups I–VI above. Such simple similarities are perhaps likely by sheer chance, but I was not really prepared to discover two complex configurations that rather closely resemble the Black-throated Green Warbler pattern. These are provided by the Least Flycatcher group and the northern members of the genus *Junco*.

The most remarkable thing about these complex patterns is their similarity in so many details. Each involves four or more taxa. When these units exceed four, the extra taxa are almost always southern, often of narrow range, and perhaps relict. The remaining, more widespread units are invariably eastern, or more accurately eastern-northern, and western in distribution. In all groups, the easternmost taxon is a taiga species with some populations in mixed forest and eastern deciduous forest centered on the Appalachians and always three major units are western. Of the three major western taxa, the southernmost (except for the juncos) tends to occur at the lowest elevations and tends to occur in environments (woodland, piñon-juniper, various scrubs) unlike the montane and coast forests, which are variously occupied by the two more northwesterly taxa. The southernmost form seems also, correspondingly, to be the most distinct member of the group; at least it is the one most prone to sympatry with other members of the group. In all of the groups except the juncos, at least three or four taxa constitute a superspecies or, as in the Least Flycatcher complex, a species group (Least and Hammond's flycatchers, within this group, may be a superspecies) with distributions and characteristics that are not incompatible with

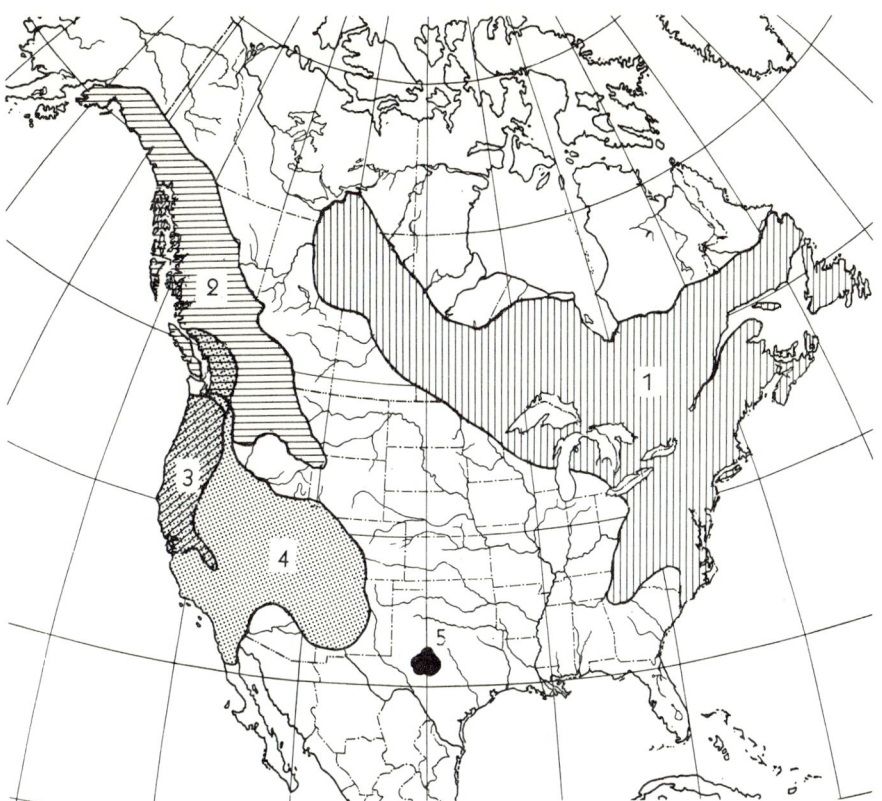

Fig. 9. Approximate breeding distributions of the Black-throated Green Warbler group. 1, Black-throated Green Warbler; 2, Townsend's Warbler; 3, Hermit Warbler; 4, Black-throated Gray Warbler; 5, Golden-cheeked Warbler. Figure reproduced by courtesy of the Cornell Laboratory of Ornithology from *The Living Bird* (vol. 3, p. 19, 1964).

descent from a not-too-ancient superspecies. Finally, again in all but the juncos, there is a broad zone of at least geographical (not necessarily ecological) overlap between the southernmost western taxon and one or more of the others, which seems to make the whole to be correctly considered as a species group.

I think the differences mentioned above indicate a somewhat different history for the junco group, similarities notwithstanding. This matter will be discussed shortly. For the rest, the fact that the distributional patterns of four or five members of three taxonomically diverse groups display so many features in common suggests a common history of distribution and species formation. To be sure, given a species group or superspecies of about the right size and range of adaptation, one might expect some similarity between its distribution and that of a comparable group simply on the basis of the limited number of major habitat blocks available; but when five forms are considered in each case, the number of possible combinations is sufficiently great so that the probability of essential duplication in major features by chance alone seems small indeed.

The Black-throated Green Warbler group.—Earlier I have discussed this group and its distributions (Fig. 9) at considerable length (Mengel, 1964, pp. 16-22, pl. I, and fig. 5). Additional comments applying to more than one group will be found in the discussion following this section. In

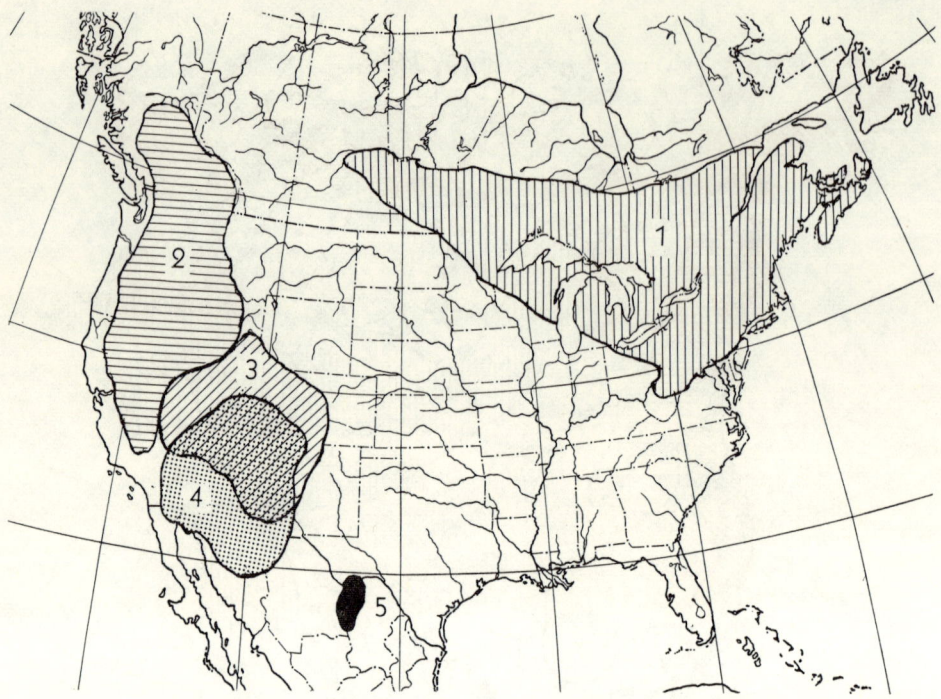

Fig. 10. Approximate breeding distributions of the Nashville Warbler group. 1, Nashville Warbler, eastern race *(Vermivora ruficapilla ruficapilla)*; 2, Nashville Warbler, western race *(V. ruficapilla ridgwayi;* 3, Virginia's Warbler; 4, Lucy's Warbler; 5, Colima Warbler. Figure reproduced by courtesy of the Cornell Laboratory of Ornithology from *The Living Bird* (vol. 3, p. 25, 1964).

brief, the eastern and northern Black-throated Green Warbler, which is more or less adapted to various mesic deciduous forest habitats in its Appalachian range, is more typical of taiga in its northern range, which does not seem to quite reach that of Townsend's Warbler. The latter may encroach into boreal forest; it is typical of montane forest in the southern parts of its range and in the southwesterly quadrant it occupies tall coastal forests where it is to some extent sympatric with the Hermit Warbler. Here the two produce at least occasional hybrids (Jewett, 1944), which resemble the Black-throated Green Warbler (see Mengel, 1964, pl. I) perhaps more than either parent! In my opinion, all in all, both more nearly resemble the Black-throated Green Warbler than either of them resembles the small Black-throated Gray Warbler, southernmost of the western forms, which, while it is partially sympatric with them in coast forests, is unique to the south in also occupying oak, piñon, juniper, and similar scrubby and woodland habitats at intermediate elevations. It is, of course, true that this difference in environment must relate to differences in the bird, which may, therefore, not all be of phylogenetic significance. Finally, the Golden-cheeked Warbler of the Edwards Plateau of Texas seems to be little more than a melanistic, relict Black-throated Green Warbler; it is interesting that other animal and plant organisms of the Edwards Plateau seem also to be relict relatives of eastern species (Blair, 1958).

The Nashville Warbler group.— This complex of distributions (Fig. 10) was likewise discussed at some length in my earlier paper (Mengel, 1964, pp. 22-25, pl. II, and fig. 6). The major difference between this pattern and the previous one is that the northwestern

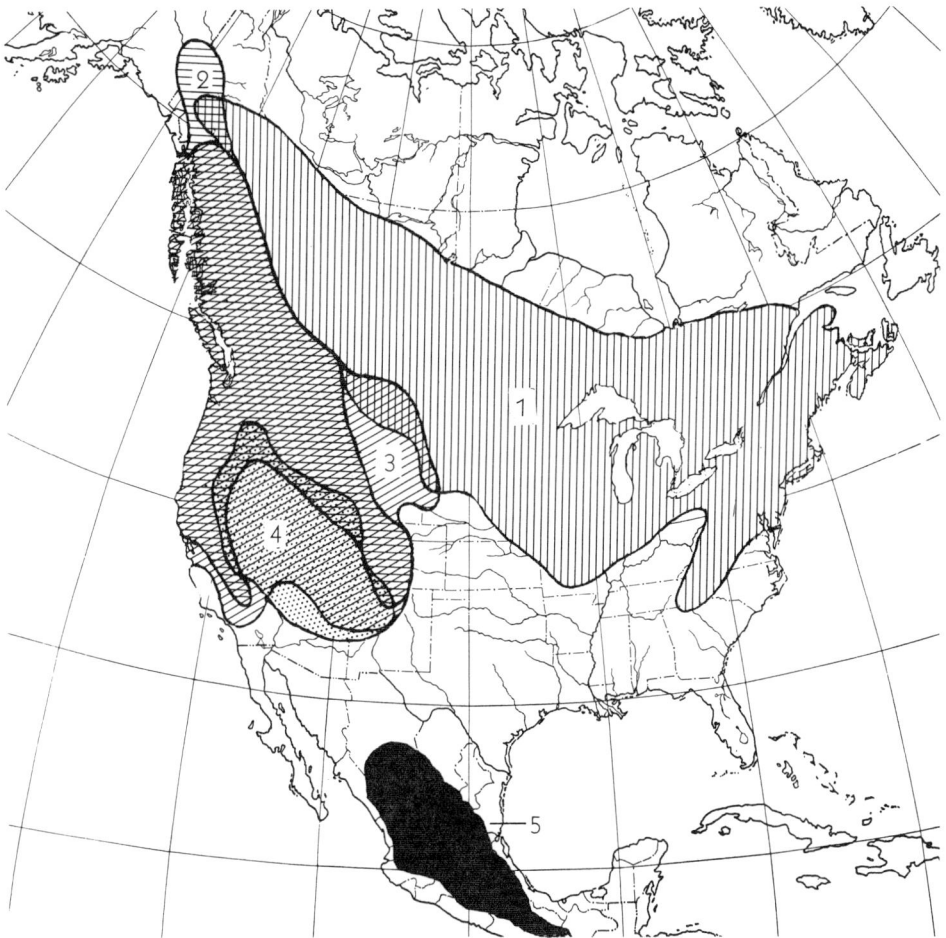

Fig. 11. Approximate breeding distributions of the Least Flycatcher group. 1, Least Flycatcher; 2, Hammond's Flycatcher; 3, Dusky Flycatcher; 4, Gray Flycatcher; 5, Pine Flycatcher.

form of the complex is here only a faintly differentiated race *(Vermivora ruficapilla ridgwayi)* of the eastern-northern Nashville Warbler—one of the rare examples of disjunct but faintly differentiated subspecies among North American birds. Otherwise the differences are overbalanced by the similarities, although the birds of the present complex are on the whole more addicted to successional stages of vegetation (which is broad-leaved, whether or not intermixed with coniferous growth). Here, again, the most distinct member of the complex seems to be the little desert-inhabiting Lucy's Warbler, southernmost of the western species.

This complex, finally, differs from the last in that the southernmost member, the Colima Warbler, appears to be closely related not to the eastern but to one of the western taxa, namely Virginia's Warbler.

The Least Flycatcher group.—The ranges of the three western members of this group have been adapted from Johnson's figure 16 (1963, p. 142), to which I have added the range of the Least Flycatcher, mapped from standard sources (Fig. 11). Also, I have added the range of the Pine Flycatcher, which seems (Johnson, 1963, p. 214) to be quite a distinct species and the southernmost member of the group.

The similarity of the apparent relations and distributions in this group to those in the two foregoing wood warbler groups is striking, the chief difference being only a somewhat greater degree of over-all sympatry in the present group. After intensive study of both the morphology and field biology of the western *Empidonaces,* Johnson concluded (1963, pp. 213-214) that Hammond's Flycatcher and the Least Flycatcher were most alike. Then, moving southward in the West, the Dusky Flycatcher, the Gray Flycatcher, and the Pine Flycatcher are progressively more distinct from the first two. The Gray Flycatcher, like the Black-throated Gray Warbler and Lucy's Warbler, whose relative geographic positions it shares, is the most prone to occur in lowland and xeric environments. The Mexican Pine Flycatcher, while differing in choice of habitat, is not unlike the Colima Warbler in its occupancy of highland pine-oak forests.

It might now be helpful to review the parallels among these three complexes.

1. Eastern-northern taxa: *Black-throated Green Warbler, Nashville Warbler* (nominate race), *Least Flycatcher*. Occupy eastern deciduous forest in its northern and Appalachian portion, mixed or lake forest, and taiga or (Nashville Warbler) its successional stages. Invariably winter farthest south and make longest migration. Much like the next following group of congeners.

2. Northernmost western taxa: *Townsend's Warbler, Nashville Warbler* (western race), *Hammond's Flycatcher*. Occupy taiga in the northern Rocky Mountains and montane forests and, to some degree, coast forest farther south, or (Nashville Warbler) their successional stages.

3. Somewhat less northern of the western taxa: *Hermit Warbler, Virginia's Warbler, Dusky Flycatcher*. Do not extend as far north as the foregoing. Variously distributed in coast and montane forests (Hermit Warbler) or montane forests and altitudinally lower woodlands and scrubs. The Virginia's Warbler and the Dusky Flycatcher, at least, tend toward partial ecological separation from their foregoing counterparts where geographical overlap occurs.

4. Southern or central western taxa: *Black-throated Gray Warbler, Lucy's Warbler, Gray Flycatcher*. Tend to occupy woodlands, scrubs, and desert or semidesert habitats and thus to be ecologically isolated from the members of the second group and at least partially isolated from members of the third group in cases of geographical overlap. More different from members of the first group than the above, and perhaps the most distinct in their species groups—as might be expected in view of the distinctiveness of their environments.

5. Southern taxa: *Golden-cheeked Warbler, Colima Warbler, Pine Flycatcher*. Southern disjuncts in plateaus and mountains of Texas or Mexico (or both). Inhabit pine forests, woodlands, or juniper habitats, and may be relictual. The Golden-cheeked Warbler is unique in occurring farther east than the others, and, quite possibly related to this, also is unique in having its nearest ties apparently with the eastern-northern instead of one of the more southerly western members of its species group.

The junco group.—The distributions of these birds are shown in Figure 12. Their evolutionary history and differentiation were considered by Miller (1941, pp. 371-372 and elsewhere). The chief differences in the distributions of this group from the others studied are three, which I think probably indicate quite different histories. The juncos are uniformly boreal or montane in distribution; there are no really desert, scrub, or low-elevation woodland forms. Second, there are among the juncos numerous disjunct, southern forms in outlying habitats such as Baja California, the volcanic belt of Mexico, and in Central America. Third, there

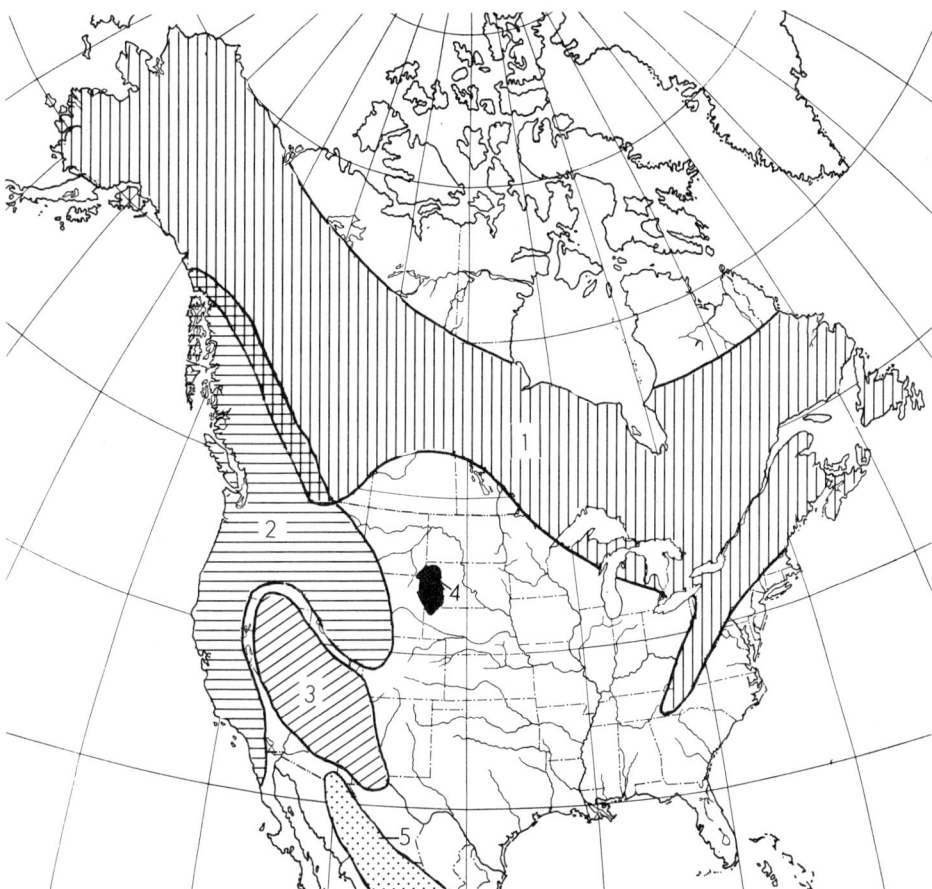

Fig. 12. Approximate breeding distributions of the juncos (southern and insular species, probably most or all of them relicts, are not shown). 1, Slate-colored Junco; 2, Oregon Junco; 3, Gray-headed Junco; 4, White-winged Junco; 5, Mexican Junco. Note the relationship of the last to the Slate-colored Junco; the Black Hills differentiate appears to be a relict related to the eastern-northern form in much the same way as the Golden-cheeked Warbler is related to the Black-throated Green Warbler (Fig. 9).

is no case in the genus of extensive sympatry between taxa.

Similarities between this distributional complex and the foregoing also exist. The eastern-northern Slate-colored Junco corresponds well with its foregoing counterparts, as does the northwestern Oregon Junco, which, further, may be most closely related to it. The Slate-colored Junco, aside from its southern-eastern race (*Junco hyemalis carolinensis*), is a bird of the boreal forest or taiga, including the northern Rocky Mountains north of the Peace River gap in east-central British Columbia. The Oregon Junco is a bird of coast forest and montane forest in the Cascades and Sierras and in the central Rocky Mountains north of the Wyoming deserts and south of the Peace River gap. The Gray-headed Junco occupies the southern Rocky Mountains between the Wyoming deserts and the Mogollon rim, while the Mexican Junco is a bird of the mountains of central and northern Mexico. These kinds correspond in number, if not in pattern, with preceding complexes, but as noted above, there is no parallel in the other groups for the several far southern, montane disjuncts found among the juncos.

The systematic treatment of the juncos in the 1957 A.O.U. Check-list is conservative, generally following the conclusions of Miller's (1941) study. There is now a large body of opinion that most junco "species" of the 1957 check-list are actually of lower taxonomic rank. Rather than forming, as do the complexes considered above, a species group consisting of a superspecies and various overlapping species, the juncos, or at least the ones with which we are concerned, appear themselves to be a superspecies consisting of but two allopatric and complexly polytypic species (K. C. Parkes, personal communication). These are the dark-eyed juncos (all of the above-named forms except the Mexican Junco and including the White-winged Junco of the Black Hills) and the yellow-eyed juncos (Mexican Junco and relatives). Within these species the kinds here named are allopatric but often contiguous, in which case they generally hybridize in the manner of semispecies. The presumptive semispecies, in turn, are themselves polytypic at a lower level, and have been divided into various named geographic races.

The juncos appear to me, as they did to Miller (1941, pp. 371-372) to be a moderately plastic group with a long boreal and montane history. Sedentary to moderately migratory, they appear likely to have differentiated *in situ* in the montane and boreal regions where they now live or in the nearest appropriate glacial refugia, with little subsequent movement resulting in allopatric distributions. While there may have been contractions of range, especially by more primitive forms, these boreal birds show little evidence of a dynamic history of the kind postulated for some of the other groups, notably wood warblers; thus their distributional resemblance to the above complexes may be, after all, essentially coincidental.

SOME NORTHERN DISTRIBUTIONS

We have already seen, in groups VI and VII, a number of distributional complexes that included a northern taiga representative, and one of the earlier groups included such distributions. It will now be instructive to survey many others as well.

By taiga distributions is here meant those that include some or all of the north-south extent of the taiga but most of which do not encroach broadly on the Central Plains to the south of the taiga, although they may variously enter the aspen-parklands ecotone just to the south. The widest encroachments beyond the taiga here allowed are shown by the Black-capped Chickadee and the Saw-whet Owl, which have obvious relationships to Old World boreal congeners. Some of the species may also extend, again variously, southward either in eastern deciduous or western montane forest. In short, we are interested in essentially forest species, or in species of successional stages of the forests in question, but not in species of wider adaptations that happen to include taiga.

It has proved instructive to divide these distributions into three classes, as shown in Table 3. (1) Distributions including much taiga, usually including some or much of Alaska, and considerable amounts of western montane and, sometimes, coastal forest. In some cases there is an eastern southward extension into mixed forest and Appalachian coniferous forest, but there is usually little adaptation to, or distribution in, the true eastern deciduous forest. (Limited exceptions are indicated in a footnote to Table 3.) (2) Species more or less widespread in eastern deciduous forest, at least in its northern parts, and extending through mixed forest into the taiga, where they occur more or less widely, but only rarely throughout (that is, to northern and western Alaska). (3) Taiga species lacking considerable distributions out-

TABLE 3.—Forest Species Important in the Taiga, Grouped in Three Distributional Classes[1]

Class 1—Taiga–Montane Forest Species[2]

B GREAT GRAY OWL +
B COMMON RAVEN +[3]
B BROWN CREEPER +[3]
B PINE GROSBEAK +
B WHITE-WINGED CROSSBILL +
B GOSHAWK ++[3]
?B PIGEON HAWK ++
B NORTHERN THREE-TOED WOOD-
　　PECKER ++
B RED CROSSBILL ++
B SAW-WHET OWL *
B BLACK-BACKED THREE-TOED WOOD-
　　PECKER *
B GRAY JAY *
B RED-BREASTED NUTHATCH *
B GOLDEN-CROWNED KINGLET *

PURPLE FINCH *[3]
B PINE SISKIN *
RUBY-CROWNED KINGLET **
B EVENING GROSBEAK **
B Bohemian Waxwing +
B Winter Wren ++
B Spruce Grouse *
Yellow-bellied Sapsucker
Olive-sided Flycatcher
Hermit Thrush I
Swainson's Thrush I
Orange-crowned Warbler II
Wilson's Warbler
White-crowned Sparrow
Fox Sparrow
Lincoln's Sparrow

Class 2—Eastern Deciduous Forest–Taiga Species

Broad-winged Hawk
Barred Owl
Eastern Phoebe
Least Flycatcher III
Red-eyed Vireo
Black-and-white Warbler
Magnolia Warbler
Black-throated Green Warbler IV
Black-throated Blue Warbler
Blackburnian Warbler
Chestnut-sided Warbler

Ovenbird
Northern Waterthrush
Mourning Warbler V
Canada Warbler
American Redstart
Scarlet Tanager VI
Rose-breasted Grosbeak VII
White-throated Sparrow
Swamp Sparrow
B Slate-colored Junco

Class 3—Species Restricted to Taiga[2]

B HAWK OWL +
B BOREAL OWL +[4]
B NORTHERN SHRIKE +
B BLACK-CAPPED CHICKADEE *
B BOREAL CHICKADEE *
Yellow-bellied Flycatcher
Gray-cheeked Thrush I
Philadelphia Vireo
Tennessee Warbler II

Cape May Warbler
Myrtle Warbler VIII
Bay-breasted Warbler IX
Blackpoll Warbler IX
Palm Warbler
Connecticut Warbler
Rusty Blackbird
Le Conte's Sparrow
B Tree Sparrow

[1] Throughout, truly boreal species are marked B; capitalized species belong to families or genera of Old World origin (for exceptions and qualifications see text).

[2] Symbols indicate the following: +, New World race(s), weakly defined; ++, New World races, well defined or numerous or both; *, closely related to an Old World species; **, closely related to an Old World genus. Roman numerals signify membership in one of nine superspecies or close species groups of summer boreals.

[3] Small numbers also breed in the northern part of the eastern deciduous forest, even in the absence of conifers.

[4] There is a small, disjunct (?) population in the Colorado Rocky Mountains (Baldwin and Koplin, 1964).

side of that formation. They may be more or less widespread, a few reaching from coast to coast. I have eliminated Eurasian species of narrow occurrence in the New World, such as the Gray-headed Chickadee. I have eliminated also, because they are too widespread to seem instructive in the present context, several species that occur in taiga but also have populations both in eastern deciduous and western montane forests (for example, Ruffed Grouse, Pileated Woodpecker, Veery, Solitary Vireo, Nashville Warbler). The three distributional patterns are shown in Figures 13-15.

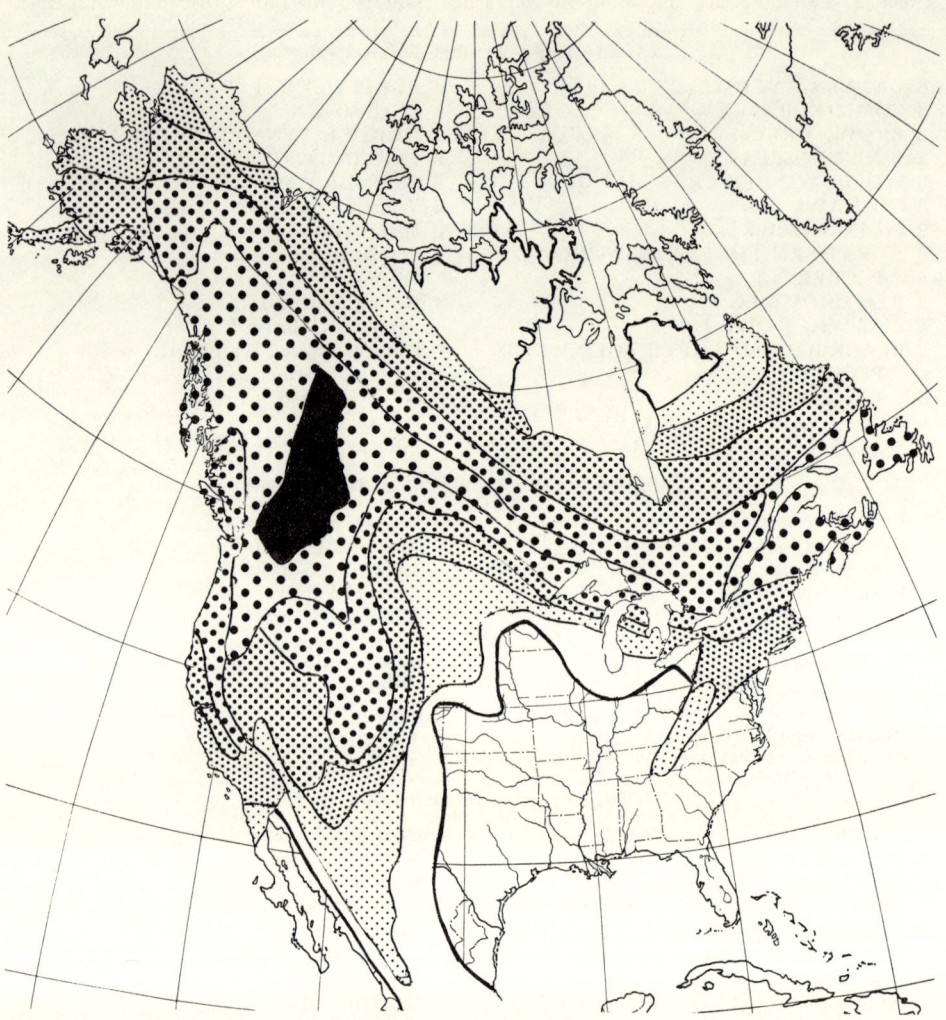

Fig. 13. The over-all distributions of the montane forest–taiga species (first group of Table 3). Species densities from the white peripheral pattern to the black center are: 1–2, 3–5, 6–10, 11–15, 16–20, 21–25, 26 or more. The wide arctic island distribution of the Common Raven is not shown.

Several things are revealed by Table 3 and Figures 13-15; but before discussing these, it is important to make a basic distinction that is often overlooked. *We should carefully avoid the tendency to confuse breeding distribution with total distribution,* which not all biogeographers have done. Two species may have identical breeding distributions, tempting us to think of them as equally "boreal," and perhaps even as having similar biogeographical histories, but they are biologically very different if one is sedentary and completely cold-adapted and the other is a long-distance migrant that travels thousands of miles to the southward well before the onset of cold weather. The first species is certainly boreal in every sense of the word; the second is never exposed to the most conspicuous aspect of boreal climate and may in fact live throughout the year in a climate that deviates but little from tropical. Such species I here call "summer boreal." That as a group they tend to have bio-

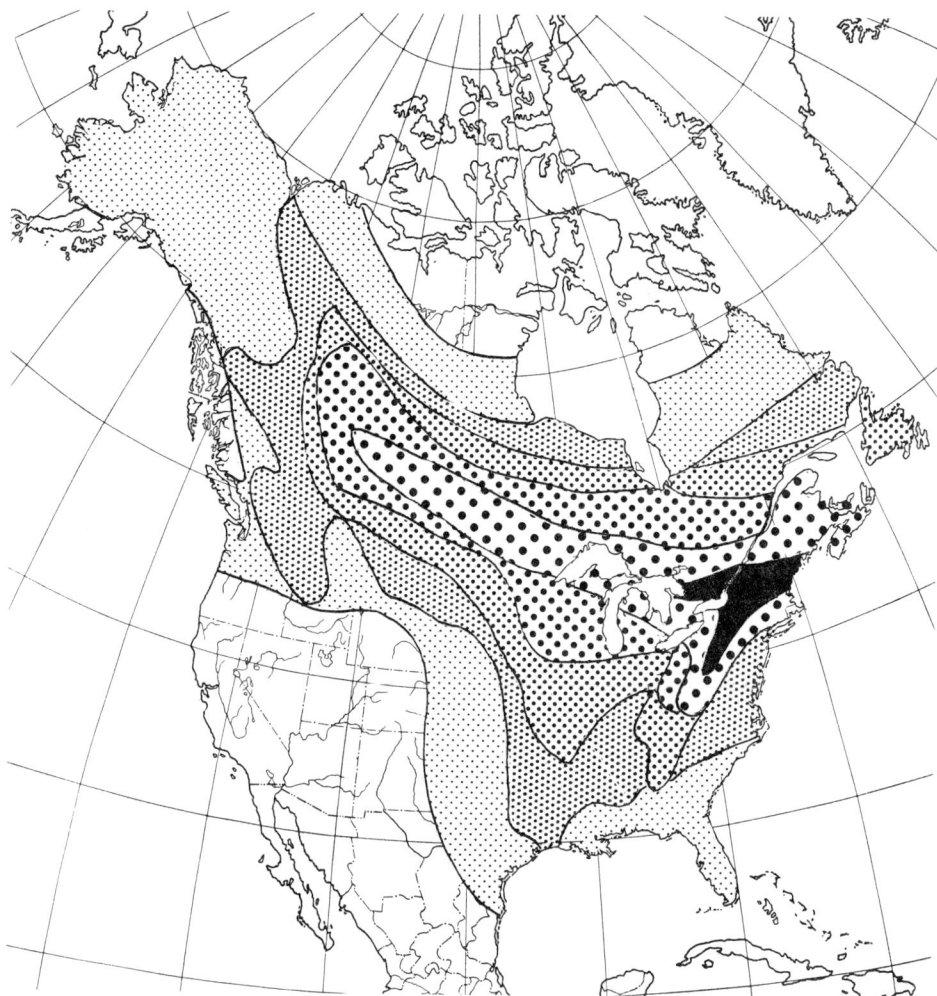

Fig. 14. The over-all distributions of the eastern deciduous forest–taiga species (second group of Table 3). Species densities from the palest peripheral pattern to the black center are: 1–2; 3–5; 6–10; 11–15; 16–20; 21 or more.

geographical histories considerably different from truly boreal species is indicated by numerous lines of evidence discussed below.

The three classes contain altogether 69 species, all terrestrial taxa of nonaquatic groups. In Table 3 I have written in all capital letters the names of species of families or subfamilies that either were thought by Mayr (1946, pp. 26-27) to be of Old World origin or else were, as are the Goshawk, Pigeon Hawk, and three-toed woodpeckers *(Picoides)*, boreal species or genera of families unanalyzed by him. Exceptions are the members of the genus *Catharus*, which, although they belong to the Old World family of thrushes, probably are northern derivatives of a secondary radiation in southern North America and which, further, are summer boreals only. All species in capital letters seem to be truly boreal birds in the sense of being more or less indifferent migrants, and they are definitely cold-adapted (the Pigeon Hawk is slightly exceptional— it probably migrates, however, only as

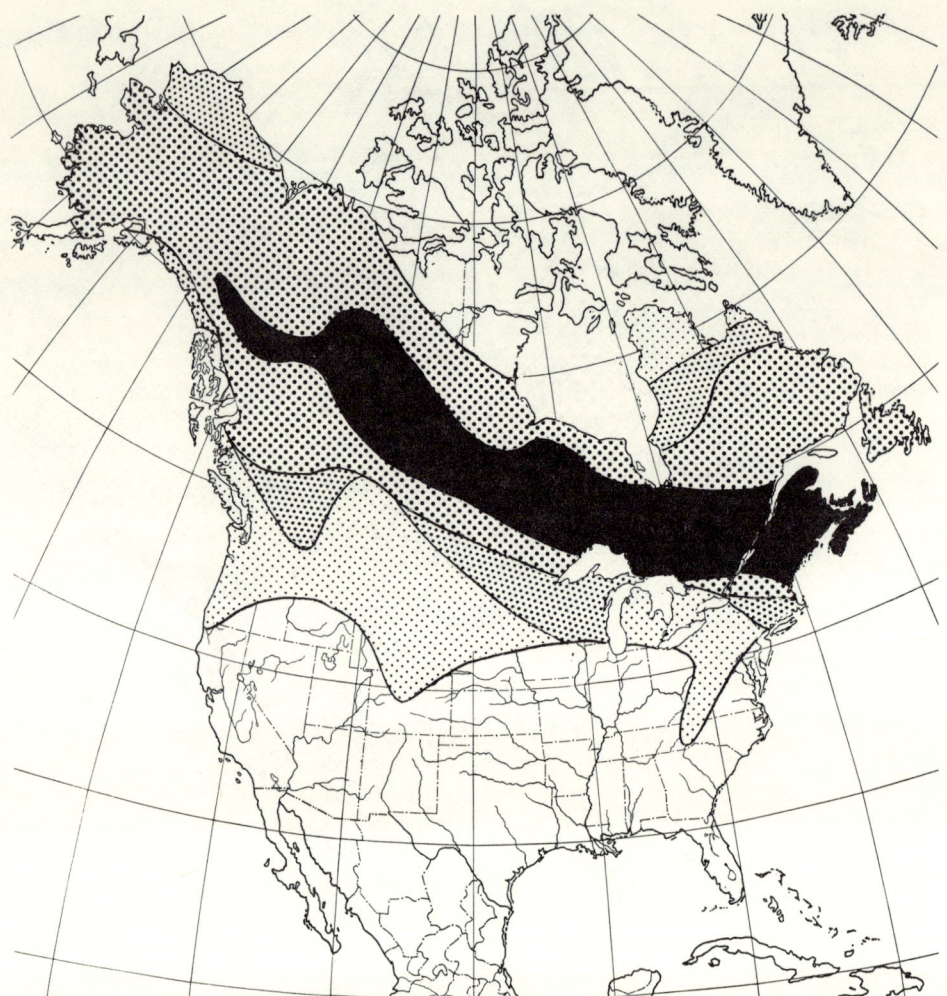

Fig. 15. The over-all distributions of the species restricted to taiga (third group of Table 3). Species densities from the palest peripheral pattern to the black center are: 1–2; 3–5; 6–10; 11–15; 16 or more.

far as it needs to in order to find an adequate food supply of small birds). These truly boreal species are marked B in Table 3. As a general criterion of cold adaptation I have used the presence of a significant wintering population in northern United States and southern Canada. Some of the species are presently Holarctic, whereas others, New World endemics, are presumably descended from such species. Within the three classes shown in Table 3, I have listed both capitalized (all caps) and uncapitalized (caps and lower case) species with close Old World relations in order of the apparent closeness of the relationship, as indicated by the symbols used in that table.

The uncapitalized species belong to families or subfamilies thought by Mayr (1946, pp. 26-27) to be of New World origin, most of them in warm-temperate or tropical environments, whether in North America (most) or farther south (blackbirds, flycatchers). Of these species only the Spruce Grouse and Bohemian Waxwing are truly boreal (and the latter is almost certainly

TABLE 4.—Features of Some Northern Distributions

Kind of distribution	Number of species	Members of Old World or unanalyzed groups (%)[1]	Members of New World groups (%)[2]
1. Montane Forest–Taiga	30	66.6 (20)	33.3 (10)
2. Eastern Deciduous Forest–Taiga	21	5.0 (1)	95.0 (20)
3. Taiga only	18	33.3 (6)	66.6 (12)

[1] Chiefly boreal species; actual numbers in parentheses.
[2] Chiefly "summer-boreal" species; actual numbers in parentheses.

a species of Old World origin as well), although other species approach the truly boreal condition in various ways, notably the Yellow-bellied Sapsucker, Winter Wren, and a few sparrows. The majority, however, are "summer boreal" only, being either moderately or exceptionally long-distance migrants without significant adaptations to cold.

Some of the statistics of the distributions of the species of Table 3 are shown in Table 4. The first point of interest shown by this table is the predominance of truly boreal species in distribution type 1 (montane forest–taiga), where they account for two-thirds of the total; their still appreciable representation in distributional type 3 (taiga only), nearly one-third; and their almost total absence from distributional type 2 (eastern deciduous forest–taiga).

Then we note that a high percentage of these truly boreal species, indeed all except the Bohemian Waxwing, Spruce Grouse, Winter Wren, and a few marginally boreal sparrows, belong either to families or subfamilies thought by Mayr (1946) to be of Old World origin or to groups not analyzed by him. Further, and perhaps of more interest in the present context than the ultimate origins of their families or subfamilies, it seems probable, on the basis of their present distributions and differentiation, that a large majority of these members of Old World or unanalyzed elements either had their origin as species in the Old World or else belong to genera that originated in the northern part of the Old World (as also may have been the case with the spruce grouse genus). *Picoides* and *Aegolius*, it is true, could be exceptions at the generic level (Stegmann, 1963, pp. 68-69), although one of the two species of each seems almost certainly to be of Old World origin.

Conversely, the summer-boreal species, with the sole exception of the three thrushes, all are members of families or subfamilies that Mayr (1946) considered to be of temperate or tropical New World origin. Interestingly, they display a distributional pattern exactly opposite to the boreal species, making up one-third of the montane forest–taiga group, two-thirds of the strictly taiga group, and virtually all of the eastern deciduous forest–taiga group.

In other words, it seems that while many species of groups originating in the tropics or temperate regions of the New World have invaded the taiga to various extents as a breeding range (class 2) and a fair number (class 3) now breed exclusively in taiga, comparatively few such birds also have significant distributions in western montane forest. Further, a significant percentage of the New World, tropically-derived groups that do occur in the western montane forest are represented there by members of superspecies, one or more of whose members have eastern or eastern-taiga distributions.

On the other hand, while the New World–derived, heat-adapted, summer-boreal species have shown considerable plasticity in adapting to new environments (taiga, for instance) for breeding, the Old World–derived, chiefly cold-adapted species have, at least recently, shown little plasticity, by failing to invade the eastern deciduous forest environments. (It is, of course, possible that some of them have been impeded in this respect by the prior presence there of congeners or relatives not otherwise considered in this analysis—for example, the White-breasted Nuthatch against the Red-breasted Nuthatch, or the Wood Thrush against the thrushes of Table 3).

The decidedly different distributional patterns displayed by the Old World and the New World elements as here noted lends considerable support to Mayr's disposition of the various groups among these elements, since these aspects of distribution were not, as such, among the criteria considered by him in arriving at his decisions.

Other aspects of these distributional patterns, and related phenomena, bear interestingly on the history of speciation on the North American continent and the role of the Central Plains in this history.

Before we go on to discussion of such matters, however, it should be mentioned that in addition to the various groups (I-VII, above) and patterns already discussed, there are still other boreal-montane groups with distributions of interest. Like some of the earlier groupings, these reveal the existence of what seem to be secondary zones of contact after periods of Pleistocene isolation. In most cases, however, the differentiation of the component taxa seems to be at the subspecific or, at most, semispecies level, rather than at the superspecific or species level. Some of the more conspicuous of these assemblages were discussed in Rand's (1948) pioneer paper on speciation and glaciation in North America—among them the races of the Gray Jay, the Yellow-bellied Sapsucker, and the White-crowned Sparrow, and the subspecies or semispecies pairs of Oregon and Slate-colored juncos, Franklin's and Spruce grouse, and Audubon's and Myrtle warblers. As many as three or four refugia could have been involved during glacial phases in the histories of some of these taxa (Rand, 1948; Hubbard, 1969, p. 428).

DISCUSSION

In reviewing groups of birds from the various formations surrounding the Central Plains, I searched for superspecies or closely related species groups whose distributional patterns were such as to suggest the geographic areas in which the several forms were differentiated. Although not exhaustive, a fairly comprehensive coverage of North American superspecies and some of the more interestingly distributed species groups resulted. While a few birds of some of the scrub environments enter the picture, the great majority of the species that provide the materials for consideration are birds of the forest, its edges, or its successional stages. This suggests that a prominent influence on the speciation of North American birds, at least from the Middle or Late Pliocene through the Quaternary, has been the periodic fractionation of woodland and forest environments.

While the records of the past are increasingly obscure with remoteness in time, in the absence of contrary evidence we must suppose that the general degree of fractionation, if not the details, in past interglacial times has been somewhat similar to that obtaining at present. That the forest and woodland environments of today are both diverse and quite fractionated is shown by Figure 16. Here we see a large deciduous forest containing an extensive southeastern pine-oak component in the east, which is in effect still larger by virtue of being connected, via the transitional lake forest, with the aspen successional stage of the taiga. The latter has a huge east-west extent, all far to the north. Finally, in the western mountains we see complexly distributed western montane and coast forests, smaller than the other forest formations and less continuous in distribution.

The stippled part of Figure 16 indicates the grassland formation and related, essentially open environments, these being oak savannah in the extreme east and various scrubs in the southwest, with cold desert in the Great Basin. Collectively these open environments may be likened to an ocean, with the deciduous forest a large eastern continent and the montane forests standing for a western complex of smaller island-continents and islands of various sizes. The role of the taiga in this analogy will be discussed below.

THE GRASSLANDS VIEWED AS A SEA

What are the present relations of this central "sea" of open land to other areas? Weaver and Clements (1938, p. 517) add relevant detail: this formation "lies in contact with more formations than any other on the continent." They go on:

> The sole complete exception is the tundra, though the contact with the subalpine and the lake forest is slight. The outstanding relations are with the deciduous and montane forests, with the woodland climax, and with the three scrub climaxes—sagebrush, chaparral, and desert scrub. The main body of the prairie touches the deciduous forest throughout the entire length of its greatly indented western border and comes in contact with the montane forest along the front ranges of the Rocky Mountains. In New Mexico and Arizona, it usually confronts the woodland climax above, as also in the southern portion of the Great Basin, while below it alternates and mingles with the desert scrub and sagebrush.

Evidently through a lapsus, this statement fails to note the long additional line of contact, from northern Minnesota through Alberta, of the grassland with the northern coniferous forest or taiga.

Since it seems to be universally true, as Chapin (1932, p. 204) long ago wrote of African birds, that "the most fundamental distinction, ecologically, is between the forest and grass-dwelling faunas," the contact between forest and grassland seems quite properly analogous to that of a sea with a continent or island, and as we have just seen, a very large part of the present boundaries of the grassland is formed by forest. However, in parts of the Southwest and elsewhere, the contact is with more similar formations of the scrub group, with which there may be more or less interdigitation or mixture, or with more or less open woodland and savannah situations, where distinctions also may become blurred. This need not detract from the utility of the analogy; that these contacts are comparatively small is of considerably less importance than the fact that they can simply be thought of as "marshy" or "estuarine" ecotones or transitional areas between hard, dry "continent" (the forest) and blue "water" (the grassland). So far as birds of strong forest affiliations are concerned, these transitional areas seem to make no difference anyway, as indicated below.

ECOLOGICAL CONTINENTS, ISLANDS, AND BRIDGES

If, with respect to American forest and forest-related birds, the grasslands and related formations function as a sea, with the forested regions serving variously as continents and islands, we have the principal requirements for the differentiation and potential speciation of isolated stocks, which, if extensive and repeated, results in evolutionary radiation.

But before we can have isolation, differentiation, and speciation we must have successful dispersal, in the sense that a colony must survive for an adequate time and multiply, a tricky and complicated matter (see, for instance, MacArthur and Wilson, 1967, pp. 92-94, 121, 122). Further, in order for

Fig. 16. Major vegetational zones of North America, modified from Aldrich (1963, fig. 1). A, tundra (arctic in the north, alpine in the cordillera). B, eastern deciduous forest, including its southeastern evergreen-forest subclimax. C, grassland and related environments, including oak savannah in the extreme east, cold desert in the Great Basin and northwest, and assorted scrubs in the southwest and western Mexico—nearly all essentially open environments. E, taiga, more open in the north, more closed in the south—overlapping hatched areas (B and E) are the ecotonal lake (or mixed) forest, and overlapping hatched and stippled (E and C) areas are the ecotonal aspen parklands. F, coast forest, or Pacific rain forest. G, montane forest (higher) and woodland (lower), being somewhat different—chiefly pine-oak except at very high elevations in the volcanic belt—in its Mexican portions. D, assorted tropical vegetation.

radiation to occur, differentiated populations must perfect the means of competing successfully—or avoiding competition—with successive invasions of the parental or other stocks (see Brown, 1957, p. 248). Finally, if the new differentiate is itself to colonize the source of its own ancestors or any other area, its chances will be materially increased by its attainment of considerable numbers and extensive geographic distribution, since both intuitively and on the basis of considerable evidence, biogeographers generally concede that more dispersal takes place from larger to smaller areas than the other way (see Darlington, 1957, pp. 540, 620). This will inevitably be more difficult if the differentiate has just such a comparatively small island as various authors (for example, Hamilton, 1962, p. 49; and Amadon, 1966, p. 246) suggest is most conducive to differentiation.

Even considerable dispersal, however, rarely seems to result in colonization by birds across water (or comparable) barriers, unless the distance is quite small. Thus, in the possibly 10,000–12,000 years since the disappearance of the Beringian land bridge (Hopkins, 1967, p. 464, fig. 4), among passerines only the Red-throated Pipit (possibly), White Wagtail (Peyton, 1963), Bluethroat, Wheatear, Yellow Wagtail, Arctic Warbler, and Gray-headed Chickadee—in order of increasing success—have crossed the Bering Strait from Siberia to Alaska. Among these species only the last two have differentiated, faintly, at the subspecific level. Conversely, only one North American species, the Gray-cheeked Thrush, has established a successful beachhead in Siberia (the attribution by Mayr, 1953, p. 391, to two wood warblers of small Siberian breeding ranges is erroneous through some lapsus; see Vaurie, 1959). Yet the water barrier, even today, is only some 75 miles wide (with islands in the middle) and must for long periods have been narrower. Further, environments on both sides are rather similar. This suggests that the rate of successful colonization across this narrow strait is less than one per 1000 years. The eastward crossing has exceeded the westward by some seven times. The former, of course, is from the larger land mass to the smaller; but to attribute much significance to this alone, when both continents are very large, might be quite wrong. However, at least in the last and probably in all glacial periods, Beringia was effectively separated by ice from temperate and tropical North America at the glacial maximum and long afterward (see Flint, 1957, pl. 3), while no such barrier divided this small area from the vast Eurasian land mass (Repenning, 1967, p. 304).

The difference between dispersal and colonization is further emphasized by the fact that dispersal over far greater distances than the width of Bering Strait, given favorable winds and other factors, is not only possible but commonplace. Thus North American passerine birds of numerous species regularly cross or island-hop the North Atlantic, appearing as a comparatively slight but nonetheless steady annual "rain" on the British Isles and the European continent (for details see Savile, 1956, pp. 442-443; and especially Parkes, 1958, p. 423). Yet there is no firm evidence that a single North American species has ever successfully colonized Europe by this route! One arctic European species, the Wheatear, has colonized northeastern North America in this way, migrating back through Europe to winter in Africa; and other long-range colonizations are known, the appearance of the Fieldfare as a breeding bird in Greenland in 1937 (Salomonsen, 1951), for example. The point, then, is not that such colonizations do not occur, but that they seem rare enough so they should not be expected, in any given major time span, to account for more than a modest percentage of all the new species formed in isolation.

The distance involved being somewhat less than that between North America and Europe, it is not surprising that the literature contains much evidence (see, for example, Zimmerman, 1969) that a similar "rain" of eastern passerine species falls constantly on the western states in migration and that some winter there. Sometimes birds are even found in the breeding season (for example, Worm-eating and Parula warblers on June 13, 1963, in the Sheep Range of Nevada—Johnson, 1965, pp. 112, 113; Rose-breasted Grosbeak in northern California on June 29, 1962—Shelton, 1963) or are actually detected breeding (for example, Parula Warbler at Point Lobos, California, two nests in 1952, also notes on other western breeding-season records—Williams et al., 1958; Chestnut-sided Warbler in Colorado in 1968—Gadd, 1969). A preliminary survey of the literature, further, suggests that this "rain" is lighter on the part of truly eastern species and may well be heavier according to the extent that the eastern or eastern-northern species concerned (Fig. 14) have pushed their breeding ranges northwestward across the taiga. (For discussion of the much stronger tendency for western birds to stray eastward in migration see McAtee et al., 1944).

Despite this constant movement of wanderers, however, little evidence has been found of the establishment in western North America of propagules by eastern species (propagule is "the minimal number of individuals of a species capable of successfully colonizing a habitable island" according to MacArthur and Wilson, 1967, p. 190). The Eastern Bluebird and Indigo Bunting may provide exceptions. This being the case, it is not surprising that, among forest species, there are today in North America south of the taiga few if any small, disjunct, little-differentiated populations suggesting a recent crossing of the Central Plains barrier by a propagule from either east or west. This is true even though, as we shall shortly see, this barrier must recently have been much narrower than it now is. Further, even among species with trans-taiga distributions, none has a disjunct population suggesting recent descent from a vagrant propagule unless it be the uniquely disjunct, slightly marked western race (Fig. 10) of the Nashville Warbler (and this seems unlikely to me). While we know that it occurs on oceanic islands, the formation on continents of avian species by the descendants of wanderers to new and distant regions would seem to be comparatively infrequent, even though it may have occurred many times in the long history of the earth.

A few kinds of birds, indeed, may provide the majority of exceptions, most being northern, "irruptive" species (see Lack, 1954, pp. 227-242, for extended discussion). For instance, the nomadic crossbills *(Loxia)* are a special case. They follow the cone crops of gymnosperms, and like the Pacific white-eyes *(Zosterops)*, tend to travel in large flocks, each a potential colony. It is of additional note that crossbills are brought into breeding condition by an abundant food supply (Tordoff and Dawson, 1965). These things may account for the facts that many Pacific islands possess endemic white-eyes and that crossbills have been notable colonizers at various points throughout their large Holarctic range and even beyond, as in the Philippines (see Brown, 1957, pp. 267-268).

On continents, in order to account for most differentiation, and especially for active radiations, we need something more than chance dispersal. What is required is a bridge, on the one hand, and a means of cutting this bridge, on the other. (We need only ask that there be a reasonable body of evidence *for* the bridge, and no strong body of evidence *against* it, to avoid a pitfall in which whiten the figurative bones of many past zoogeographers.)

In this way whole, vigorous populations rather than a few individuals can be introduced into new areas, and then isolated when the bridge is cut. Differentiation can then occur during the isolation phase, and the newly differentiated populations can be brought into sympatry upon any ensuing reestablishment of the bridge. A great body of paleontological and biogeographical evidence attests to the efficacy of bridges in the history of proliferation of the world fauna.

In our present analogy, if the grasslands be viewed as a "sea" and the western montane forests as "islands" of various sizes, then it does not take much imagination to see that the taiga, for species with appropriate adaptations, is no less than a broad "land bridge" tending to link these "islands" with the "continent" of the eastern deciduous forest. The taiga itself, at least temporarily, is also of "continental" proportions. I say temporarily because the entire taiga lies on glaciated terrain and cannot be more than a few thousand years old where we now know it. In the event of another glaciation comparable to the four or more continental advances in the past (for details see Flint, 1947, 1957; and Wright and Frey, 1965), it would gradually be obliterated in its present range, and such of it as survived would necessarily retreat far to the south and be subject to numerous vicissitudes. Any populations of birds retreating with the taiga, whether or not it was then fractionated, would be brought into new ranges and, to some degree, new environments to which they would be forced to adapt if they were to survive. Let us examine this concept further.

THE TAIGA AS A BRIDGE

Earlier, I (Mengel, 1964) suggested that some 25 wood warblers have breeding distributions in the taiga that are extensive enough to be in danger of disjunction by another major glaciation. Consideration of other groups has extended this number to about 70 species; more could probably be added by consideration of additional widespread species that also occur in taiga (for example, the Hairy Woodpecker).

Glaciation as an isolating mechanism.—Biogeographers and taxonomists have sought, often persuasively, to explain observed patterns of speciation and distribution by postulating glacial isolation in separate "refugia." For birds, pioneer zoogeographic efforts were Adams's (1905) and Rand's (1948) in North America and Stresemann's (1919) and Salomonsen's (1930) in Europe, the latter two surprisingly much overlooked in subsequent literature. The efforts of systematists are numerous; one may cite Pitelka's works on the dowitchers (1950) and scrub jays (1951). Most of the considerable North American literature was summarized by Selander (1965). The evidence for refugia is of various kinds and should, of course, be viewed on its merits in individual cases. The geological and especially the paleobotanical evidence, although improving, is still often inadequate, and a good deal of our evidence for refugia comes from the relationships and distributions of organisms themselves. Circularity, therefore, if it cannot always be avoided, must be recognized.

Various respected colleagues have occasionally suggested to me that glaciation is over-used in biology and systematics. Nevertheless, it is difficult to imagine that continental ice and mountain glaciers thousands of feet thick could have occupied more than half of North America no more than 20,000 years ago—to say nothing of at least three times previously—without having some effects on the number and distribution of northern species. As the timing and relations of the glacial advances and periglacial conditions are more and more elucidated, it becomes increasingly difficult to ignore such relationships between geology, paleoecology, and biogeography as appear to be significant. As Deevey has said (1961,

p. 605), no matter what the limitations of biogeography, "the recognition of historic process as a determinant of modern ranges is a long step in the right direction."

Let us, then, try to imagine the history of widespread taiga species upon the advent of the Wisconsin glaciation and thereafter.

First, we know that elements of an earlier taiga survived and must, therefore, have retreated southward before the slowly advancing ice and in response to a cooling climate. In places where advance of the ice was comparatively rapid, forest was occasionally overrun. The extent of advance necessitated a considerable compression of vegetational zones, the details of which, however, are still unclear. It seems probable that the tundra border to the ice, when present, was narrow, especially in post-glacial phases (see Watts, 1967, pp. 93-94; Wayne, 1967, p. 393; and for further arguments see Martin, 1958; and Deevey, 1961, pp. 605-607). While some of the facts of zonation are controversial, we know in any case that during full glaciation, boreal coniferous trees ranged far south of their present position in the east (Whitehead, 1967) and far lower than their present elevations in the west (Martin and Mehringer, 1965). A narrower taiga (which I think all indications suggest was likely) is not in itself enough to isolate displaced taiga species into the separate refugia necessary for differentiation. Yet, if we are to think of the full-glacial period as an isolation phase of the cycle (though not, as we shall see, the only isolation phase), we must somehow break the displaced taiga bridge. In earlier papers various authors (see Rand, 1948, p. 320; Hall, 1958, p. 372; and Mengel, 1964, pp. 13-15, 17, and elsewhere) invoked treeless steppes south of the glacial boundary to break the taiga for this purpose. However, paleobotanical and other evidence recently has been accumulating (see Wendorf, 1961; Martin, 1963; and Wright, this volume) that suggests much of the present Central Plains and grassland were occupied variously by savannah, parkland, woodland, and forest at the Wisconsin full glacial, the grasslands being contracted to an unknown size and displaced southward to an unknown degree, and that at the least there was a significant coniferous element in the eastern parts of the Northern Plains (Wright, this volume).

That this taiga element was transcontinental at any given time, however, has not been demonstrated and may well not be true. For one thing, there is considerable evidence (Flint, 1947, pp. 134, 176) that broad outwash plains were present where the Mississippi River drained the giant continental Laurentide Ice Sheet, with blowing dust (immense deposits of loess resulted) and inclement conditions in general. Such a huge valley train could well have tended to isolate small passerine birds occurring in periglacial taiga, particularly if the taiga was scattered and narrow in the vicinity. Second, a very interesting hypothesis put forward by Smith (1965) and discussed further by Wright (this volume) suggests that the large region of the present Dakota Badlands and Nebraska Sandhills was occupied by a windy, virtually treeless desert and by dunes caused by intense winds blowing off of the James River Lobe of the Wisconsin ice. This large area, which is thought to have persisted during much of Wisconsin time—and perhaps earlier glacial periods—existed while a spruce forest occurred 250 miles to the southeast in what is now Kansas. Whether or not there was again periglacial taiga northwest of the Sandhills, this great forbidding area of cold, strong winds, deserts, and blowing dust seems likely to have provided an appreciable disjunction in the taiga environment.

Lastly, a further barrier may have existed in the form of Rocky Mountain and Cascade–Sierra Nevadan mountain glaciers extending from the Cordilleran ice sheets in present northern Montana, Idaho, and Washington intermittently south to New Mexico and California. Hubbard (1969) has invoked this barrier to account for the differentiation of the Myrtle and Audubon's warblers (Fig. 8, A). However, his maps (Hubbard, 1969, figs. 7, 8), although admittedly diagrammatic, suggest a mountain glacial barrier broader and more continuous than any that seems actually to have existed, and even broader alpine tundra belts for which there is no evidence whatever either way. A more accurate indication of mountain glaciation south of the ice sheets is provided by Flint (1947, pl. 5), whose detailed map indicates all glaciated terrain, not just the Wisconsin (see his pl. 3). It is, therefore, not certain that Rocky Mountain glaciers and tundra were ever continuous enough at any one time fully to separate coniferous forest-adapted populations. The Cascade-Sierran glaciers were somewhat more continuous, being closer to a source of precipitation (Flint, 1947, p. 217), and in fact may have been more effective in separating bird populations.

Regardless of how effective montane glaciers were as isolating agents, other interesting developments took place during Late Wisconsin deglaciation, from about 12,000 BP until the present. These are as follows: the northern Rocky Mountains were covered by the Cordilleran Ice Sheet, while to the east the lower ground was covered by the Laurentide Ice Sheet, the two coalescing along the foot of the Rockies in present Alberta, British Columbia, and Yukon (Flint, 1947, chapter 12). There is considerable evidence that as these two sheets contracted, a long arm of tundra and then of taiga (successively) invaded the lower ground from southern Alberta to the Mackenzie River delta (see Banfield, 1961, pp. 104, 105, writing of differentiation in caribou). The northwestward orientation of this corridor is clearly suggested by the isochrones in Bryson et al. (1969, figs. 1-2), who wrote (p. 5):

> . . . the corridor between the Cordillera and the retreating ice front is of great interest to the anthropologist, biologist, and climatologist. Assuming that the isochrone map is roughly correct, it indicates that between 11,000 BP and 8,500 BP a broad corridor opened along the Mackenzie River and through Alberta. What had been a broad, high, east-west saddle between the Cordilleran and Laurentide ice centers became a low north-south corridor from the Arctic to the Great Plains.

This evidence tends to explain a peculiar and recurrent feature in the distribution of North American birds and numerous other animals, namely the strong tendency for essentially eastern taxa that had adapted to the taiga and its successional stages to occur northwestward to, or nearly to, Alaska at the expense of western montane kinds that had adapted to montane coniferous forest and its successional stages. This distributional pattern is shown by all of the nearly 40 species of Figs. 14-15. Collectively these birds form an avifauna that quite clearly got there first, filling a majority of the niches, while the montane forest avifauna tended to be blocked by the persistent but dwindling Cordilleran Ice Sheet (see also Hubbard, 1969, p. 426). Significantly, the few really obvious western exceptions are provided by species like Steller's Jay, Western Tanager (Fig. 3, B and D), Western Wood Pewee (Fig. 7, A), and Townsend's Warbler (Fig. 9) whose most likely eastern congeneric competitors appear to have few or imperfect adaptations to boreal forest.

Finally, several coniferous forest-adapted, differentiated east-west semispecies pairs come together and hybridize exactly where they would be expected to have met upon the final disappearance of the Cordilleran ice barrier, namely in the eastern part of the Rocky Mountains centering more or less on the Peace River gap. These are the Yellow-shafted and Red-shafted flickers (Fig. 4, D), the well-marked subspecies of the Yellow-bellied Sapsucker (Howell, 1952), Myrtle and Audubon's warblers (Fig. 8, A), Slate-colored and Oregon juncos (Fig. 12), and the well-marked northern ("Spruce Grouse") and montane ("Franklin's Grouse") races of the Spruce Grouse (not figured). Thus, while some degree of separation—perhaps considerable—may have been imposed on various birds by Rocky Mountain glaciers and by the "badlands" during full glaciation, it also seems probable that these separations were reinforced in Late Wisconsin time. Then, as the montane taxa worked their way back, both northward and up into the higher elevations behind the shrinking Cordilleran Ice Sheet, simultaneously the taiga birds were moving rapidly northwestward up the fast-opening Laurentide-Cordilleran corridor (for example, see Short, 1965, p. 410, concerning flickers).

Now let us assume that various more or less trans-taiga species have in fact been separated, have survived, and have differentiated in two or more refugia in the east and west. The first phase in the differentiation process—for taiga species—may then be thought of as a full-glacial separation phase. If differentiation to the specific level does occur, as I suggest that it often has, then this very occurrence shows that the expansion (see p. 320) of woodland, savannah, or parkland into much of the present grassland in the Central Plains in full-glacial time and during pluvial conditions has not prevented differentiation in these forest-related species. In other words, regardless of what grows there, the Central Plains have not provided an environment suitable for effectively rejoining disjunct species either in full-glacial or interglacial times. The analogy of the isolating "sea" has held up, even if the sea has intermittently become "shallow" or "marshy."

During the first phase of differentiation we may expect the eastern differentiates to have occupied a southward-depressed and perhaps somewhat modified area of taiga adjacent to compacted eastern deciduous forest and at least partly at low elevations. Simultaneously the western differentiates should have been comparatively widespread in a mixture of southward-displaced taiga and coastal and montane forest to which they had been obliged to adapt. With the exception of being incorporated into the montane forest and being generally at somewhat greater elevations, the western differentiates at this stage should have experienced selective pressures little different from those affecting the eastern isolates.

Isolation in interglacial times.—Upon withdrawal of the glacier, the western differentiates should have had the opportunity, varying with the sizes and positions of their full-glacial distributions, of withdrawing upward into one or more forested mountain "islands." These vary greatly in size, from small ranges scattered through the Great Basin and Wyoming deserts and grasslands to huge massifs like the Yellowstone–Teton–Wind River complex and the Canadian Rockies. Some are more, and some less, continuous and connected, but all have the capacity to isolate western differentiates more thoroughly, perhaps, than they were isolated in full-glacial times and certainly to reduce the sizes of their populations in a way conducive to rapid differentiation. Further, Hibbard (1960, p. 25) has amassed considerable evidence which he interprets as suggesting that the climate of each major interglacial was warmer and drier than the present one, if indeed we are now experiencing another interglacial. If this is so, the western montane islands may have grown still smaller, higher, and more widely separated from one another by grassland, scrub, and desert than they are at present, and may do so again. The further differentiation of their insular populations would thus be enhanced, while their capacity to colonize other environments should simultaneously be decreased. There should be, therefore, decreasing likelihood—even given no significant adaptation to montane forest conditions—of their reinvasion of the taiga environment and reestablishment of a trans-taiga range from west to east.

Another factor perhaps tending to inhibit insular montane forest differentiates from reinvasion of taiga is the likelihood of their having to compete successfully, as they expanded eastward, with an increasing number of their eastern-northern congeners and other relatives. Similarly, eastern-northern forms expanding across the taiga may have encountered increasing competition for niches from progressively greater numbers of true boreal, chiefly Old World birds (see Fig. 13 and, particularly, the considerable evidence accumulated by MacArthur, 1959, fig. 1 and table 1, showing the density gradients in question). This could be one reason—the easier access of Beringia to Eurasia than to North America during glacial maxima being another (see Flint, 1957, pl. 3)—why no North American summer-boreal species save the Gray-cheeked Thrush appears to have established a foothold in Siberia.

The eastern differentiates during stage two are confronted with a different situation from that of those in the west. They have few mountains of consequence and none of great size in which to become isolated, and no grasslands to effect such isolation (although deciduous forest, in theory, could do so). On the other hand, they have available for occupancy the whole rapidly re-expanding taiga, which seems clearly to have a unique capacity rapidly to preempt recently glaciated terrain (Braun, 1950, p. 521), and they can travel via the southern Alberta–Mackenzie Delta corridor to reach the northwest early (as we have seen), well in advance of many of their trapped (and perhaps specialized) western vicariants. Thus the second phase of differentiation is also probably often a new expansion (bridge-crossing) phase for the eastern isolates and perhaps too—though much more rarely—for some western ones.

Repeatability of the process—a model sequence.—In theory, any group of birds properly situated in time and space could have repeated the entire cycle of isolation and differentiation just postulated as many as several times, since there have been at least four major glaciations in North America, each with minor divisions, and perhaps several more (Deevey, 1961, pp. 600-604).

With this thought in mind, I constructed a model predicting what might result, given the kind of isolation cycle just postulated, if a hypothetical wood warbler first became adapted to taiga with the Nebraskan glaciation (Mengel, 1964, pp. 13-16, fig. 4). I postulated this adaptation in the Nebraskan because of my opinion (1964, p. 38) that the wood warblers arose in warm-temperate broad-leaved forest, hence south of boreal vegetation until the latter was forced southward by Pleistocene cooling. This timing, however, is not essential—it would make no difference to the model if this wood warbler had earlier and more gradually acquired northern adaptations. But in any case, if it were then subjected to three additional glacial-interglacial cycles, what would result? The model involves four expansion phases for a widespread, "continental" eastern species (three for the three interglacials and one for the Recent), which establishes in each case a trans-taiga distribution, and three isolation sequences (one for each of the last three glaciations), each resulting in the separation of the eastern-northern taxon into two populations and the subsequent differentiation of a new, less widely distributed "insular" species in the west. We are left with four species—one widespread lowland eastern one and three less widely distributed western species that are restricted to various mountain ranges and systems. If we predict some competition between western isolates, resulting in geographic displacement southward of the earlier-differentiated, essentially insular western species (theoretically weaker competitors), and continued evolution of the intermittently expanding, essentially "continental" eastern-northern species (theoretically the stronger competitor), we end up with the eastern species, on the whole, being progressively more like those in the west as we proceed from south to north.

That something like this may in fact have happened is indicated by the distributions and characters of several species groups. The situation predicted by the model seems to be precisely the case with the species of the Black-throated Green Warbler complex (Fig. 9), with the further provision that the southern Golden-cheeked Warbler of the Edwards Plateau of Texas must additionally be accounted for. The latter appears to be simply a melanistic close relative of the eastern Black-throated Green Warbler, quite possibly a relict pinched off during a recent trend toward warming and drying of the Central Plains. Also, as we have seen, similar configurations are shown by the Nashville Warbler group and the Least Flycatcher group, and there are some interesting similarities in the distributions of some of the juncos—although the over-all history of these essentially boreal birds is probably different from that of the summer-boreal warblers, flycatchers, and other species.

Not only do the Black-throated Green and Nashville warbler groups conform closely to the model, suggesting similar histories, but also there are numerous, less complex groups that show partial conformity with it. That is, they conform to what would be expected if a species covered only part of the sequence or, alternatively, had covered all of it but had part of the record erased by extinction or movement, or both, of critical populations. (And extinctions have probably been significant: witness the minute sizes today of populations of Kirtland's, Golden-cheeked, Bachman's, and perhaps Colima warblers.) Among the wood warblers such groups are provided by the three species of the Mourning Warbler group (Fig. 8, E), by the Myrtle and Audubon's warblers (Fig. 8, A), and the Tennessee and Orange-crowned warblers (Fig. 8, C).

It also is possible that the seven or eight wood warblers, some obviously rather closely related, that have sympatric breeding ranges confined strictly to taiga (Table 3) are perhaps unanalyzable relicts of such histories. An alternative has been postulated by Hubbard (1969, p. 428), who proposed differentiation of stocks leading to the formation of two of these species, the Bay-breasted and Blackpoll warblers, by isolation of the latter in the well-known glacial refugium in Alaska, probably during the Illinoian (or penultimate) glaciation. This seems unlikely to me for two reasons. First, the ancestral Blackpoll Warbler was presumably a summer-boreal, warmth-adapted bird and a long-distance migrant, like the present one; yet this hypothesis calls for such a 12-gram bird to migrate twice annually across an ice sheet more than a thousand miles wide at its narrowest point and estimated (Flint, 1947, p. 221) to have been 10,000 feet thick (thus, possibly 12,000–13,000 feet high) in the region of northern British Columbia. Second, should the Blackpoll Warbler nonetheless have originated in Beringia, it seems at least possible that we might find it in Siberia today; of the Beringian mammal fauna envisioned by Macpherson (1965, p. 170), all are (or were) Holarctic.

My model had one more feature: it provided for the possible reinvasion of the taiga by western derivatives, while proposing—as has been emphasized—that this should be comparatively rare. The distributions of the Orange-crowned and Wilson's warblers suggest reinvasion of the taiga by montane forest species, while the Myrtle and Mourning warblers may possibly be derived from such species. Interestingly, all but the Myrtle are birds of shrubby environments in open forest or forest edge, presumably a more ubiquitous habitat than any forest climax.

Finally, still other groups of birds display distributional patterns and phenetic resemblances suggesting a history compatible with the suggested importance of the taiga bridge. Among these are the purple finches (Fig. 8, B) and the Acadian–Yellow-bellied–Western Flycatcher group (Fig. 8, F), and there are certainly others. For example, the interrelationships of the Yellow-throated and Solitary vireos and the Red-eyed, Philadelphia, and Warbling vireos are suggestive, but their ranges are a little too extensively sympatric to invite analysis. The present relations between the Yellow-shafted and Red-shafted flickers (Fig. 6, D) might also suggest such a history, at least in part, and indeed did so to Short (1965, pp. 409-411). Flickers, however, are not restricted to forests, but rather are restricted only by the availability of trees for nesting and hence are difficult to isolate. They seem to me quite possibly to have a history that is somewhat different, albeit difficult to interpret, which will be discussed later.

Exceptions to the model and variations on its theme.—Before passing on from the model, I should emphasize, as I have before (Mengel, 1964, p. 16), that the exact agreement of the model with the Black-throated Green Warbler group is because it was in fact constructed with that group in mind. It also conforms, however, with a simplified outline of glacial events and with certain biogeographical principles, such as more frequent movement from large areas to small.

One possible weakness of the model as originally constructed is that it assumed that earlier glaciations, at least in their general features, were much like the Wisconsin. While there seems to be no reason to think this was not true of the ice sheets themselves (see Flint, 1947, pp. 281, 283, pl. 3), there is some evidence, patiently marshaled by Hibbard (1960), that full-glacial climates differed—those of the Nebraskan and Kansan glacial maxima being warmer than those of the Illinoian and, especially, Wisconsin, particularly with respect to winter climates. To an unknown degree this could, of course, have affected the size, position, and completeness of isolation of periglacial refugia for taiga-adapted species. In view of new knowledge about the length and complexity of the Wisconsin (Deevey, 1961) and the apparent differentiation rates of at least some passerine birds (Johnston and Selander, 1964), it is even conceivable that radiations comparable to that of the model could be accomplished within the Wisconsin and its interstadials. After all, as Deevey noted (1961, p. 604), the biogeographer's dimensions are limited to "present" and "past." Such rapid radiation would, however, run counter to long-established orthodox theory concerning the rate and timing of bird evolution (see Moreau, 1966a, for review).

Some other ways in which real histories are likely to differ from the model are: (1) The original east-west separation may have been before the Pleistocene, and may in some cases have involved an ancestor adapted to sclerophyllous woodland vegetation of southwestern, perhaps Madro-Tertiary, origin. The eastern differentiate would then pursue all or part of the course of the model. (2) Two or more western differentiates—say one in the Rocky Mountains and one in the coast forests —could result from isolated derivatives of a single invasion by an eastern ancestor. (3) One western differentiate could, especially at full-glacial time, expand and then subsequently be isolated in separate "islands," becoming two or more species. Radiations of birds within island groups following probably single colonizations are well known—for example, certain Tristan da Cunha finches; the Geospizinae, or Darwin's finches, in the Galápagos; and the Drepaniidae, or Hawaiian honeycreepers (Lack, 1947, p. 162). (4) The eastern form of a complex of species, instead of being ancestral to the western taxa, as in the model, may have been derived from a western species in comparatively recent times; this may actually have happened with fair frequency among essentially western or Mexican-western groups and among boreal birds of, mostly, Old World origin. But this really amounts to a different model. It has, I think, happened rarely in New World groups of tropical origin, generally, and wood warblers, in particular.

My main reason for thinking so is that the forest-related, Nearctic members of the Parulidae are today uniquely and conspicuously an eastern, eastern-northern, and taiga group. Among families of American and presumably tropical or warm-temperate origin, no other North American group remotely approaches the wood warblers in number of species north of Mexico, average size of genera (related to recent radiations), preponderance of chiefly eastern over essentially western forms, or in the extent of invasion of the boreal regions for breeding. Hence the purely biogeographical evidence suggests strongly that the family had as the main staging area for its more recent, northern evolution the eastern part of the contracting Arcto-Tertiary forest (ancestral, but perhaps not equal, to the eastern deciduous forest) of the Late Tertiary and Quaternary. This should mean that by late in the Pliocene, when the principal surviving unit of this forest was in the east (see MacGinitie, 1958, p. 70), most of the ancestral wood warblers in question were there too.

Here is the hard evidence in favor of the deciduous-forested East as the principal source of birds for the radiations of wood warblers to the north and west (Mengel, 1964, p. 34 and table 1). (1) Of some 46 forest-related wood warblers important north of Mexico, only 10 are endemic to western montane regions (not all adapted to western montane forest). (2) Two more species, shared with the taiga, bring the total indigenous to the west to 12. (3) Compare this with the eastern deciduous forest, which possesses 11 endemic species and shares some 14 more with the taiga for a total of 25 that are indigenous (four of which belong to monotypic endemic genera). (4) Most of seven or eight strictly taiga species are more widespread in the eastern than in the western taiga and, further, are long-distance eastern migrants, suggesting eastern origins if, as is widely held, migration tends to be conservative. (5) Most of the movement seems to be from the larger to the smaller environments. (6) Most importantly of all, of the 12 endemic and/or probably autochthonous species of wood warblers of the western montane forests, only Wilson's Warbler is without an obvious vicariant in the eastern deciduous or northern coniferous forest, and even Wilson's Warbler has two congeners there. (7) Conversely, only one eastern species that has not acquired extensive coniferous forest adaptations possesses a vicariant or vicariants in the western montane forests—the Yellow-throated Warbler, which is discussed below. (8) Among these birds, finally, there is a trend in the degree of adaptive specialization from deciduous forest to mixed (lake) forest to taiga to western montane forest formations. So much for the wood warblers, then; these eight points tend to suggest an eastern origin for many of them. Other groups, although none with so many members, show some similar distributional patterns at the level of superspecies or species groups. It will now be interesting to compare all of these birds with others of presumably different history.

Summer-boreal and boreal birds of the taiga bridge.—The summer boreals, nearly all strongly migratory, are members of families or subfamilies of tropical or warm-temperate American origins that have attained significant distributions in the taiga. They include: numerous wood warblers (Parulidae); several vireos (Vireonidae); several flycatchers (Tyrannidae); at least one grosbeak, the Rose-breasted (Cardinalinae); and one tanager, the Western (Thraupidae). The truly boreal species of the taiga are mostly sedentary species or weak migrants and are mostly derived from Old World families, subfamilies, or genera.

It is revealing to contrast the patterns of speciation displayed by these two assemblages. We are concerned with most of the 70-odd species of Tables 3 and 4 and Figures 13-15. Of summer boreals, excluding non-passerines (three species) and the weakly migratory sparrows and Rusty Blackbird (seven species), there are 31 species important as breeding birds in the taiga. Of these, at least 13 (42 percent) are members of nine superspecies or close-knit species groups (numbered I-IX in Table 3), which collectively contain 27 species (average, three per group), 14 of which occur elsewhere than in taiga. Others (the Yellow-bellied Flycatcher, Philadelphia Vireo, and so forth) belong to still other species groups not quite so homogenous as the foregoing. In other words, the members of this group of 31 species, most of which are strongly migratory, have collectively been involved in comparatively recent radiations resulting in the differentiation—at the least—of nine ancestral into 27 descendent species.

In marked contrast is the situation with the true-boreal group. Actually counted, again excluding sparrows, were those species of Table 3 marked B, 26 in all. These 26 species have only two (8 percent) members that belong to superspecies or species groups (which contain only six species in all, of which four occur outside of the taiga). Thus in this group, two ancestral taxa have been involved in radiations resulting in at most six derivatives, suggesting that the rate of radiation of boreal birds has been less than one-fourth as great as that of summer boreals in the same environment. The members of the groups concerned are the Carolina and Mexican chickadees (with the Black-capped, Fig. 8, D) and the Cassin's Finch (with the Purple and House finches, Fig. 8, B). Also worthy of mention are the two well-marked hybridizing forms of the Spruce Grouse. All have distributions that could have resulted from a history of taiga interruption. The reasons for the apparent scarcity of this kind of pattern with these birds do not seem particularly clear, but the phenomenon is intriguing.

Of course the two *Aegolius* owls and the two three-toed woodpeckers are themselves small species groups, and the more southerly Loggerhead Shrike and Cedar Waxwing qualify as members of groups with the Northern Shrike and Bohemian Waxwing. However, the speciation patterns involved here are not the intracontinental kind with which we have been concerned, but rather intercontinental patterns involving movements across the Bering bridge prior to isolation by water. (This could also be true of parts of the chickadee and purple finch groups mentioned above.)

The species groups in question have been discussed, variously, by Parkes (1958, pp. 429-430—waxwings and shrikes) and Stegmann (1963, pp. 68-69—owls, woodpeckers, and waxwings). These birds have much in common: occurring and widespread in the taiga of the Old and New worlds are the Boreal Owl, Northern Three-toed Woodpecker, Northern Shrike, and Bohemian Waxwing; widespread but more southerly, in the New World only, are the related Saw-whet Owl (southern part of the taiga and south of the taiga), Black-backed Three-toed Woodpecker (southern part of the taiga and montane forest), Loggerhead Shrike (widespread south of the taiga), and Cedar Waxwing (widespread in the southernmost part of the taiga and south of the taiga). The only variation on this theme is provided by the waxwings, which have an additional, localized species in southeastern Siberia, the Japanese Waxwing.

Stegmann favored a New World origin for all of these groups on the grounds that the southern New World species is more "primitive." He thought its ancestors crossed Bering Strait, differentiated, and returned as the present northern species. Parkes inclined toward an Old World origin, as do I. It seems to me more logical, and compatible with the principle of movement from a larger to a smaller land mass, that the ancestor crossed the Bering Strait from west to east and was then disjoined and differentiated. The New World derivative was then forced south by glaciation, which, because of the different and discontinuous pattern of Old World glaciation (see Flint, 1947, pl. 3), need not have happened in the Old World. Some of the New World taxa, at this point, successfully adapted to deciduous forest and other environments south of the taiga, something few boreal birds have apparently accomplished, at least recently. Finally, upon reestablishment of a Bering bridge, a later, more advanced Old World descendant of the original colonist again crossed into the New World, where in each case it has now evolved to the subspecific level. This pattern, seen also in other, older taxa, is compatible with the sequence of events postulated in the "wood warbler" model discussed above.

In any case, these patterns suggest that at least some of the boreal birds are susceptible to isolation by a comparatively narrow (water) gap and are quite capable of speciation. This makes their relatively low degree of differentiation within the New World even more interesting, because all of them but the Bohemian Waxwing have broad, trans-taiga distributions and hence should be vulnerable to the same processes of glacial and interglacial isolation as the summer-boreal birds. I have thought about this a great deal and confess that no really satisfactory revelation has come to me so far. A few comments may be ventured, however, some of them volunteered by R. S. Hoffmann and K. C. Parkes (personal communications).

1. If a periglacial taiga strip, newly connecting southeastern and Rocky Mountain refugia in late-glacial time, were narrow and close to the ice field for any long period, it is possible that cold itself could preclude the presence, as breeding birds, of some summer-boreal species, while failing to inhibit cold-adapted true boreals. Thus, partially differentiated boreals would experience a much briefer period of complete isolation than would differentiating summer boreals. Suggesting the same, Hoffmann also observes that many true boreals today do not occur very far south in the Rocky Mountains. Thus, it might be that some boreal differentiates did in fact reach species level in mountain regions, but perished in the warm conditions of the Hypsithermal or some comparable period. Along similar lines, montane boreal species might, because of their adaptations to cold, be capable of occupying cold intermontane valley forests and basins during full-glacial periods, at which time their partly differentiated forms might lose genetic differences built up during interglacial or interstadial isolation in montane islands.

2. It is curious and probably relevant that, considering all North American passerines, the sedentary species display several times more infraspecific (that is, named subspecific) variation than do either strong or weak migrants, while partial migrants (here used to mean species in which some populations are migratory and some sedentary, with perhaps still others partially migratory in that not all individuals migrate) display the most of any category (Mengel and Jenkinson, manuscript in preparation). Why the resi-

dents, which of course include many of the true boreals, should display several times more infraspecific variation than migrants, while simultaneously undergoing several times less speciation, is an interesting point and one that has suggested to various authors that most continental "subspeciation" does not lead to species formation. On this general point Parkes thought as follows (letter):

> . . . sedentary species may more readily form mostly inbreeding micropopulations, with small isolating barriers (minor habitat gaps, etc.) that are crossed often enough to provide limited gene flow, if any—but these species do not "pioneer." The migratory species, on the other hand, may be subject to a greater possibility of panmixia (notably in species with a strong flocking tendency postbreeding), but are also more prone to "pioneer" and establish new populations in peripheral or isolated areas where relatively rapid speciation may take place.

Another partial explanation of this peculiar difference in patterns and type of geographic variation may be found in the probability that strongly migratory summer-boreal birds, whose breeding ranges were divided by glaciation in North America, have been especially prone to become markedly allohiemic (Salomonsen, 1955)—that is, to have the separated breeding populations also wintering in distinct and separate areas. Because environmental stresses are often greater in winter than in summer, a species that has genetically isolated breeding populations and is also subjected to markedly different conditions in separate and unlike winter quarters should differentiate much more rapidly than one whose winter populations either are not separate or occupy similar regions. We have just seen that truly boreal forms may have been more difficult to separate by glacial isolation than summer boreals; more importantly, however, because of their failure to migrate, even if separate, they would occupy winter quarters quite similar in climate and aspect.

As to how summer boreals may have tended to become effectively allohiemic, consider the following hypothesis. It is frequently stated (for example, see Dorst, 1962, p. 372), and has been discussed at length by Mayr (1953, p. 391), that migration tends to be conservative, being "originally perhaps always a backtracking of the route of immigration." Certainly the eastern summer boreals, breeding ranges of which thrust northwestward into the taiga, tend to return to their winter ranges via eastern migration routes. Conversely, the few summer boreals that have ranges confined to the taiga and the western montane forest and that may have invaded the former from centers of differentiation in the latter, tend to migrate by more westerly routes to more nearby wintering grounds. It is extremely interesting that the shortest migrations displayed by any group of summer boreals, and the most westerly, are almost all those of the western species, many of which were suggested above to be western differentiates from originally eastern-northern stock.

Now, noting the characteristic wedge-shaped outline of the maximum penetrations of the several glacial advances (Flint, 1947, pp. 281, 283, pl. 3), imagine that a population of a summer-boreal species breeds in the northwest and performs a conservative migration, necessarily hundreds of miles longer than formerly, in order to avoid traversing the advancing glacier. At some point, as must have happened in the case of every migrant invader of a truly new environment, as of North America from Eurasia, the conservative migration must have become so costly that selection became strong against individuals that persisted in it. Simultaneously, any individuals with a hereditary mechanism (formerly a liability) favoring the adoption of a new migration route and a new winter range would have been favored by selection. I have already mentioned the "rain" of eastern-northern summer boreals on the Pacific Coast and in the southwest. Such individuals could have supplied the material for such an

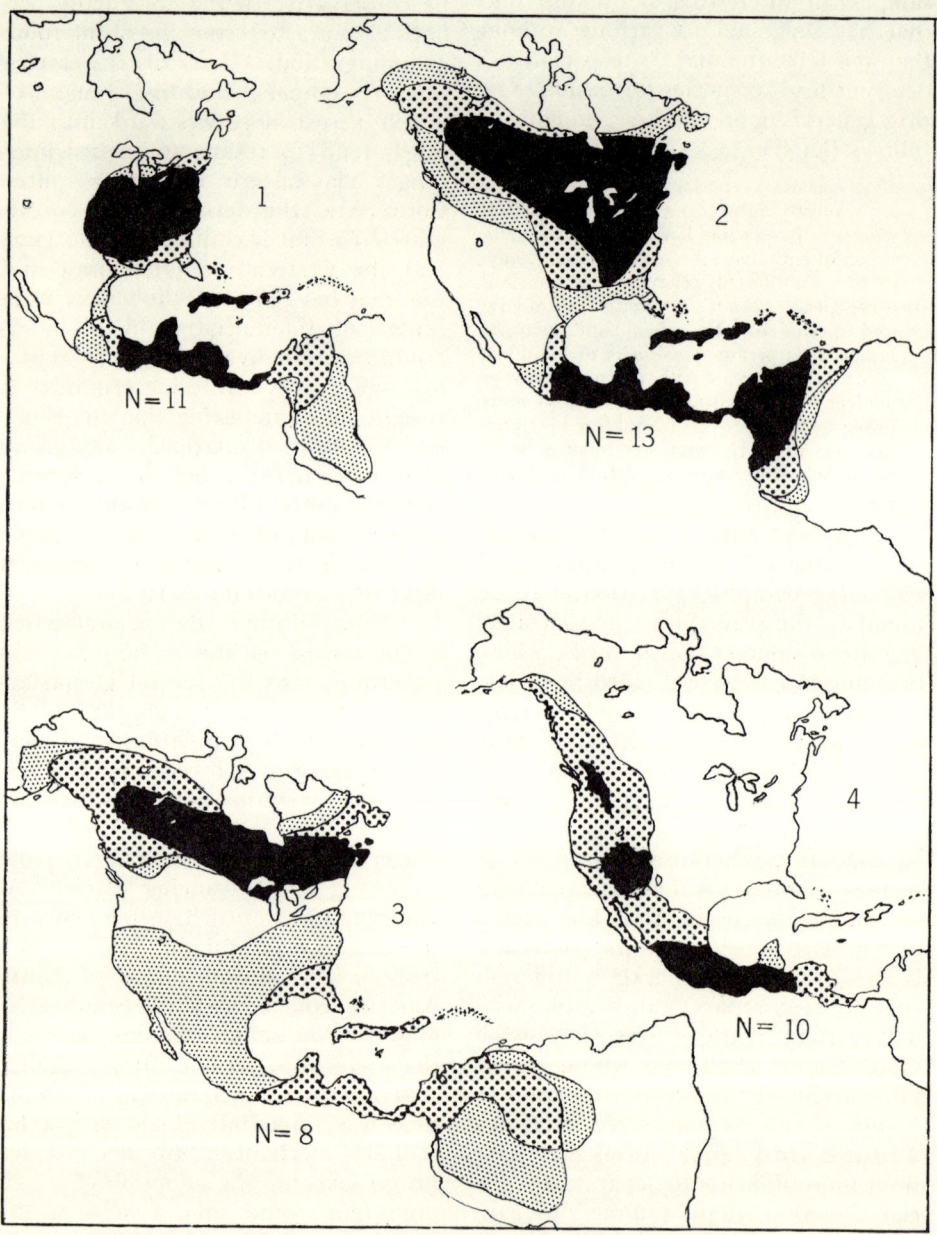

Fig. 17. Breeding and winter distributions of 42 species of northern, forest-related wood warblers. 1, eastern deciduous forest species; 2, eastern deciduous forest–taiga species; 3, species restricted to taiga; 4, western montane species. From the palest peripheral pattern to the black centers, species densities are: 1–3; 4–6; 7 or more. Note the essentially unique and quite wide overlap of breeding and wintering ranges on the part of western species.

adaptive shift in migration routes. In any case, upon the adoption by a northwestern population of a new, western winter range and shortened migration route, the condition of allohiemy would have been obtained and differentiation of the western derivative would theoretically have been facilitated. The winter quarters today of the western wood warblers, for example, center on western Mexico and the Central American highlands (Skutch, 1957, p. 270), areas that provide habitats quite different from the generally tropical Caribbean, Central American, and northern South American regions occupied by the eastern-northern and eastern wood warblers (Fig. 17).

In all, Figure 17 shows the generalized breeding ranges and winter ranges of 42 species of wood warblers divided into four groups. It will immediately be seen that species of the eastern deciduous forest have comparatively short average migrations and their winter ranges cluster around the Caribbean; eastern-northern species often migrate much farther and display a tendency towards "leapfrog" migration, suggesting that the slight land area of Central America and the West Indies is saturated in winter with a combination of resident and migrant passerines. Populations of many of the summer-boreal birds are immense. Further, a tendency to interspecific territoriality on migration and in winter, noted for at least two species of wood warbler—Myrtle and Cape May (Woolfenden, 1962; Kale, 1967)—may be rather general and, if so, might well contribute to winter spacing of populations. In any case, the longer migrations performed by the more northern species of wood warblers would seem to be further evidence for the later origins of the more northern breeding ranges.

BRIDGES SOUTH OF THE TAIGA

The observations I can properly venture about the ornithological history of the Central Plains and surrounding regions south of the taiga will be limited. At present, the grasslands cut straight to the Gulf of Mexico on the coasts of Texas and Tamaulipas, and in so doing, separate the distributions of a number of similar and obviously related pairs of species or incipient species. Some of the forms of groups I (Fig. 3, C), II (Fig. 4, A), III (Fig. 5, E), and V (Fig. 7, B), and all of those of Group IV (Fig. 6) are paired on opposite sides of this grassland barrier. Represented in all are some 10 pairs of taxa. Not mapped, but possibly related to some or all of these groups historically, are the Mississippi and Plumbeous kites.

The biological test of sympatry is lacking in most cases, as in the Florida and western populations of the Scrub Jay, Brown and Long-billed thrashers, Bachman's and Botteri's sparrows, Indigo and Varied buntings, eastern and southwestern Whip-poor-wills, Red-eyed and Yellow-green vireos, and Parula and Olive-backed warblers, where differentiation could be either at the subspecies (especially jays and whip-poor-wills), semispecies, or species level. It is available in limited areas, as we have seen above, for others, such as the Red-bellied and Golden-fronted woodpeckers and the Great-tailed and Boat-tailed grackles, which appear to have achieved reproductive isolation, and in the Black-crested and Tufted titmice, which have not.

It is for the paleobotanist rather than the biogeographer to provide the critical vegetational history and climatic evidence, but the biogeographical evidence rather strongly suggests that woodland connections between Mexico and the southeastern United States have existed in earlier times, perhaps in the Pliocene and probably more than once in the Pleistocene. Such connections permitted the distribution around the Gulf of Mexico of forms subsequently disjoined by grassland, or by retreat into Pleistocene refugia in Mexico and the southeastern United States, or both. Reestablishment of sundered woodland connections may at times have permitted intergradation of imperfect differentiates, as in the titmice today, and the secondary contact of others after reproductive isolation had been attained, as in the woodpeckers and grackles. It is interesting, however, that none of these birds has established wide sympatry with another. Concerning these groups, it seems unwise to suggest more than this at present.

Across the contemporary grasslands there stretch today narrow bands of riparian woodland and, sometimes, impoverished lowland forest. These, according to the early explorers, were there upon their arrival (for review see Short, 1965, p. 409). These riparian threads may be thought of as narrow bridges for certain woodland or forest-edge species (for example, Baltimore and Bullock's orioles). It is not certain, however, that such bridges—even as augmented by human plantings in the Central Plains—would ever permit gene exchange adequate to prevent differentiation in the absence of other contact (see Wright, 1946). Further, it is probable that even these narrow bands disappeared during the longest of the warm and dry parts of interglacial periods.

On the other hand, as Wendorf *et al.* (1961), Martin (1963), and others have recently shown, from full Wisconsin glaciation until around 10,000 years BP a pluvial situation existed in the Southwest and the Southern Plains that resulted in the occupancy of the grasslands and much of the deserts by savannah, pine parkland, and in higher areas, boreal montane forest. We may suppose that for the species of Group II (Fig. 4) that are woodland and forest-edge tolerant or edge-obligate, the isolation phase in the speciation process may not have been full-glacial time, as many authors have supposed (for example, Short, 1965, p. 409), but rather the interglacial intervals, particularly their extremes. Thus, the generally narrow contacts present now in riparian situations in the Central Plains, often alluded to as "secondary," are more likely remnantal contacts soon to vanish if Hibbard's prophecy (1960, p. 25) that the warmest and driest post-Wisconsin climates are still to come should prove true.

This would also perhaps explain why none of these species has achieved full specific distinctness, with the consequent ability to establish sympatry; this would be made extremely difficult if there were broad "swamping" in the Central Plains during each full glacial and complete isolation only during maximally xeric interglacial times. The bridge, for these edge and woodland species would, unlike that provided by the taiga for forest species, be an obstacle to full speciation rather than an aid. The differentiates of the truly forest species connected with the taiga theoretically enjoyed considerable to total isolation during both glacial and interglacial times.

On the other hand, it is entirely possible that pluvial occupancy of the Central Plains by savannah-woodland-park has permitted some species to gain transcontinental distribution and subsequent isolation by interglacial aridity, with the opportunity for species formation if subsequent pluvial-period swamping could be avoided. I have not made a thorough search for such species but would guess that perhaps some of the sparrows, possibly Chipping and Brewer's, may have arisen from common ancestral stock by some such process.

Some east-west pairs of species that are presently distributed wholly or partly south of the taiga appear to owe their differentiation to Pleistocene events without any indication of ever having had an ancestor with a trans-taiga distribution. Obviously, most of these must descend, if not from taiga species, then from woodland species, and it is possible that some of them owe their original disjunction to Pliocene fractionation of woodland. These are the species of Group I (Fig. 3), all but the first two of which (Chimney and Vaux's swifts) are forest or forest-edge birds: Blue and Steller's jays, Tufted and Plain titmice, and Scarlet and Western tanagers. All save possibly the swifts appear safely distinct at the species level, and all, today, are separated by the grasslands. Presumably, if they came together at all during the last pluvial period, they already enjoyed genetic isolation.

Finally, there are the species pairs and groups found in Group III (Fig. 5, A-D) whose histories, while obscure, seem somehow to be involved with the history of a pine-oak sclerophyll woodland and scrub element, once probably continuous around the Gulf Coast (Pitelka, 1951, pp. 383-384), and which are thought by some to be ancestral to an element of the present eastern deciduous forest (see Kendeigh, 1961, p. 298). This element probably descended from one of the Madro-Tertiary forest geoflora (discussed by Axelrod, 1958). At some time in the Late Cenozoic, probably Early Pleistocene, this sclerophyll belt was apparently sundered by the advance of the developing grasslands to the Gulf Coast and with it the populations of a number of bird species and perhaps other animals.

It is interesting that except for the Barred Owl, no birds highly typical of the eastern deciduous forest (descended from the temperate Arcto-Tertiary forest) are extensively shared with the related temperate mesophytic forests of Mexico, while various reptiles, plethodontid salamanders, and two small mammals are held in common by these two areas (Martin and Harrell, 1957). Either such birds have become extinct in the Mexican forests—or never did occur that far south—or else the disjunction of the deciduous forest environment that must once have occurred predated their existence as species.

ACKNOWLEDGMENTS

For helpful suggestions tendered by correspondence I am indebted to Claude W. Hibbard. John P. Hubbard, Kenneth C. Parkes, and Robert S. Hoffmann were kind enough to read and criticize the manuscript carefully and extensively, to my immense benefit (though any errors of fact or opinion remain strictly my own), and I have further enjoyed long discussions with the last. I have benefitted also by questions directed to me at various times by William L. Brown, Jr., Richard Selander, and Thomas Lovejoy, which helped focus my thinking on various points. Philip S. Humphrey has made useful suggestions; and references have been brought to my attention by a number of kind people, including Reginald E. Moreau, Max C. Thompson, and Gary D. Schnell. I thank O. S. Pettingill, Jr., and the Cornell Laboratory of Ornithology for permission to reproduce material from *The Living Bird*, and McGraw-Hill Book Company for permission to quote from Weaver and Clements' *Plant Ecology*. Also, the publishers of *Arctic and Alpine Research* kindly permitted quotation of a short passage from the first volume of that excellent journal. Wakefield Dort, Jr., and J. Knox Jones, Jr., the capable organizers of this symposium, have been more than kind in their patience with regard to delays in transmittal of the manuscript. Finally, my wife, Marion J. Mengel (whose publication name is Marion Anne Jenkinson), not only made most of the illustrations (I am indebted to Thomas Swearingen for one of them), but also did much typing and rigorous and skillful editing and performed other tedious tasks in a most efficient and patient way.

LITERATURE CITED

Adams, C. C.
1905. The postglacial dispersal of the American biota. Biol. Bull., 9:53-71.

Aldrich, J. W.
1963. Geographic orientation of American Tetraonidae. Jour. Wildlife Mgt., 27:528-545.

Allen, A. A.
1939. [Account of the Ivory-billed Woodpecker.] Pp. 1-12, in Life histories of North American woodpeckers (A. C. Bent, comp.), Bull. U.S. Nat. Mus., 174:viii+322 pp.

Amadon, D.
1950. The Hawaiian honeycreepers (Aves, Drepaniidae). Bull. Amer. Mus. Nat. Hist., 95:153-262.
1966. The superspecies concept. Syst. Zool., 15:245-249.

American Ornithologists' Union
1957. Check-list of North American birds. American Ornithologists' Union, Lord Baltimore Press, Inc., Baltimore, Maryland, x+691 pp.

Axelrod, D. I.
1950. Studies in Late Tertiary paleobotany. VI. Evolution of desert vegetation in western North America. Publ. Carnegie Inst. Washington, 380:215-306.
1952. A theory of angiosperm evolution. Evolution, 6:29-60.
1958. Evolution of the Madro-Tertiary geoflora. Bot. Rev., 24:433-509.

Baldwin, P. H.
1953. Annual cycle, environment and evolution in the Hawaiian honeycreepers (Aves: Drepaniidae). Univ. California Publ. Zool., 52:285-398.

Baldwin, P. H., and J. R. Koplin
1964. The Boreal Owl and its status in Colorado. Jour. Colorado-Wyoming Acad. Sci., 5:46-47 (abst.).

Banfield, A. W. F.
1961. A revision of the reindeer and caribou, genus *Rangifer*. Bull. Nat. Mus. Canada, 177:vi+137.

Behle, W. H.
1942. Distribution and variation of the Horned Larks *(Otocoris alpestris)* of western North America. Univ. California Publ. Zool., 46:205-316.

Bent, A. C.
1948. Life histories of North American nuthatches, wrens, thrashers and their allies. Bull. U.S. Nat. Mus., 195:xii+459.
1958. Life histories of North American blackbirds, orioles, tanagers, and allies. Bull. U.S. Nat. Mus., 211:ix+549.

Bigelow, R. S.
1965. Hybrid zones and reproductive isolation. Evolution, 19:449-458.

Blair, W. F.
1958. Distributional patterns of vertebrates in the southern United States in relation to past and present environments. Pp. 433-468, in Zoogeography (C. L. Hubbs, ed.), Publ. Amer. Assoc. Advance. Sci., Washington, D.C., 51:x+509.

Blake, E. R.
1968. [Family Vireonidae.] Pp. 103-138, in ["Peters"] Check-list of birds of the world, vol. XIV (R. A. Paynter, Jr., ed.), Mus. Comp. Zool., Cambridge, Massachusetts, x+433 pp.

Bowman, R. I.
1961. Morphological differentiation and adaptation in the Galápagos finches. Univ. California Publ. Zool., 58:viii+302.

Braun, E. L.
1950. Deciduous forests of eastern North America. Blakiston Company, Philadelphia, xiv+596 pp.

Brewer, R.
1963. Ecological and reproductive relationships of Black-capped and Carolina chickadees. Auk, 80:9-47.

Brodkorb, P.
1960. How many species of birds have existed? Bull. Florida State Mus. Biol. Sci., 5:41-53.

Brown, W. L., Jr.
1957. Centrifugal speciation. Quart. Rev. Biol., 32:247-277.

Bryson, R. A., W. M. Wendland, J. D. Ives, and J. T. Andrews
1969. Radiocarbon isochrones on the disintegration of the Laurentide ice sheet. Arctic and Alpine Res., 1:1-13.

Chapin, J. P.
1932. The birds of the Belgian Congo, Part 1. Bull. Amer. Mus. Nat. Hist., 65:x+756.

Cody, M. L.
1968. On the methods of resource division in grassland bird communities. Amer. Nat., 102:107-147.

Darlington, P. J., Jr.
1957. Zoogeography: the geographical distribution of animals. Chapman & Hall, Limited, London, xiii+675 pp.

Deevey, E. S.
1961. Recent advances in Pleistocene stratigraphy and biogeography. Pp. 594-623, in Vertebrate speciation (W. F. Blair, ed.), Univ. Texas Press, Austin, xvi+642 pp.

Dixon, K. L.
1955. An ecological analysis of the interbreeding of crested titmice in Texas. Univ. California Publ. Zool., 54:125-206.

Dorst, J.
1962. The migrations of birds. Heinemann, London, xix+476 pp.

Enderson, J. H.
1964. A study of the Prairie Falcon in the central Rocky Mountain region. Auk, 81:332-352.

Fenneman, N. M.
　1931. Physiography of western United States. McGraw-Hill Book Co., Inc., New York, x+534 pp.
Flint, R. F.
　1947. Glacial geology and the Pleistocene epoch. John Wiley & Sons, New York, xviii+589 pp.
　1957. Glacial and Pleistocene geology. John Wiley & Sons, New York, xiv+553 pp.
Gadd, S.
　1969. Chestnut-sided Warbler breeds in Colorado. Auk, 86:552-553.
Haffer, J.
　1967. Zoogeographical notes on the "non-forest" lowland bird faunas of northwestern South America. El Hornero, 10:315-333.
Hall, B. P.
　1961. The taxonomy and identification of pipits (genus *Anthus*). Bull. British Mus. (Nat. Hist.), Zool., 7:243-289.
Hall, E. R.
　1958. Introduction [to "Geographic distribution of contemporary organisms"]. Pp. 371-373, in Zoogeography (C. L. Hubbs, ed.), Publ. Amer. Assoc. Advance. Sci., Washington, D.C., 51:x+509.
Hamilton, T. H.
　1962. Species relationships and adaptations for sympatry in the avian genus *Vireo*. Condor, 64:40-68.
Hibbard, C. W.
　1960. An interpretation of Pliocene and Pleistocene climates in North America. Ann. Rep. Michigan Acad. Sci., 62:5-30.
Hoffmann, R. S.
　1958. The meaning of the word "taiga." Ecology, 39:540-541.
Hoffmann, R. S., and R. D. Taber
　1967. Origin and history of Holarctic tundra ecosystems, with special reference to their vertebrate faunas. Pp. 143-170, in Arctic and alpine environments (H. E. Wright, Jr., and W. H. Osburn, eds.), Indiana Univ. Press, Bloomington, xii+308 pp.
Hopkins, D. M.
　1967. The Cenozoic history of Beringia—a synthesis. Pp. 451-484, in The Bering land bridge (D. M. Hopkins, ed.), Stanford Univ. Press, Stanford, California, xvi+495 pp.
Howell, T. R.
　1952. Natural history and differentiation in the Yellow-bellied Sapsucker. Condor, 54:237-282.
Hubbard, J. P.
　1969. The relationships and evolution of the *Dendroica coronata* complex. Auk, 86:393-432.
Jewett, S. G.
　1944. Hybridization of Hermit and Townsend Warblers. Condor, 46:23-24.

Johnson, N. K.
　1963. Biosystematics of sibling species of flycatchers in the *Empidonax hammondii–oberholseri–wrightii* complex. Univ. California Publ. Zool., 66:79-238.
　1965. The breeding avifaunas of the Sheep and Spring ranges in southern Nevada. Condor, 67:93-124.
　1969. Review: three papers on variation in flickers (*Colaptes*) by Lester L. Short, Jr. Wilson Bull., 81:225-230.
Johnston, R. F., and R. K. Selander
　1964. House Sparrows: rapid evolution of races in North America. Science, 144:548-550.
Kale, H. W.
　1967. Aggressive behavior by a migrating Cape May Warbler. Auk, 84:120-121.
Kendeigh, S. C.
　1961. Animal ecology. Prentice-Hall, Inc., Englewood Cliffs, New Jersey, x+468 pp.
Lack, D.
　1947. Darwin's finches. Cambridge Univ. Press, Cambridge, x+204 pp.
　1954. The natural regulation of animal numbers. Clarendon Press, Oxford, viii+343 pp.
　1956. A review of the genera and nesting habits of swifts. Auk, 73:1-32.
Lanyon, W. E.
　1957. The comparative biology of the meadowlarks (*Sturnella*) in Wisconsin. Publ. Nuttall Ornith. Club, 1:vi+67.
　1966. Hybridization in meadowlarks. Bull. Amer. Mus. Nat. Hist., 134:1-26.
Lowery, G. H., Jr., and B. L. Monroe, Jr.
　1968. [Family Parulidae.] Pp. 3-93, in ["Peters"] Check-list of birds of the world, vol. XIV (R. A. Paynter, Jr., ed.), Mus. Comp. Zool., Cambridge, Massachusetts, x+433 pp.
MacArthur, R. H.
　1959. On the breeding distribution pattern of North American migrant birds. Auk, 76:318-325.
MacArthur, R. H., and J. W. MacArthur
　1961. On bird species diversity. Ecology, 42:594-598.
MacArthur, R. H., and E. O. Wilson
　1967. The theory of island biogeography. Princeton Univ. Press, Princeton, New Jersey, xii+203 pp.
MacGinitie, H. D.
　1953. Fossil plants of the Florissant beds, Colorado. Publ. Carnegie Inst. Washington, 599:iii+198.
　1958. Climate since the Late Cretaceous. Pp. 61-79, in Zoogeography (C. L. Hubbs, ed.), Publ. Amer. Assoc. Advance. Sci., Washington, D.C., 51:x+509.
Macpherson, A. H.
　1965. The origin of diversity in mammals of the Canadian arctic tundra. Syst. Zool., 14:153-173.

Marshall, J. T., Jr.
  1957. Birds of pine-oak woodland in southern Arizona and adjacent New Mexico. Pacific Coast Avifauna, 32:125 pp.

Martin, P. S.
  1958. Pleistocene ecology and biogeography of North America. Pp. 375-420, *in* Zoogeography (C. L. Hubbs, ed.), Publ. Amer. Assoc. Advance. Sci., Washington, D.C., 51:x+509.
  1963. The last 10,000 years. Univ. Arizona Press, Tucson, viii+87 pp.

Martin, P. S., and B. E. Harrell
  1957. The Pleistocene history of temperate biotas in Mexico and eastern United States. Ecology, 38:468-480.

Martin, P. S., and P. J. Mehringer, Jr.
  1965. Pleistocene pollen analysis and biogeography of the Southwest. Pp. 433-451, *in* The Quaternary of the United States (H. E. Wright, Jr., and D. G. Frey, eds.), Princeton Univ. Press, Princeton, New Jersey, x+922 pp.

Mayfield, H.
  1965. The brown-headed Cowbird, with old and new hosts. Living Bird, 4:13-28.

Mayr, E.
  1942. Systematics and the origin of species. Columbia Univ. Press, New York, xiv+334 pp.
  1946. History of the North American bird fauna. Wilson Bull., 58:3-41.
  1953. On the origin of bird migration in the Pacific. Pp. 387-407, *in* Proc. Seventh Pacific Sci. Cong., Pacific Sci. Assoc., Whitcombe and Tombs, Ltd., Aukland.
  1963. Animal species and evolution. Harvard Univ. Press, Cambridge, Massachusetts, xvi+797 pp.
  1969. Principles of systematic zoology. McGraw-Hill Book Co., New York, xi+428 pp.

McAtee, W. L., T. D. Burleigh, G. H. Lowery, Jr., and H. L. Stoddard
  1944. Eastward migration through the gulf states. Wilson Bull., 56:152-160.

Mengel, R. M.
  1963. A second probable hybrid between the Scarlet and Western tanagers. Wilson Bull., 75:201-203.
  1964. The probable history of species formation in some northern wood warblers (Parulidae). Living Bird, 3:9-43.

Miller, A. H.
  1941. Speciation in the avian genus *Junco*. Univ. California Publ. Zoo., 44:173-434.
  1951. An analysis of the distribution of the birds of California. Univ. California Publ. Zool., 50:531-644.

Moreau, R. E.
  1954. The main vicissitudes of the European avifauna since the Pliocene. Ibis, 96:411-431.
  1966a. On estimates of the past numbers and of the average longevity of avian species. Auk, 83:403-415.
  1966b. The bird faunas of Africa and its islands. Academic Press, New York, x+424 pp.

Murray, B. G., Jr.
  1969. A comparative study of the Le Conte's and Sharp-tailed sparrows. Auk, 86:199-231.

Nickell, W. P.
  1968. Return of northern migrants to tropical winter quarters and banded birds recovered in the United States. Bird-Banding, 39:107-116.

Norris, R. A.
  1958. Comparative biosystematics and life history of the nuthatches *Sitta pygmaea* and *Sitta pusilla*. Univ. California Publ. Zool., 56:119-300.

Parkes, K. C.
  1958. The Palaearctic element in the New World avifauna. Pp. 421-432, *in* Zoogeography (C. L. Hubbs, ed.), Publ. Amer. Assoc. Advance. Sci., Washington, D.C., 51:x+509 pp.

Paynter, R. A., Jr.
  1955. The ornithogeography of the Yucatan Peninsula. Bull. Peabody Mus. Nat. Hist., Yale Univ., 9:ii+347.

Peters, J. L., and many others
  1931-. Check-list of birds of the world. Various vols., Mus. Comp. Zool., Cambridge, Massachusetts [to be 15 volumes when completed].

Peyton, L. J.
  1963. Nesting and occurrence of White Wagtails in Alaska. Condor, 65:232-235.

Phillips, A. R., and L. L. Short, Jr.
  1968. A probable intrageneric hybrid pewee (Tyrannidae: *Contopus*) from Mexico. Bull. British Orn. Club, 88:90-93.

Pitelka, F. A.
  1941. Distribution of birds in relation to major biotic communities. Amer. Midland Nat., 25:113-137.
  1950. Geographic variation and the species problem in the shore-bird genus *Limnodromus*. Univ. California Publ. Zool., 50:1-108.
  1951. Speciation and ecologic distribution in American jays of the genus *Aphelocoma*. Univ. California Publ. Zool., 50:195-464.

Rand, A. L.
  1948. Glaciation, an isolating factor in speciation. Evolution, 2:314-321.

Rausch, R. L.
  1963. A review of the distribution of Holarctic Recent mammals. Pp. 29-43, *in* Pacific basin biogeography (J. L. Gressitt, ed.), Bishop Museum Press, Honolulu, x+563 pp.

Repenning, C. A.
  1967. Palearctic-Nearctic mammalian dispersal in the Late Cenozoic. Pp. 288-311, *in*

The Bering land bridge (D. M. Hopkins, ed.), Stanford Univ. Press, Stanford, California, xiii+451 pp.

Rising, J. D.
1968. A multivariate assessment of interbreeding between the chickadees, *Parus atricapillus* and *P. carolinensis*. Syst. Zool., 17:160-169.

Salomonsen, F.
1930. Diluviale Isolation und Artenbildung. Proc. Seventh Int. Ornith. Cong., pp. 413-438.
1951. The immigration and breeding of the Fieldfare (*Turdus pilaris* L.) in Greenland. Proc. Tenth Int. Ornith. Cong., pp. 515-526.
1955. The evolutionary significance of bird-migration. Dan. Biol. Medd., 22 (6):1-62.

Savile, D. B. O.
1956. Known dispersal rates and migratory potentials as clues to the origin of the North American biota. Amer. Midland Nat., 56:434-453.

Selander, R. K.
1965. Avian speciation in the Quaternary. Pp. 527-542, in The Quaternary of the United States (H. E. Wright, Jr., and D. G. Frey, eds.), Princeton Univ. Press, Princeton, New Jersey, x+922 pp.

Selander, R. K., and D. R. Giller
1961. Analysis of sympatry of Great-tailed and Boat-tailed grackles. Condor, 63:29-86.
1963. Species limits in the woodpecker genus *Centurus* (Aves). Bull. Amer. Mus. Nat. Hist., 124:213-274.

Shelford, V. E.
1963. The ecology of North America. Univ. Illinois Press, Urbana, xxii+610 pp.

Shelton, L. A.
1963. A further record of a Rose-breasted Grosbeak in northern California. Condor, 65:241.

Short, L. L., Jr.
1965. Hybridization in the flickers *(Colaptes)* of North America. Bull. Amer. Mus. Nat. Hist., 129:307-428.
1967. A review of the genera of grouse (Aves, Tetraoninae). Amer. Mus. Novit., 2289:1-39.
1968. Sympatry of red-breasted meadowlarks in Argentina, and the taxonomy of meadowlarks (Aves: *Leistes, Pezites,* and *Sturnella*). Amer. Mus. Novit., 2349:1-30.
1969. Taxonomic aspects of avian hybridization. Auk, 86:84-105.

Sibley, C. G.
1954. Hybridization in the red-eyed towhees of Mexico. Evolution, 8:252-290.

Sibley, C. G., and L. L. Short, Jr.
1959. Hybridization in the buntings *(Passerina)* of the Great Plains. Auk, 76:443-463.
1964. Hybridization in the orioles of the Great Plains. Condor, 66:130-150.

Sibley, C. G., and F. C. Sibley
1964. Hybridization in the red-eyed towhees of Mexico: the populations of the southeastern plateau region. Auk, 81:479-504.

Sibley, C. G., and D. A. West
1959. Hybridization in the Rufous-sided Towhees of the Great Plains. Auk, 76:326-338.

Skutch, A. F.
1957. Migrant wood warblers in their Central American homes. Pp. 269-274, in The warblers of America (L. Griscom, A. Sprunt, et al.), Devin-Adair Co., New York, xii+356 pp.

Smith, H. T. U.
1965. Dune morphology and chronology in central and western Nebraska. Jour. Geol., 73:557-578.

Stegmann, B.
1963. The problem of the Beringian continental land connection in the light of ornithogeography. Pp. 65-78, in Pacific basin biogeography (J. L. Gressitt, ed.), Bishop Museum Press, Honolulu, x+563 pp.

Storer, R. W.
1955. A preliminary survey of the sparrows of the genus *Aimophila*. Condor, 57:193-201.
1961. A hybrid between the Painted and Varied buntings. Wilson Bull., 73:209.

Stresemann, E.
1919. Über die europäischen Gimpel *(Pyrhula)*. München, Beitr. Zoogeogr. pal. Reg., 1:25-56.

Sutton, G. M.
1938. Oddly plumaged orioles from western Oklahoma. Auk, 55:1-6.
1951. Mexican birds. Univ. Oklahoma Press, Norman, xvi+282 pp.

Tanner, J. T.
1952. Black-capped and Carolina chickadees in the southern Appalachian Mountains. Auk, 69:407-424.

Tordoff, H. B.
1950. A hybrid tanager from Minnesota. Wilson Bull., 62:3-4.

Tordoff, H. B., and W. R. Dawson
1965. The influence of daylength on reproductive timing in the Red Crossbill. Condor, 67:416-422.

Udvardy, M. D. F.
1958. Ecological and distributional analysis of North American birds. Condor, 60:50-66.
1963. Bird faunas of North America. Proc. Thirteenth Int. Ornith. Cong., pp. 1147-1167.

Vaurie, C.
1959. The birds of the Palearctic fauna . . . Order Passeriformes. H. F. G. Witherby, Limited, London, xii+762 pp.

Verbeek, N. A. M.
  1967. Breeding biology and ecology of the Horned Lark in alpine tundra. Wilson Bull., 79:208-218.
Voous, K. H., Jr.
  1947. On the history of the distribution of the genus *Dendrocopos*. Limosa, 20:1-142.
  1960. Atlas of European birds. Thomas Nelson and Sons, Ltd., London, 284 pp.
Watts, W. A.
  1967. Late-glacial plant macrofossils from Minnesota. Pp. 89-97, *in* Quaternary paleoecology (E. J. Cushing and H. E. Wright, Jr., eds.), Yale Univ. Press, New Haven, Connecticut, viii+433 pp.
Wayne, W. J.
  1967. Periglacial features and climatic gradient in Illinois, Indiana, and western Ohio, east-central United States. Pp. 393-414, *in* Quaternary paleoecology (E. J. Cushing and H. E. Wright, Jr., eds.), Yale Univ. Press, New Haven, Connecticut, viii+433 pp.
Weaver, J. E., and F. E. Clements
  1938. Plant ecology. McGraw-Hill Book Company, Inc., New York, xxii+601 pp.
Wendorf, F. (ed.)
  1961. Paleoecology of the Llano Estacado. Mus. New Mexico, Santa Fe, 144 pp.
West, D. A.
  1962. Hybridization in grosbeaks *(Pheucticus)* of the Great Plains. Auk, 79:399-424.
Whitehead, D. R.
  1967. Studies of full-glacial vegetation and climate in southeastern United States. Pp. 237-248, *in* Quaternary paleoecology (E. J. Cushing and H. E. Wright, Jr., eds.), Yale Univ. Press, New Haven, Connecticut, viii+433 pp.
Williams, L., K. Legg, and F. S. L. Williamson
  1958. Breeding of the Parula Warbler at Point Lobos, California. Condor, 60:345-354.
Woolfenden, G. E.
  1962. Aggressive behavior by a wintering Myrtle Warbler. Auk, 79:713-714.
Wright, H. E., Jr., and D. G. Frey (eds.)
  1965. The Quaternary of the United States. Princeton Univ. Press, Princeton, New Jersey, x+922 pp.
Wright, S.
  1946. Isolation by distance under diverse systems of mating. Genetics, 31:39-59.
Zimmerman, D. A.
  1969. New records of wood warblers in New Mexico. Auk, 86:346-347.

APPENDIXES

APPENDIX A. — SCIENTIFIC NAMES OF SPECIES MENTIONED IN TEXT (BASED ON A.O.U. CHECK-LIST, 1957)

Blackbird,
  Red-winged, *Agelaius phoeniceus*
  Rusty, *Euphagus carolinus*
Bluebird,
  Eastern, *Sialia sialis*
  Western, *Sialia mexicana*
Bluethroat, *Luscinia svecica*
Bobolink, *Dolichonyx oryzivorus*
Bunting,
  Indigo, *Passerina cyanea*
  Lark, *Calamospiza melanocorys*
  Lazuli, *Passerina amoena*
  Painted, *Passerina ciris*
  Varied, *Passerina versicolor*
Buzzard,
  Long-legged, *Buteo rufinus*
  Upland, *Buteo hemilasius*
Chickadee,
  Black-capped, *Parus atricapillus*
  Boreal, *Parus hudsonicus*
  Carolina, *Parus carolinensis*
  Gray-headed, *Parus cinctus*
  Mexican, *Parus sclateri*
Cowbird, Brown-headed, *Molothrus ater*
Creeper, Brown, *Certhia familiaris*
Crossbill,
  Red, *Loxia curvirostra*
  White-winged, *Loxia leucoptera*
Curlew, Long-billed, *Numenius americanus*
Dickcissel, *Spiza americana*
Dove, Mourning, *Zenaidura macroura*
Falcon,
  Altai, *Falco altaicus*
  Prairie, *Falco mexicanus*
  Saker, *Falco cherrug*
Fieldfare, *Turdus pilaris*
Finch,
  Cassin's, *Carpodacus cassinii*
  Purple, *Carpodacus purpureus*
Flicker,
  Gilded, *Colaptes chrysoides*
  Red-shafted, *Colaptes cafer*
  Yellow-shafted, *Colaptes auratus*
Flycatcher,
  Acadian, *Empidonax virescens*
  Coues', *Contopus pertinax*
  Dusky, *Empidonax oberholseri*
  Gray, *Empidonax wrightii*
  Hammond's, *Empidonax hammondii*
  Least, *Empidonax minimus*
  Olive-sided, *Nuttallornis borealis*
  Pine, *Empidonax affinis*
  Western, *Empidonax difficilis*
  Yellow-bellied, *Empidonax flaviventris*
Godwit, Marbled, *Limosa fedoa*
Goshawk, *Accipiter gentilis*
Grackle,
  Boat-tailed, *Cassidix mexicanus*
  Great-tailed, *Cassidix major*

Grosbeak,
 Black-headed, *Pheucticus melanocephalus*
 Evening, *Hesperiphona vespertina*
 Pine, *Pinicola enucleator*
 Rose-breasted, *Pheucticus ludovicianus*
Grouse,
 Ruffed, *Bonasa umbellus*
 Sage, *Centrocercus urophasianus*
 Sharp-tailed, *Pedioecetes phasianellus*
 Spruce, *Canachites canadensis*
Gull, Franklin's, *Larus pipixcan*
Hawk,
 Broad-winged, *Buteo platypterus*
 Ferruginous, *Buteo regalis*
 Marsh, *Circus cyaneus*
 Pigeon, *Falco columbarius*
 Red-tailed, *Buteo jamaicensis*
 Swainson's, *Buteo swainsoni*
Jay,
 Blue, *Cyanocitta cristata*
 Gray, *Perisoreus canadensis*
 Mexican, *Aphelocoma ultramarina*
 Scrub, *Aphelocoma coerulescens*
 Steller's, *Cyanocitta stelleri*
Junco,
 Gray-headed, *Junco caniceps*
 Mexican, *Junco phaeonotus*
 Oregon, *Junco oreganus*
 Slate-colored, *Junco hyemalis*
 White-winged, *Junco aikeni*
Kinglet,
 Golden-crowned, *Regulus satrapa*
 Ruby-crowned, *Regulus calendula*
Kite,
 Mississippi, *Ictinia misisippiensis*
 Plumbeous, *Ictinia plumbea*
Lark, Horned, *Eremophila alpestris*
Longspur,
 Chestnut-collared, *Calcarius ornatus*
 McCown's, *Rhynchophanes mccownii*
Marsh Wren,
 Long-billed, *Telmatodytes palustris*
 Short-billed, *Cistothorus platensis*
Meadowlark,
 Eastern, *Sturnella magna*
 Western, *Sturnella neglecta*
Nighthawk, Common, *Chordeiles minor*
Nuthatch,
 Brown-headed, *Sitta pusilla*
 Pygmy, *Sitta pygmea*
 Red-breasted, *Sitta canadensis*
 White-breasted, *Sitta carolinensis*
Oriole,
 Baltimore, *Icterus galbula*
 Bullock's, *Icterus bullockii*
Ovenbird, *Seiurus aurocapillus*
Owl,
 Barred, *Strix varia*
 Boreal, *Aegolius funereus*
 Burrowing, *Speotyto cunicularia*
 Great Gray, *Strix nebulosa*
 Hawk, *Surnia ulula*
 Little, *Athene noctua*
 Saw-whet, *Aegolius acadicus*
 Short-eared, *Asio flammeus*
 Spotted, *Strix occidentalis*
Pewee (see also Wood Pewee),
 Tropical, *Contopus cinereus*
Phalarope, Wilson's, *Steganopus tricolor*
Phoebe, Eastern, *Sayornis phoebe*
Pipit,
 Meadow, *Anthus pratensis*
 Red-throated, *Anthus cervinus*
 Short-clawed, *Anthus furcatus*
 Sprague's, *Anthus spraguei*
 Water, *Anthus spinoletta*
Plover,
 Mountain, *Eupoda montana*
 Upland, *Bartramia longicauda*
Prairie Chicken,
 Greater, *Tympanuchus cupido*
 Lesser, *Tympanuchus pallidicinctus*
Raven,
 Common, *Corvus corax*
 White-necked, *Corvus cryptoleucus*
Redstart, American, *Setophaga ruticilla*
Sapsucker, Yellow-bellied, *Sphyrapicus varius*
Shrike,
 Loggerhead, *Lanius ludovicianus*
 Northern, *Lanius excubitor*
Siskin, Pine, *Spinus pinus*
Sparrow,
 Bachman's, *Aimophila aestivalis*
 Baird's, *Ammodramus bairdii*
 Botteri's, *Aimophila botterii*
 Brewer's, *Spizella breweri*
 Cassin's, *Aimophila cassinii*
 Chipping, *Spizella passerina*
 Clay-colored, *Spizella pallida*
 Fox, *Passerella iliaca*
 Grasshopper, *Ammodramus savannarum*
 Henslow's, *Passerherbulus henslowii*
 Lark, *Chondestes grammacus*
 Le Conte's, *Passerherbulus caudacutus*
 Lincoln's, *Melospiza lincolnii*
 Rufous-winged, *Aimophila carpalis*
 Sage, *Amphispiza belli*
 Savannah, *Passerculus sandwichensis*
 Sharp-tailed, *Ammospiza caudacuta*
 Sierra Madre, *Xenospiza baileyi*
 Striped, *Oriturus superciliosus*
 Swamp, *Melospiza georgiana*
 Tree, *Spizella arborea*
 Vesper, *Pooecetes gramineus*
 White-crowned, *Zonotrichia leucophrys*
 White-throated, *Zonotrichia albicollis*
Swift,
 Chimney, *Chaetura pelagica*
 Vaux's, *Chaetura vauxi*
Tanager,
 Scarlet, *Piranga olivacea*
 Western, *Piranga ludoviciana*
Thrasher,
 Brown, *Toxostoma rufum*
 Cozumel, *Toxostoma guttatum*
 Long-billed, *Toxostoma longirostre*
 Sage, *Oreoscoptes montanus*
Thrush,
 Gray-cheeked, *Hylocichla minima*
 Hermit, *Hylocichla guttata*

Swainson's, *Hylocichla ustulata*
Wood, *Hylocichla mustelina*
Titmouse,
 Black-crested, *Parus atricristatus*
 Plain, *Parus inornatus*
 Tufted, *Parus bicolor*
Towhee,
 Collared, *Pipilo ocai*
 Green-tailed, *Chlorura chlorura*
 Rufous-sided, *Pipilo erythrophthalmus*
Veery, *Hylocichla fuscescens*
Vireo,
 Black-whiskered, *Vireo altiloquus*
 Philadelphia, *Vireo philadelphicus*
 Red-eyed, *Vireo olivaceus*
 Solitary, *Vireo solitarius*
 Warbling, *Vireo gilvus*
 Yellow-green, *Vireo flavoviridis*
 Yellow-throated, *Vireo flavifrons*
 Yucatan, *Vireo magister*
Wagtail,
 White, *Motacilla alba*
 Yellow, *Motacilla flava*
Warbler,
 Adelaide's, *Dendroica adelaidi*
 Arctic, *Phylloscopus borealis*
 Audubon's, *Dendroica auduboni*
 Bachman's, *Vermivora bachmani*
 Bay-breasted, *Dendroica castanea*
 Blackburnian, *Dendroica fusca*
 Blackpoll, *Dendroica striata*
 Black-and-white, *Mniotilta varia*
 Black-throated Blue, *Dendroica caerulescens*
 Black-throated Gray, *Dendroica nigrescens*
 Black-throated Green, *Dendroica virens*
 Cape May, *Dendroica tigrina*
 Chestnut-sided, *Dendroica pensylvanica*
 Colima, *Vermivora crissalis*
 Connecticut, *Oporornis agilis*
 Golden-cheeked, *Dendroica chrysoparia*
 Grace's, *Dendroica graciae*
 Hartlaub's, *Vermivora superciliosa*
 Hermit, *Dendroica occidentalis*
 Kirtland's, *Dendroica kirtlandii*
 Lucy's, *Vermivora luciae*
 MacGillivray's, *Oporornis tolmiei*
 Magnolia, *Dendroica magnolia*
 Mourning, *Oporornis philadelphia*
 Myrtle, *Dendroica coronata*
 Nashville, *Vermivora ruficapilla*
 Olive-backed, *Parula pitiayumi*
 Olive-capped, *Dendroica pityophila*
 Orange-crowned, *Vermivora celata*
 Palm, *Dendroica palmarum*
 Parula, *Parula americana*
 Socorro, *Parula graysoni*
 Tennessee, *Vermivora peregrina*
 Townsend's, *Dendroica townsendi*
 Virginia's, *Vermivora virginiae*
 Wilson's, *Wilsonia pusilla*
 Worm-eating, *Helmitheros vermivorous*
 Yellow-throated, *Dendroica dominica*
Waterthrush, Northern, *Seiurus noveboracensis*
Waxwing,
 Bohemian, *Bombycilla garrulus*
 Cedar, *Bombycilla cedrorum*
 Japanese, *Bombycilla japonica*
Wheatear, *Oenanthe oenanthe*
Whip-poor-will, *Caprimulgus vociferus*
Willet, *Catoptrophorus semipalmatus*
Woodpecker,
 Black-backed Three-toed, *Picoides arcticus*
 Gila, *Centurus uropygialis*
 Golden-fronted, *Centurus aurifrons*
 Imperial, *Campephilus imperialis*
 Ivory-billed, *Campephilus principalis*
 Ladder-backed, *Dendroica scalaris*
 Northern Three-toed, *Picoides tridactylus*
 Nuttall's, *Dendrocopos nuttallii*
 Pileated, *Dryocopus pileatus*
 Red-bellied, *Centurus carolinus*
 Red-cockaded, *Dendrocopos borealis*
Wood Pewee,
 Eastern, *Contopus virens*
 Western, *Contopus pertinax*
Wren (see also Marsh Wren),
 Winter, *Troglodytes troglodytes*
Yellowthroat, *Oporornis trichas*

APPENDIX B.—GENERA USED IN CONSTRUCTING TABLE 2 (NON-GRASSLAND) AND THE NUMBER OF NORTH AMERICAN NON-GRASSLAND SPECIES IN EACH

(M indicates a monotypic species)

Tyrannus, 5; *Muscivora*, 1; *Myiarchus*, 5; *Sayornis*, 3; *Empidonax*, 9; *Contopus*, 3; *Nuttallornis*, 1 (M); *Pyrocephalus*, 1 (M); *Perisoreus*, 1 (M); *Cyanocitta*, 2; *Aphelocoma*, 2; *Pica*, 2; *Corvus*, 5; *Gymnorhinus*, 1 (M); *Nucifraga*, 1; *Parus*, 11; *Auriparus*, 1 (M); *Psaltriparus*, 2; *Sitta*, 4; *Certhia*, 1; *Chamaea*, 1 (M); *Troglodytes*, 3; *Thryomanes*, 1; *Thryothorus*, 1; *Campylorhynchus*, 1; *Salpinctes*, 2; *Mimus*, 1; *Dumetella*, 1 (M); *Toxostoma*, 8; *Turdus*, 2; *Ixoreus*, 1 (M); *Catharus*, 5; *Sialia*, 3; *Hylocichla*, 1 (M); *Myadestes*, 1; *Polioptila*, 2; *Regulus*, 2; *Bombycilla*, 2; *Phainopepla*, 1 (M); *Lanius*, 2; *Vireo*, 12; *Mniotilta*, 1 (M); *Protonotaria*, 1 (M); *Limnothlypis*, 1 (M); *Helmitheros*, 1 (M); *Vermivora*, 9; *Parula*, 1; *Dendroica*, 21; *Seiurus*, 3; *Oporornis*, 4; *Icteria*, 1 (M); *Cardellina*, 1 (M); *Wilsonia*, 3; *Setophaga*, 1; *Icterus*, 6; *Euphagus*, 2; *Quiscalus*, 2; *Molothrus*, 2; *Piranga*, 4; *Richmondena*, 1 (M); *Pyrrhuloxia*, 1 (M); *Pheucticus*, 2; *Guiraca*, 1 (M); *Passerina*, 4; *Hesperiphona*, 1 (M); *Carpodacus*, 5; *Pinicola*, 1; *Acanthis*, 2; *Spinus*, 4; *Loxia*, 2; *Pipilo*, 3; *Aimophila*, 4; *Amphispiza*, 2; *Junco*, 5; *Spizella*, 5; *Zonotrichia*, 4; *Passerella*, 1 (M); *Melospiza*, 3.

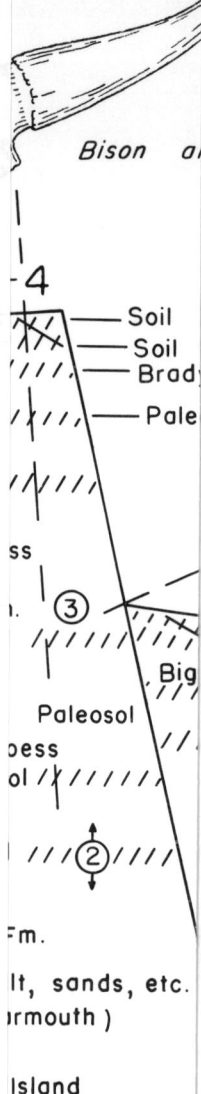

e-2 complex is 12,0
ociated with Paleo
became extinct sor
terstadial Paleosol
rather than Paleos
n recorded from the

e of bison thus far.
omplex in the uppe

Quaternary Mammalian Sequence in the Central Great Plains

C. Bertrand Schultz and Larry D. Martin

ABSTRACT

Recent recovery and study of fossils of small vertebrates from important Quaternary localities in Nebraska, where large species previously had been collected in abundance, have provided supplementary evidence concerning the stratigraphic sequence in various phylogenetic lines of mammals. Chief among the fossiliferous localities are Medial Pleistocene quarries found in Sheridan County (northwestern Nebraska), Cherry County (north-central Nebraska), and Harlan and Nuckolls counties (southern Nebraska). The proximity of the localities to the midcontinent glaciated areas is an important factor in providing better stratigraphic control of the various phylogenetic sequences. Geomorphologic, stratigraphic, and paleoecologic studies in the areas of the fossil localities and in adjacent regions aid research workers in dating and interpreting the deposits. Abundant remains of fish, amphibians, reptiles, and birds are associated with the mammalian bones and teeth, thus assisting in environmental interpretations.

Fossils have been collected at the type locality of the Sappa Formation where fossiliferous sediments occur above and below the Pearlette Volcanic Ash. More geologic documentation has been obtained concerning the occurrence of faunas from deposits attributed to the Yarmouth and Sangamon interglacial and the Illinoian glacial periods. An increasing number of Early, Middle, and Late Pleistocene, as well as Holocene, vertebrate localities provide reliable information for establishing a mammalian sequence for the entire Quaternary of the Central Great Plains of North America. New evidence concerning the evolution of the bison and mammoths is summarized.

Three new geologic names are proposed, namely, (1) *Gothenburg Member* for the upper unit of the Loveland Formation and (2) *Buzzard's Roost Paleosol Complex* and (3) *Ingham Paleosol* for the upper and lower buried soil layers in the Loveland.

The Central Great Plains have long been noted as an important source of Quaternary vertebrates. No other place in North America has produced such an abundance of vertebrate fossils from deposits ranging in age from Early to Late Pleistocene. The sequence of vertebrate remains from the Holocene (or Recent) also is becoming known, and is providing evidence of the migrations and evolution of mammals during the past 8000 years.

The Central Great Plains region provides an excellent place for the establishment of a faunal sequence for the Quaternary. This region is closely associated with both the extensive continental glaciated region on the east and the glaciated areas in the Rocky Mountains to the west. Thus, there is an opportunity to obtain precise geologic evidence for dating the various fossiliferous deposits. It should be pointed out, however, that there is much yet to be learned about the geologic sequence of the Quaternary in the Central Great Plains. The stratigraphic record is complex, and the problems of the Quaternary can be solved only by integrated research programs using all available evidence. The present paper will touch upon some of the problems, and also may provide preliminary evidence concerning eventual establishment of an acceptable mammalian sequence for the Quaternary of this region.

We use and recommend the use of the age terms Early, Middle, and Late Pleistocene, and Holocene for the divisions of the Quaternary (Reed *et al.*, 1965, p. 202; Schultz, 1968, p. 115). Early Pleistocene means Early Nebraskan to Middle Kansan; Middle Pleistocene includes Late Kansan to Late Illinoian; Late Pleistocene designates Sangamon to Late Wisconsin; and Holocene refers to approximately the past 8000 years, which is often considered to be the Recent.

The University of Nebraska State Museum field parties have been making a systematic stratigraphic collection of Quaternary fossils since the early 1930's, and have brought together the largest stratigraphic Quaternary collection in North America. As usual, the larger the collections, the more complex the problems become, but some trends of development in certain phylogenetic lines of mammals are becoming apparent. The first attempt to establish a faunal sequence of Pleistocene mammals in the Central Great Plains region was made in the mid-1930's by Lugn and Schultz (1934). Up until that time, vertebrate paleontologists in general, including Hay (1914, p. 25, and 1924, pp. 278, 302) and Matthew (1902, p. 317, and 1918, p. 227), had considered many Pleistocene fossils from Iowa, Nebraska, and surrounding areas to be associated with Aftonian interglacial deposits. Since that time it has become evident that rather few of the specimens attributed to the Aftonian were of that age, most having been derived from much later deposits (Barbour and Schultz, 1937; Schultz and Stout, 1948, 1961).

The well preserved regional terraces (see Fig. 1) in the Central Great Plains region have aided greatly in establishing faunal sequences. Nowhere in North America are terrace-fills so well exposed. Many geologists (Bryan and Ray, 1940; Bryan, 1941, 1950; Schultz and Stout, 1945, 1948; Lueninghoener, 1947; Frankforter, 1950; Schultz *et al.*, 1951; Schultz and Tanner, 1957) have considered that geomorphology is an important key to correlating Quaternary deposits. Certainly geomorphologic evidence must be considered when stratigraphic, petrographic, or paleontologic studies are undertaken. Geomorphology can provide many clues to the solving of problems concerning the establishment of faunal sequences.

Charts such as the one published by Flint (1955, p. 42) for eastern Nebraska and adjacent areas make certain Pleistocene sequences look too simple. As Dreeszen noted earlier in this volume, the Nebraskan may be composed of several tills, and the Kansan probably has at least two glacial tills (Stout *et al.*, 1965, p. 12). Furthermore, the Illinoian also must have had at least three advances of the glaciers (Frye *et al.*, 1965, pp. 46, 50). All this complicates the problem of dating local faunas. Frequently, paleoecological evidence gained from the study of fossil remains is of great aid in evaluating the problems at hand, but geomorphic, stratigraphic, pedologic, and other evidence also must be considered. The faunal shifts indicated by Hibbard *et al.* (1965, p. 514) must be far more complicated than were shown, and some of the local faunas should be reexamined in so far as the geologic designations are concerned. The present writers are reconsidering all of the stratigraphic positions of the local faunas in the Nebraska region. The new evidence concerning interstadials and interglacials must be taken into consideration. Certainly the assumption that all faunas showing "interglacial affinities" must pertain to the Aftonian, Yarmouth, or Sangamon is doubtful, and this may have resulted in the miscorrelation of faunas with these events. The argument has been presented by various workers that the interstadials do not represent enough time to permit the readjustment of animal and

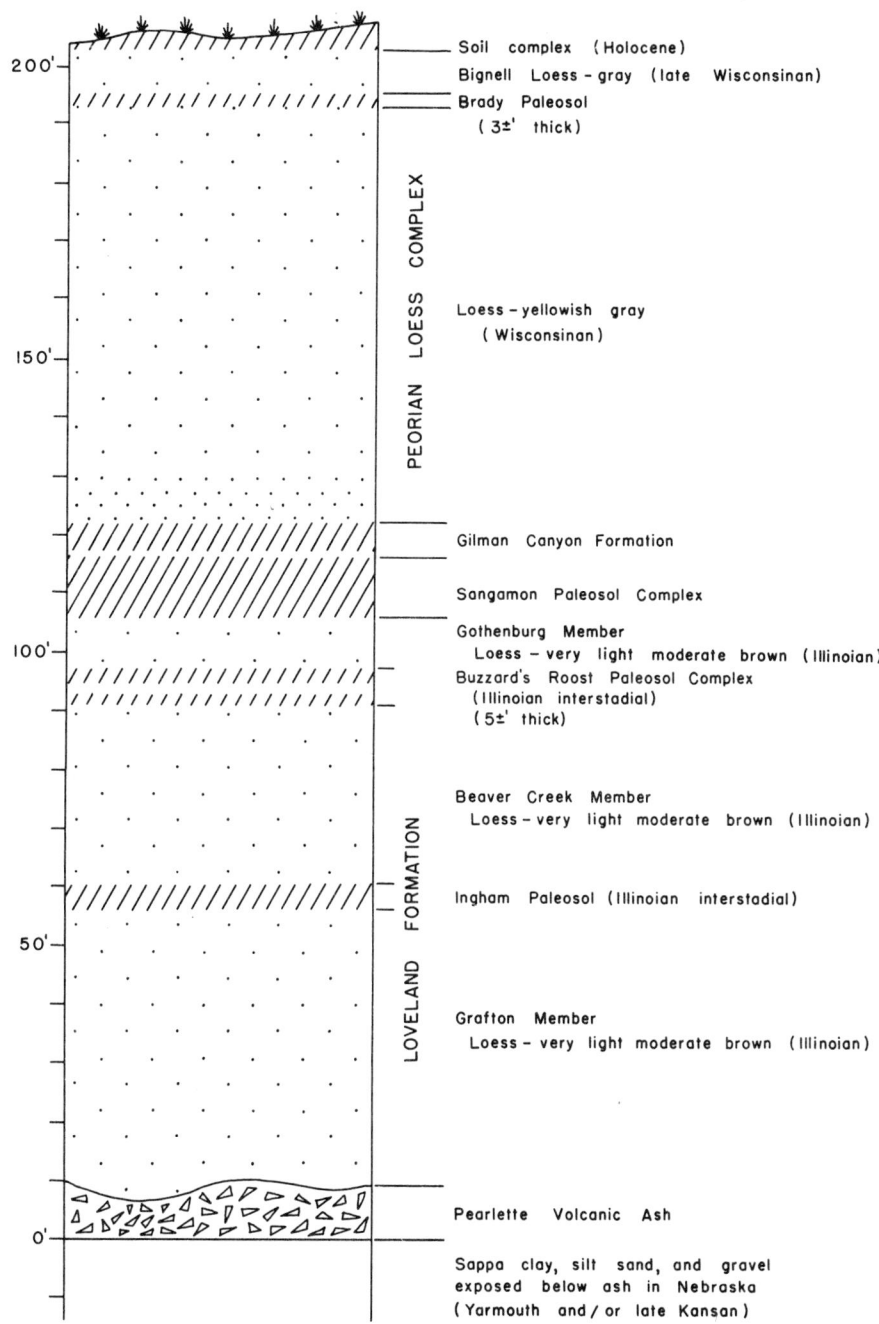

FIG. 2. Buzzard's Roost section, showing thicknesses of Quaternary loesses and paleosols in Terrace-4 fill, 6 mi. S and 8 mi. W Gothenburg, W ½, SE ¼, sec. 7, T. 10 N, R. 26 S, Lincoln County, Nebraska. Type locality for the following new geologic names: *Gothenburg Member* of Loveland Formation; *Buzzard's Roost Paleosol Complex;* and *Ingham Paleosol*. Minor paleosols also occur in the Peorian Loess in localities near this section. Fossil vertebrates are frequently found associated with the paleosols, usually at the top of the A-horizons, and rarely occur in the massive loess deposits. (Figure modified from Schultz and Tanner, 1957, fig. 7, and Schultz, 1968, fig. 8-1.)

plant ranges to the changing climatic conditions. However, this argument does not seem to be valid, because there are many examples of adjustments of animal ranges that have taken place during the latest Holocene, that is, even during the past two centuries (Jones, 1964, p. 31).

The loess canyon area south and east of North Platte has many paleosols, or buried soils, which aid in establishing a more complete sequence of climatic events in the western part of the Central Great Plains. There are two paleosols (see Figs. 1 and 2) in the Loveland, which are exposed in many localities in Nebraska. The paleosols indicate that the climate fluctuated at frequent intervals, geologically speaking, suggesting that the advances and retreats of the Illinoian ice sheets influenced the environment several hundred miles to the west and south. Perhaps the loesses can be correlated with the Liman (the lower), Jacksonville (middle), and Buffalo Hart (upper) tills of the Illinoian in the Mississippi Valley (Frye et al., 1968, p. 8).

The Loveland could be considered a group, but the present writers prefer to call it a formation, at least in the Iowa and Nebraska region. Schultz (1968, p. 118) divided the Loveland into three members. The Lower Member of the Loveland appears to be the Grafton Loess of Reed and Dreeszen (1965, p. 38), and the Middle Member may be equated with their Beaver Creek Loess (see Fig. 2). The buried soil (equal to "Yarmouth Soil development" of Reed and Dreeszen, 1965, p. 62) on top of the Lower Member at Buzzard's Roost (Univ. Nebraska State Museum locality Ln-103, 6 mi. S and 8 mi. W Gothenburg, in W $\frac{1}{2}$, SE $\frac{1}{4}$, sec. 7, T. 10 N, R. 26 S, Lincoln County, Nebraska) is the *Ingham Paleosol* (new name), and is named after the abandoned town of Ingham in southeastern Lincoln County, where formerly there was extensive mining of the Pearlette Volcanic Ash (Barbour, 1916, pp. 372-373). The buried soil complex (see Fig. 2) on top of the Middle Member at Buzzard's Roost is the *Buzzard's Roost Paleosol Complex* (new name). This paleosol complex frequently shows two closely associated soil profiles, and forms a distinctive marker in many of the loess areas of Nebraska. Above the Buzzard's Roost Paleosol Complex is the Upper Member (restricted "Loveland Loess" of Reed and Dreeszen, 1965, p. 40) of the Loveland Loess. The present writers propose that the Upper Member of the Loveland Loess be called the *Gothenburg Member* (new name), rather than restricting the Loveland to the upper portion of the Illinoian loess in central and western Nebraska. The Gothenburg Member is named after the town of Gothenburg. The Sangamon Soil Complex, some 12 feet in thickness, is developed on top of the Gothenburg Member of the Loveland. The type locality for the Ingham Paleosol, the Buzzard's Roost Paleosol Complex, and the Gothenburg Member of the Loveland is in the Terrace-4 fill at Buzzard's Roost, but all three are well exposed in many localities in the canyons in central and southwestern Nebraska.

The Buzzard's Roost recorded geologic section (see Fig. 2; also Frankel, 1956, pp. 53-55; Reed and Dreeszen, 1965, p. 62) is as shown in Table 1.

Important fossil assemblages have been collected in association with the Brady Paleosol at the top of the Peorian Loess. It is apparent that the Brady Paleosol is younger than anticipated by Schultz et al. (1951). This buried soil seems to be equivalent to the paleosol designated as YY in the base of Terrace-2, Fill A (see Fig. 1), rather than Paleosol X associated with the Peorian Loess. Reed and Dreeszen (1965, p. 1) provided a carbon-14 date of 9160 ± 250 years for the age of the Brady Paleosol, although they mentioned that the date might be a minimal one (see legend, Fig. 1, for carbon-14 dates for soil YY in base of Terrace-4; also Schultz and Tanner, 1965, p. 85,

TABLE 1. Buzzard's Roost Recorded Geologic Section

	Thickness in Feet
HOLOCENE	
Soil complex	2
Bignell Loess: Silt, gray	10
PLEISTOCENE	
Brady Paleosol: Buried soil complex, carbonaceous	3±
Peorian Loess: Silt, sandy to clayey; aeolian, light buff brown, massive, columnar jointing; calcareous, secondary-lime pellets at several levels, disseminated carbon flakes	70
Gilman Canyon Formation: Silt, dark gray, carbonaceous in upper 1 to 2 feet	6±
Sangamon Paleosol Complex: Silt, sandy to clayey, dark brownish gray to reddish brown; granular, calcareous, few secondary-lime pellets; many disseminated carbon flakes; many fossil rodent burrows	10±
Loveland Formation: Loess and paleosols	
Gothenburg Member: Loess-silt, sandy, very light moderate brown, massive; columnar jointing, slightly calcareous, few secondary-lime pellets, disseminated carbon flakes	8
Buzzard's Roost Paleosol Complex: Silt and clayey silt, double soil development —upper 1.5 feet, sandy silt, medium to dark brownish gray, slightly calcareous, many disseminated carbon flakes; next lower 1.5 feet, clayey silt, light reddish brown, massive, slightly calcareous, disseminated carbon flakes; lower 3 feet, silt, clayey, dark brownish gray, granular, slightly calcareous, many disseminated carbon flakes	6
Beaver Creek Member: Loess-Silt, sandy, very light moderate brown, massive; columnar jointing, slightly calcareous, secondary-lime pellets near top and middle of Member, disseminated carbon flakes	30
Ingham Paleosol: Silt and clayey silt, soil development—upper 1.5 feet, clayey silt, dark brownish gray, granular, calcareous, secondary-lime pellets, many disseminated carbon flakes (? A-horizon); next lower 2 feet, moderately clayey silt, medium dark brownish gray, calcareous, many secondary-lime pellets, disseminated carbon flakes	4.5
Grafton Member: Loess-silt, sandy, very light moderate brown, massive; slightly calcareous with few secondary-lime pellets in upper 4 feet, disseminated carbon flakes. (Lower portion poorly exposed in canyon area)	45
Pearlette Volcanic Ash	10±
Sappa Formation: Clay, silt, and sand (only a few feet exposed at base of section, but exposed in other parts of the canyon)	

and Ruhe, 1968, p. 61, and this volume). It is significant that the Bignell Loess, the Brady Paleosol, and Paleosol YY have not produced any mammoth remains, whereas Paleosol X often yields mammoth bones and teeth.

A large collection of fossil mammals is being assembled from the Gilman Canyon Formation (equal to "*Citellus* Zone" in part), which lies on top of the Sangamon Paleosol Complex. The Gilman Canyon Formation appears to be equivalent to most, if not all, of the Altonian (Frye *et al.*, 1968, p. 16) of Illinois.

One of the problems of the Quaternary that deserves attention is the relationship or correlation of the time of accumulation of the Middle and Late Pleistocene loesses with the time of the structural development of the Rocky Mountains since the Late Kansan glaciation. Moist conditions seem to have existed in the Central Plains during most of the first half of the Quaternary and there was little loess deposition, but since the final retreat of the Kansan, an ever-increasing amount of loess has been deposited. Schultz *et al.* (1951) postulated that the Middle Pleistocene uplift of the Rocky Mountains in the Colorado-Wyoming region and in the mountain areas westward to the Pacific coast may well have produced a physiographic barrier or "rain shadow" effect on the Nebraska region, where the largest sand dune area in North America was developed and where the most extensive, thick loess deposits on the continent accumulated. Certainly the blowing dust and sand must have had a great influence on the distribution of vertebrate animals, especially the larger ones, during the latter part of the Quaternary. The absence or scarcity of vertebrate remains in the various major loess deposits strongly suggests that the animals did not live continuously in the regions where the sand and dust were blowing. In a recent publication, Smith (1968, p. 46) reported: "A distinguishing characteristic of the Nebraska area is a multicycle dune history in which successive episodes of clearly defined eolian activity alternated with intervals of stabilization." He went on to point out that during one interval "desert conditions fully comparable to those of the present day Sahara were prevalent." Today, of course, the sand dunes in Nebraska are stabilized and are covered with vegetation. It is evident that it is necessary to more precisely correlate the main periods of blowing sands and stabilization with the loesses, silts, and paleosols of the terrace fills of the adjacent areas. Fortunately, vertebrate fossils now are being discovered in association with the major paleosols of the Medial and Late Pleistocene, and also of the Holocene. A. L. Lugn (1962, 1968) felt that the Sandhills were a major source for the loess in the areas adjacent to the dunes.

The Terrace-4 fill of Schultz and Stout (1945, 1948) is a critical one, and problems of extinction and migration of mammals are involved in it. Here, we find the earliest evidence of mammoths, bovids, certain cervids, and other mammals (Schultz, 1964, 1968). These mammals had migrated from Asia during late Kansan or early Yarmouthian times.

Some of the best known and key fossil localities involved in the Terrace-4 fill are to be found south of the Niobrara River in Sheridan County, Nebraska. These localities have been worked extensively since the 1880's by many museums. Unfortunately the collections made prior to the late 1920's have no stratigraphic data associated with the specimens. The University of Nebraska State Museum has worked the old localities on various occasions since 1930, and has developed new sites in the area. More work is being planned for the next five years in this important area.

The Rushville Quarries, University of Nebraska State Museum locality Sh-3, have provided important faunal and stratigraphic data. Here the faunal evidence is the same as at the Hay Springs Quarries (UNSM locality Sh-1) but the larger vertebrate fossils are more abundant. The fossils are in the base of the terrace-fill and also are found in sediments above. The local fauna from the base is tentatively associated with the Yarmouth, and the faunal evidence from some 20 feet higher in exposure appears to be Early Illinoian.

The Hay Springs local fauna from Sheridan County in northwestern Nebraska has been regarded by Hibbard (1956, p. 1265), as being Late Illinoian or Sangamon in age. This determination was based on the presence of *Microtus pennsylvanicus* (Ord) in this local fauna, and the advanced nature of the muskrat, *Ondatra nebrascensis*, from the Hay Springs deposits. However, a small muskrat, *Ondatra* cf. *annectens* (Brown), also occurs with *O. nebrascensis* in these quarries, and *Microtus pennsylvanicus* occurs stratigraphically below the Loveland Loess and the Sangamon Paleosol in University of Nebraska State Museum locality No-101, located near Angus in Nuckolls County, south-central Nebraska. The Angus local fauna resembles Hibbard's Cragin and Jinglebob local faunas (Hibbard, 1955) from Kansas as well as the Middle Pleistocene faunas from Sheridan County, Nebraska. The following mammals have been collected from the Angus locality: ground sloth (genus and species undetermined); *Lepus* sp., jackrabbit; *Silvilagus* sp., cottontail; bat (genus and species undetermined); *Sorex cinereus* Kerr, masked shrew; *Blarina brevicauda* (Say), short-tailed shrew; *Scalopus* sp., mole; *Spermophilus* cf. *tridecemlineatus* (Mitchell), thirteen-lined ground squirrel; *S.* cf. *richardsoni* (Sabine), Richardson's ground squirrel; *Cynomys niobrarius* Hay, Niobrara prairie dog; *Geomys bursarius* (Shaw), plains pocket gopher; *Perognathus hispidus* Baird, hispid pocket mouse; *Perognathus* sp., pocket mouse; *Castor* sp., beaver; *Castoroides* sp., giant beaver; *Onychomys* sp., grasshopper mouse; *Reithrodontomys* sp., harvest mouse; *Peromyscus* sp., white-footed mouse; *Neotoma* sp., woodrat; *Synaptomys* sp., bog lemming; *Microtus ochrogaster* (Wagner), prairie vole; *M. pennsylvanicus* (Ord), meadow vole; *Ondatra nebrascensis* (Hollister), muskrat; *Canis* cf. *latrans* (Say), coyote; *C. dirus nebrascensis* Frick, Nebraska dire wolf; *Vulpes velox* (Say), fox; *Mustela frenata* Leichtenstein, long-tailed weasel; *Taxidea* cf. *taxus* (Schreber), badger; *Mammuthus (Archidiskodon) imperator* (Leidy), imperial mammoth; mastodont (genus and species undetermined); *Mylohyus browni* Gidley, peccary; *Platygonus* sp., peccary; *Camelops kansanus* Leidy, camel; *Odocoileus sheridanus* Frick, deer; *Capromeryx fucifer* Matthew, subpronghorn; ?*Stockoceros* sp., four-horned pronghorn; *Equus excelsus* Leidy, horse; and *E. calobatus* Troxell, stilt-legged horse. Descendants of most of the smaller mammals in the Angus Local Fauna can still be found in southern and west-central Nebraska today. This suggests that climatic conditions were no more severe at the time of deposition of the Angus quarry deposits than they are now in central Nebraska.

Another Middle Pleistocene collecting site has been developed at the type locality of the Sappa Formation in Harlan County, south-central Nebraska, University of Nebraska State Museum locality Hn-102. Here, fossils have been collected from both above and below the Pearlette Volcanic Ash bed. The fauna from below the ash bed includes fishes, amphibians, reptiles, birds, and the following mammals: *Spermophilus* sp., ground squirrel; *Cynomys* sp., prairie dog; *Geomys* sp., pocket gopher; *Perognathus* sp., pocket mouse; *Reithrodontomys* sp., harvest mouse; *Onychomys* sp., grasshopper mouse; *Peromyscus* sp., white-footed mouse; *Synaptomys meltoni* Paulson, Melton's bog lemming; *Microtus* cf. *llanensis* Hibbard, upland vole; *Ondatra* sp., muskrat; *Zapus sandersi* Hibbard, Sanders' jumping mouse. Specimens of mastodont and horse *(Equus)* were also recovered. This fauna is similar to the Cudahy fauna in Kansas, and is quite different from the small animals from Sheridan County (UNSM localities Sh-1, Sh-9, Sh-3, and Sh-4) and Nuckolls County (UNSM locality No-101), Nebraska. Unfortunately, the Pearlette Ash does not appear to be present at any of the Sheridan County localities, although it has been reported. The Pleistocene ash problem is a difficult one to solve, as has been pointed out by V. H. Dreeszen of the Nebraska Geological Survey. How many ash falls actually took place? What is the age of the Pearlette Ash—Late Kansan, or Yarmouth, or Early Illinoian? Perhaps there were two or more ash falls. Much more extensive work will have to be done.

On the North Prong of the Middle Loup River in southern Cherry County, we have a series of quarries, UNSM localities Cr-10, Cr-102, and others, in which we find a cold or glacial climate assemblage, including musk oxen *(Symbos cavifrons)*, with warm interstadial forms directly above. Also there is considerable faunal evidence indicating a warm climate below the *Symbos* layer. The field evidence here is being reinvestigated at the present time. New techniques in stratigraphic collecting were developed from 1962 to 1967 (Schultz et al., 1963, 1967) at Big

Bone Lick, Kentucky, by paleontologists and geologists; and it is certain that more precise stratigraphic and paleoecological data now can be obtained at the Cherry County localities than Jakway (1962) was able to assemble in the early 1960's. Unfortunately there had been a certain amount of reworking and mixing of specimens at the latter localities during Middle Pleistocene times, and this made the problems of association more complex. The same problems plagued the field parties at Big Bone Lick.

At the Rushville quarries there is an abundance of the remains of the mammoth *Mammuthus (Archidiskodon) imperator*. The teeth, that is the structure of the plates and the thickness of the enamel bands, indicate that these were among the most primitive mammoths in North America and among the first to reach the Great Plains (see Fig. 3). The taxonomy of the North American mammoths is in need of complete revision and this is one of the research projects being carried on at the University of Nebraska State Museum at the present time. A large amount of material is needed to accomplish the revision, and a concentrated effort to gather additional specimens associated with stratigraphic data has been progressing since the early 1930's. Erwin H. Barbour and his associates had been encouraged by Henry Fairfield Osborn to try to make a stratigraphic collection; Osborn (1942) did not have such a collection available when he did his extensive research on the proboscideans. Most museums have shied away from collecting fossils as large as mammoths. Specimens of this kind take considerable storage space and also are costly to collect. However, funds have been available to continue the collecting, preservation, and storage at the University of Nebraska State Museum. At present a new and more adequate facility is being constructed for research and storage of the Museum's systematic biological collection.

Thus, the mammoths soon will be more accessible for research.

Three mammoths, *Mammuthus (Archidiskodon) hayi* (Barbour, 1915), *M. (A.) scotti* (Barbour, 1925), and *M. (A.) haroldcooki* (Hay and Cook, 1928, p. 33, and 1930, p. 32) have been considered primitive, perhaps the most primitive mammoths from North America. Many authors (Osborn, 1942) have listed the holotypes as having been derived from the Aftonian. Actually, judging by the mandibles, the animals all appear to have been extremely old when they died, because the last molars (m3) were worn almost down to the roots. In *scotti* and *haroldcooki*, the broadness of the enamel bands and primitive look of the plates are due to the obliquity of wear. The plates slant pronouncedly backward. One must use extreme care in measuring mammoth teeth and must consider the stage of wear before making comparisons. During the Late Pleistocene, mammoth teeth developed thinner plates, and the enamel also was thinner and more crenulated. In Figure 3 a bar graph illustrates the decrease of thickness of enamel in M3 and m3 in the *Mammuthus (Archidiskodon)* lineage from Yarmouth Interglacial to Early Wisconsinan. This preliminary study was based on mammoth teeth from three stratigraphically documented local faunas. There appear to have been at least two phylogenetic lines of mammoths paralleling each other in the Late Pleistocene, the *Mammuthus (Archidiskodon)* and the *M. (Parelephas)*. Utmost care must be taken in the measuring of mammoth teeth or erroneous measurements will result. The front enamel bands in the plates are usually thinner than those at the rear, and the enamel at the center of the teeth usually measures less than at the sides.

One primitive mammoth tooth (UNSM 2655), right m3, is strikingly different from any of the other specimens in our collection, or from any other reported material from the

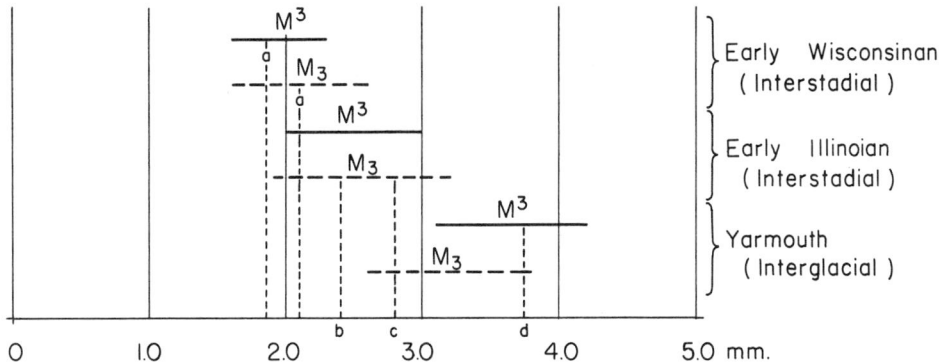

Fig. 3. Bar graph of enamel thickness in M3 and m3 in the *Mammuthus (Archidiskodon)* lineage from Yarmouth Interglacial to Early Wisconsinan. Specimens of Yarmouth age are from University of Nebraska State Museum locality Sh-3, south of Rushville, Sheridan County, Nebraska; Early Illinoian specimens are from UNSM localities Cr-102 and Cr-11, North Prong of Middle Loup River, northwest of Mullen, Cherry County, Nebraska; and Early Wisconsinan ("*Citellus* Zone") specimens are from Franklin and Lincoln counties, Nebraska. The range midpoints of enamel thicknesses in the holotypes of *M. (A.) maibeni* (a), *M. (A.) scotti* (b), *M. (A.) hayi* (c), and *M. (A.) imperator* (d) also are included. Measurements in millimeters.

United States. It has only three and one-half plates per 10 centimeters, measured perpendicular to the plates, and is reminiscent of the Early Pleistocene or Villafranchian mammoth teeth of Spain, France, and Italy. This might suggest that mammoths could have migrated to the Central Great Plains prior to the Yarmouth, perhaps during an interstadial in the Kansan. This specimen came from a locality in northern Butler County, near localities where intertill sediments of Kansan age are known. Unfortunately, the exact provenience of the Butler County tooth is not known. Mammoths continued to evolve thinner plates and enamel until some 12,000 years ago, at which time they became extinct.

A complete revision of the bison based on precise stratigraphic control also is badly needed. Here again, as is the case with the mammoths, only a preliminary progress report can be made at this time concerning trends.

Bison bison and *B. latifrons* remains have not been found associated together in the Nebraska area. In fact, no *Bison antiquus* and *Bison latifrons* have been found together (Schultz and Frankforter, 1946; Schultz, 1968). In the Sangamon, thus far only *B. anti-*

quus barbouri has been recorded and no *B. latifrons* has been found. The *B. latifrons* remains in Nebraska and adjacent areas all are Middle Pleistocene in age. No remains of *B. latifrons* or any other species of bison thus far have been found at Hay Springs, Rushville, or Angus, but it is possible that remains will be encountered when further excavations are made. It must be remembered that these huge beasts were not plentiful, as were the horses and camels. The evidence at hand indicates that they did not reach the Central Great Plains area from Asia, via the trans-Bering land bridge until Medial Pleistocene times, and then not in great numbers as was the case of the mammoths. At Big Bone Lick, Kentucky, some 700 miles east of Nebraska, *B. bison, B. antiquus,* and *B. latifrons* have been frequently reported to be associated with each other. Intensive excavations, research, and mapping by the University of Nebraska State Museum field parties in cooperation with U.S. Geological Survey scientists have failed to find this to be true (Schultz et al., 1963, 1967). Although the holotype of *B. latifrons* came from Boone County, the county in which Big Bone

Lick is located, no specimens of this species were encountered at Big Bone Lick.

Although there are many references in the literature to the occurrence of *Bison latifrons* at Big Bone Lick, a careful study of these particular specimens preserved in museums in London, Paris, Philadelphia, and Cambridge, Massachusetts, shows that these were incorrectly identified, and in most cases they are simply modern *Bison bison*, not more than a few hundred years old.

Recently *B. latifrons* has been reported from the Rancho La Brea tar pits of Los Angeles, California (Miller, 1968, p. 4), but the specimens reported should be further investigated before they are accepted as conclusive evidence of *B. latifrons* in the Late Pleistocene. More stratigraphic data, as well as specimens, are being accumulated from various parts of North America so that a revision of the bison, as well as the musk oxen and other bovids, can be made in the near future.

Evidence found in southeastern New Mexico in two dust caves of the Guadalupe Mountains (just west of the Southern Great Plains region) suggests that some of the forms that disappeared from the Central Great Plains some 12,000 years ago may have lingered on for thousands of years to the south. The following evidence for this was summarized by Schultz (1968, p. 125). In Burnet Cave (Schultz and Howard, 1935, pp. 273-298, pls. 11-16), horse and camel bones (found buried in the dust) date as recently as 7800 to 8000 years ago. Two species of horse (Schultz and Howard, 1935, p. 285, pl. 15, figs. 1-8), a large extinct form (*Equus* cf. *excelsus* Leidy) and a small one (*Equus* cf. *tau* Owen), were found in the same zone.

Associated with the bones was a Plainview-type dart point (Howard, 1935, p. 69, pl. 35, fig. 3; Schultz and Howard, 1935, pl. 11, fig. 2), which usually dates much more recently than Folsom- and Scottsbluff-type points found elsewhere in the Central Great Plains. It is not a Clovis-type point as has been suggested (Hester, 1967, pp. 183-184; Martin and Guilday, 1967, pp. 59-60). Folsom- and Scottsbluff-type points appear to date between 9500 and 10,000 BP, so it would seem reasonable to consider that the makers of the Plainview point could have lived as recently as 8000 years ago. Howard (1935, p. 69) first reported the provenience of the dart point in the following manner: "It was five feet seven inches below the [undisturbed] surface, also it was directly under a rock, the underside of which was covered with charcoal and ashes as though it [the rock] had fallen directly from the roof or wall upon the remains of a fire." The charcoal dated by Libby (sample C823) was at a slightly lower level than the dart point, so all of the animal bones reported by Schultz and Howard (1935) must be slightly more recent.

Other extinct animals from Burnet Cave, which are more recent than charcoal sample C823, are as follows (see Schultz and Howard, 1935): giant bear, *Arctodus*; large camel, *Camelops* sp.; large deer, *Sangamona fricki* Schultz and Howard; prong buck, *Tetrameryx onusrosagris* Roosevelt and Burden; shrub ox, *Euceratherium collinum morrisi* Schultz and Howard; shrub ox, *Preptoceras sinclairi neomexicana* Schultz and Howard; and large bison, *Bison antiquus* Leidy.

Lundelius (1967, p. 287) reported that a fauna consisting of proboscidians, sloths, glyptodonts, saber-toothed cats, *Bison antiquus, B. occidentalis, Camelops, Tanupolama*, horses, and peccaries persisted in central Texas until approximately 8000 years ago.

Hermit's Cave, which is located in Lincoln National Forest on the eastern slope of the Guadalupe Mountains, was worked in 1938 by the University of Nebraska State Museum in association with the New Mexico State Museum, and again in 1955 by the University of Nebraska in association with the West Texas Museum (Texas Technological College). Haynes (1967, p. 278) did not regard radiocarbon dates from this cave as relating to human occupation; however, in 1955 fossil mammals were found associated with fireplaces of early man. The dating of charcoal and wood (by M. Rubin of the U.S. Geological Survey's Low-Radiation Laboratory) indicates that mammoths and dire wolves lived as recently as 12,000 years ago in that area, which is adjacent to the Great Plains region. Sample W498 (University of Nebraska State Museum, no. 5762) was dated 11,850 ±350 BP, and sample W499 (UNSM, no. 5763) was dated 12,270±450 BP. Both samples were associated with the mammoth (*Mammuthus* sp.) and dire wolf (*Canis dirus* Leidy) remains that were found below a layer of dust.

An increasing number of Early, Middle, and Late Pleistocene fossiliferous localities and various new buried sites in Holocene deposits in the Great Plains region and adjacent areas are providing reliable information that can be used for the development of a stratigraphically documented sequence of the vertebrates.

ACKNOWLEDGMENTS

The writers are grateful to their colleagues Vincent Dreeszen, Mylan Stout, and Lloyd Tanner for their counsel and other assistance, and to Dwight Brennfoerder for the preparation of the illustration.

LITERATURE CITED

Barbour, E. H.
    1915. A new primitive mammoth, *Elephas hayi*. Amer. Jour. Sci., ser. 4, 40:129-134.
    1916. Nebraska pumicite. Publ. Nebraska Geol. Surv., ser. 1, 4:355-401.
    1925. *Elephas scotti*, a new primitive mammoth from Nebraska. Bull. Nebraska State Mus., 1:21-24.

Barbour, E. H., and C. B. Schultz
    1937. An early Pleistocene fauna from Nebraska. Amer. Mus. Novit., 942:1-10.

Bryan, K.
    1941. Correlations of the deposits of Sandia Cave, New Mexico, with glaciology chronology. Smithsonian Misc. Collections, 99(23):45-64.
    1950. The geology and fossil vertebrates of Ventana Cave: Geological interpretation of the deposits. Pp. 75-126, 4 tables (pt. 3), *in* The stratigraphy and archaeology of Ventana Cave, Arizona (E. W. Haury, ed.), Univ. Arizona Press and Univ. New Mexico Press, 599 pp.

Bryan, K., and L. L. Ray
    1940. Geologic antiquity of the Lindenmeir site in Colorado. Smithsonian Misc. Collections, 99(2):1-76.

Flint, R. F.
    1955. Pleistocene geology of eastern South Dakota. U.S. Geol. Surv. Prof. Paper, 262:1-173.

Frankel, L.
    1956. Pleistocene geology and paleoecology of parts of Nebraska and adjacent areas. Unpubl. Ph.D. thesis, Univ. Nebraska, 297 pp.

Frankforter, W. D.
    1950. The Pleistocene geology of the middle portion of the Elkhorn River Valley. Univ. Nebraska Studies (n.s.), 5:1-46.

Frye, J. C., H. B. Willman, and R. F. Black
    1965. Outline of glacial geology of Illinois and Wisconsin. Pp. 43-61, *in* The Quaternary of the United States (H. E. Wright, Jr., and D. G. Frey, eds.), Princeton Univ. Press, Princeton, New Jersey, x+922 pp.

Frye, J. C., H. B. Willman, and H. D. Glass
    1968. Correlation of Midwestern loesses with the glacial succession. Pp. 3-21, *in* Loess and related eolian deposits of the world (C. B. Schultz and J. C. Frye, eds.), Univ. Nebraska Press, Lincoln, 369 pp.

Hay, O. P.
    1914. The Pleistocene mammals of Iowa. Bull. Iowa Geol. Surv., 23:1-662.
    1924. The Pleistocene of the middle region of North America and its vertebrated animals. Publ. Carnegie Inst. Washington, 322A:1-385.

Hay, O. P., and H. J. Cook
    1928. Preliminary descriptions of fossil mammals recently discovered in Oklahoma, Texas and New Mexico. Proc. Colorado Mus. Nat. Hist., 8:33.
    1930. Fossil vertebrates collected near, or in association with, human artifacts at localities near Colorado, Texas; Frederick, Oklahoma; and Folsom, New Mexico. Proc. Colorado Mus. Nat. Hist., 9:4-40.

Haynes, C. V., Jr.
    1967. Carbon-14 dates and early man in the New World. Pp. 267-286, *in* Pleistocene

extinctions: the search for a cause (P. S. Martin and H. E. Wright, Jr., eds.), Yale Univ. Press, New Haven, Connecticut, x+453 pp.

Hester, J. J.
1967. The agency of man in animal extinctions. Pp. 169-200, *in* Pleistocene extinctions: the search for a cause (P. S. Martin and H. E. Wright, Jr., eds.), Yale Univ. Press, New Haven, Connecticut, x+453 pp.

Hibbard, C. W.
1955. The Jinglebob Interglacial (Sangamon?) Fauna from Kansas and its climatic significance. Contrib. Mus. Paleontol., Univ. Michigan, 12:179-228.
1956. *Microtus pennsylvanicus* (Ord) from the Hay Springs Local Fauna of Nebraska. Jour. Paleontol., 30:1263-1266.

Hibbard, C. W., C. E. Ray, D. E. Savage, D. W. Taylor, and J. E. Guilday
1965. Quaternary mammals of North America. Pp. 509-525, *in* The Quaternary of the United States (H. E. Wright, Jr., and D. G. Frey, eds.), Princeton Univ. Press, Princeton, New Jersey, x+922 pp.

Howard, E. B.
1935. Evidence of early man in North America. Mus. Jour., Univ. Pennsylvania, 24:61-171.

Jakway, G. E.
1962. The Pleistocene faunal assemblages of the Middle Loup River terrace-fills of Nebraska. Unpubl. Ph.D. thesis, Univ. Nebraska, 43 pp.

Jones, J. K., Jr.
1964. Distribution and taxonomy of mammals of Nebraska. Univ. Kansas Publ., Mus. Nat. Hist., 16:1-356.

Lueninghoener, G. C.
1947. The post-Kansan geologic history of the Lower Platte Valley area. Univ. Nebraska Studies (n.s.), 2:1-82.

Lugn, A. L.
1962. The origin and sources of loess. Univ. Nebraska Studies (n.s.), 26:xi+1-105.
1968. The origin of loesses and their relation to the Great Plains in North America. Pp. 139-182, *in* Loess and related eolian deposits of the world (C. B. Schultz and J. C. Frye, eds.), Univ. Nebraska Press, Lincoln, 369 pp.

Lugn, A. L., and C. B. Schultz
1934. The geology and mammalian fauna of the Pleistocene of Nebraska [part I by Lugn, part II by Schultz]. Bull. Nebraska State Mus., 1:319-393.

Lundelius, E. L., Jr.
1967. Late-Pleistocene and Holocene faunal history of central Texas. Pp. 287-319, *in* Pleistocene extinctions: the search for a cause (P. S. Martin and H. E. Wright, Jr., eds.), Yale Univ. Press, New Haven, Connecticut, x+453 pp.

Martin, P. S., and J. E. Guilday
1967. A bestiary for Pleistocene biologists. Pp. 1-62, *in* Pleistocene extinctions: the search for a cause (P. S. Martin and H. E. Wright, Jr., eds.), Yale Univ. Press, New Haven, Connecticut, x+453 pp.

Matthew, W. D.
1902. List of the Pleistocene fauna from Hay Springs, Nebraska. Bull., Amer. Mus. Nat. Hist., 16:317-322.
1918. Contributions to the Snake Creek Fauna with notes upon the Pleistocene of Western Nebraska American Museum Expedition of 1916. Bull., Amer. Mus. Nat. Hist., 38:226-229.

Miller, W. E.
1968. Occurrence of a giant bison, *Bison latifrons*, and a slender-limbed camel, *Tanupolama*, at Rancho La Brea. Contrib. Sci., Los Angeles County Mus., 147:1-9.

Osborn, H. F.
1942: Proboscidea: A monograph of the discovery, evolution, migration and extinction of the mastodonts and elephants of the world. Amer. Mus. Nat. Hist., Amer. Mus. Press, 2 (Stegodontoidea, Elephantoidea): xxvii+805-1675.

Reed, E. C., and V. H. Dreeszen
1965. Revision of the classification of the Pleistocene deposits of Nebraska. Bull. Nebraska Geol. Surv., 23:1-65.

Reed, E. C., V. H. Dreeszen, C. K. Bayne, and C. B. Schultz
1965. The Pleistocene in Nebraska and northern Kansas. Pp. 187-202, *in* the Quaternary of the United States (H. E. Wright, Jr., and J. C. Frey, eds.), Princeton Univ. Press, Princeton, New Jersey, x+922 pp.

Ruhe, R. V.
1968. Identification of paleosols in loess deposits in the United States. Pp. 49-65, *in* Loess and related eolian deposits of the world (C. B. Schultz and J. C. Frye, eds.), Univ. Nebraska Press, Lincoln, 369 pp.

Schultz, C. B.
1964. Quaternary vertebrate paleontology and stratigraphy of the Central Great Plains. Rep. VI Int. Cong. Quaternary, Paleozoological Sec., 2:583-589.
1968. The stratigraphic distribution of vertebrate fossils in Quaternary eolian deposits in the midcontinent region of North America. Pp. 115-138, *in* Loess and related eolian deposits of the world (C. B. Schultz and J. C. Frye, eds.), Univ. Nebraska Press, Lincoln, 369 pp.

Schultz, C. B., and W. D. Frankforter
1946. The geologic history of the bison in the Great Plains, a preliminary report. Bull. Univ. Nebraska State Mus., 3:1-10.

Schultz, C. B., and E. B. Howard
1935. The fauna of Burnet Cave, Guadalupe Mountains, New Mexico. Proc. Acad. Nat. Sci. Philadelphia, 87:273:298.

Schultz, C. B., G. C. Lueninghoener, and W. D. Frankforter

1951. A graphic resume of the Pleistocene of Nebraska (with notes on the fossil mammalian remains). Bull. Univ. Nebraska State Mus., 3:1-41.

Schultz, C. B., and T. M. Stout
1945. The Pleistocene loess deposits of Nebraska. Amer. Jour. Sci., 243:231-244.
1948. Pleistocene mammals and terraces in the Great Plains. Bull. Geol. Soc. Amer., 59:553-588.
1961. Field Conference on the Tertiary and Pleistocene of Western Nebraska. Spec. Publ. Univ. Nebraska State Mus., 2:1-54.

Schultz, C. B., and L. G. Tanner
1957. Medial Pleistocene fossil vertebrate localities in Nebraska. Bull. Univ. Nebraska State Mus., 4:59-81.
1965. The Medicine Creek artifact sites. Pp. 84-89, in Guidebook for [INQUA] Field Conference D, Central Great Plains (C. B. Schultz and H. T. U. Smith, eds.), Nebraska Academy of Sciences, Lincoln, 123 pp.

Schultz, C. B., L. G. Tanner, F. C. Whitmore, Jr., L. L. Ray, and E. C. Crawford
1963. Paleontologic investigation at Big Bone Lick State Park, Kentucky: a preliminary report. Science, 142:1167-1169.
1967. Big Bone Lick, Kentucky: a pictorial story of the paleontological excavations at this famous fossil locality from 1962 to 1966. Mus. Notes, Univ. Nebraska State Mus., 33:1-12.

Smith, H. T. U.
1968. Nebraska dunes compared with those of North Africa and other regions. Pp. 29-46, in Loess and related eolian deposits of the world (C. B. Schultz and J. C. Frye, eds.), Univ. Nebraska Press, Lincoln, 369 pp.

Stout, T. M., V. H. Dreeszen, C. B. Schultz, and C. K. Bayne
1965. Pleistocene classifications. Pp. 11-15, in Guidebook for [INQUA] Field Conference D, Central Great Plains (C. B. Schultz and H. T. U. Smith, eds.), Nebraska Acad. Sci., Lincoln, 123 pp.

Influence of Late-Glacial and Post-Glacial Events on the Distribution of Recent Mammals on the Northern Great Plains

ROBERT S. HOFFMANN AND J. KNOX JONES, JR.

ABSTRACT

Recent mammals that occur in the northern part of the Great Plains include: (1) grassland (steppe) species of the Great Plains proper; (2) steppe or desert species that have invaded the plains from the southwest (Sonoran region) or west (Great Basin); (3) coniferous forest species that also occur in the Rocky Mountains or boreal forest; (4) deciduous forest species that reach their western limits on the plains; (5) a few basically Neotropical species that reach their northern limits on the plains; and (6) widespread North American or Pan-American elements. All mammals in the first, second, fifth, and sixth groups have fundamentally contiguous ranges, whereas boreomontane species and those associated with the deciduous forest may have either contiguous or discontinuous distributions.

The existence of true grassland and steppe species, with relatively narrow habitat requirements and largely concordant patterns of distribution, argues for the continuous existence of a true steppe or savannah environment on the Great Plains throughout the Late Pleistocene, albeit shifting in location. Southwestern steppe and desert mammals may have spread northward in post-Wisconsin time, probably mostly during drier periods; some species of Sonoran or Neotropical affinities may be moving northward presently in response to a recent continental warming trend. Steppe and desert mammals of the Great Basin pushed eastward onto the Great Plains in interglacial and post-Wisconsin periods through low passes across the Continental Divide in Montana and Wyoming; glacial periods split these taxa into two groups, separated by the Rocky Mountains, and promoted speciation in some cases.

Discontinuous ranges of certain mammals reflect the presently discontinuous distribution of both coniferous and deciduous forests in the Northern Great Plains region. Species inhabiting boreal and montane coniferous forests now occur on isolated mountain ranges such as the Bighorns and Black Hills, and suggest that coniferous forest connected these areas with the Rocky Mountains to the west and the boreal taiga to the northeast during Late Wisconsin and early post-glacial times. The absence of many boreal and montane species on these isolated ranges may be a function of their relatively small areas and lack of environmental diversity on one hand, and of the influence of post-glacial climates that were warmer or drier (or both) than at present, on the other. Similarly, some species of the eastern deciduous forests are now discontinuously distributed in patches of suitable habitat along rivers in the western part of the Great Plains. Post-glacial conditions moister than at present probably permitted major westward expansion of deciduous forest; a drier phase that followed caused the forest to be broken up and restricted to favorable microenvironments. More recent reinvasion of the eastern part of the Great Plains by many mammals of the deciduous forest is further postulated; reinvasion was not possible for boreal and montane species, because suitable habitat failed to reappear in the northern and western parts of the region.

INTRODUCTION

The mammalian fauna of the Northern Great Plains is, with some exceptions, reasonably well known in terms of the distributions and habitats of the 113 species of mammals known to occur there—see especially Bailey (1927); Warren (1942); Over and Churchill (1945); Jones (1964); Long (1965); and Hoffmann and Pattie (1968). No analysis of the distributional patterns of Recent mammals has been published previously for this region, which, aside from widespread species and those typical of the interior grasslands of North America, contains elements associated with the Rocky Mountains and Great Basin to the west, boreal areas to the north, deciduous forest to the east, and Chihuahuan and Sonoran deserts to the southwest, as well as a few invaders of tropical affinities from the south.

A biogeographic analysis of the Northern Plains fauna may provide evidence concerning the nature of Pleistocene and early Holocene environments. On the other hand, data on late-glacial and post-glacial environments may assist in the interpretation of geographic and ecologic distributions of Recent mammals.

Circular reasoning must be avoided, however, when dealing with mammalian distributions and Pleistocene-Holocene environments in this manner. Consider, for example, the presence of a relict population of a boreal species, such as the red squirrel *(Tamiasciurus hudsonicus)*, in a small area on the Great Plains that is surrounded by habitat unsuitable to survival of the squirrel and disjunct from the main range of the species. For *Tamiasciurus*, which lacks means of long-distance dispersal of the kind available to winged or wind-blown organisms, such a relict population constitutes prima facie evidence that the environment in the presently uninhabited area intervening between the relict population and the contiguous range of the species was at some time in the past suitable for occupancy, presumably by being cooler or moister, or both. Similarly, palynological or pedological evidence for a coniferous forest in the area in question also is clear evidence for a different climate at some time in the past. In the absence of mammalian fossils of known stratigraphy, and of time control of the pollen or soil evidence, it is unwarranted to assume that the coniferous forest suggested by one set of data provided the suitable habitat necessary for the boreal species to disperse to the point occupied by the present relict population. The squirrel may have occupied that place prior to the occurrence of the coniferous forest documented by pollen, or it may have dispersed through suitable habitat that developed later in time. If, however, stratigraphic evidence from fossil mammals, or radiometrically determined pollen or soil dates are available, the temporal correspondence between the past vegetation and dispersal of the mammalian species is on more secure ground, although prior or subsequent dispersal in response to other, unrecorded vegetational events still cannot be ruled out.

A second sort of zoogeographic argument is based upon groupings of ecogeographically similar species. For example, mammals that occur in the Great Basin and that are adapted to life in arid-steppe and sagebrush-semi-desert environments may be associated together in a single "Great Basin" unit. This sort of approach has been used by botanists and ornithologists (Hultén, 1937; Stegmann, 1938; Udvardy, 1963). Such groupings are subjective, unless arrived at by newer techniques using computers (Smith and Fisher, this volume), but not necessarily invalidated thereby. The occurrence of a species belonging to such an ecogeographic group in another geographic area, such as a Great Basin species in the Northern Great Plains, is evidence that the species dispersed from its original geographic center to the other area at some time in the past. It may be argued that the geographic range of the mammal, of itself, provides only equivocal evidence for area of origin and direction of dispersal, but its ecology provides further proof. The sagebrush vole, a Great Basin species, is closely restricted to habitats dominated by various species of sagebrush (*Artemesia*), not only in the Great Basin portion of its range, but also on the Great Plains. The fact that sagebrush is widespread and dominant over most of the Great Basin, but scattered and not dominant in most of the Northern Great Plains, strongly suggests that both the vole and its vegetational habitat originated in the Great Basin and subsequently dispersed to the Great Plains. Even so, it could be postulated that both the sagebrush and the vole were initially not members of the Great Basin biota, but dispersed there from the Great Plains where they persist only as relict populations; a final appeal from this possibility can be based only upon dated paleontological evidence.

It seems to us desirable, in the discussion that follows, to use these two sorts of biogeographic evidence, even though they are not infallible.

PRESENT ENVIRONMENTS OF THE NORTHERN GREAT PLAINS

We arbitrarily define the Northern Great Plains as the region north of 40° north latitude, west of the Missouri, Sioux, and Red rivers, east of the Continental Divide and its contiguous mountain ranges, and south of the aspen parklands—the ecotone between the boreal coniferous forest (or taiga, *fide* Hoffmann, 1958) and the grasslands of the plains. We further restrict ourselves to the most recent major period of Pleistocene glaciation (Wisconsin) and subsequent Holocene time, with one exception.

Within the limits of the area defined above, there presently exist four major biomes (Shelford, 1963): (1) grassland, or steppe; (2) deciduous forest; (3) coniferous forest; and (4) alpine. The latter two are included by virtue of the existence of isolated ranges of hills or mountains in the Great Plains area, which are surrounded by grasslands (see map, Fig. 1). Deciduous forests are found mostly along the eastern fringe of the area and in river valleys. Grassland plant associations at present far exceed all the others in areal extent.

STEPPES

Four main types of steppe vegetation have been recognized in the Northern Great Plains (Weaver and Albertson, 1956; Shelford, 1963). On the east, merging with the deciduous forest ecotone, is tall-grass steppe, also called "true prairie." Farther west is a zone of mixed-grass steppe, in the midst of which lie the Sandhills of Nebraska, a unique area that combines tall and mixed grasses with, in certain restricted areas, deciduous and coniferous forest elements. Still farther west is the short-grass steppe, which abuts on the montane coniferous forest in the foothills of the Rocky Mountains. Finally, bunch-grass steppe occurs east of the Continental Divide in central and southwestern Montana, and perhaps in central Wyoming. On the north, tall-, mixed-, and short-grass steppes form a complex ecotone, the "Aspen parkland," with the taiga.

CONIFEROUS FORESTS

The two major divisions of coniferous forest in the Northern Great Plains region are the taiga, or boreal coniferous forest, and the Rocky Mountain coniferous forests (Oosting, 1950; Shelford, 1963). The boreal taiga is questionably represented in the Turtle Mountains of North Dakota. This low range (to 2500 feet in elevation) supports a forest of aspen *(Populus tremu-*

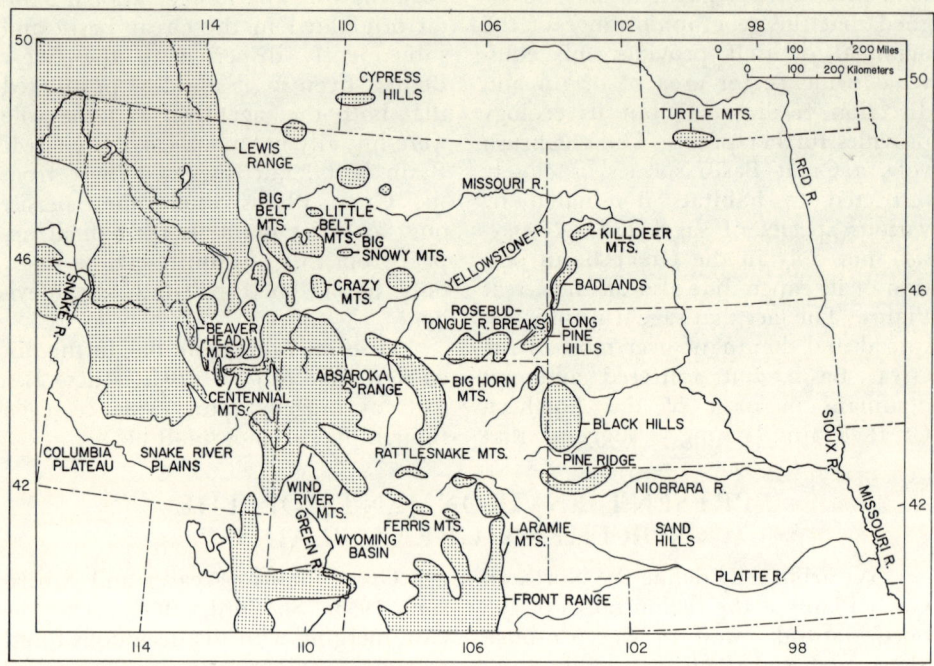

Fig. 1. Map of the Northern Great Plains, central and northern Rocky Mountains, and Great Basin region, showing places referred to in the text. Stippled areas indicate approximate distribution of montane areas.

loides), balsam poplar *(P. balsamifera)*, and paper birch *(Betula papyrifera)*, successional and subdominant species in the taiga. However, coniferous elements of the taiga are absent (Rudd, 1951). Rocky Mountain coniferous forest covers most of the isolated mountain ranges of Montana and Wyoming. Lower elevations support yellow pine *(Pinus ponderosa)* and occasionally Douglas-fir *(Pseudotsuga menziesii)* as dominants. At higher elevations, subalpine forests of spruce *(Picea* sp.) and fir *(Abies lasiocarpa)* occur (Daubenmire, 1943). Not all of the isolated mountain ranges are sufficiently high in elevation or latitude to support subalpine forests, but these do occur in the Big Belt, Little Belt, Big Snowy, and Crazy mountains of central Montana, and in other ranges to the southwest (but east of the Continental Divide). In Wyoming, subalpine forests are widespread in the Bighorn Mountains and occur also in smaller mountainous areas, including the northern, isolated ranges of the Laramie Mountains (Laramie Peak, for example), the Ferris Mountains (and probably the adjacent Green and Seminoe mountains), and possibly the Rattlesnake Range (Cary, 1917; Findley and Anderson, 1956). Montane coniferous forests occupy belts below subalpine forests in all of these ranges, and are the only coniferous forests in some others, including the Rosebud and Tongue River breaks and the Long Pine Hills of southeastern Montana (and the associated woodlands of adjacent South Dakota); the Killdeer Mountains and Badlands of western North Dakota; the Fox Ridge and Badlands area of South Dakota; the Pine Ridge and Niobrara escarpments of South Dakota and Nebraska, and the Wildcat and Bighorn ridges and other areas in the latter state (Wells, 1965). A number of other small, low ranges in Wyoming and central Montana are poorly known, but probably support only montane coniferous forests.

Two large isolated ranges stand out as special cases. The Black Hills and adjacent Bear Lodge Mountains of South Dakota and Wyoming contain a mixture of floral associations. The coniferous forest of the Black Hills has a montane belt of yellow pine with Rocky Mountain affinities, and a subalpine belt containing white spruce *(Picea glauca)*, paper birch, and various understory plants with northern affinities. In drier areas, eastern deciduous elements such as bur oak *(Quercus macrocarpa)*, green ash *(Fraxinus pennsylvanica)*, American elm *(Ulmus americana)*, and ironwood *(Ostrya virginiana)* occur (Wetmore, 1967). Farther north, the Cypress Hills of Saskatchewan and Alberta have coniferous forests of lodgepole pine *(Pinus contorta* var. *latifolia)*, white spruce *(P. glauca* var. *albertiana)*, and paper birch *(B. papyrifera* var. *subcordata)*. At the varietal level these taxa indicate northern Rocky Mountain affinities (Newsome and Dix, 1968), thus differing from other isolated subalpine spruce forests, which contain Englemann spruce *(Picea engelmanni)*, of the central Rocky Mountains.

ALPINE

A few peaks attain sufficient altitude to project above timberline, and are capped by alpine tundra. These include small areas in the Big Belt, Little Belt, Big Snowy, and Crazy mountains, and a more extensive area in the Big Horn Mountains. Other areas, such as Laramie Peak, approach timberline, but do not have an alpine belt at present.

DECIDUOUS FORESTS

Along the southeastern edge of the Northern Great Plains, as here defined, is the fringe of oak-hickory forest and woodland, with some oaks *(Q. borealis, Q. macrocarpa)* and basswood *(Tilia americana)* occurring farther north (Küchler, 1964; Shelford, 1963). This, along with the adjacent deciduous forest-grassland ecotone, provides habitats for a number of typical deciduous forest mammals. Another sort of deciduous forest, the so-called northern floodplain forest, dominated by cottonwoods *(Populus* sp.), willows *(Salix* sp.), and elms, forms gallery forest and woodland stretching far westward in the major river valleys. In some cases these narrow, dendritic deciduous forests stretch more or less continuously across the Northern Great Plains, as along the upper Missouri and Yellowstone rivers, providing narrow avenues of westward dispersal for eastern mammals.

LATE PLEISTOCENE AND HOLOCENE CLIMATIC AND VEGETATIONAL PATTERNS

Pollen profiles, plant macrofossils, and paleosols and other geological data provide means of determining the soils and vegetations at various points in the Northern Great Plains during late-glacial and post-glacial times. From these data, climatic inferences can be made. Also, recent attempts to reconstruct regional climatic complexes, or "climata" (Bryson and Wendland, 1966; Bryson *et al.*, this volume), independently derived from meteorological principles, provide additional data. The following summary draws upon these to sketch the general features of climate and vegetation in the Northern Great Plains from the time of the Late Wisconsin glacial maximum until the present. Chronology and terminology follow Bryson *et al.* (this volume).

Full-glacial (to approximately 13,000 BP).—Dillon (1956), Dorf (1959), and others have implied that open tundra or tundra-like vegetation occupied the Northern Great Plains south of the continental glacier, with "Canadian" or "subarctic" conditions, still presumably more or less treeless, in Wyoming and Nebraska. However, more recent studies (Kapp, Wright, both this volume) indicate that a boreal spruce forest occupied the Northern Great Plains in parts of Kansas, Nebraska, and South Dakota (and possibly also in parts of North Dakota) back at least to 23,000 BP. Whether a zone of tundra existed between the taiga and the glacial front has not been proved, but possibly such a zone did exist, by analogy with southern Minnesota, where a tundra vegetation was established following glacial retreat (12,650±350 BP, Watts, 1967). Typical steppe vegetation may have been absent during this time from the Northern Great Plains, or it may have existed in arid basins in the rain shadow of the Rocky Mountains such as in the Wyoming Basin (Bryson and Wendland, 1966). Also, an open landscape with sparse vegetation extending from western Nebraska northward to the ice edge (Wright, this volume) is postulated. Farther to the south, steppe or pine savannah supporting a patchy steppe vegetation may have existed in eastern Colorado and western Kansas (Bryson et al., this volume), as is postulated by Martin and Mehringer (1965) for eastern New Mexico and western Texas. Whereas cool summers are suggested by the spruce forest in the Northern Great Plains, winters may not have been as cold as at present, due to the barrier of the continental glacier (Bryson and Wendland, 1966).

Late-glacial (13,000 to 10,500 BP). —During this period the mean position of the continental ice had retreated northeastward, with the southern edge exhibiting minor phases of retreat and readvance (Two Creeks, Valders, and others—see Terasmae, 1967; Cushing, 1967). A boreal spruce forest still occupied much of the Northern Plains, probably including those parts of Nebraska and the Dakotas that had been treeless at the height of Wisconsin glaciation (Wright, this volume; Watts and Wright, 1966). As in earlier Wisconsin time, the question of tundra between the forest and the ice edge cannot yet be answered. Steppe, or a mosaic or savannah of pine and steppe vegetation, still was mostly south of the Northern Great Plains, though perhaps occupying southwestern Nebraska, northeastern Colorado, and the Wyoming Basin. Summers were cooler, and winters probably warmer, than at present.

Pre-Boreal (10,500 to 9140 BP) *and Boreal* (9140 to 8450 BP).—About 10,500 BP there was a sudden shift in pollen spectra from the Northern Great Plains and adjacent areas (Ogden, 1967, and others) suggesting an abrupt change in climate away from the atmospheric circulation pattern of the late-glacial (Bryson and Wendland, 1966). This resulted in replacement of the boreal spruce forest by steppes throughout the Northern Great Plains (Wright, 1968a, 1968b) as far north as southwestern Manitoba. The continental ice sheet melted back rapidly under the influence of this climatic change, which may be considered to mark the end of the Pleistocene and start of the Holocene, and the Pre-Boreal period. Between 10,500 and 9000 BP a low corridor between the Arctic and the Northern Great Plains was gradually opened by retreat of the ice, and cold Arctic air was able to flow southward onto the plains, with increasing frequency and intensity as the corridor widened (Bryson et al., this volume) to initiate the Boreal period. Grassland could then have spread northwestward south of the ice edge behind the boreal forest, which was colonizing newly exposed soil (perhaps following a pioneer herbaceous "tundra" stage). Climate during this time was probably increasingly continental with warmer summers, but colder winters. Strong westerlies would have spread a wedge of dry air eastward to a line running through Manitoba, Minnesota, Wisconsin, and Illinois (Bryson and Wendland, 1966).

Atlantic (8450 to 4680 BP).—The Cochrane readvance may have resulted in a slowing or temporary cessation of melting of the continental glacier; but this was shortlived, and the glacier must have continued to melt rapidly *in situ* after about 8000 BP (Bryson and Wendland, 1966). The thinning of the continental ice permitted more southward flow of Arctic air, and strong westerlies continued, resulting in a maximum eastward penetration of grassland about 7000 BP. However, warmer summers permitted an expansion of species of the coniferous and deciduous forests northward of their present limits. Later in this period, the Northern Plains may have become wetter.

Sub-Boreal (4680 to 2690 BP).—By this time steppes had retreated from their easternmost maximum to approximately their present position. A southward shift in the winter, and especially summer, Arctic frontal zone about 3500 BP is suggested by a southward shift in the northern and southern limits of the boreal forest at this time. A cooling of the climate of the Northern Great Plains from the Atlantic maximum may have accompanied this shift. At the end of the Sub-Boreal the climate may have been similar to that at present.

Sub-Atlantic (2890 to 1690 BP).—Bryson and Wendland (1966) suggested that a shift of the upper-air anticyclonic eddy northeastward from the Great Basin would have resulted in a considerably wetter climate in the Northern Plains region during this period.

Scandic (1690 to 1100 BP).—During this transition period, conditions began to return toward those of early Atlantic time, becoming warmer and probably drier on the Northern Great Plains.

Neo-Atlantic (1000 to 760 BP).—Warmer climate continued during this period, but greater moisture probably was available in the region of the Great Plains (Bryson and Wendland, 1966).

Pacific (760 to 410 BP).—Around the year 1200 A.D., a shift to drier conditions on the Northern Great Plains is noted both in pollen spectra and archaeological data (Baerreis and Bryson, 1967).

Neo-Boreal (410 to 115 BP).—Colder, moister conditions returned to the Northern Plains in this period.

Recent (115 BP to present).—The Neo-Boreal was brought to a close about the year 1850 A.D. by an increase in strength of dry westerlies, resulting in warmer, drier conditions on the Northern Great Plains, and our present climate.

This summary, based on the chronology and terminology advocated by Bryson *et al.* (this volume), gives a progressively more detailed picture of climatic and vegetational fluctuations in the Northern Plains region. The greater frequency of climatic shifts as one approaches the present is a result, however, of increase in availability of data, and one must assume similar low-magnitude shifts to have occurred in the longer earlier periods. The traditional post-glacial climatic sequence of Hypsithermal, with warm, wet (Climatic Optimum) and warm, dry (Xerothermic) phases, followed by return, through a cooler, moister Sub-Atlantic period to the Recent climate, can now be replaced by the more detailed model. Chronologically, the Climatic Optimum is approximately equivalent to the last half of the Atlantic, and the Xerothermic to the Sub-Boreal. The apparent contradictions in this correlation are probably accountable in terms of regional differences in climate. Thus, low-index summer circulation might bring drier conditions to the eastern deciduous forest, while providing more summer rainfall to the Northern Great Plains (Bryson and Wendland, 1966, fig. 13). Since the "Xerothermic" was originally defined by replacement of mesic by more xeric tree species in areas east of the Mississippi, the Sub-Boreal period might then be a period of greater dryness in one area, and more moisture in an adjacent area. These regional variations add to the complexities of interpreting past environments and their influence on mammalian distributions.

PAST MAMMALIAN FAUNAS ON THE PLAINS

The few Wisconsin faunas that have been published on indicate only in the most general way the past distributions of the species comprising the present mammalian fauna of the Northern Great Plains (see Hibbard, this volume). Many species ranged far south of their present southern limits during and after the Wisconsin, and it thus is appropriate to examine first Wisconsin faunas from the Southern Great Plains.

Full-glacial.—The full-glacial period is represented by a fauna from central Texas dated 25,000 to 40,000 BP (Slaughter and Ritchie, 1963). It is a typical steppe fauna in which the cotton rat *(Sigmodon hispidus)* is represented, but it also includes the black-tailed prairie dog *(Cynomys ludovicianus)* and the prairie vole *(Microtus ochrogaster),* the present southern limits of which are farther west and north, respectively. Another fauna (Dalquest, 1962), which is undated and perhaps earlier, is generally similar but included *Microtus pennsylvanicus* and *Blarina brevicauda,* the present limits of which are farther north and east, respectively. A third fauna, also undated, but probably "between 20,000 and 30,000" BP (Dalquest, 1964) contained several plains species and also *Microtus pennsylvanicus* and remains of the bog lemming, *Synaptomys cooperi.*

In the same area as the two faunas last mentioned, but dated at 16,775 ±565 BP, is a fauna that included *Blarina brevicauda, Microtus pennsylvanicus, Synaptomys cooperi, Sigmodon hispidus,* and possibly *Microtus ochrogaster* and *Spermophilus richardsonii* (Dalquest, 1965). However, also included among the species from this site were *Sorex cinereus, S. palustris,* and *Thomomys talpoides,* the present ranges of which lie far to the northwest. This fauna indicates that full-glacial conditions still existed at this time on the Southern Plains, and that conditions probably were somewhat more boreal than 10,000 years earlier. However, the presence of *Sigmodon* and *Oryzomys* does not suggest extremely cold winters, as was also true of the earlier Texas deposits.

In contrast to the Texas faunas, the full-glacial Jones fauna (Hibbard and Taylor, 1960) from southwestern Kansas was composed almost solely of steppe or savannah species, lacking the mammals associated with eastern deciduous forests and also southern elements. However, some boreal species *(Sorex cinereus, Microtus pennsylvanicus)* were present, as in the Texas sites. Mid-Wisconsin faunas from Nebraska have been incompletely reported, but in addition to steppe species *(Cynomys, Antilocapra),* there are boreal-forest *(Rangifer)* and shrub-tundra *(Ovibos)* mammals present (Schultz et al., 1951; Banfield, 1962; Hibbard et al., 1965). These faunas, when compared with the climatic and vegetational data summarized above, strongly suggest that during the Wisconsin glacial period the mammals of the present Northern Great Plains fauna occurred under steppe or savannah conditions from southern Nebraska, western Kansas, and eastern Colorado into central Texas and eastern New Mexico (excluding the Llano Estacado plateau—see Wendorf, 1961), and probably even farther south. The area of the present Northern Plains was occupied by a boreal woodland biota or by cold loess steppe and tundra.

Late-glacial.—During the late-glacial period (13,000 to 10,500 BP), there is evidence of continuing cool climates on the Southern Great Plains (summers, at least) by the continued presence of *Sorex cinereus* and *Microtus pennsylvanicus.* *Mustela erminea, Spermophilus franklinii,* and *Synaptomys cooperi* were present and also attest to boreal influences (Slaughter and Hoover, 1963; Lundelius, 1967). At the same time, the presence of *Sig-*

modon indicates probable absence of harsh winters, and elements of the eastern deciduous forest continued to be represented. Mammals of the late-glacial in southwestern Kansas indicate continuing cool steppe conditions (*Sorex cinereus, Sorex palustris, Microtus pennsylvanicus, Synaptomys cooperi, Zapus hudsonius*), but *Blarina* represents a deciduous forest element not present in full-glacial times (Schultz, 1967). Published data from Nebraska for "Late-Wisconsin" indicate the continued presence of steppe species and the apparent disappearance of tundra-related mammals, but details are thus far lacking (Schultz and Frankforter, 1948).

Farther to the west, late-glacial faunas from eastern Wyoming reveal a complex of steppe, taiga, and tundra species, probably overlapping in time due to the high relief offered by the Laramie Mountains. It is possible to imagine steppes at the base of mountains, with *Antilocapra, Lagurus,* and *Mustela nigripes;* then montane and subalpine forest, with *Microsorex, Clethrionomys,* and *Martes;* and finally alpine tundra, with *Ochotona, Oreamnos,* and *Dicrostonyx*. An eastern species, *Cryptotis parva,* also was present (Guilday *et al.,* 1967; Anderson, 1968). A similar faunal complex from west of the Continental Divide in Idaho is dated 10,370 to 11,580 BP (Guilday and Adam, 1967).

Holocene.—Other faunas, from Pre-Boreal to early Atlantic time from the Southern and Central Great Plains, suggest a warmer or drier (or both) climate, at least in summer. By that time *Sorex cinereus* and *Microtus pennsylvanicus* no longer were represented in Texas faunas, although *Synaptomys* persisted until at least 7300 BP (Lundelius, 1967). At this same time a mixed faunal complex can be identified, including species of the subalpine and montane forests and also the arid steppe, at Burnet Cave in the mountains of southeastern New Mexico (Murray, 1957).

Numerous cave faunas from central Texas indicate a continued trend to a warmer and drier climate on the Southern Great Plains from later Atlantic time to about 1000 BP, when a modern faunal composition was achieved (Lundelius, 1967). Sampling frequency provided by the faunas may, of course, be too low to illustrate minor oscillations in climate. Unfortunately, the Northern Plains region has not yet yielded faunas from this period, and interpretations of faunal and climatic shifts must be based on indirect biogeographic evidence.

Farther to the east there is scattered fossil evidence for an eastward extension of plains mammals during the late Pleistocene or Holocene. *Spermophilus tridecemlineatus* has been found in several sites in Pennsylvania and Virginia, and Guilday *et al.* (1964) believed dispersal there of this species to have been in late-glacial times (after 11,000 BP). *Microtus ochrogaster* is recorded in northern Michigan at 750 BP (Pruitt, 1954), and the spotted skunk (*Spilogale putorius interrupta*) occurred in Illinois between 6500 and 3500 BP (Parmalee and Hoffmeister, 1957; Parmalee, 1967, 1968). However, most interpretations of climatic oscillations in this period are based on distributional data (Smith, 1957; Jones, 1964).

PRESENT DISTRIBUTION PATTERNS

STEPPE SPECIES

Thirteen species that occur on the Great Plains are intimately associated with the interior grasslands of central North America (Table 1). A few are endemic to the region, but most occur also to the west or south of the plains, and several (*Spermophilus tridecemlineatus* and *Microtus ochrogaster,* for example) range far to the east. In addi-

TABLE 1. Mammals of the Northern Great Plains, Listed by Faunal Units as Discussed in the Text

Steppe Species (13)
    Lepus townsendii, White-tailed jackrabbit
    Cynomys ludovicianus, Black-tailed prairie dog
    Spermophilus franklinii, Franklin's ground squirrel
    Spermophilus tridecemlineatus, Thirteen-lined ground squirrel
    Geomys bursarius, Plains pocket gopher
    Perognathus fasciatus, Olive-backed pocket mouse
    Perognathus flavescens, Plains pocket mouse
    Perognathus hispidus, Hispid pocket mouse
    Reithrodontomys montanus, Plains harvest mouse
    Microtus ochrogaster, Prairie vole
    Vulpes velox, Swift fox
    Mustela nigripes, Black-footed ferret
    *Spilogale putorius (subspecies *interrupta*), Spotted skunk
Invaders from the Southwest (9)
    Myotis thysanodes, Fringe-tailed bat
    Euderma maculatum, Spotted bat
    Sylvilagus audubonii, Desert cottontail
    Lepus californicus, Black-tailed jackrabbit
    Spermophilus spilosoma, Spotted ground squirrel
    Perognathus flavus, Silky pocket mouse
    Dipodomys ordii, Ord's kangaroo rat
    Reithrodontomys megalotis, Western harvest mouse
    Onychomys leucogaster, Northern grasshopper mouse
Invaders from the Great Basin (8)
    Sorex merriami, Merriam's shrew
    Sorex preblei, Preble's shrew
    Sylvilagus idahoensis, Pygmy rabbit
    *Eutamias minimus, Least chipmunk (karyotype B)
    Spermophilus richardsonii, Richardson's ground squirrel
    Perognathus parvus, Great Basin pocket mouse
    Lagurus curtatus, Sagebrush vole
    *Spilogale putorius (subspecies *gracilis*), Spotted skunk
Boreal or Montane Species (37)
    Sorex arcticus, Arctic shrew
    Sorex cinereus, Masked shrew
    Sorex palustris, Water shrew
    Sorex vagrans, Vagrant shrew
    Sorex nanus, Dwarf shrew
    Microsorex hoyi, Pygmy shrew
    Myotis evotis, Long-eared bat
    Myotis volans, Long-legged bat
    Plecotus townsendii, Western big-eared bat
    Ochotona princeps, Pika
    Sylvilagus nuttallii, Mountain cottontail
    Lepus americanus, Snowshoe hare
    Eutamias amoenus, Yellow-pine chipmunk
    *Eutamias minimus, Least chipmunk (karyotype A)
    Marmota flaviventris, Yellow-bellied marmot
    Spermophilus lateralis, Golden-mantled ground squirrel
    Tamiasciurus hudsonicus, Red squirrel
    Glaucomys sabrinus, Northern flying squirrel
    Thomomys talpoides, Northern pocket gopher
    Neotoma cinerea, Bushy-tailed woodrat
    Clethrionomys gapperi, Red-backed vole
    Phenacomys intermedius, Montane heather vole
    Microtus longicaudus, Long-tailed vole
    Microtus montanus, Montane vole
    Microtus pennsylvanicus, Meadow vole

* Two species are listed in each of two groups; see text for discussion.

TABLE 1. (Continued)

    Arvicola richardsoni, Water vole
    Synaptomys cooperi, Southern bog lemming
    Zapus hudsonius, Meadow jumping mouse
    Zapus princeps, Western jumping mouse
    Martes americana, Marten
    Mustela erminea, Ermine
    Mustela nivalis, Least weasel
    Gulo gulo, Wolverine
    Lynx canadensis, Lynx
    Alces alces, Moose
    Rangifer tarandus, Caribou
    Ovis canadensis, Bighorn sheep

Deciduous Forest Species (18)
    Didelphis marsupialis, Opossum
    Blarina brevicauda, Short-tailed shrew
    Cryptotis parva, Least shrew
    Scalopus aquaticus, Eastern mole
    Myotis keenii (subspecies *septentrionalis*), Keen's bat
    Pipistrellus subflavus, Eastern pipistrelle
    Lasiurus borealis (subspecies *borealis*), Red bat
    Nycticeius humeralis, Evening bat
    Sylvilagus floridanus, Eastern cottontail
    Tamias striatus, Eastern chipmunk
    Marmota monax, Woodchuck
    Sciurus carolinensis, Gray squirrel
    Sciurus niger, Fox squirrel
    Glaucomys volans, Southern flying squirrel
    Peromyscus leucopus, White-footed mouse
    Neotoma floridana, Eastern woodrat
    Microtus pinetorum, Woodland vole
    Urocyon cinereoargenteus, Gray fox

Invaders from the South (3)
    Tadarida brasiliensis, Free-tailed bat
    Dasypus novemcinctus, Nine-banded armadillo
    Sigmodon hispidus, Hispid cotton rat

Widespread Species (27)
    Myotis leibii, Masked bat
    Myotis lucifugus, Little brown bat
    Eptesicus fuscus, Big brown bat
    Lasionycteris noctivagans, Silver-haired bat
    Lasiurus cinereus, Hoary bat
    Castor canadensis, Beaver
    Peromyscus maniculatus, Deer mouse
    Ondatra zibethicus, Muskrat
    Erethizon dorsatum, Porcupine
    Canis latrans, Coyote
    Canis lupus, Wolf
    Vulpes vulpes, Red fox
    Ursus americanus, Black bear
    Ursus arctos, Grizzly bear
    Procyon lotor, Raccoon
    Mustela frenata, Long-tailed weasel
    Mustela vison, Mink
    Taxidea taxus, Badger
    Mephitis mephitis, Striped skunk
    Lutra canadensis, Otter
    Felis concolor, Cougar
    Lynx rufus, Bobcat
    Cervus canadensis, Wapiti
    Odocoileus hemionus, Mule deer
    Odocoileus virginianus, White-tailed deer
    Antilocapra americana, Pronghorn
    Bison bison, Bison

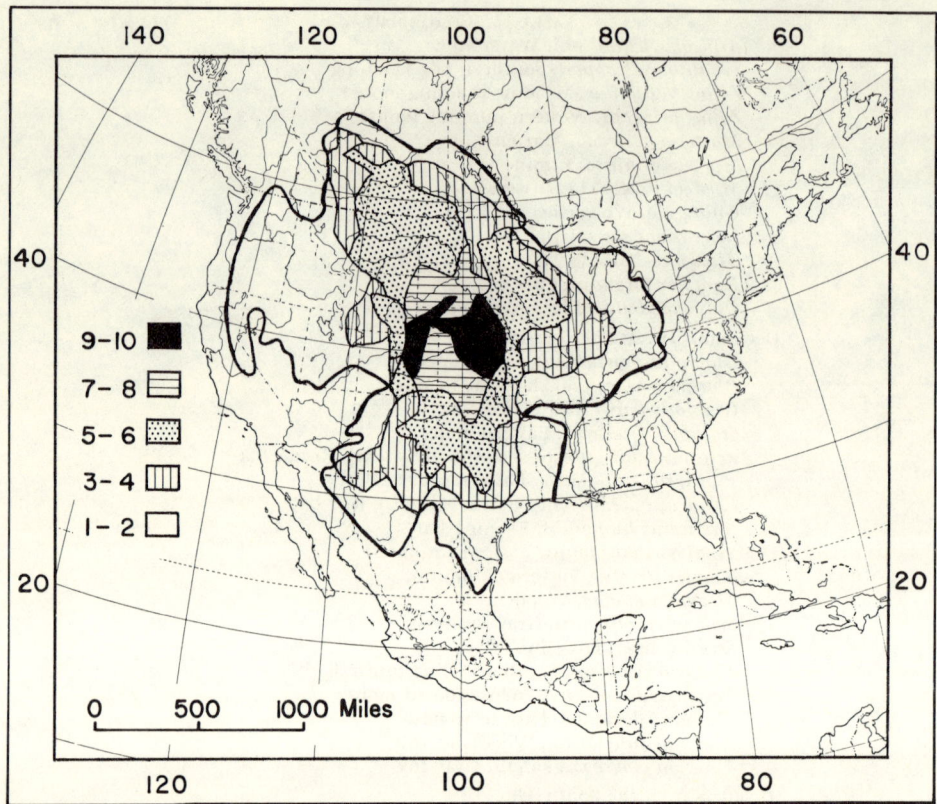

FIG. 2. Superimposed distribution patterns for 10 species of steppe mammals (carnivores excluded—see Table 1). Center of maximum species diversity is in the Northern and Central Great Plains.

tion to the 13, two species that nominally are classed as members of some other faunal assemblage have well-defined subspecies in the region. These are the ground squirrel *Spermophilus richardsonii* (subspecies *richardsonii*), which is primarily a Great Basin species, and *Spilogale putorius* (subspecies *interrupta*), which may in fact be specifically distinct from the spotted skunks of the western United States. Finally, there are species such as the northern grasshopper mouse *(Onychomys leucogaster)*, pronghorn *(Antilocapra americana)*, and bison *(Bison bison)*, which are typically grassland inhabitants but have distributional patterns that dictate their placement in another faunal grouping.

The existence of true steppe species, with relatively narrow habitat requirements and largely concordant patterns of distribution (see Fig. 2), argues for the continuous existence of a steppe, or at least savannah, environment on the plains in the Late Pleistocene, albeit shifting in location. Judging from paleozoological and botanical evidence (see above), most of the grassland species had more southerly distributions in Late Pleistocene (glacial) times than now. The prairie vole, *Microtus ochrogaster*, and black-tailed prairie dog, *Cynomys ludovicianus*, for example, "left" relics (*M. ludovicianus* of western Louisiana and eastern Texas and *C. mexicanus* in northern Mexico, respectively) as evidence of a more southerly extent of range in that period. The black-footed ferret, *Mustela nigripes*, closely associated geographically with its principal prey species, the black-tailed prairie dog, recently has been reported from a Late

Pleistocene cave fauna in Lehmi County, Idaho, west of its present distributional limits, on the west side of the Continental Divide, and at a place where *Cynomys* was apparently absent (Guilday and Adam, 1967).

Most of the grassland species are typical of the High Plains and areas of short and mixed grasses, rather than of the tall-grass steppe (*Spermophilus franklinii* being the only exception). A relatively large number (nine) are, at the latitude of Nebraska and southern South Dakota, restricted to the west of the Missouri River and its deciduous riparian association. Some of these (*Lepus townsendii, Perognathus fasciatus,* and *Perognathus flavescens*) were able to cross the Missouri farther northward, however, as were certain elements of other faunal units (*Onychomys leucogaster, Antilocapra americana,* and *Odocoileus hemionus* are examples); several then moved southward in the grasslands on the east side of the river. In this manner, some dispersed at least as far south as the northwestern part of the present state of Missouri. The hispid pocket mouse (*Perognathus hispidus*) has, perhaps rather recently, reached the flood plain of the Missouri River in southeastern Nebraska, but evidently has been unable to cross this barrier. It is of note that *P. hispidus* dispersed eastward in late-glacial or early Holocene times, because Oesch (1967) and Parmalee *et al.* (1969) have reported remains of this mouse from Crankshaft Pit, in Jefferson County of eastern Missouri (but south of the Missouri River).

INVADERS FROM THE SOUTH

Three species, the free-tailed bat (*Tadarida brasiliensis*), armadillo (*Dasypus novemcinctus*), and hispid cotton rat (*Sigmodon hispidus*), recently have invaded the southern part of the Northern Plains region (see Fig. 3) and are forerunners of a potentially larger invasion of essentially Neotropical species should the current warming trend on the North American continent continue. Even now, to the south of the region under current consideration, species such as the pygmy mouse (*Baiomys taylori*), rice rat (*Oryzomys palustris*), and ring-tailed cat (*Bassariscus astutus*) may be involved in northward dispersal. At least two other species (here classed as invaders to the Northern Plains from the east, because their arrival in the area evidently was by that route in Late Pleistocene or earliest Holocene times), *Didelphis marsupialis* and *Urocyon cinereoargenteus,* also can be thought of as excurrent species from the Neotropical Region.

Of the three mammals here considered, the hispid cotton rat has had the most spectacular dispersal into the region. As late as 30 years ago this species was known no farther north than central Kansas, but now occurs over much of southeastern Nebraska and in northwestern Missouri.

The free-tailed bat has been taken but three times in Nebraska, in each instance at Lincoln: a male on August 15, 1913; a pregnant female on June 27, 1931; and a male on August 27, 1956. The two males were young of the year that evidently wandered northward prior to seasonal migration, from colonies to the south in which they were reared. Many such individuals are now known from late summer and early autumn recoveries in western and central Kansas (see Jones *et al.*, 1967). The adult female taken in June may well have represented an individual that "overshot" (in spring migration) the northernmost maternity colonies of this bat in extreme south-central Kansas and adjacent Oklahoma.

The armadillo is less well known in the Northern Plains region than the other two species here recorded, but evidently is relatively well established, albeit rare, as far north as the Kansas River and some of its major tributaries to the west in Kansas (Smith and Lawlor, 1964). The as yet unpublished report of a specimen taken several years ago in southern Nebraska has prompted us to include *Dasypus* here.

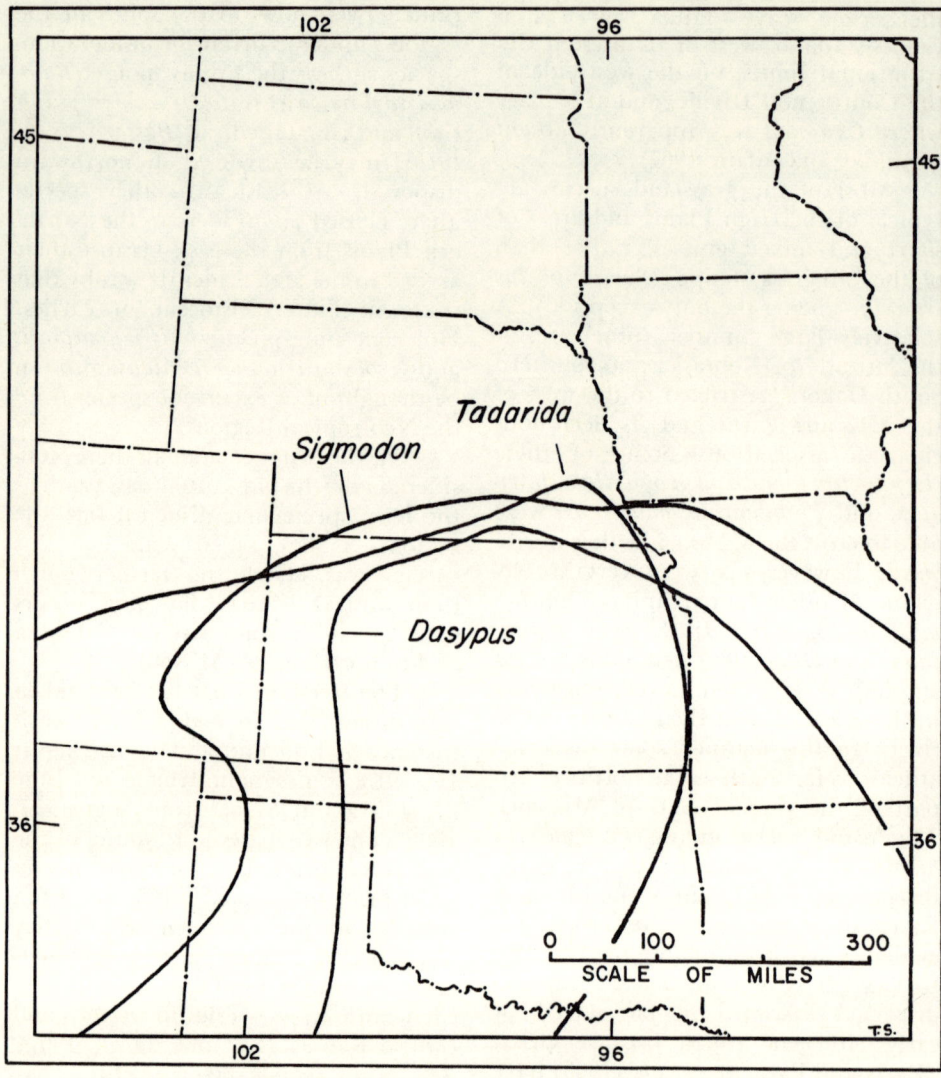

FIG. 3. Northern limits of distribution of three species of Neotropical mammals (*Tadarida brasiliensis, Dasypus novemcinctus, Sigmodon hispidus*).

INVADERS FROM THE SOUTHWEST

Nine species of mammals (Table 1) that occur on the Northern Plains have their origins in the Chihuahuan-Sonoran Region to the southwest (Fig. 4). Two bats, *Myotis thysanodes* and *Euderma maculatum,* are in this grouping; the former is represented by an isolated (and presumably relict) population on the Black Hills, and the latter is known from one specimen from Billings, Montana, and another from Byron, in the Big Horn Basin of Wyoming.

Of the seven terrestrial species, four *(Sylvilagus audubonii, Spermophilus spilosoma, Perognathus flavus,* and *Dipodomys ordii)* are more or less limited to the western part of the plains. *S. spilosoma* and *P. flavus* occur northward in this region only to about the level of the 43rd parallel, whereas the cottontail and kangaroo rat occur on the High Plains well to the north of that latitude. The other three, *Lepus*

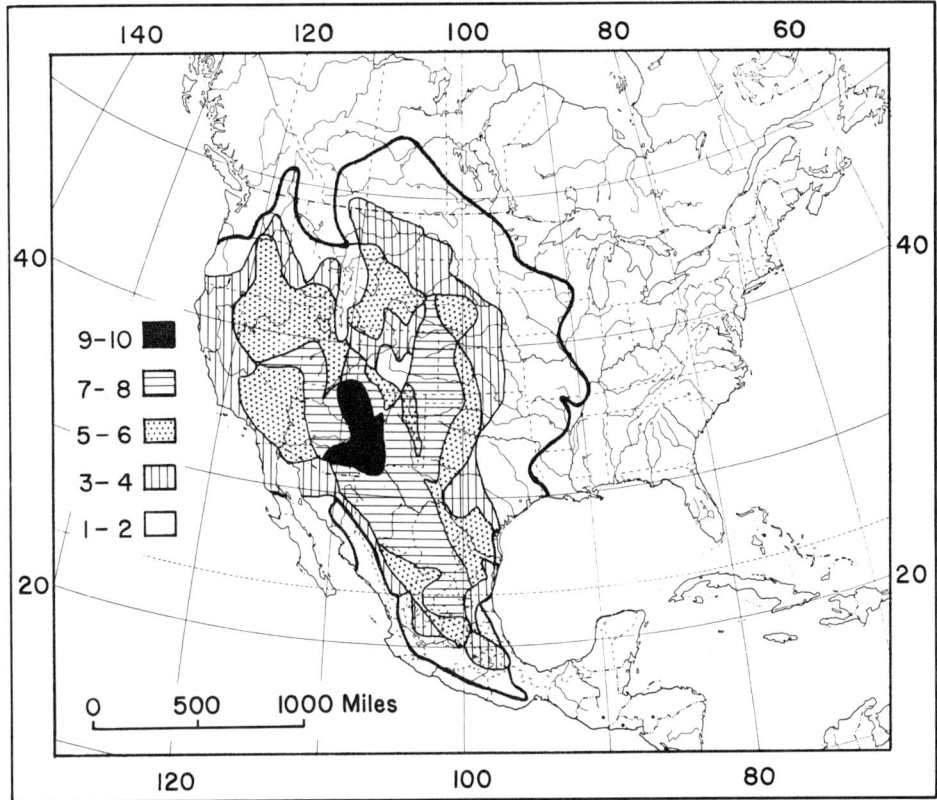

Fig. 4. Superimposed distribution patterns for nine species of southwestern steppe and desert mammals (see Table 1). Center of maximum species diversity is at the northern edge of the Chihuahuan-Sonoran desert regions.

californicus, Reithrodontomys megalotis, and *Onychomys leucogaster,* all have extensive distributions in the grasslands, the harvest mouse having reached eastward to Wisconsin and Indiana, and the grasshopper mouse occurring as far east as northern Iowa and adjacent Minnesota. The last two species mentioned are placed in the southwestern faunal unit, rather than being classed as grassland species, principally because of their extensive distributions in the Southwest and Great Basin regions. Fossil evidence indicates that in the Wisconsin, as at present, many of these species were associated with typical grassland species in Texas.

INVADERS FROM THE GREAT BASIN

Eight Great Plains species (Table 1) appear to be geographically associated primarily with the Great Basin and adjacent cold-desert and semidesert areas (Snake River Plains and Columbia Plateau of Idaho, Oregon, and Washington). They also have an affinity for the sagebrush (*Artemesia tridentata*)–grass communities (Shelford, 1963), and within the Great Plains are restricted to, or most common in, habitats that include sagebrush as an important constituent. The superimposed distributions of these species is shown in Fig. 5. A Great Basin center is apparent, and two principal routes of penetration eastward onto the Northern Great Plains can be seen. One of these is from southeastern Idaho eastward through the Wyoming Basin, and the second from eastern Idaho northeastward through the southwestern corner of Montana. In both of these areas relatively low passes, covered by sage-

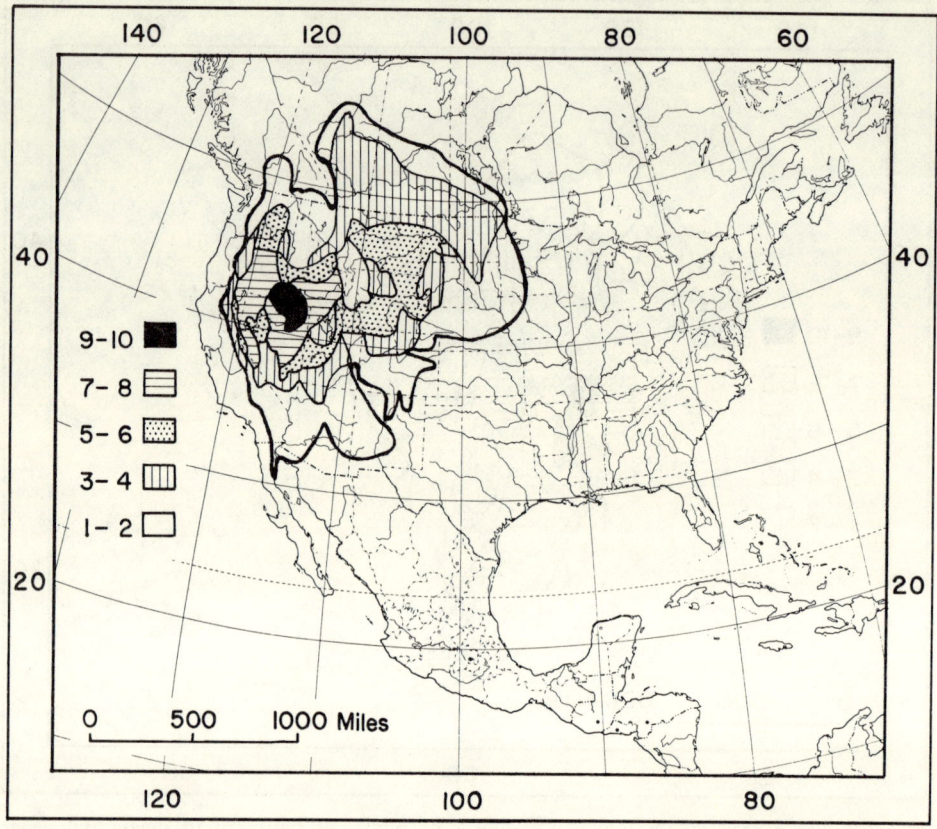

Fig. 5. Superimposed distribution pattern for eight species of Great Basin steppe and desert mammals (see Table 1), plus *Lepus californicus deserticola*. Center of maximum species diversity is in the northern Great Basin.

brush and grass, cross the Continental Divide.

Some of these mammals have only recently spread into the extreme western edge of the Northern Great Plains *(sensu lato)* by crossing the Divide. Thus, in Montana, *Sylvilagus idahoensis, Lepus californicus deserticola,* and *Perognathus parvus* are restricted to southwestern Montana (Fig. 6), and do not occur elsewhere on the plains (except *L. californicus*, which is principally an invader of the Great Plains from the Southwest). In fact, the jackrabbit may possibly have spread into southwestern Montana only within the last half-century (Hoffmann *et al.,* 1969). The Great Basin pocket mouse has penetrated a short distance into Wyoming but has not crossed the Green River into the Wyoming Basin, possibly because of the presence there of *P. fasciatus.* The sagebrush vole, *Lagurus curtatus,* also occurs in southwestern Montana, but is much more widespread, and has also crossed the Wyoming Basin to occupy a large area of the Northern Plains (Fig. 6).

A different pattern is shown by Merriam's shrew, *Sorex merriami,* which is the only shrew normally found in arid sagebrush habitats. It appears to be widespread in the Great Basin, but has been captured so infrequently that the limits of its range cannot be said to be well known. However, its absence from Idaho and southwestern Montana and its occurrence in southwestern Wyoming suggest that this shrew moved through the Wyoming Basin into the Northern Plains region. There, it is widespread in Montana

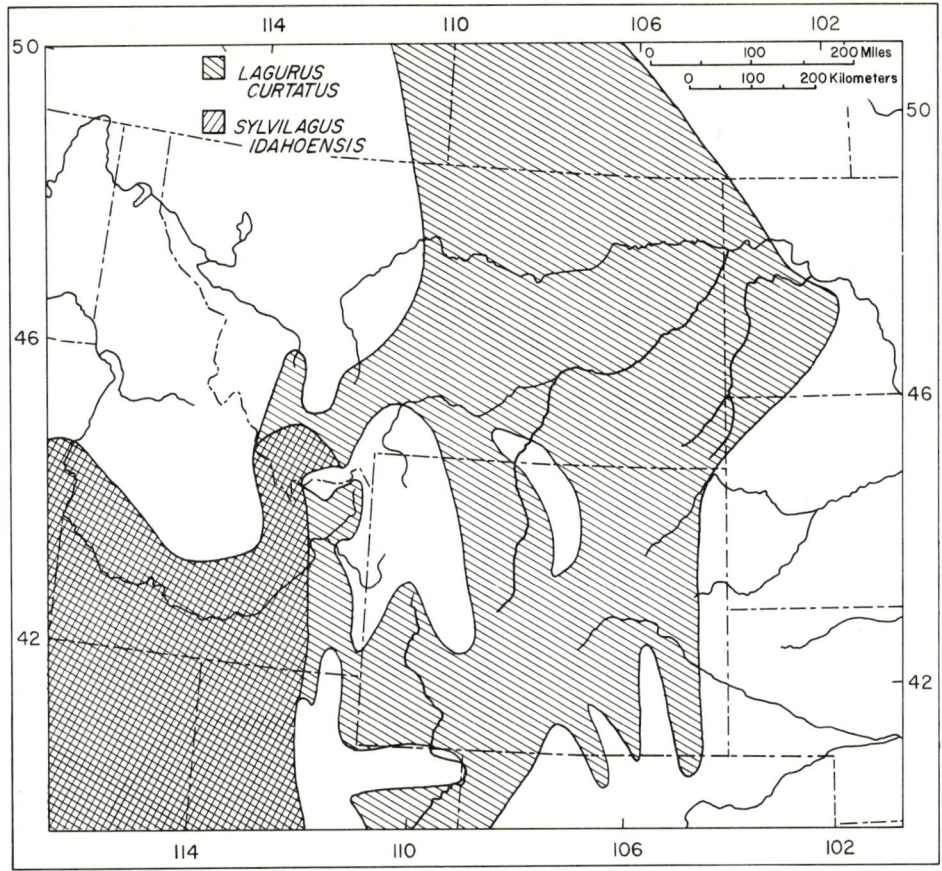

FIG. 6. Two patterns of distribution of Great Basin mammals on the Northern Great Plains. *S. idahoensis* is restricted to southwestern Montana, whereas *L. curtatus* is widespread on the Northern Plains.

and Wyoming, and extends eastward to western Nebraska and the Dakotas, but records are scarce.

Recent chromosome studies (Sutton and Nadler, 1969) suggest that populations of the least chipmunk, *Eutamias minimus*, of the Great Basin *(scrutator, pictus)* and Northern Great Plains *(pallidus, cacodemus)* have a different karyotype ("B") than those of least chipmunks of the boreal forest and most populations in the central and southern Rocky Mountains (karyotype "A"). Subspecies of *E. minimus* with karyotype "B" thus resemble the distribution of *Sorex merriami*.

None of the above species has significantly differentiated in separate Great Basin and Great Plains centers. However, some mammals in this group do exhibit such a pattern. For example, Richardson's ground squirrel *(Spermophilus richardsonii)* presently occupies three disjunct areas: *S. r. nevadensis* and *S. r. aureus* in the Great Basin; *S. r. elegans* in Wyoming and Colorado; and *S. r. richardsonii* in Montana, the Dakotas, and elsewhere on the Northern Great Plains. Only the last-mentioned race is typical of the steppe (Hoffmann and Pattie, 1968); elsewhere the species is mostly confined to semidesert or montane grassland (Durrant and Hansen, 1954).

Also, *Sorex preblei* may be a Great Basin derivative of *Sorex cinereus* that has reinvaded the Northern Plains through southwestern Montana in post-

Wisconsin times. However, as in the case of Merriam's shrew, records are scarce. Finally, the Great Basin spotted skunk, *Spilogale putorius gracilis*, and the Great Plains race, *S. p. interrupta*, may reflect Pleistocene isolation in these two centers.

BOREAL OR MONTANE SPECIES

The Great Plains is bordered on its western edge by the Rocky Mountains and on the north by the boreal forest, or taiga. Some of this perimeter is sharply defined by mountains that rise abruptly from the steppe, as does Colorado's Front Range. Elsewhere, however, isolated mountain ranges occur far out onto the Great Plains, many of them sufficiently elevated to support coniferous forest and even alpine communities. Thirty-seven species of mammals with boreal or montane distributions (Table 1) are found within the grassland, many of them isolated on ecological "islands." Some of the species in this category are represented on the Northern Great Plains by populations of both boreal (north) and montane (west) affinities, and must be thought of in this dual context.

Boreal taxa.—Some mammals having boreal affinities are of more or less peripheral occurrence on the Northern Plains, being restricted to the eastern Dakotas in the northeastern part of the region (Table 2). These include *Sorex arcticus, Sorex palustris hydrobadistes, Microsorex hoyi hoyi, Glaucomys sabrinus canescens, Clethrionomys gapperi loringi, Martes americana americana, Mustela erminea bangsi, Alces alces andersoni,* and *Rangifer tarandus.* Other boreal taxa, however, occur much farther to the south and west. For example, the red squirrels of the Black Hills and the north end of the Laramie Mountains (*Tamiasciurus hudsonicus dakotensis*) appear more closely allied to the reddish races of the boreal forest than to the dark-colored Rocky Mountain subspecies (Fig. 7). Moreover, they occur in pine hills of eastern Montana and the western Dakotas, along a possible corridor connecting the Black Hills with the boreal forest after late-glacial time, when such forests disappeared from the Northern Great Plains (see map, Fig. 1). The boreal forest subspecies of least chipmunk (*Eutamias minimus borealis*) is replaced on the Black Hills by *E. m. silvaticus*, a race once thought to be "indistinguishable" from *borealis* (Howell, 1929). Sutton and Nadler (1969) have shown that Black Hills populations have karyotype "B," as does *Eutamias minimus confinus* of the Bighorn Mountains. In the Laramie Mountains, and elsewhere in the central and southern Rockies, however, races of *E. minimus* with the "A" karyotype occur, suggesting their boreomontane affinities.

The masked shrew, *Sorex cinereus*, meadow jumping mouse, *Zapus hudsonius*, and meadow vole, *Microtus pennsylvanicus*, presently are fairly widespread on the Northern Great Plains, but appear to be allied with boreal rather than montane faunas. Aside from the oft-times localized populations of these species on the plains, meadow voles reach the Bighorns, and jumping mice reach both the Bighorn and Laramie mountains (Fig. 8). There is some question, however, concerning the specific identity of *Zapus* from these two ranges (Paul B. Robertson, personal communication). A puzzling distribution is that of the snowshoe hare, in that *Lepus americanus seclusus* of the Bighorn Mountains is believed to be more closely related to the boreal *L. a. americanus* than to *L. a. bairdii* of the Rocky Mountains (Baker and Hankins, 1950).

The southern bog lemming (*Synaptomys cooperi*) presents a somewhat different problem. Together with its largely allopatric congener, the northern bog lemming (*S. borealis*), it is widely distributed in boreal and mixed forests and in the northern half of the deciduous forest. On the Northern and Central plains its distribution (at the southwestern edge of its range) is discontinuous like that of the meadow vole. It is here restricted to grassy riparian areas (Jones, 1964), probably with cool microclimates.

Montane taxa.—Other mammals inhabiting coniferous forests have penetrated the plains from the west and occur, at least in part, in "islands" on isolated mountain ranges in the Northern Great Plains (Table 2). Some that have reached the Black Hills include

TABLE 2. Boreomontane Faunas of Selected Isolated Mountain Ranges in the Northern Great Plains*

Species	Big Belt and Crazy Mts.	Little Belt and Big Snowy Mts.	Laramie Mountains	Bighorn Mountains	Black Hills	Turtle Mountains
Sorex arcticus	—	—	—	—	—	B
Sorex cinereus	M	M	M	M	B	B
Sorex nanus	M	M	M	—	M	—
Sorex palustris	M	M	M	M	—	—
Sorex vagrans	M	M	M	M	—	—
Microsorex hoyi	—	—	—	—	—	B
Myotis evotis	M	M	M	M	M	—
Myotis volans	M	M	M	M	M	—
Plecotus townsendii	M	M	M	M	M	—
Ochotona princeps	M	M	—	M	—	—
Sylvilagus nuttallii	M	M	M	M	M	—
Lepus americanus	M	M	—	B	—	B
Eutamias amoenus	M	M	—	—	—	—
Eutamias minimus (karyotype A)	—	B	M	—	—	B
Marmota flaviventris	M	M	M	M	M	—
Spermophilus lateralis	M	—	M	—	—	—
Tamiasciurus hudsonicus	M	M	M,B	M	B	B
Glaucomys sabrinus	M	—	—	—	M	B
Thomomys talpoides	M	M	M	M	M	—
Neotoma cinerea	M	M	M	M	M	—
Clethrionomys gapperi	M	M	M	M	M	B
Phenacomys intermedius	M	M	M	—	—	—
Microtus longicaudus	M	M	M	M	M	—
Microtus montanus	M	M	M	M	—	—
Microtus pennsylvanicus	M	M	—	B	B	B
Arvicola richardsoni	M	—	—	M	—	—
Synaptomys cooperi	—	—	—	—	—	—
Zapus hudsonius	—	—	B	B	B	B
Zapus princeps	M	M	M	M	—	—
Martes americana	M	M	M	—	?	B
Mustela erminea	—	—	M	—	M	B
Mustela nivalis	B	B	—	—	—	B
Gulo gulo	B-M	B-M	B-M	B-M	B-M	B-M
Lynx canadensis	B-M	B-M	B-M	B-M	B-M	B-M
Alces alces	M	—	—	—	—	B
Rangifer tarandus	—	—	—	—	—	B
Ovis canadensis	M	M	M	M	M	—
Total boreomontane species	30	27	25	23	20	17
Percent boreal affinities (B)	3	7	6	13	20	88
Percent montane affinities (M)	90	86	86	78	65	0
Percent boreomontane (B-M) or questionable	7	7	8	9	15	12

* Taxa with montane (M) or boreal (B) affinities are indicated; in a few instances, different populations of the same species fall into different categories. Taxa occurring without essential differentiation in both boreal and Rocky Mountain coniferous forest are designated B-M. Species listed are those identified as boreomontane in Table 1.

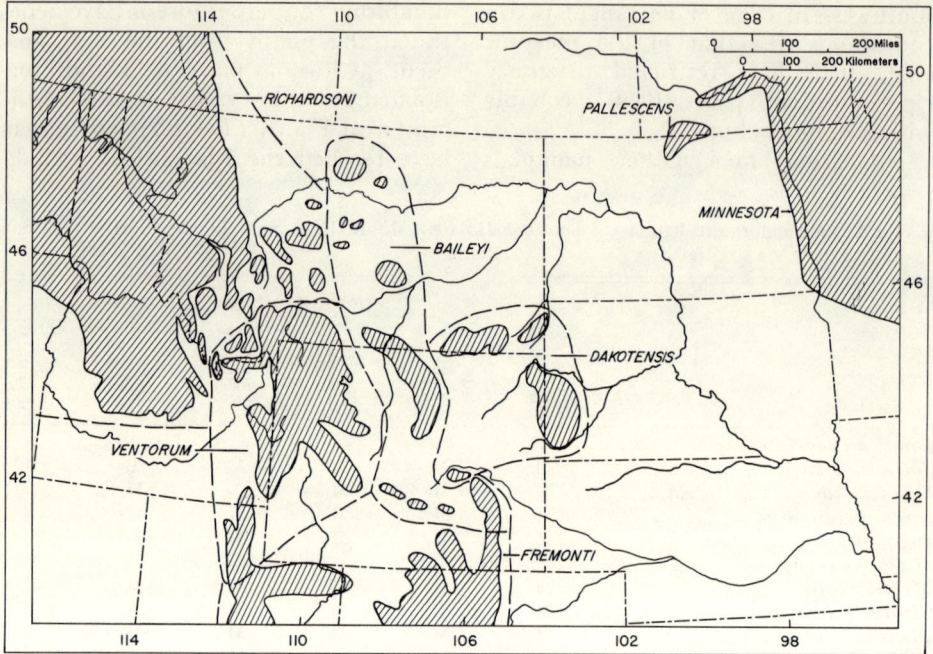

Fig. 7. Approximate distributions of some of the races of *Tamiasciurus hudsonicus* on isolated ranges in the Northern Plains. Our allocation of red squirrels from the northern Laramie Mountains to *T. h. dakotensis* and of various other isolated populations to the subspecies *richardsoni, baileyi,* and *ventorum* is tentative.

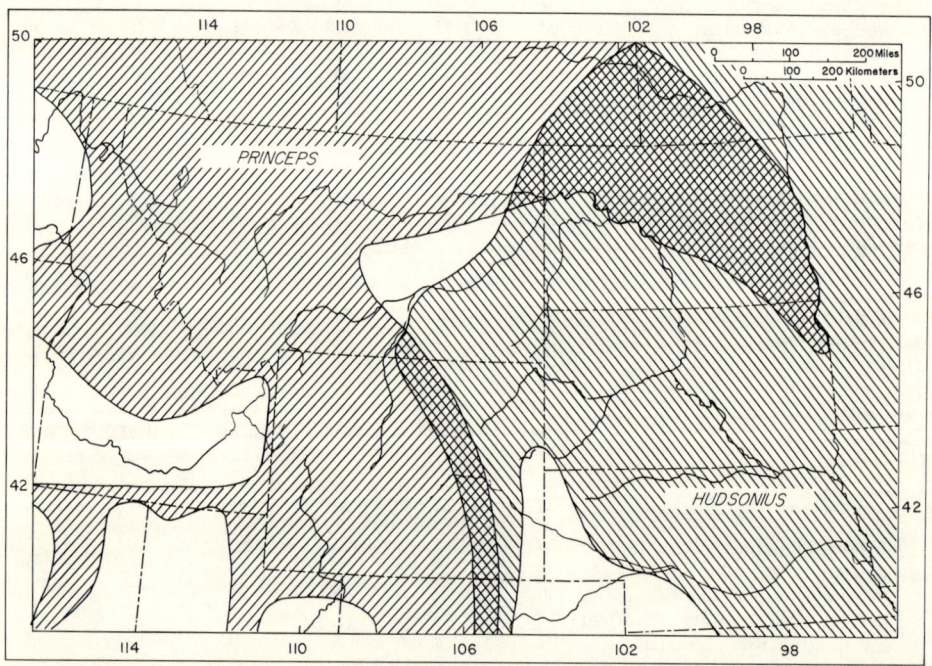

Fig. 8. Distributions of two species of jumping mice, *Zapus*, on the Northern Great Plains. *Z. hudsonius* represents a boreal species widespread on the Northern Plains, whereas *Z. princeps* is a western, montane species.

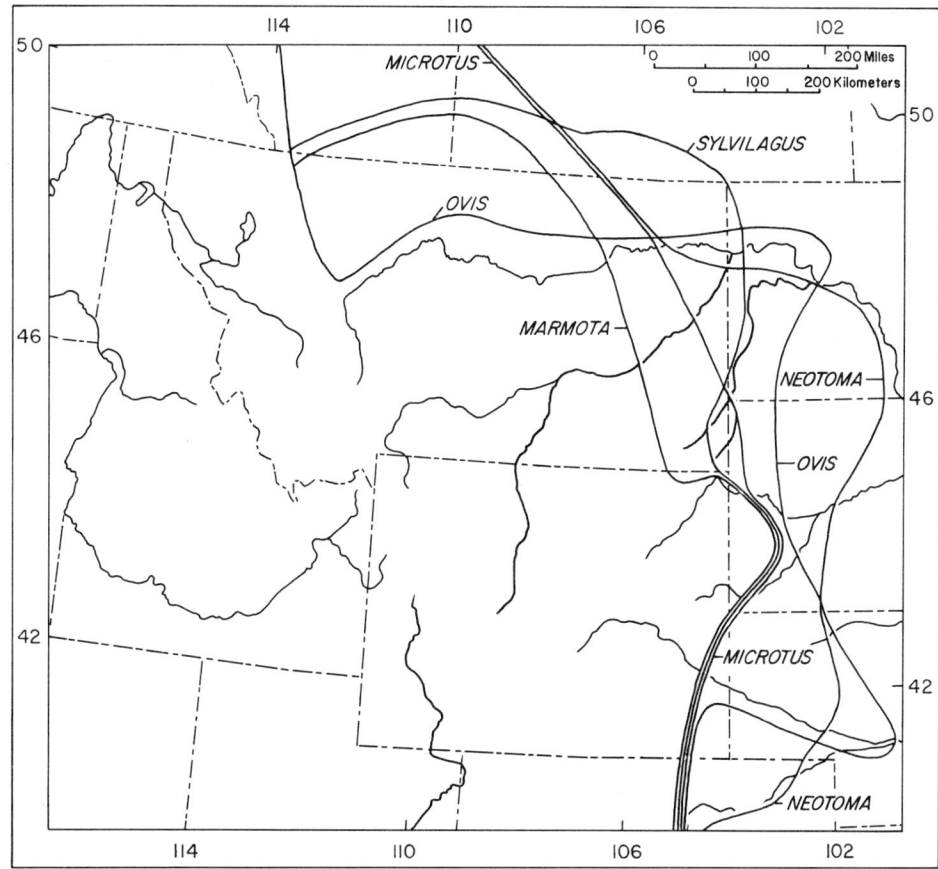

Fig. 9. Eastern limits of distribution of five species of montane mammals (*Sylvilagus nuttallii, Marmota flaviventris, Neotoma cinerea, Microtus longicaudus, Ovis canadensis*).

Sylvilagus nuttallii, Marmota flaviventris, Clethrionomys gapperi brevicaudus, and *Microtus longicaudus* (Fig. 9). All save *Clethrionomys* have rather broad ecological "niches" and are not confined strictly to coniferous forests. Other elements that have montane affinities, including *Sorex nanus, Glaucomys sabrinus bangsi* (Fig. 10), and *Mustela erminea muricus*, presently are known from the Black Hills, but are missing (at least unreported) from some of the ranges, such as the Bighorns, intervening between the Black Hills and the Rockies. These species all have rather narrow ecological "niches," and perhaps this is causally related to their distributions. Redbacked voles constitute an intermediate state. Although *C. gapperi* is generally confined to coniferous forests, it has dispersed as far as the Black Hills, and has survived there and on intervening mountain ranges. The allocation of *C. g. brevicaudus* to the montane group rather than the boreal group is based on the assessment that "it resembles *G. c. galei* more than it does any other named kind" (Cockrum and Fitch, 1952). Yet it is possible that, like the red squirrel and northern jumping mouse, *brevicaudus* has affinities with the boreal fauna. As in the case of the snowshoe hare of the Bighorn Mountains, more data are needed.

Aside from the kinds mentioned above, several mammals of western affinity occur not only as far east as the Black Hills, but are variously distributed in rocky areas or montane forest

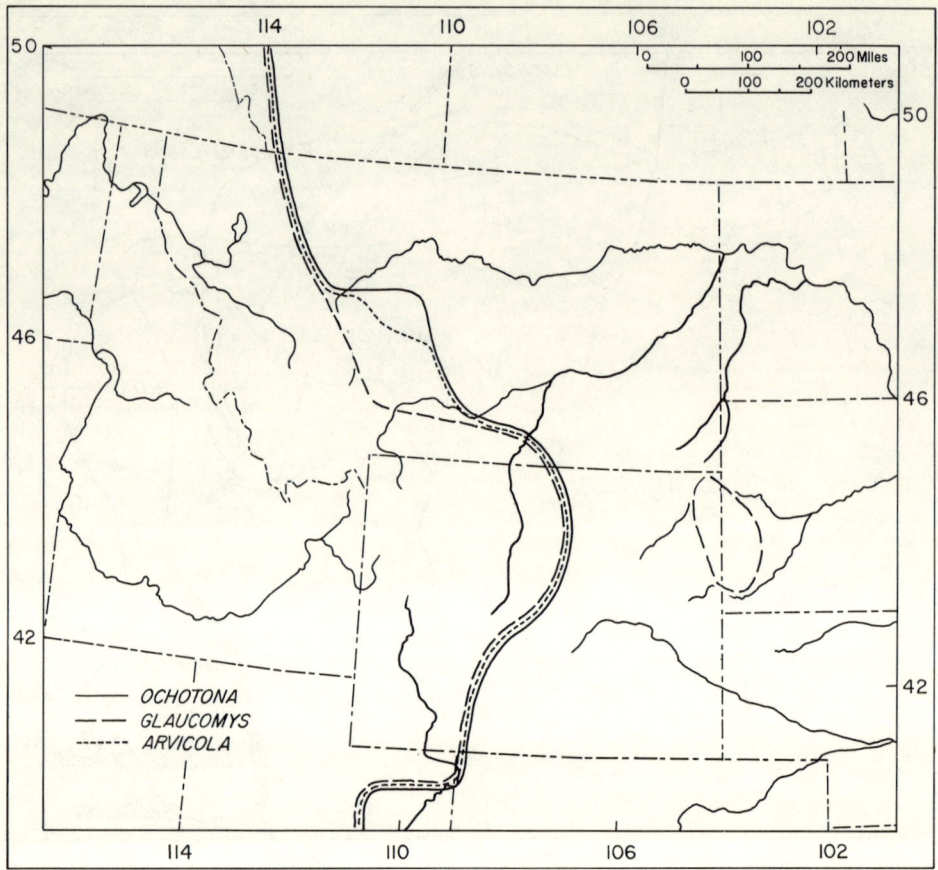

Fig. 10. Eastern limits of distribution of three species of montane mammals (*Ochotona princeps, Glaucomys sabrinus, Arvicola richardsoni*).

from the Platte Valley of western Nebraska northward to the Badlands and adjacent areas of North Dakota. These include three bats (*Myotis evotis, Myotis volans, Plecotus townsendii*), as well as *Neotoma cinerea* and *Ovis canadensis*, the latter extinct within historic times (Fig. 9). Two other species fall more or less into the same category —*Thomomys talpoides*, which has a broader distribution on the Northern Great Plains than those mammals mentioned above, and *Myotis leibii*, listed among the widespread taxa in Table 1 but a saxicolus species that in the Northern Plains is essentially limited to the western part of the region.

Another group of species does not occur as far east as the Black Hills; either they failed to reach the Hills in glacial times, or else did so but subsequently became extinct. Those found in both the Bighorn and Laramie mountains include *Sorex vagrans, Microtus montanus,* and *Zapus princeps,* as well as subspecies of other boreomontane or widespread taxa that are distinct from those elsewhere on the Northern Great Plains—*Sorex cinereus cinereus, Sorex palustris navigator,* and probably *Bison bison athabascae,* for example. These species have, or had, discontinuous distributions along the eastern margins of their ranges, being mostly restricted to montane environments. This might be questioned for the vagrant shrew and bison, which usually are considered as eurytopic, but there is evidence for considering at least some of the montane populations as ecologically and taxonomically distinct. Montane vagrant shrews *(S. v. obscurus)* were considered a subspecies of *Sorex vagrans* by Findley (1955). Typical vagrant shrews are western lowland species, occurring only as far east as western Montana and Wyoming. Montane populations, formerly placed in the species *S. obscurus* (Jackson, 1928), are the taxa contributing to the montane element in the fauna of the isolated ranges of the Northern Great Plains. Moreover, recent work (Darwen N. Hennings, personal communication) suggests that *vagrans* and *obscurus* rarely if ever intergrade where they are sympatric in western Montana.

The occurrence of bison within the Rocky Mountains is well documented (Fryxell, 1928; Beidelman, 1955; G. M. Christman, personal communication); the forest subspecies *(B. b. athabascae)* apparently occurred regularly above timberline on isolated ranges in the plains and may represent a montane element at this taxonomic level.

Some montane taxa occur in the Bighorn and adjacent ranges or in the Laramie Mountains, but not in both. Those occurring in the Bighorns include *Ochotona princeps, Tamiasciurus hudsonicus baileyi,* and *Arvicola richardsoni* (Figs. 7, 10). *Lepus americanus seclusus* of the Bighorns is thought to represent a boreal element, but the possibility that it is related more closely to Rocky Mountain populations than to those of the boreal forest should be considered because other mammals with the same distributional pattern have montane affinities. Taxa absent in the Bighorns, but present in the Laramies, include *Spermophilus lateralis lateralis, Tamiasciurus hudsonicus fremonti,* and *Phenacomys intermedius.* All of these, with the exception of the mantled ground squirrel, are restricted to montane coniferous forest or subalpine to alpine habitats *(Ochotona, Arvicola).* They are thus comparable to other species in the Bighorn and Laramie mountains, discussed above.

A final group of mammals include those that do not occupy the Laramie or Bighorn mountains, but have dispersed to some extent east from the main ranges of the Rockies in Montana. These include *Eutamias amoenus, Spermophilus lateralis cinerascens,* and *Alces alces shirasi,* which occur in the Big Belt and Crazy mountains.

DECIDUOUS FOREST SPECIES

Eighteen species (Table 1) have invaded the Northern Plains from deciduous forests to the east or southeast. Mammals in this grouping fall into two general, evenly divided categories: (1) species more or less restricted to temperate eastern North America, and (2) species that are distributed in the temperate and tropical regions to the east and south of plains, but that definitely reached the area from the east. Some kinds in the latter category occur southward well into Middle America, and in several instances all the way to South America.

Kinds that fall into the first grouping are *Blarina brevicauda, Scalopus aquaticus, Myotis keenii* (subspecies *septentrionalis*), *Tamias striatus, Marmota monax, Sciurus carolinensis, Sciurus niger, Neotoma floridana,* and *Microtus pinetorum.* Species that are more widely distributed but that evidently invaded the plains from an easterly direction are (those distributed as far south as South America marked with an asterisk): **Didelphis marsupialis, Cryptotis parva, Pipistrellus subflavus, *Lasiurus borealis, Nycticeius humeralis, *Sylvilagus floridanus, Glaucomys volans, Peromyscus leucopus,* and **Urocyon cinereoargenteus.* Of those species that reach South America, only *D. marsupialis* and probably *U. cinereoargenteus* originated there.

Looking at the distributional picture from another point of view, eastern invaders of the Northern Plains region also can be characterized as to

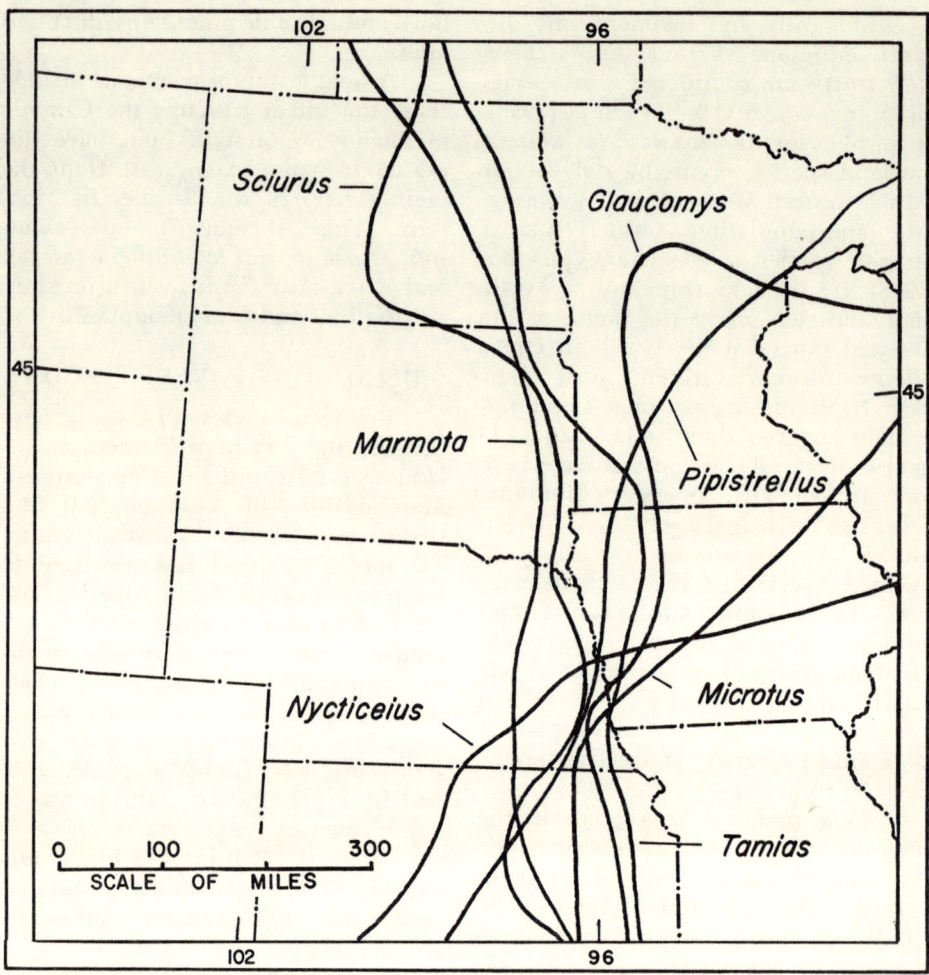

Fig. 11. Western limits of distribution of seven deciduous forest species of mammals restricted to the eastern part of the Northern Plains *(Nycticeius humeralis, Pipistrellus subflavus, Tamias striatus, Sciurus carolinensis, Glaucomys volans, Marmota monax, Microtus pinetorum)*.

whether or not the western limits of their ranges are more or less coincident with, or do not extend westward much beyond, the western limits of hardwood forest. Figure 11 indicates the approximate western boundary of the ranges of such species—*N. humeralis, P. subflavus, T. striatus, S. carolinensis, G. volans, M. monax,* and *M. pinetorum*.

Other invaders from the east are essentially inhabitants of relatively mesic grasslands or, to the west of the hardwood forest, riparian communities, and they occur far onto the plains along the many eastward-flowing tributaries of the Missouri River system (see Fig. 12). Most appear to have continuous dendritic distributions, but at least three (*M. keenii, P. leucopus, N. floridana*—see Figs. 15-17) have isolated western populations. Two, the fox squirrel (*S. niger*) and the opossum (*D. marsupialis*), have extended their ranges northward and westward markedly in historic time, partly as a result of natural dispersal and partly as a result of introduction.

The fox squirrel, not found on the Missouri above the mouth of the Niobrara by Lewis and Clark, now occurs along that river and adjacent parts of its tributaries all through the Dakotas

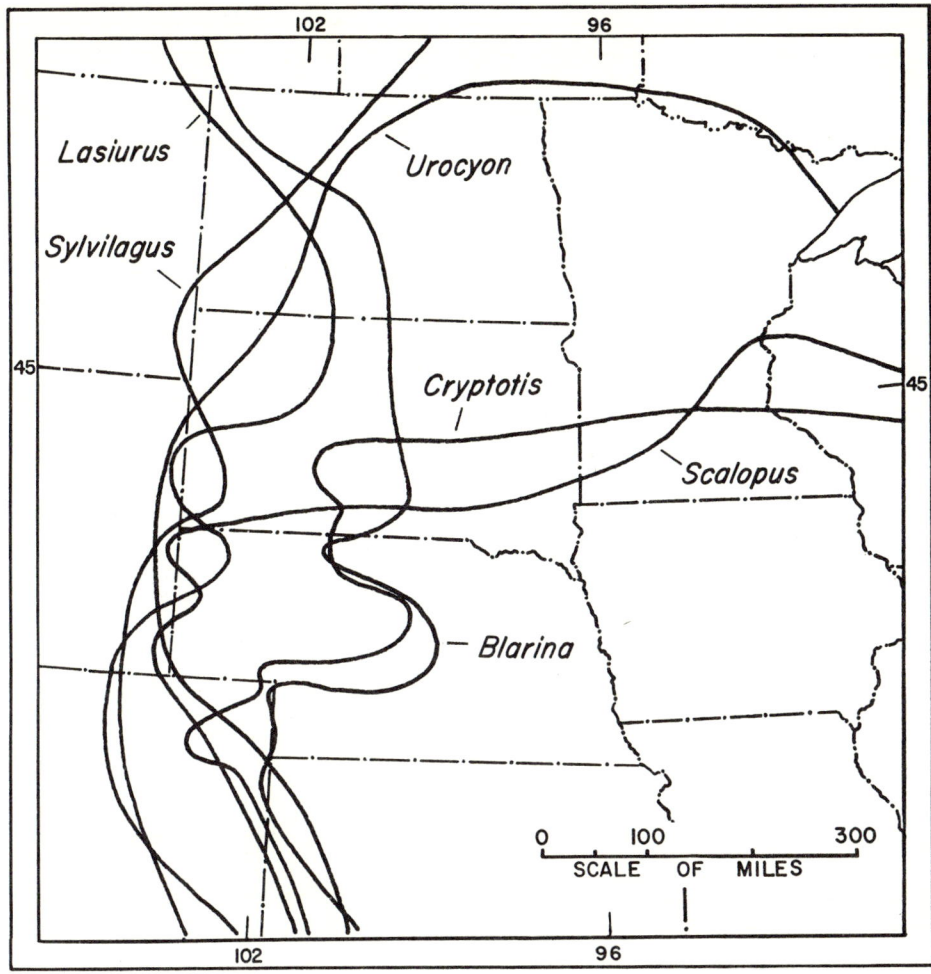

Fig. 12. Western limits of distribution of six species of eastern affinities *(Blarina brevicauda, Cryptotis parva, Scalopus aquaticus, Lasiurus borealis, Sylvilagus floridanus, Urocyon cinereoargenteus)*.

and even up the Yellowstone as far as south-central Montana (Hoffmann et al., 1969). It also is known westward along the Niobrara and in the White River drainage and Hat Creek Basin in northwestern Nebraska; along the North Platte the species has reached extreme eastern Wyoming (Long, 1965). E. A. Hibbard (1957) in North Dakota, Hoover and Yeager (1953) in Colorado, and Packard (1956) in Kansas all have recorded westward extension of the range of the fox squirrel in recent years.

The opossum, a Neotropical species that invaded the plains region from the east, has not extended its range as far northward as has *Sciurus niger*, but has been almost as successful to the west (see Fig. 13). Evidently limited to the woodlands of southeastern Nebraska at the time of arrival of European man, this species now occurs northward well into South Dakota, out the Niobrara system in Nebraska almost to the western boundary of the state (Jones, 1964), along the North Platte through western Nebraska to adjacent Wyoming (Brown, 1965), and out the South Platte into Colorado.

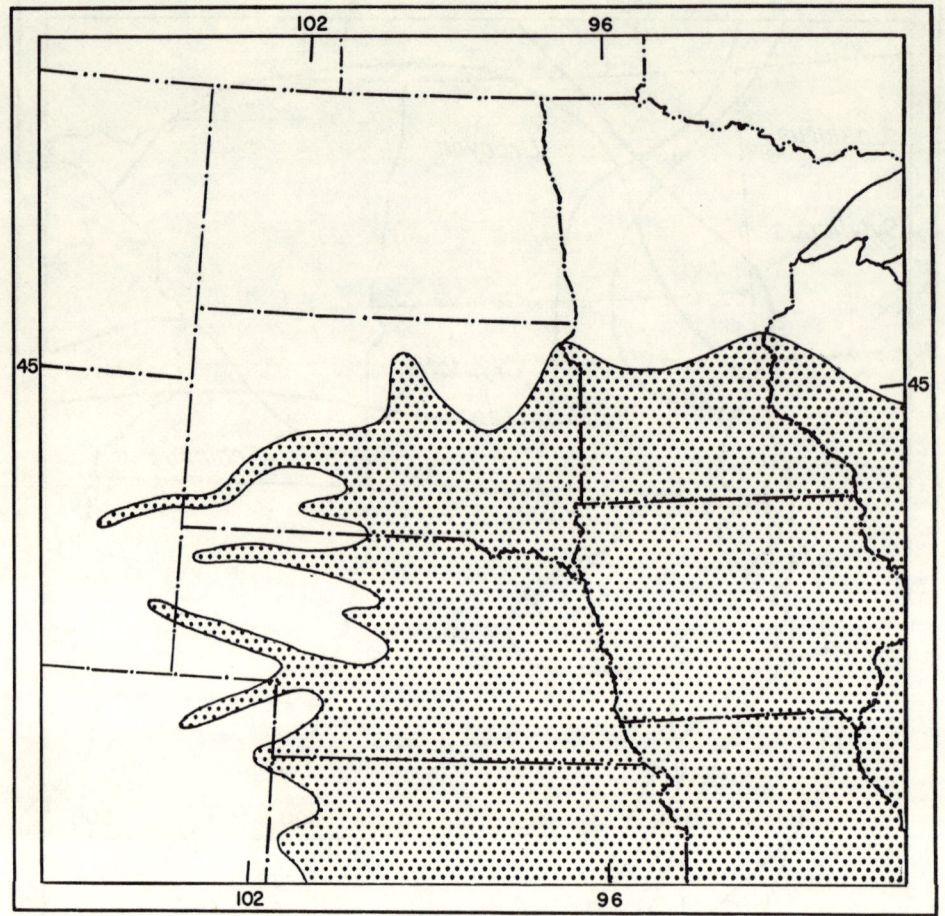

Fig. 13. Present distribution of the opossum *(Didelphis marsupialis)* on the Northern Great Plains. Northwestern extensions of range represent recently invaded areas.

Judging from its remarkable dispersal in the past half-century, it may be supposed that this unique mammal has yet to reach the ultimate limits of its distribution on the Northern Great Plains.

Considering this matter, it is well to note that the valleys of many plains river systems probably are better timbered now than at any other period since warm, moist post-glacial times, owing to control of prairie fires and the planting and protection of trees by man, and also that a source of food (mostly in the form of grain) now is present in valleys where few fruit- and nut-bearing trees occurred previously.

WIDESPREAD SPECIES

As is true in all major biotic communities, there are some species on the plains that are tolerant of a wide range of environmental conditions or have rather specialized requirements that are met in several biotopes. Such widespread mammals in the fauna of the Northern Great Plains include 27 species (Table 1). Among eurytopic taxa, ungulates and especially carnivores are well represented in this widespread group. The coyote, wolf, red fox, grizzly bear, long-tailed weasel, striped skunk, cougar, and bobcat are exemplary carnivores in this regard, as are the wapiti and bison among the ungulates. Among the small, terrestrial rodents only the deer mouse falls in this category. Finally, a number of bats appear to belong among the eurytopic taxa.

Certain other species have more narrowly defined niches. Aquatic or semiaquatic mammals with specialized niche requirements are found throughout the Great Plains wherever suitable aquatic habitats are present. These include the beaver, muskrat, mink, and otter. The black bear and porcupine are primarily woodland species, although they once occurred in gallery forests and scarp woodlands throughout the plains. The raccoon is more or less restricted to gallery forests and other riparian plant communities, and its general distribution closely approaches that of the gray fox, which is here treated as a deciduous forest element. The division between them is necessarily arbitrary. The white-tailed deer also is primarily associated with deciduous forest. The badger and pronghorn, although found in a variety of habitats, are most abundant in nonforested areas and are primarily western in distribution. Their ranges and, to a lesser extent, habitat preferences parallel those of the mule deer.

The spotted skunk (*Spilogale putorius*), two subspecies of which are here included with the Great Plains and Great Basin faunas (*S. p. interrupta* and *S. p. gracilis*, respectively), would, if considered a single species (Van Gelder, 1959), undoubtedly be placed in the widespread group. However, in the plains region, where their ranges meet, these two act as good species (Long, 1965; Mead, 1968); we believe the burden of proof now lies with those who would unite these skunks at the specific level.

ANALYSIS AND DISCUSSION

STEPPE SPECIES

The available paleozoological and botanical data suggest a movement of steppe mammals into the Northern Great Plains region after the late-glacial period, probably from a glacial refugium to the south and west. As the climate became warmer and drier, species of the arid Southwest also penetrated the Northern Plains. Warm, dry conditions to the east of the plains also permitted a few steppe mammals to disperse eastward across the Missouri River and the Mississippi and, in the case of *Spermophilus tridecemlineatus*, as far as eastern Pennsylvania. When, and how many times, such eastward movements of steppe mammals occurred is debatable. Guilday *et al.* (1964) argued that *Spermophilus* moved eastward in late-glacial time, whereas Smith (1957) and others have proposed the post-glacial "Xerothermic" period, when the "Prairie Peninsula" was presumed to have been formed, as the time of eastward spread of these mammals.

Whether the trans-Missouri and, in some cases, trans-Mississippi dispersal of typically plain species occurred in a dry post-glacial period or earlier is unknown. It is certain, however, that some grassland species did spread eastward—for example, the plains pocket gopher (*Geomys bursarius*), ground squirrels (*Spermophilus tridecemlineatus* and *S. franklinii*), and the prairie vole (*Microtus ochrogaster*). Certain steppe-associated mammals of other faunal units, such as the badger, *Taxidea taxus* (Clelland, 1966), and the western harvest mouse (*Reithrodontomys megalotis*), also have invaded the tall-grass prairie to the east of the plains. Too, the middle part of the post-glacial period is the only time when the spotted skunk (*Spilogale putorius interrupta*), now restricted to the western side of the Mississippi River, ranged eastward across the Mississippi, as evidenced by remains found in rock shelters and caves in western Illinois (Parmalee and Hoffmeister, 1957; Parmalee, 1967, 1968).

Although dated evidence of the eastward surges of grassland species in late-glacial and post-glacial times is scanty, it now seems probable that such movements occurred several times, for differing distances, depending on the frequency and duration of dry climatic regimes.

INVADERS FROM THE SOUTH AND SOUTHWEST

As pointed out in an earlier section, it is apparent that many species of southern or southwestern affinities invaded the Central and Northern plains in late-glacial or post-glacial times. Some may have dispersed farther north and east than is indicated by their present ranges in warm, dry periods of the Holocene, perhaps in company with still other species that since have retreated from the region. There is some evidence that at least two mammals, *Sylvilagus audubonii* and *Perognathus flavus,* were distributed to the east of presently occupied ranges during the "Great Drought" of a generation ago.

Only three species, *Lepus californicus, Sigmodon hispidus,* and possibly *Dasypus novemcinctus,* actually seem to be invading new areas of the Northern Plains at present. Within historic time the black-tailed jackrabbit has occupied the area north of the Platte River in Nebraska and now occurs northward into South Dakota. This invasion has caused it to overlap broadly the range of the related white-tailed jackrabbit *(L. townsendii),* and blacktails have replaced whitetails in the southern part of the newly occupied range. However, at the same time, *Lepus townsendii* seems to be shifting its range northward. Whether dispersal to the north and northeast in the last century by this lagomorph was the result of intensive agricultural practices in areas formerly occupied to the south (in western Kansas and southern Nebraska, for example), of competition with the invading black-tailed jackrabbit, or a response to the recent gradual warming trend on the North American continent has been argued (see Jones, 1964). We incline toward the last explanation, noting that *L. townsendii* is essentially an inhabitant of the "northern grasslands" and that it has for many years successfully occupied parts of Nebraska that (1) are heavily cultivated and (2) support substantial populations of black-tailed jackrabbits. Admittedly, however, the problem of competition between the two species has not as yet been studied in detail and may also prove to be an important factor in this situation. It is of interest to note that the black-tailed jackrabbit also has recently invaded southwestern Montana from the Great Basin.

The current warming trend probably accounts also for the spectacular northward spread in the last century of the cotton rat. From central Kansas this species extended its distribution northward between 1933 and 1947 in that state at an average rate of seven miles per year (Cockrum, 1948). *S. hispidus* first was taken in Nebraska in 1958 (Jones, 1960), in the extreme southeastern part of the state; since that time it has dispersed northward at least to the vicinity of Holstein, Adams County. It probably now occupies much of the area drained by the Blue and Republican rivers in the southeastern part of the state. According to Genoways and Schlitter (1967), recent northward movement of the species to Nebraska has proceeded at an average rate of about five and one-half miles per year. It is of note that these same authors reported the cotton rat for the first time to the north of the Missouri River in northwestern Missouri. Since their report, the species has been found to be relatively widely distributed there and likely has reached the southern border of Iowa. Factors that may limit the northward dispersal of *Sigmodon* are poorly understood,

INVADERS FROM THE GREAT BASIN

During the Wisconsin, a barrier of boreal forest evidently existed between the Great Basin and Wyoming Basin (Findley and Anderson, 1956), separating the former from the Northern Great Plains. Late-glacial faunas from Wyoming (see above) indicate that species of the Rocky Mountain coniferous forest then occurred at lower elevations. With the northward and upward retreat of coniferous forest in the Holocene, nonforested connections between the Great Basin and the Northern Great Plains were established. Great Basin species such as *Sylvilagus idahoensis, Perognathus parvus, Lagurus curtatus,* and *Sorex merriami,* along with *Lepus californicus deserticola* and perhaps subspecies of *Eutamias minimus* and *Thomomys talpoides,* were then able to spread to varying degrees onto the Northern Plains. Other taxa, which already had differentiated from closely related Great Basin populations in pre-Wisconsin time, also regained access to the Northern Plains.

The clearest example of this is in the *Spermophilus richardsonii* complex. Earlier in the Pleistocene this species probably was confined to one of the three areas that it now occupies (see Fig. 14). If, as Durrant and Hansen (1954) suggested, *S. r. nevadensis* is closest to the ancestral stock, then *S. r. elegans* may have evolved after the species dispersed eastward into Wyoming and Colorado, through the Wyoming Basin, in interglacial times and was isolated in a subsequent glacial period. The earliest North American ground squirrels of the advanced subgenus *Spermophilus* are known from the Late Pliocene in the Columbia Plateau region of Oregon (Black, 1963). *S. r. elegans* is first reported from deposits of Kansan age in southwestern Kansas (Hibbard, 1937, and personal communication). These records do not contradict the proposed dispersal pattern. As in the case of *S. r. elegans,* interglacial conditions would have permitted ground squirrels to disperse northeastward from the Snake River Plains into Montana, and glacial conditions would have interposed a boreal forest barrier along the Continental Divide, permitting differentiation of *S. r. richardsonii* on the Northern Great Plains.

Nadler (1964, 1966) has shown that *S. r. nevadensis* and *S. r. elegans* have similar karyotypes and chromosome numbers ($2n=34$), whereas *S. r. richardsonii* has a different karyotype and a diploid number of 36 (see Fig. 14). He suggested that because centric fusion is considered the most plausible form of rearrangement leading to reduced chromosome number, *S. r. richardsonii* is likely to be the ancestral form and to have given rise to *S. r. elegans* and *S. r. nevadensis,* in contrast to the hypothesis of Durrant and Hansen cited previously. In either case, interglacial dispersal along routes through the Wyoming Basin and southwestern Montana, with subsequent glacial isolation, is a tenable hypothesis for the mechanism of speciation in these ground squirrels. Nadler (1966) further recognized the possibility that centric fission, leading to higher chromosome numbers, might be involved in ground squirrel evolution, in which case *S. r. nevadensis* would be ancestral. He subsequently demonstrated (1968) that in karyotype and chromosome number *S. r. aureus* was indistinguishable from *elegans* and *nevadensis*. On the basis of present chromosomal evidence, *S. r. aureus* and *S. r. richardsonii* show no intergradation, but instead occupy mutually exclusive areas on opposite sides of the Madison River in Montana (Hoffmann et al., 1969). A second post-Wisconsin dispersal of the Great Basin population northeastward

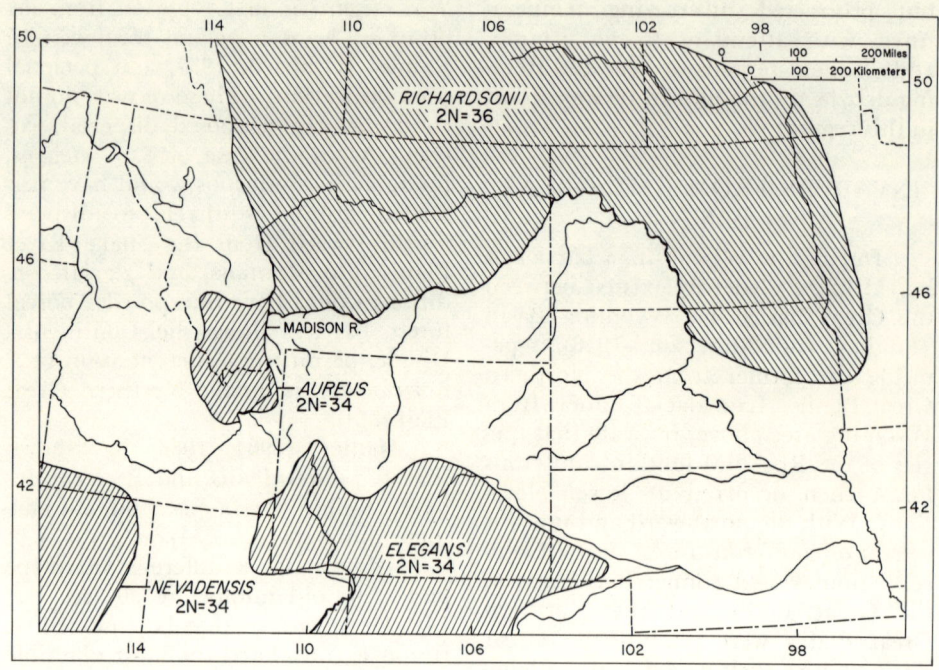

FIG. 14. Distribution (in part) of the four subspecies of *Spermophilus richardsonii*. The Great Plains race (*S. r. richardsonii*, 2n=36) is contrasted with the other three races (2n=34).

over the Continental Divide would produce such a pattern and would provide partial support for the centric fission mechanism. The alternative, that *S. r. aureus* is derived from *S. r. richardsonii*, requires the Madison River to have prevented gene flow for long enough to produce the major divergence between the populations presently divided by the river. The Madison does not appear to be a barrier of this magnitude. [Since this manuscript was completed, further field work has shown that the zone of contact between the two races is not along, but is somewhat to the west of, the Madison River. Thus, no geographic barrier separates the two subspecies. Even so, hybridization is rare, lending further support to the idea that the contact between the two came about after a prolonged period of glacial isolation.]

Another possible example of this Great Basin–Great Plains pattern of differentiation involves a series of shrews of the *Sorex cinereus* group.

These shrews are principally boreal-taiga and tundra mammals, widespread in North America and with a limited range in eastern Siberia (Hoffmann and Peterson, 1967). Several small shrews (*S. lyelli*, *S. preblei*, *S. milleri*), all with limited disjunct ranges, have been regarded as allopatric species in the *cinereus* superspecies. *S. preblei* is now believed to be partly sympatric with *S. cinereus*; they occur together in western Montana, with habitat segregation (Hoffmann et al., 1969), and also in eastern Montana, where *S. preblei* is difficult to distinguish from the small *S. c. haydeni*. Preble's shrew may have been isolated originally from typical *cinereus* in the northern Great Basin–Columbia Plateau region during a glacial period. In isolation it became adapted to more arid steppe and semidesert conditions as well as to coniferous forest habitats. Then an interglacial environment permitted it to disperse northeastward through southwestern Montana and onto the Northern Great Plains, where it occupied the arid grassland habitats, meeting populations of *S. c. haydeni.*

Three other mammals, although fairly widespread geographically, appear to conform in part to the Great Basin–Northern Great Plains pattern. There are, however, unsolved taxonomic problems that prevent straightforward interpretation of these situations. The least chipmunk *(Eutamias minimus)* occurs throughout much of the Great Basin, Rocky Mountains, and Northern Great Plains. Within this large range, two main sorts of habitats are utilized—desert to arid steppe and coniferous forest. Least chipmunks from the boreal forest (subspecies *borealis* and *neglectus*) and from Rocky Mountain coniferous forests (subspecies *arizonensis, consobrinus,* and *operarius*) were shown by Sutton and Nadler (1969) to have a distinctive karyotype ("A"), found only in *E. minimus* and in *E. cinereicollis* (a chipmunk of forests of the extreme southern Rocky Mountains). The races *E. m. minimus* and *E. m. caryi,* inhabiting the Wyoming Basin and San Luis Valley of Colorado, respectively, also possess karyotype "A," but differ from the above-mentioned subspecies in that they are desert or arid-steppe inhabitants.

In contrast, least chipmunks from the Great Basin and Northern Great Plains (subspecies *scrutator, pictus, pallidus,* and *cacodemus,* and at least one steppe population considered to be *borealis*) have a different karyotype ("B"), a configuration found in at least 13 other species of North American *Eutamias.* Two montane races, *E. m. confinis* of the Bighorn Mountains and *E. m. silvaticus* of the Black Hills, are also "B" karyotypes, and they appear to be derived from adjacent plains populations rather than from boreal or Rocky Mountain ancestors. The possibility deserves consideration that these Great Basin–Northern Great Plains chipmunks have had an evolutionary history similar to that of the *Sorex cinereus–Sorex preblei* complex and have been divorced for some time from most populations of *E. minimus* of the boreal forest and Rocky Mountains. Recent studies of the northern pocket gopher, *Thomomys talpoides,* provide additional support for this same concept. This pocket gopher has a wide geographic range and broad ecological tolerance, living from desert through alpine conditions. It has been shown (Long, 1965) that in some places different subspecies of *talpoides* occur in contiguous allopatry without intergrading, separated by distinct habitat preferences; more recently, Thaeler (1968) has found within the "species" *talpoides* a number of populations with different chromosome numbers and karyotypes. It is possible that the desert and steppe populations of this pocket gopher also will conform to the Great Basin–Great Plains pattern herein discussed. Spotted skunks *(Spilogale),* mentioned previously, also may fit the pattern.

BOREAL AND MONTANE SPECIES

The presence of 37 species of boreomontane mammals, one-third of the total fauna of the Northern Plains, many of which are isolated on ecological "islands," is of biogeographic significance. The ecological requirements of the smaller insectivores, lagomorphs, and rodents are such that a continuous coniferous forest environment must have been present to link the now isolated mountain ranges on the Northern Plains with boreal forests, and especially Rocky Mountain coniferous forests, at some times during late-glacial or post-glacial periods.

The small, low Turtle Mountains of North Dakota, separated from the present edge of the boreal forest by only a short distance, have a fauna that is exclusively boreal in affinity. Conversely, the isolated mountain ranges of central Montana (Big Belt, Crazy, Little Belt, Big Snowy, and others) and the Laramie Mountains of southeastern

Wyoming are predominately montane in faunal relationships, reflecting their proximity to the main ranges of the Rocky Mountains (Table 2).

The two most isolated mountainous areas of the Northern Plains, the Bighorn Mountains and the Black Hills, are intermediate, but not equally so, between the conditions described above. Moreover, each of these ranges, and the Laramie Mountains as well, lacks boreomontane species for which suitable niches may be present. Thus, the Black Hills are apparently not inhabited at present by such species as the vagrant shrew, water shrew, pika, snowshoe hare, golden-mantled ground squirrel, montane vole, water vole, and heather vole, whereas the Bighorns lack the golden-mantled ground squirrel, northern flying squirrel, heather vole, and perhaps marten, ermine, and the dwarf shrew. Pikas, snowshoe hares, flying squirrels, and meadow voles (*M. pennsylvanicus*) have not yet been found in the Laramie Mountains; similar gaps exist also in the isolated montane faunas of central Montana. It is, of course, possible that additional collecting will reveal that some of the species mentioned inhabit these mountain ranges, but the unfilled niches are most likely to be the result of periods of post-glacial climatic stress (warmer or drier, or both, than at present), perhaps reinforced by random extinction.

Another cause of the absence of a species in a particular range is the presence of a closely related congener, occupying the ecological niche and competitively excluding the other. This appears to be the case with the chipmunks, where either a boreal or montane derivative is present, but not both. A clear example of this is seen in central Montana, where the Little Belt Mountains are occupied by the montane *Eutamias amoenus*, but the Big Snowy Mountains, 15 to 20 miles to the east, are inhabited by *E. minimus borealis* (Table 2). The small weasels, *Mustela erminea* and *M. nivalis*, also appear to be competitively exclusive (Hoffmann and Pattie, 1968). The jumping mice present a currently unresolved problem. Both species are reported from the Bighorn and Laramie mountains (see Fig. 8). However, the presence of *Zapus hudsonius* in the Bighorns is based on only two specimens, and there are some doubts concerning the specific identity of *Z. hudsonius* from the Laramie Mountains. Should both boreal and montane jumping mice occur in these areas, habitat segregation and other ecological niche differences may be expected (Taber and Hoffmann, 1963).

Boreal-forest mammals probably maintained continuity with populations on the Black Hills, Bighorns, and Laramie Mountains until well into the post-glacial period as the boreal forest of Nebraska and the Dakotas disappeared, via a forest corridor extending across the plains from northeast to southwest. Remnants of this are to be found today as a series of scarp woodlands (Wells, 1965) running from the Turtle Mountains and Pembina escarpment through the Killdeer Mountains and Long Pine Hills and their adjacent "badlands," to the Black Hills and Pine Ridge. Another corridor of possible importance in the post-glacial period may have been from east to west along the escarpments paralleling the Niobrara River. Connections between the Rocky Mountain coniferous forests and the isolated ranges of central Montana probably were maintained from southwest to northeast, from a center between the Lewis Range and the Absaroka Range. The Bighorn Mountains probably had contact with the Rockies through the Owl Creek Range, and directly eastward with the boreal forest via the Black Hills, or indirectly via the northern Laramie ranges (Casper, Haystacks), the Hartville Uplift, and the Pine Ridge escarpment. The isolated sections of the Laramie ranges, in addition to the boreal conection just mentioned, were in contact to the south with the Rocky Mountains via the Front Range, but may also have had a connection to the northwest, through the Freezout, Ferris, and Green mountains to the Wind River Mountains.

Ecological analysis shows that the majority of boreomontane species have preferred habitats in coniferous forests dominated by spruce and fir—that is, subalpine forest or boreal taiga. The only exceptions appear to be the water vole *(Arvicola)* and pika *(Ochotona)*, which are presently restricted to high subalpine and alpine habitats and which occur in the Bighorn Mountains and in some of the more westerly ranges in Montana. It is possible, however, that *Arvicola* might have a broader ecological niche under different environmental conditions, as its Old World congener is widespread at low elevations, and the pika in certain areas of the Great Basin presently occurs in fairly arid rocky areas.

Negative evidence worthy of note is the absence of indigenous populations of two stenotopic alpine mammals of the Rocky Mountains, the hoary marmot *(Marmota caligata)* and the mountain goat *(Oreamnos americanus)*, even though suitable alpine habitats exist in some of the isolated mountain ranges under consideration. This is demonstrated by the successful introduction by man of mountain goats in most of these areas. The explanation of their absence is probably two-fold: rocky tundra environments did not stretch continuously from the main ranges of the Rockies out to all of the isolated ranges, and warmer periods in the post-glacial period caused extermination of local populations through elevation of the timberline. The late-glacial presence of *Oreamnos* in the Laramie Range and its present absence from Wyoming and Colorado suggest operation of the second factor (Hoffmann and Taber, 1967).

DECIDUOUS FOREST SPECIES

In contrast to the movements of steppe mammals, deciduous forest elements of the Northern Great Plains would be expected to retreat eastward in drier periods, but to spread westward onto the plains when the climate was wetter, such as in the Sub-Atlantic or Neo-Atlantic periods. Mammalian fossils *(Blarina, Cryptotis, Scalopus, Microtus pinetorum)* provide some evidence for this in the late-glacial period on the Southern Plains, as does the presence of the two shrews in postglacial deposits farther to the north. Mammals associated primarily with the eastern deciduous forest may have reached limits far beyond those of the same species today, only to be displaced eastward again in drier times. Little concrete evidence, however, is available of post-glacial mammalian movements on the plains, principally for two reasons: few fossil remains, at least of small mammals that are sound zoogeographic indicators, are available from that period, and the absence of suitable refugia for mammals with eastern affinities on the plains during the dry periods resulted in relatively few relict populations to document wider dispersal.

At least three species—*Myotis keenii, Peromyscus leucopus,* and *Neotoma floridana*—have present-day geographic ranges seemingly related to earlier, more widespread distributions (Figs. 15-17). The contiguous distribution of Keen's bat reaches westward in forested situations in the central part of the plains only to approximately the 98th meridian, although records to the north of the region place it farther to the west. A relict population on the Black Hills, subspecifically indistinguishable from *M. k. septentrionalis*, attests to a more widespread distribution formerly, at least along the eastward-flowing tributaries of the Missouri, that allowed *M. keenii* to reach the Hills, whence it was isolated (Fig. 15). Two specimens from northwest-

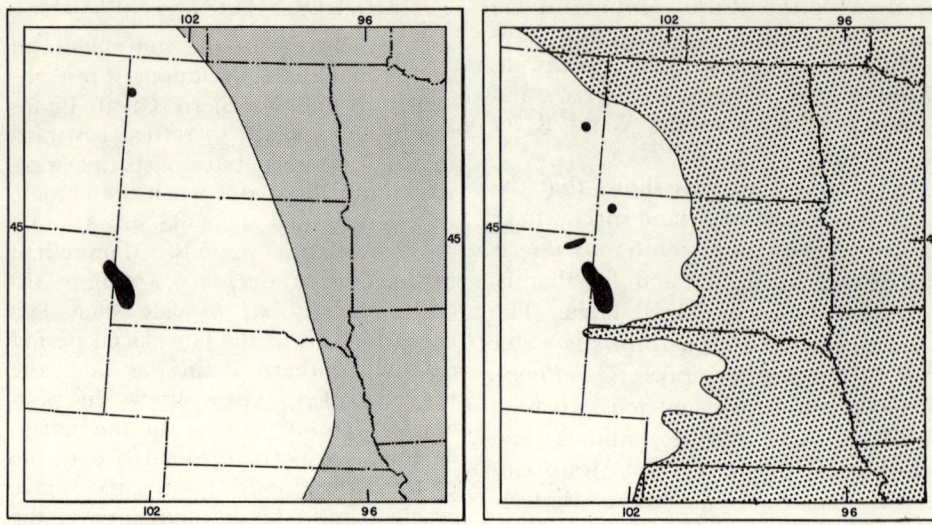

FIGS. 15-16. Distribution of Keen's bat *(Myotis keenii)*, left, and the white-footed mouse *(Peromyscus leucopus)*, right, on the Northern Great Plains. Contiguous distribution of each species is stippled, and isolated relict populations are black. Other more westerly isolated populations of *Peromyscus* in Montana and Alberta are not shown.

ern North Dakota (Miller and Allen, 1928), both taken many years ago, also may represent a relict population, or they may indicate instead that this species still occurs sparingly that far west along the Missouri River.

The white-footed mouse, *Peromyscus leucopus*, typically an inhabitant of deciduous woodlands in the northern part of its range, now occurs in a number of semi-isolated or isolated areas from northwestern Nebraska to Alberta that are associated with isolated stands of deciduous timber. The Black Hills, restricted areas along the Little Missouri and Yellowstone rivers, and the Slim Buttes of northwestern South Dakota are known areas of isolation (Fig. 16). If these relict populations had resulted from dispersal in the wake of the retreating Wisconsin ice sheet, it would be expected that *P. leucopus* now would have a more northerly distribution in the plains region. The only logical explanation, therefore, is that the species dispersed northwestward along river systems in company with deciduous timber during a postglacial period of wetter climate. Subsequent drier periods destroyed the continuous dendritic extensions of eastern woodland, but left isolated patches in certain favored areas, in some of which white-footed mice also survived. Relict populations of the meadow jumping mouse, *Zapus hudsonius*, in the same region, on the other hand, most likely date from the time following glacial retreat.

Another rodent, the eastern woodrat *(Neotoma floridana)*, has a relict population (*N. f. baileyi*) in northern Nebraska that apparently is restricted to the relatively cool canyons of the Niobrara River and a few of its tributaries. *N. floridana* could have reached this area only at a time when deciduous woodlands were more extensive, both geographically and in number of species, than now—in other words, during a wetter, possibly warmer climate. Drier periods then excluded this rat from other parts of northern and eastern Nebraska that formerly were occupied, leaving the relict subspecies, *baileyi*, stranded along the Niobrara (Fig. 17). Presumably the relict distribution of such trees as the cork elm *(Ulmus thomasi)* in the same area supports this explanation, as does the fact

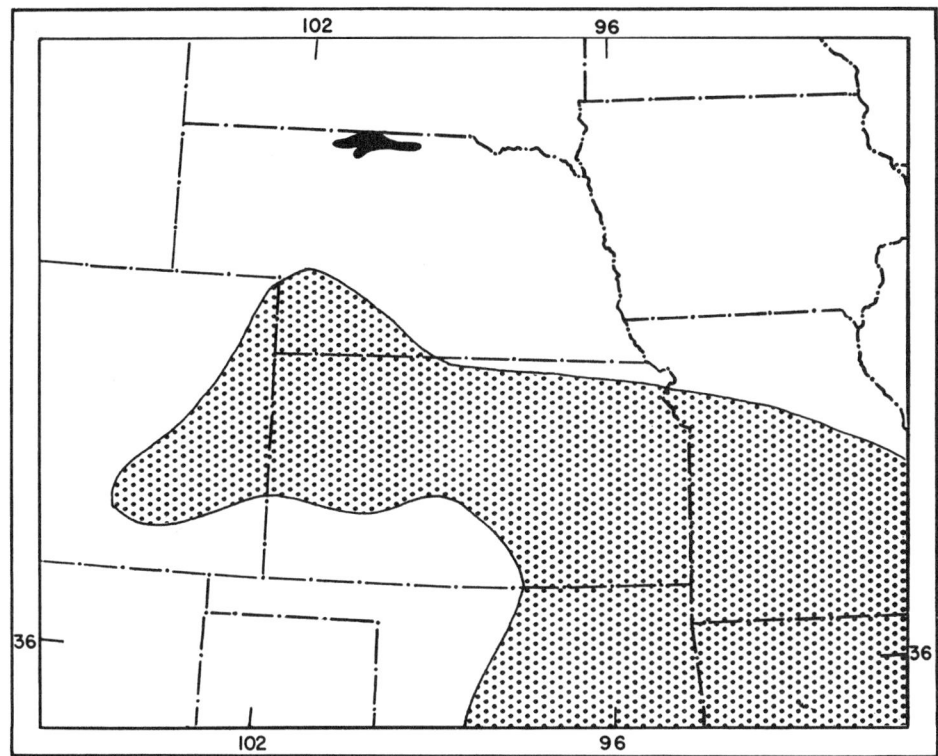

Fig. 17. Distribution of the eastern woodrat *(Neotoma floridana)* on the Northern Great Plains. Contiguous distribution of this rat is stippled, whereas an isolated and relict population *(N. f. baileyi)* is black.

that *N. f. baileyi* most closely resembles *N. f. osagensis*, the subspecies of eastern Kansas and adjacent regions (Elmer C. Birney, personal communication). The more westerly distribution of another subspecies of this woodrat, *N. f. campestris*, which occurs today in southwestern Nebraska and adjacent parts of Colorado and Kansas, is more difficult to explain; the distribution of *campestris* is spotty and irregular throughout its range, and a few small populations that were isolated in xeric times may have dispersed since, to a greater or lesser degree.

The distributional history of the short-tailed shrew, *Blarina brevicauda*, also is instructive. This insectivore evidently invaded the plains early in post-glacial time, as evidenced by remains found in association with an early man site in southwestern Nebraska that has been dated as about 9000 BP. In warm, wet, post-glacial interludes the short-tailed shrew may have had a more extensive distribution than now on the Great Plains. With the advent of periods of arid climate, the contiguous distribution of the species probably was displaced far to the eastward, perhaps to the Mississippi or beyond, leaving isolated populations in the wake of retreat. How many of these relict groups survived drier postglacial events and subsequent genetic "swamping" by reinvading stocks is a moot point, but current evidence indicates that at least a few may have done so.

Jones (1964) described the possible routes of reinvasion of the Great Plains by stocks of *Blarina* from the northeast and the southeast, and noted that the two stocks acted essentially as full species where they met in a fairly well-defined secondary zone of contact in southern Nebraska—a zone now known to extend eastward into Iowa and beyond. He supposed that the two stocks, the larger (northern) *B. b. brevicauda* and the smaller *B. b. carolinensis,* represented disjunct segments of a formerly continuous cline in size from north to south in the species, a cline still evident in most of the western range of the genus (Jones and Glass, 1960). Recently, however, it has been shown that the northern and southern segments of *Blarina* differ both in fundamental number and diploid number of chromosomes (M. R. Lee and E. C. Birney, personal communications), suggesting the possibility that two species may be involved, rather than one clinally variable species (see also Hibbard, this volume, who treats the two as distinct species in Late Pleistocene faunas; Oesch, 1967; Parmalee, 1967). However that may be, it does not necessarily argue against the thesis of effects of post-Wisconsin climatic shifts on the distribution of these shrews, particularly when the unique distributional pattern in central and western Nebraska is considered.

In summary, then, it can be seen that some species with eastern affinities now occur westward in the Northern Plains region only approximately to the margin of continuous hardwood forest, whereas others have pushed far westward onto the plains, especially as dendritic segments along the major eastward-flowing river systems. The limited fossil record, together with the occurrence of a few mammals in favorable ecological situations on the Central and Northern Great Plains, points to the fact that many eastern species had more extensive distributions during portions of Holocene time than today.

CONCLUSIONS

The general environment of the Northern Great Plains in the Late Wisconsin, as inferred from the evidence of mammalian distributions, must have been one of coniferous forest, connecting the boreal coniferous forest south of the glacial ice and montane coniferous forests extending from the Rocky Mountains eastward, and including the isolated ranges. Such forests may have been widespread on the plains, or they may have been restricted to ecologically suitable corridors; paleobotanical evidence points to the former. A zone of tundra or wooded tundra may have been present between the forests and the ice edge; there is no pollen evidence for the Great Plains such as exists for the Great Lakes region and the Northeast (Davis, 1967; Terasmae, 1967; Watts, 1967), but the tundra-dwelling *Ovibos* occurred from eastern Montana to eastern Iowa (Hibbard *et al.,* 1965; Hibbard, personal communication; Charlesworth, 1957). Steppe or savannah probably was restricted to a limited area east of the Rocky Mountains, southward from eastern Colorado and western Kansas, and perhaps in the Wyoming Basin.

An expansion of coniferous forests downward in the Rocky Mountains would have had the effect of making confluent the coniferous forests of the central Rockies (Montana to northern Wyoming) with those of the southern Rocky Mountains (Utah and southern Wyoming southward) (Findley and Anderson, 1956). Closure of nonforested gaps west of the Wyoming Basin and between the Beaverhead and Centennial mountains in the area of southwestern Montana and northeastern Idaho also would have isolated Great Basin–Snake River Plains populations of steppe and semidesert mammals from formerly contiguous segments of the Northern Great Plains. This isolation during glacial periods, together with associated environmental changes in both the Great Basin and Great Plains, could well have resulted in speciation involving several kinds of mammals.

With the onset of drier, warmer climates in the post-glacial period, coniferous forests retreated northward and upward, again opening nonforested corridors between the Great Basin and Great Plains. Steppe and desert species spread northward in the wake of the retreating forest. Contact was restored between those taxa isolated during the Wisconsin, with introgression between those for which differentiation remained below the species level. Upon coming into contact, species from either area appear in most cases to be competitively exclusive—for example, the kit and swift foxes *(Vulpes)*, the western and eastern spotted skunks *(Spilogale)*, and the pocket mice *(Perognathus parvus, apache,* and *fasciatus)*. Some coniferous forest species were isolated on montane islands of suitable habitat that were surrounded by the encroaching steppe.

Finally, as conditions became warmer, many deciduous forest species spread westward onto the plains, especially along riparian gallery forests, only to withdraw eastward again with the onset of drier conditions, but leaving behind, in some cases, relics. Drier conditions in turn favored the northward movement of both southwestern and Great Basin taxa, but likely excluded some boreomontane elements from the plains. Later, Neotropical species spread northward due to postglacial warming.

Such a sequence of distributional events beginning about 10,000 years ago could have resulted in the present distributional patterns exhibited by mammals on the Northern Great Plains.

LITERATURE CITED

Anderson, E.
   1968. Fauna of the Little Box Elder Cave, Converse County, Wyoming. The Carnivora. Univ. Colorado Studies, Earth Sci. Ser., 6:1-59.

Baerreis, D. A., and R. A. Bryson (eds.)
   1967. Climatic change and the Mill Creek culture of Iowa. Arch. Archaeol., no. 29, Soc. Amer. Archaeol., 673 pp.

Bailey, V.
   1927. A biological survey of North Dakota. N. Amer. Fauna, 49:vi+1-226.

Baker, R. H., and R. M. Hankins
   1950. A new subspecies of snowshoe rabbit from Wyoming. Proc. Biol. Soc. Washington, 63:63-64.

Banfield, A. W. F.
   1962. A revision of the reindeer and caribou, genus *Rangifer*. Bull. Nat. Mus. Canada, 177:vi+1-137.

Beidleman, R. G.
   1955. An altitudinal record for bison in northern Colorado. Jour. Mamm., 36:470-471.

Black, C. C.
   1963. A review of the North American Tertiary Sciuridae. Bull. Mus. Comp. Zool., 130:109-248.

Brown, L. N.
   1965. Status of the opossum, *Didelphis marsupialis,* in Wyoming. Southwestern Nat., 10:142-143.

Bryson, R. A., and W. M. Wendland
   1966. Tentative climatic patterns for some late-glacial and post-glacial episodes in central North America. Mimeographed, Univ. Wisconsin, Madison, 33 pp.

Cary, M.
   1917. Life zone investigations in Wyoming. N. Amer. Fauna, 42:1-95.

Charlesworth, J. K.
   1957. The Quaternary era. Edward Arnold, London, 2 vols., 1700 pp.

Clelland, C. E.
   1966. The prehistoric animal ecology and ethnozoology of the upper Great Lakes region. Anthro. Papers, Mus. Anthro., Univ. Michigan, 29:x+1-294.

Cockrum, E. L.
   1948. The distribution of the hispid cotton rat in Kansas. Trans. Kansas Acad. Sci., 51:306-312.

Cockrum, E. L., and K. L. Fitch
   1952. Geographic variation in red-backed mice (genus *Clethrionomys*) of the southern Rocky Mountain region. Univ. Kansas Publ., Mus. Nat. Hist., 5:281-292.

Cushing, E. J.
   1967. Late-Wisconsin pollen stratigraphy and the glacial sequence in Minnesota.

Pp. 59-88, *in* Quarternary paleoecology (E. J. Cushing and H. E. Wright, Jr., eds.), Yale Univ. Press, New Haven, Connecticut, vii+433 pp.

Dalquest, W. W.
1962. The Good Creek formation, Pleistocene of Texas, and its fauna. Jour. Paleont., 36:568-582.
1964. A new Pleistocene local fauna from Motley County, Texas. Trans. Kansas Acad. Sci., 67:499-505.
1965. New Pleistocene formation and local fauna from Hardeman County, Texas. Jour. Paleont., 39:63-79.

Daubenmire, R. F.
1943. Vegetational zonation in the Rocky Mountains. Bot. Rev., 9:325-393.

Davis, M. B.
1967. Late-glacial climate in northern United States: a comparison of New England and the Great Lakes region. Pp. 11-43, *in* Quaternary paleoecology (E. J. Cushing and H. E. Wright, Jr., eds.), Yale Univ. Press, New Haven, Connecticut, vii+433 pp.

Dillon, L. S.
1956. Wisconsin climate and life zones in North America. Science, 123:167-176.

Dorf, E.
1959. Climatic changes of the past and present. Contrib. Mus. Paleo., Univ. Michigan, 13:181-210.

Durrant, S. D., and R. M. Hansen
1954. Distribution patterns and phylogeny of some western ground squirrels. Syst. Zool., 3:82-85.

Findley, J. S.
1955. Speciation of the wandering shrew. Univ. Kansas Publ., Mus. Nat. Hist., 9:1-68.

Findley, J. S., and S. Anderson
1956. Zoogeography of the montane mammals of Colorado. Jour. Mamm., 37:80-82.

Fryxell, F. M.
1928. The former range of the bison in the Rocky Mountains. Jour. Mamm., 9:129-139.

Genoways, H. H., and D. A. Schlitter
1967. Northward dispersal of the hispid cotton rat in Nebraska and Missouri. Trans. Kansas Acad. Sci., 69:356-357.

Guilday, J. E., and E. K. Adam
1967. Small mammal remains from Jaguar Cave, Lemhi County, Idaho. Tebiwa, 10:26-36.

Guilday, J. E., H. W. Hamilton, and E. K. Adam
1967. Animal remains from Horned Owl Cave, Albany County, Wyoming. Geol. Contrib., Univ. Wyoming, 6:97-99.

Guilday, J. E., P. S. Martin, and A. D. McCrady
1964. New Paris No. 4: a Pleistocene cave deposit in Bedford County, Pennsylvania. Bull. Nat. Speleological Soc., 26:121-194.

Hibbard, C. W.
1937. Notes on some vertebrates from the Pleistocene of Kansas. Trans. Kansas Acad. Sci., 40:233-237.

Hibbard, C. W., D. E. Ray, D. E. Savage, D. W. Taylor, and J. E. Guilday
1965. Quaternary mammals of North America. Pp. 509-525, *in* The Quaternary of the United States (H. E. Wright, Jr., and D. G. Frey, eds.), Princeton Univ. Press, Princeton, New Jersey, x+922 pp.

Hibbard, C. W., and D. W. Taylor
1960. Two late Pleistocene faunas from southwestern Kansas. Contrib. Mus. Paleo., Univ. Michigan, 16:1-223.

Hibbard, E. A.
1957. Range and spread of the gray and fox squirrels in North Dakota. Jour. Mamm., 37:525-531.

Hoffmann, R. S.
1958. The meaning of the word "taiga." Ecology, 39:540-541.

Hoffmann, R. S., and D. L. Pattie
1968. A guide to Montana mammals. Univ. Montana Printing Service, Missoula, x+133 pp.

Hoffmann, R. S., and R. S. Peterson
1967. Systematics and zoogeography of *Sorex* in the Bering Strait area. Syst. Zool., 16:127-136.

Hoffmann, R. S., and R. D. Taber
1967. Origin and history of Holarctic tundra ecosystems, with special reference to their vertebrate faunas. Pp. 143-170, *in* Arctic and alpine environments (H. E. Wright, Jr., and W. H. Osburn, eds.), Indiana Univ. Press, Bloomington, xii+308 pp.

Hoffmann, R. S., P. L. Wright, and F. E. Newby.
1969. The distribution of some mammals in Montana. I. Mammals other than bats. Jour. Mamm., 50:579-604.

Hoover, R. L., and L. E. Yeager
1953. Status of the fox squirrel in northeastern Colorado. Jour. Mamm., 34:359-365.

Howell, A. H.
1929. Revision of the American chipmunks (genera Tamias and Eutamias). N. Amer. Fauna, 52:1-157.

Hultén, E.
1937. Outline of the history of arctic and boreal biota during the Quaternary period. Aktiebolaget Thule, Stockholm.

Jackson, H. H. T.
1928. A taxonomic review of the American long-tailed shrews (genera *Sorex* and *Microsorex*). N. Amer. Fauna, 51:vi+1-238.

Jones, J. K., Jr.
1960. The hispid cotton rat in Nebraska. Jour. Mamm., 41:132.
1964. Distribution and taxonomy of mam-

mals of Nebraska. Univ. Kansas Publ., Mus. Nat. Hist., 16:1-356.

Jones, J. K., Jr., E. D. Fleharty, and P. B. Dunnigan
 1967. Distributional status of bats in Kansas. Misc. Publ. Mus. Nat. Hist., Univ. Kansas, 46:1-33.

Jones, J. K., Jr., and B. P. Glass
 1960. The short-tailed shrew, *Blarina brevicauda*, in Oklahoma. Southwestern Nat., 5:136-142.

Küchler, A. W.
 1964. Potential natural vegetation of the conterminous United States. Map, scale 1:3,168,000, Amer. Geogr. Soc., New York.

Long, C. A.
 1965. The mammals of Wyoming. Univ. Kansas Publ., Mus. Nat. Hist., 14:493-758.

Lundelius, E. L., Jr.
 1967. Late-Pleistocene and Holocene faunal history of central Texas. Pp. 287-319, in Pleistocene extinctions: the search for a cause (P. S. Martin and H. E. Wright, Jr., eds.), Yale Univ. Press, New Haven, Connecticut, x+453 pp.

Martin, P. S., and P. J. Mehringer, Jr.
 1965. Pleistocene pollen analysis and biogeography of the Southwest. Pp. 433-451, in The Quaternary of the United States (H. E. Wright, Jr., and D. G. Frey, eds.), Princeton Univ. Press, Princeton, New Jersey, x+922 pp.

Mead, R. A.
 1968. Reproduction in western forms of the spotted skunk (genus *Spilogale*). Jour. Mamm., 49:373-390.

Miller, G. S., Jr., and G. M. Allen
 1928. The American bats of the genera *Myotis* and *Pizonyx*. Bull. U.S. Nat. Mus., 144:viii+1-218.

Murray, K. F.
 1957. Pleistocene climate and the fauna of Burnet Cave, New Mexico. Ecology, 38:129-132.

Nadler, C. F.
 1964. Chromosomes and evolution of the ground squirrel, *Spermophilus richardsonii*. Chromosoma, 15:289-299.
 1966. Chromosomes and systematics of American ground squirrels of the subgenus *Spermophilus*. Jour. Mamm., 47:579-596.
 1968. Chromosomes of the ground squirrel, *Spermophilus richardsonii aureus* (Davis). Jour. Mamm., 49:312-314.

Newsome, R. D., and R. L. Dix
 1968. The forests of the Cypress Hills, Alberta and Saskatchewan, Canada. Amer. Midland Nat., 80:118-185.

Oesch, R. D.
 1967. A preliminary investigation of a Pleistocene vertebrate fauna from Crankshaft Pit, Jefferson County, Missouri. Bull. Nat. Speleological Soc., 29:163-185.

Ogden, J. G., III
 1967. Radiocarbon and pollen evidence for a sudden change in climate in the Great Lakes region approximately 10,000 years ago. Pp. 117-127, in Quaternary paleoecology (E. J. Cushing and H. E. Wright, Jr., eds.), Yale Univ. Press, New Haven, Connecticut, vii+433 pp.

Oosting, H. J.
 1950. The study of plant communities. W. H. Freeman and Co., San Francisco, 389 pp.

Over, W. H., and E. P. Churchill
 1945. Mammals of South Dakota. Univ. South Dakota Mus., 62 pp., mimeographed.

Packard, R. L.
 1956. The tree squirrels of Kansas. Misc. Publ. Mus. Nat. Hist., Univ. Kansas, 11:1-67.

Parmalee, P. W.
 1967. A Recent cave bone deposit in southwestern Illinois. Bull. Nat. Speleological Soc., 29:119-147.
 1968. Cave and archaeological faunal deposits as indicators of post-Pleistocene animal populations and distribution in Illinois. Univ. Illinois Coll. Agric., Spec. Publ., 14:104-113.

Parmalee, P. W., and D. F. Hoffmeister
 1957. Archaeozoological evidence of the spotted skunk in Illinois. Jour. Mamm., 38:261.

Parmalee, P. W., R. D. Oesch, and J. E. Guilday
 1969. Pleistocene and Recent vertebrate faunas from Crankshaft Cave, Missouri. Rept. Investigations, Illinois State Mus., 14:iv+1-37.

Pruitt, W. O.
 1954. Additional animal remains from under Sleeping Bear Dune, Leelanau County, Michigan. Papers Michigan Acad. Sci., Arts, Letters, 39:253-256.

Rudd, V. E.
 1951. Geographic affinities of the flora of North Dakota. Amer. Midland Nat., 45:722-739.

Schultz, C. B., and W. D. Frankforter
 1948. Preliminary report on the Lime Creek sites: new evidence of early man in southwestern Nebraska. Bull. Univ. Nebraska State Mus., 3:43-62.

Schultz, C. B., G. C. Lueninghoerner, and W. D. Frankforter
 1951. A graphic résumé of the Pleistocene of Nebraska (with notes on the fossil mammalian remains). Bull. Univ. Nebraska State Mus., 3:1-41.

Schultz, G. E.
 1967. Four superimposed late-Pleistocene vertebrate faunas from southwest Kansas. Pp. 321-336, in Pleistocene extinctions: the search for a cause (P. S. Martin and H. E. Wright, Jr., eds.), Yale Univ. Press, New Haven, Connecticut, x+453 pp.

Shelford, V. E.
　1963. The ecology of North America. Univ. Illinois Press, Urbana, xxii+610 pp.

Slaughter, B. H., and B. R. Hoover
　1963. Sulpher River formation and the Pleistocene mammals of the Ben Franklin local fauna. Jour. Grad. Res. Cent., 31:132-148.

Slaughter, B. H., and R. Ritchie
　1963. Pleistocene mammals of the Clear Creek local fauna, Denton County, Texas. Jour. Grad. Res. Cent., 31:117-131.

Smith, J. D., and T. E. Lawlor
　1964. Additional records of the armadillo in Kansas. Southwestern Nat., 9:48-49.

Smith, P. W.
　1957. An analysis of post-Wisconsin biogeography of the prairie peninsula region based on distributional phenomena among terrestrial vertebrate populations. Ecology, 38:205-218.

Stegmann [=Shtegman], B.
　1938. Osnovi ornitogeograficheskovo deleniya Palearctiki. Fauna SSSR, Ptitsi, 1 (2), 156 pp., in Russian and German.

Sutton, D. A., and C. F. Nadler
　1969. Chromosomes of North American chipmunks of the genus *Eutamias*. Jour. Mamm., 50:524-535.

Taber, R. D., and R. S. Hoffmann
　1963. Behavioral adaptations of mammals to mountain environments. Proc. 16th Intern. Congr. Zool., 3:54.

Terasmae, J.
　1967. Postglacial chronology and forest history in the northern Lake Huron and Lake Superior regions. Pp. 45-58, *in* Quaternary paleoecology (E. J. Cushing and H. E. Wright, Jr., eds.), Yale Univ. Press, New Haven, Connecticut, vii+433 pp.

Thaeler, C. S.
　1968. Karyotypes of sixteen populations of the *Thomomys talpoides* complex of pocket gophers (Rodentia—Geomyidae). Chromosoma, 25:172-183.

Udvardy, M. D. F.
　1963. Bird faunas of North America. Proc. 13th Intern. Ornith. Congr., pp. 1147-1167.

Van Gelder, R. G.
　1959. A taxonomic revision of the spotted skunks (genus *Spilogale*). Bull. Amer. Mus. Nat. Hist., 117:229-392.

Warren, E. R.
　1942. The mammals of Colorado. . . . Univ. Oklahoma Press, Norman, xviii+330 pp.

Watts, W. A.
　1967. Late-glacial plant macrofossils from Minnesota. Pp. 89-97, *in* Quaternary paleoecology (E. J. Cushing and H. E. Wright, Jr., eds.), Yale Univ. Press, New Haven, Connecticut, vii+433 pp.

Watts, W. A., and H. E. Wright, Jr.
　1966. Late-Wisconsin pollen and seed analysis from the Nebraska Sand Hills. Ecology, 47:202-210.

Weaver, J. E., and F. W. Albertson
　1956. Grasslands of the Great Plains. Johnson Publ. Co., Lincoln, Nebraska, ix+395 pp.

Wells, P. V.
　1965. Scarp woodlands, transported grassland soils, and concept of grassland climate in the Great Plains region. Science, 148:246-249.

Wendorf, F.
　1961. An interpretation of Late Pleistocene environments of the Llano Estacado. Pp. 115-133, *in* Paleoecology of the Llano Estacado, Publ. Ft. Burgwin Res. Center, 1:1-144.

Wetmore, C. M.
　1967. Lichens of the Black Hills of South Dakota and Wyoming. Biol. Ser., The Museum, Michigan State Univ., 3:209-464.

Wright, H. E., Jr.
　1968a. The roles of pine and spruce in the forest history of Minnesota and adjacent areas. Ecology, 49:937-955.
　1968b. History of the Prairie Peninsula. Pp. 78-88, *in* The Quaternary of Illinois (R. E. Bergstrom, ed.), Spec. Publ. Univ. Illinois Coll. Agric., 14:1-179.

Pleistocene Mammalian Local Faunas from the Great Plains and Central Lowland Provinces of the United States

CLAUDE W. HIBBARD

ABSTRACT

The Pleistocene mammalian faunal sequence of the physiographic divisions of Texas, New Mexico, Oklahoma, Kansas, and Nebraska is listed and a bibliography is given for each fauna or biota. The 37 faunas treated have been recovered south of the nonglaciated region. They are tentatively correlated with the events in the glaciated region.

The taxa are assigned to each glacial or interglacial interval of time. The first and last appearances of genera are given. Geographical shifts of distribution are noted, as are the retraction and expansion of ranges of genera. Taxa that were sympatric in the past, but that now are allopatric, are discussed on the basis of climate at the time they lived together. An interpretation of the environment as suggested by the mammals, as well as the entire biota or fauna, is given.

There is evidence of a slight cooling in temperature in Meade County, Kansas, at an elevation of 2500 feet, between the time the Upper Pliocene Rexroad local fauna lived and the arrival of the Nebraskan Dixon local fauna in Kingman County, Kansas, elevation of 1580 feet. There is no faunal evidence to support a cold climate in the Great Plains Province south and southwest of the Nebraskan glacial region during Nebraskan time.

Faunal evidence shows that Kansan time was slightly cooler than the Nebraskan. The Illinoian glacial interval was cooler than the Kansan, but not as cold as the later part of the Wisconsin or the Recent. Large land tortoises are known from deposits of both Yarmouth and Sangamon interglacials in Kansas. The sharp zonation of our present climate is a product of the latter part of the Wisconsin.

INTRODUCTION

A summary of Pleistocene local mammalian faunas from the Great Plains and Central Lowland provinces (Fig. 1, after Fenneman, 1931) has been compiled from published reports cited in the bibliography. All publications bearing directly on a fauna or biota from this region have been listed.

Late Cenozoic deposits in southwestern Kansas and northwestern Oklahoma have yielded the largest number of successive faunas known from any given local geographical area in the world. These faunas provide data that can be analyzed to show faunal and biotic shifts and retraction of ranges during the Pleistocene. These deposits and their contained faunas are in the type area of nonglaciated Pleistocene deposits of North America.

In the analysis of the mammalian faunas, the Rexroad local fauna of Upper Pliocene age from the southern Great Plains Province is used to demonstrate the change in that area from a Pliocene to a Pleistocene mammalian fauna. In previous works I have drawn heavily from the entire biota or fauna from a given site for the interpretation of Pleistocene climates.

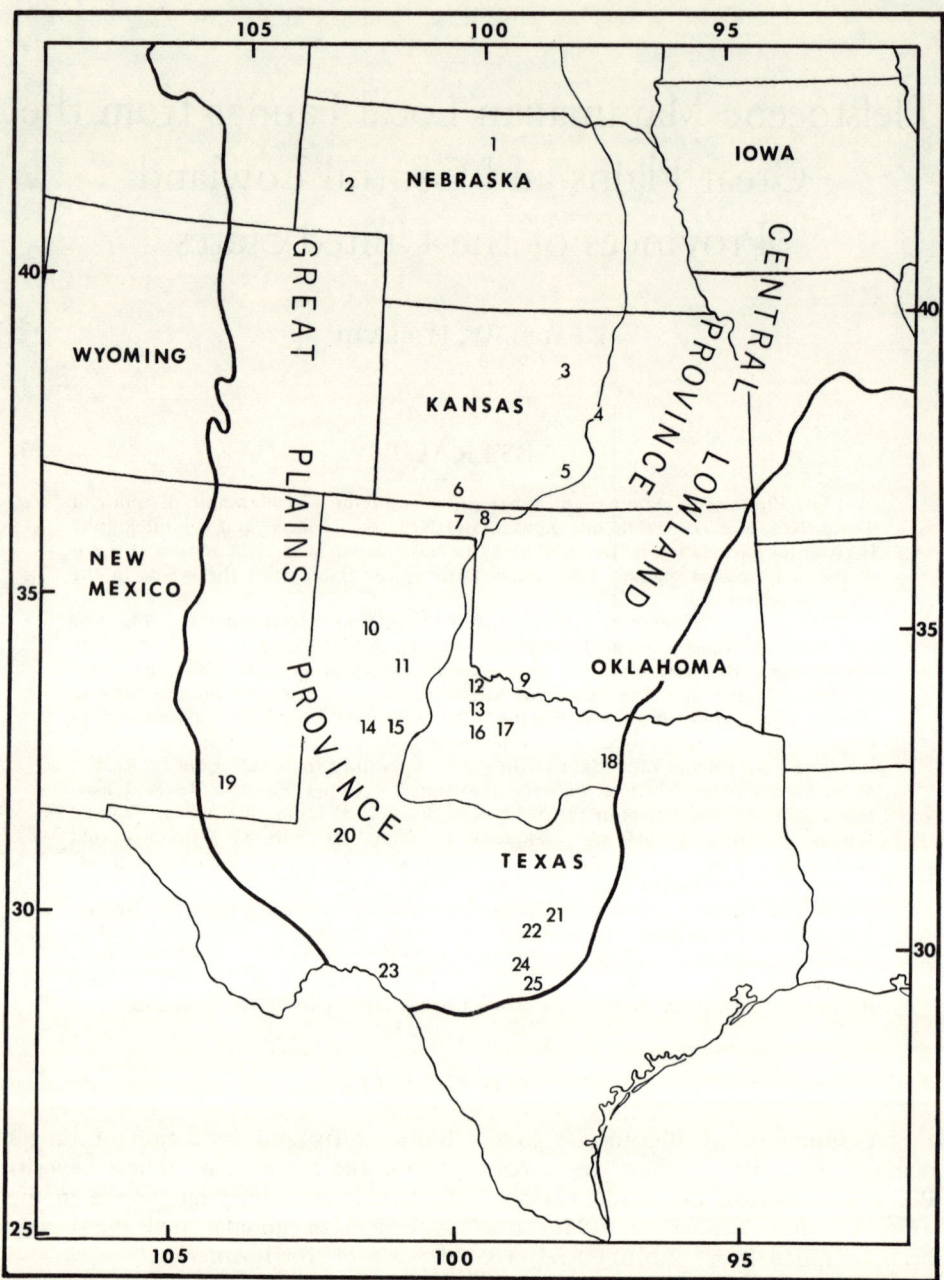

FIG. 1. List of local faunas from the Great Plains and Central Lowland provinces and their geographical location. 1, Brown County, Nebraska (Sand Draw); 2, Morrill County, Nebraska (Broadwater-Lisco); 3, Lincoln County, Kansas (Rezabek); 4, McPherson County, Kansas (Sandahl); 5, Kingman County, Kansas (Dixon); 6, Meade County, Kansas (Adams, Borchers, Butler Spring, Cragin Quarry, Cudahy, Deer Park, Jinglebob, Jones, Mt. Scott, Rexroad, Robert, Sanders, and Seger Gravel Pit); 7, Beaver County, Oklahoma (Berends); 8, Harper County, Oklahoma (Doby Springs and Bar-M); 9, Tillman County, Oklahoma (Holloman); 10, Randall County, Texas (Cita Canyon and Cudahy); 11, Briscoe County, Texas (Rock Creek); 12, Hardeman County, Texas (Howard Ranch); 13, Foard County, Texas (Easley Ranch); 14, Lubbock County, Texas (Slaton); 15, Crosby County, Texas (Blanco); 16, Knox County, Texas (Gilliland and Cudahy); 17, Baylor County, Texas (Gilliland and Cudahy); 18, Denton County,

It should be clearly understood that I do not consider the Rexroad local fauna as the latest Upper Pliocene fauna. It is older than the Hagerman local fauna of Idaho, which has a potassium-argon radiogenic date of $3.48 \pm 0.27 \times 10^6$ years (Evernden et al., 1964; see Hibbard and Zakrzewski, 1967; Bjork, 1970; Zakrzewski, 1969). So far, no fauna has been recognized in the Plains Province that is comparable in age to the Hagerman fauna. I suggested (1956) that this time may be represented in Meade County, Kansas, by the time interval between the Rexroad Formation and the Meade Formation (Ballard Formation, Hibbard, 1958a).

Previously, I assigned the Ballard Formation to Nebraskan and Aftonian time; the Angell Member has been referred to pre-Nebraskan gravels. In this discussion, the Deer Park and Sanders local faunas from the upper part of this formation are treated as Aftonian faunas. Taylor (1966) considered these two faunas and the Spring Creek local faunas of Meade County (Berry and Miller, 1966) as assemblages from the interval prior to the first major continental glaciation, but from later than an alpine or limited continental glaciation. The faunas from this area can be tentatively correlated with the events in the glaciated region on the assumption that there were only four major continental glaciations and three major interglacial intervals in North America (Hibbard, 1958b).

Hibbard and Taylor (1960) stated:

The southwestern Kansas–northwestern Oklahoma area is so situated that it has recorded the marked climatic changes of the Pleistocene epoch. It is on the boundary between the dry subhumid and semi-arid climatic types (Thornthwaite, 1948); and, fossil assemblages indicate former dryer as well as moister conditions. The Pleistocene faunas include many species which now live hundreds of miles to the north or at much higher elevations, and others now restricted to the south and east. These significant changes in distribution suggest considerable north-south shifts of isotherms, and east-west changes of moisture belts. Probably, few other areas in the Great Plains could record these changes to the same extent, for the shifts in isohyets might be rather inconspicuous westward farther into the rainshadow of the Rocky Mountains, or eastward, into what is now a moister climate. Although the exact amplitude of the north-south range shifts is not known, presumably both to the north and south at some place they would be less apparent.

The main reason why the shifts in distribution of species are so conspicuous here is that the area studied lies in the Great Plains—a tremendous expanse of relatively low relief that forms a broad pathway for ready dispersal. Furthermore, the Plains lie east of the Rocky Mountains and, as a result, are beyond the influence of westerly winds from the Pacific Ocean. They are a part of that great interior lowland of North America, between the Rocky Mountains and Appalachians, over which air masses from the arctic or Gulf of Mexico can pass with little obstruction by mountains. These unimpeded air masses vitally affect all life on the Plains. Thus, from the successive faunal differences in a local area, it is entirely possible to infer climatic changes over a large part of the continent.

UPPER PLIOCENE (REXROAD LOCAL FAUNA)

The Upper Pliocene Rexroad local fauna numbers 52 taxa of mammals (Table 1), consisting of two insectivores, one bat, 16 carnivores, 20 rodents, one sloth, one mastodon, four rabbits, five artiodactyls, and two horses. All belong to extinct species except one carnivore, which has been assigned to *Taxidea taxus* (Schreber).

The following six genera have not

Texas (Clear Creek); 19, Eddy County, New Mexico (Burnet Cave); 20, Midland County, Texas (Scharbaur); 21, Burnet County, Texas (Longhorn Cave); 22, Llano County, Texas (Millers Cave); 23, Val Verde County, Texas (Centipede and Damp Cave); 24, Kendall County, Texas (cave without a name); 25, Bexar County, Texas (Friesenhahn).

been reported from the Pleistocene: *Hesperoscalops* (a mole), *Dipoides* (a beaver with ever-growing teeth, which gave rise to *Paradipoides stovalli* Rinker and Hibbard), *Symmetrodontomys* (a peromyscine Cricetinae), and three rabbits, *Notolagus*, *Pratilepus* (which give rise to *?Sylvilagus bensonensis* Gazin), and *Nekrolagus* (which gave rise to *Sylvilagus*, *Oryctolagus*, and *Lepus*—see Hibbard, 1963). Twenty-seven species of mammals are known only from the Upper Pliocene Rexroad and older Pliocene faunas.

Megalonyx (a sloth), *Procyon* (a raccoon), and *Sigmodon* (a cotton rat) are South American genera. *Megalonyx* or a closely related sloth is known from Hemphillian faunas. *Procyon* and *Sigmodon* appear for the first time in the Rexroad fauna and seem to have made their way northward after North and South America were connected in the Pliocene.

Inferences about conditions under which the fauna lived can be derived from the sediments containing fossils, the fossils themselves, and from the habitats of some of the close living relatives of extinct members of the fauna.

The presence of *Stegomastodon*, a semibrowser, and two other genera of mastodons known from these deposits, as well as *Megalonyx*, suggests a greater abundance of broad-leafed shrubs and trees than now occurs in the region.

Wetmore (1944), after a study of the fossil birds from localities KU 2 and 3, stated: "Of the identified specimens more than one-half belong to aquatic species that live in and around marshes, streams and ponds. Remains of turkeys represent birds of wooded areas, while parrots, pigeons and quail are species of forest, or regions where thickets and groves grow amid plains, prairies or savannas." The turkey has been recently identified by Brodkorb (1964) as a species of the ocellated turkey group. Collins (1964), after the study of four genera of fossil ibises from Locality 3, stated: "Ecological information derived from these ibises and previous work indicates that this area probably had a warm, moist, frost-free, tropical climate as is found today in parts of northern South America where ibises of these genera are sympatric." Kansas at this time was probably the wintering ground for many of the more northern populations of birds that now migrate much farther southward during the winter months.

The giant land tortoise (*Geochelone rexroadensis* Oelrich), a member of this fauna, indicates frost-free winters.

I picture the region at the time the fauna lived to have had well-developed gallery forests along the banks of braided streams. The valleys contained some marsh areas, and the remaining valley areas were savanna with tall grasses and scattered groves of trees and shrubs. Short grass would have occurred only on the driest of upland areas. Unfortunately, no fossil plants or pollen are known from these deposits. Lignitic deposits are present at KU Locality 3.

A knowledge of the Upper Pliocene fauna and the climate in which it lived is important to the understanding of the later Pleistocene climates of the Great Plains Province.

NEBRASKAN (THE DIXON LOCAL FAUNA)

The fauna representing the Nebraskan interval of time is poorly understood in the Great Plains and Central Lowland provinces.

Only the Dixon local fauna (*sensu stricto*), recovered from five tons of matrix, is assigned to this interval. Restriction of the Dixon fauna excludes *Cryptotis kansasensis* Hibbard, *Sigmodon* sp., and the mollusks reported from Locality 2 (Taylor, 1960, pp. 37-38). The fossils from Locality 2 were recovered from approximately one-half ton of matrix taken 1¾ mi. S and 1 mi. E Norwich, Kingman County, Kansas. This locality is 20 miles south of the Dixon faunal site. I consider the fauna from this site as having lived in that region during Early Aftonian.

Fourteen taxa of mammals (Table 2) are known from the Nebraskan interval. The following genera from the Rexroad fauna are known from the Great Plains Province during Nebraskan time: *Sorex, Spermophilus, Geomys, Procastoroides, Peromyscus,* and *Nebraskomys.*

The following four genera of microtines make their first appearance in the Great Plains Province during the Nebraskan (see Table 2): *Pliopotamys, Ophiomys, Pliolemmus,* and *Synaptomys. Pliopotamys* (ancestral muskrat) and *Ophiomys* (a small microtine with rooted teeth) occur in the Upper Pliocene Hagerman local fauna of Idaho.

Pliolemmus (a vole) and *Synaptomys* (a lemming), both with evergrowing teeth, probably developed from microtine stocks in northern North America. They were able to extend their ranges southward or southwestward into an area that is considered to have had a climate with cooler summers and a more effective rainfall than existed at the time the Rexroad fauna lived. Two species of *Sorex* are only known to occur in the Dixon local fauna.

The southward retraction of the ranges of the genera *Urocyon, Procyon, Buisnictis, Spilogale, Liomys, Prodipodomys, Baiomys, Onychomys,* and *Sigmodon* during glaciation is as important as the first appearance of advanced microtines. *Baiomys* and *Liomys* never have been found in post-Pliocene deposits in southwestern Kansas after the retraction of their ranges at the close of the Pliocene.

The climate is considered to have been more of the maritime type at this time in the Southern Great Plains Province.

AFTONIAN

The six faunas assigned to the Aftonian (see Table 3) show the return of southern genera such as *Paenamarmota, Perognathus, Bensonomys,* and *Sigmodon* to the Central Great Plains Province. The Sand Draw, Deer Park, Cita Canyon, and Blanco local faunas contain the remains of the giant land tortoise, *Geochelone.* These tortoises lived as far north as Brown County, Nebraska, at an elevation of 2425 feet.

Sixty-three different taxa of mammals are known from these faunas. The following genera appear for the first time in the Great Plains Province: *?Ursus, Canimartes, Lynx, Panthera, Smilodon, Chasmaporthetes, Glyptotherium, Rhynchotherium,* and *Capromeryx.*

These 22 genera make their last appearance: *Borophagus, Satherium, Canimartes, Trigonictis, Buisnictis, Chasmaporthetes, Ischyrosmilus, Paenamarmota, Prodipodomys, Procastoroides, Bensonomys, Pliopotamys, Nebraskomys, Ogmodontomys, Pliophenacomys, Ophiomys, Pliolemmus, Glyptotherium, Rhynchotherium, Serobelodon, Pliauchenia,* and *Pliohippus (Astrohippus).*

Borophagus, Chasmaporthetes, and *Ischyrosmilus* seem to have been displaced by the more progressive *Smilodon* and *Homotherium (Dinobastis)* and the short-faced bears. *Prodipodomys* was replaced by more advanced kangaroo rats, and *Procastoroides* was replaced by *Castoroides.*

The genus *Bensonomys* Gazin (1942) was erected for *Eligmodontia arizonae* Gidley from the Benson local fauna of Arizona. *Bensonomys arizonae, B. pliocaenicus* (Hibbard), *B. eliasi* (Hibbard), and *B. meadensis* are not closely related to the South American cricetid *Eligmodontia* or its ancestral stock. *Eligmodontia* has four well-developed roots on m1. The similarity in characteristics between *Eligmodontia* and *Bensonomys* are considered as parallel development.

The microtines *Ogmodontomys*, *Ophiomys*, and *Pliolemmus* appear to have been replaced in the plains region by more progressive microtines, such as *Microtus* and *Pitymys* from Eurasia, which make their first appearance in the Kansan fauna. The mastodons *Rhynchotherium* and *Serbelodon?* seem to have given way to *Mammut*, *Cuvieronius*, and *Stegomastodon*.

Three microtine genera, *Nebraskomys* (gave rise to *Atopomys*), *Pliophenacomys* (gave rise to *Phenacomys*), and *Pliopotamys* (gave rise to *Ondatra*, the muskrat), are examples of phyletic replacement. The loss of the 22 genera has been chiefly by replacement with more progressive taxa.

The presence of two genera of sloths and three genera of mastodons in western Nebraska, southwestern Kansas, and northwestern Texas indicates the presence of gallery forests and savannas during the time they inhabited the region.

Warmer winters than occurred during the Nebraskan are indicated by the return of *Sigmodon* (cf. *S. intermedius*), a cotton rat, as far north as Brown County, Nebraska. I have assigned the Sand Draw mammals listed in Table 3 to the Aftonian because they were taken in the area of the Frick Quarry, which was worked by Morris Skinner. This quarry yielded remains of large land tortoises of the genus *Geochelone*. The vertebrates reported occur at this stratigraphic level or higher in the geological section. This is the area of McGrew's (1944, fig. 14) type locality of the Sand Draw local fauna. The additions to the faunal list are based on specimens collected during the summers of 1967 and 1968.

Mild winters are further indicated by (1) the abundance of *Biomphalaria kansasensis* Berry, a subtropical-tropical gastropod from Meade County, Kansas (Berry and Miller, 1966); (2) the occurrence of large species of the land tortoise *(Geochelone)* in both Brown County, Nebraska, and Meade County, Kansas; and (3) the occurrence of the large tortoise (*Gopherus canyonensis* Johnston), the ocellated turkey (*Agriocharis leopoldi* Miller and Bowman), and the glyptodon (*Glyptotherium* cf. *texanum* Osborn) from the Cita Canyon local fauna of Randall County, Texas.

At the time the giant land tortoises lived in Nebraska, Kansas, and Texas, there must have been a complete withdrawal of glacial ice from the Arctic Ocean, since these large land tortoises lived in a frost-free climate.

The Sanders local fauna of Meade County occurs above a thick bed of caliche. This fauna is Late Aftonian in age, and the fauna indicates the return of a subhumid maritime climate to this region after the semiarid interval during which the caliche was developed. It is not known what effect this semiarid interval had upon the fauna of southwestern Kansas and northwestern Texas; but during this interval there would have been a greater range in daily temperatures, and the rainfall for this region probably dropped to 15 to 18 inches for the year.

KANSAN LOCAL FAUNAS

All evidence points to a slightly cooler climate during Kansan time than during Nebraskan time in southwestern Kansas. Present northern species of *Sorex*, *Microtus*, *Synaptomys*, and *Phenacomys* were well represented in the Cudahy fauna.

The five local faunas (see Fig. 1 and Table 4) are considered to span Kansan time and include latest Blancan and earliest Irvingtonian mammal ages (Savage, 1951; Hibbard, 1958b; Hibbard *et al.*, 1965).

The Seger Gravel Pit local fauna, Meade County, Kansas, from the Stump Arroyo (sand and gravel) Member of the Crooked Creek Formation is the only Late Blancan fauna known from the Great Plains Province of this age.

The Holloman local fauna of Oklahoma and the Gilliland local fauna of Texas (elevation approximately 1400 feet), from the Seymour Formation, are pre-Cudahy faunas and occur in the Osage Plains area of the Central Lowland Province. They are the earliest Irvingtonian faunas recognized in this area. Hibbard and Dalquest (1966) assigned the Rock Creek local fauna of Texas from the Great Plains Province to this interval of time. Until the exact stratigraphic relationship of the Pearlette Volcanic Ash is known to this fauna, it might have been best to have considered it as a post-Cudahy fauna and a pre-Borchers fauna (Yarmouth) on the basis of the known vertebrates. It seems equivalent to the Arkalon local fauna of Seward County, Kansas, which is post-Pearlette Volcanic Ash fauna and a pre-Borchers fauna (Hibbard, 1953).

The Cudahy fauna, which occurs just below the Pearlette ash, is known from Kansas, Oklahoma, and Texas. The Kansas and Oklahoma part of the fauna occurs in the Great Plains Province, whereas the Vera faunule from Knox and Baylor counties, Texas, is from the Osage Plains of the Central Lowland Province. The composition of the faunas from the two regions reflects both the difference in elevation and their geographical locations. Sixty-two taxa of mammals are known.

The following genera are known for the first time from the Great Plains Province (including the Rock Creek local fauna): *Microsorex, Arctodus, Scalopus, Thomomys, Ondatra, Phenacomys, Microtus, Pitymys, Synaptomys (Mictomys), Mammuthus, Sylvilagus, Preptoceros, Equus (Asinus),* and *Equus (Hemionus).*

Those known for the first time from the Central Lowland Province are: *Dinobastis, Ondatra, Microtus, Nothrotherium, Chlamytherium, Glyptodon, Cuvieronius, Mammuthus, Sylvilagus, Tetrameryx, Capromeryx,* and *Equus (Asinus).* Two genera, *Stegomastodon,* which was crowded out by *Mammuthus,* and the camel *Titanotylopus,* make their last appearance. *Microtus* and *Pitymys,* of Eurasian origin, make their first appearance. The mammalian fauna begins to take on a modern aspect, since 11 of the fossil remains are assigned or referred to Recent species.

In the Central Lowland Province were recovered the remains of two sloths *(Nothrotherium* and *Paramylodon),* the giant armadillo *(Chlamytherium),* the large glyptodon *(Glyptodon),* two mastodons *(Stegomastodon, Cuvieronius),* a mammoth *(Mammuthus),* five species of camels, two pronghorns, three species of the horse *(Equus),* the tapir *(Tapirus),* the ocellated turkey, and the giant land tortoise *(Geochelone).* These vertebrates are part of the Gilliland fauna and occur below the Vera faunule of the Cudahy fauna in the Seymour Formation. The diversity of large grazers in the fauna indicates the presence of tall grasses in the valleys. The presence of the browsers indicates the occurrence of shrubs and a good stand of small trees (Hibbard and Dalquest, 1966). The fauna indicates more effective moisture and temperatures not below freezing at the time it lived in that area of Texas. Blair and Hubbell (1938) assigned this region to the Mesquite Plains biotic district. Blair (1950 and 1954) gave a good description of the present biota of this district in Texas and Oklahoma. The present mean annual precipitation (prior to 1953) in that region was 25.58 inches; the maximum temperature was 120°F, and the minimum, −14°F.

The following taxa, closely related to northern species, are members of the

Cudahy fauna of the High Plains Region: *Sorex cinereus, S. lacustris, S. megapalustris, Microsorex pratensis, Mustela* (cf. *erminea*), *Spermophilus richardsonii, Spermophilus* (cf. *franklinii*), *Phenacomys, Microtus paroperarius, Pitymys meadensis, Synaptomys (Mictomys) meltoni,* and *Zapus sandersi.* These mammals were found directly below the Pearlette Volcanic Ash in Kansas.

The earliest appearance of *Equus (E. giganteus* and *E. scotti)* is in Late Kansan deposits (Irvingtonian). The places of origin of *Equus (Equus), Equus (Asinus),* and *Equus (Hemionus)* are unknown. Their origin could have been in southern North America as well as in Asia. Therefore, the southern shift of the ranges of *Nannippus* and *Plesippus* during Kansan time could have brought them into close competition with these other three groups of horses, which were able to displace them in the Great Plains Province during Kansan time. Regardless of where these groups originated, they were successful and underwent considerable adaptive radiation in North America during the rest of the Pleistocene.

The Seger Gravel Pit local fauna was recovered from the Stump Arroyo Member (sand and gravels) of the Crooked Creek Formation (Hibbard, 1951). The latest remains of *Nannippus phlegon* and *Plesippus* (a zebrine horse) known from the Great Plains Province occur in this deposit. Their remains have never been found in deposits above the top of the Stump Arroyo Member of the Crooked Creek Formation in this region. The top of this member in the Great Plains region is considered as the boundary between Late Blancan and the Irvingtonian mammal ages.

YARMOUTH (THE BORCHERS LOCAL FAUNA)

This interval of time was the longest of all interglacials, and one would expect it to yield the greatest number of faunas. It appears, however, to have been a stable tectonic time in the Rocky Mountains west of the Great Plains Province. It was a time of deep soil development, filling of valleys, and the development of deep caliches in the south and southwest. The stable surface, capped by the caliche, forms part of the High Plains and some of the Lowland Plains surfaces upon which is developed much of our present drainage systems.

I have searched for the past 27 years in Kansas, Oklahoma, and Texas for an equivalent to the Borchers local fauna. It is evident that the Arkalon vertebrates taken from above the sand and gravel that contained the pure volcanic ash (Hibbard, 1953) are part of an Early Yarmouth local fauna of pre-Borchers local fauna age. I have found that this interval of time in these areas is represented chiefly by caliche and deep-weathered zones. Gerald R. Paulson, presently working along the Little Sioux River in western Iowa, has found what appears to be one of the first molluscan faunas of an Early Yarmouth age in silts above the Pearlette Volcanic Ash. Deposits in Nebraska probably contain fossils of this interval. If it were not for the Crooked Creek fault and the Meade Basin there would be no record of the Borchers local fauna.

Twenty-five different taxa of mammals are known from this fauna (see Table 5). They were recovered from just above the weathered Pearlette Volcanic Ash. In this same area, elements of the Cudahy fauna are recovered from just below the ash. Two genera of mammals are new to the area, *Etadonomys,* a kangaroo rat that is related to *Prodipodomys,* and *Lepus.* Large premolars and molars of a rabbit have been recovered from earlier deposits but no p3's or P2's, which would allow for determination of the genus present, have been taken. Only four of the species recovered are considered the same as, or closely related to, living species. The fauna, as a whole, is southern. Only the genus *Etadonomys* makes its last appearance in this time interval.

The most common taxon in the fauna is *Onychomys fossilis*, the second is *Sigmodon hilli*, which is derived from *Sigmodon intermedius* of the Upper Pliocene Rexroad fauna. *Sigmodon hilli* has the third root on the labial (external) side of m1 greatly reduced or absent. *S. hispidus*, a later invader from the American tropics, has four well-developed roots on m1. I am unable to account for the absence of *Peromyscus*.

Three of the species known from the Rexroad fauna *(Sorex taylori, Perognathus pearlettensis,* and *P. gidleyi)* are recorded for the first time in this area since the retraction of the Rexroad fauna. It is of interest that a gray fox (*Urocyon atwaterensis* Getz) appears in the fauna. The *Synaptomys* known from the fauna is a side branch of the subgenus *Synaptomys*. This lineage, *Synaptomys vetus* Wilson and *S. landesi* Hibbard, is known to occur only in the Grand View local fauna of Idaho and the Borchers. The presence of *Synaptomys landesi, Zapus burti,* and the copperhead, *Agkistrodon contortrix* Linnaeus (see Brattstrom, 1967), indicates a more effective moisture then than now. Those interested in the habitat differences between the Cudahy and Borchers faunas should note the difference between the two snake faunas of the same region that lived under different climatic conditions. Present are the giant land tortoise *Geochelone*, the southern *Sigmodon hilli, Neotoma taylori* (closest living relative is *Neotoma alleni*), and the small *Spilogale* (cf. *putoris ambarvalis*), which are considered to be "mild winter" forms (Hibbard *et al.*, 1965). The climate is considered to have been of the maritime type, with frost-free winters and with the Arctic Ocean not frozen over. Some time after the Borchers fauna lived, there was a shift to a semiarid climate, and during that time the extensive caliche was developed that caps the Crooked Creek Formation. It is upon this surface that many of the present drainage systems were developed in the Great Plains Province.

ILLINOIAN LOCAL FAUNAS

Most of these faunas are known from Beaver and Harper counties, Oklahoma, and Meade County, Kansas (see Fig. 1 and Table 6), in the Plains Province. The Sandahl local fauna from McPherson County, Kansas, was taken on the western edge of the Lowland Province, but it clearly shows some of the faunal differences between the two physiographic and biotic provinces (Dice, 1943). All seven Illinoian faunas in the region were taken from basin fills or terraces along streams that have entrenched below the upland surface of the region.

Fifty-seven different kinds of mammals are known from these faunas, of which 53 are known from the Early Illinoian and 33 from the Late Illinoian. A faunal shift occurred in Meade County after the Early Illinoian, and more southern elements entered the region during Late Illinoian time as shown by the Mt. Scott local fauna. For example, *Blarina brevicauda fossilis* occurs in the Early Illinoian faunas and is replaced by *B. carolinensis* in the Late Illinoian faunas. The study of pollen from the Oklahoma and Kansas faunal sites by Kapp (1965) has contributed greatly to the understanding of the climates at this time. Kapp also shows a shift in the flora in this region during the Illinoian. The molluscan species studied by Miller (1966) also show a slight shift in ranges during Illinoian time in this area.

The following genera are known for the first time from the Early Illinoian of the Plains Province: *Burosor, Paradipoides, Castor, Neofiber* (which is known from the earlier Port Kennedy Cave fauna of Pennsylvania), and *Bison*. Only one genus, *Oryzomys* (the rice rat), is new for the region in the Late Illinoian, but it appears with the large southern *Terrapene* in the Mt. Scott fauna. This is the earliest record for this South American rodent in the United States. The m1 has four, three, or two roots, depending upon the species (observations in part and written communication, October 27, 1967, from Robert T. Orr).

Mylohyus (the long-nosed peccary) makes its first appearance along the borders of the Central Lowland and Great Plains provinces. It is of interest to note the presence of the American mastodon in the Sandahl fauna of McPherson County, Kansas. The mastodon indicates the greater abundance of trees and shrubs for browsing than in the Great Plains Province.

Twenty-four of the taxa are referred to Recent species. The three species of *Sorex* now occur to the north and west of the area of recovery. *Spermophilus richardsonii, Cynomys vetus* (of the white-tailed prairie dog group), and *Thomomys* are western and northwestern taxa. *Blarina brevicauda* and *Zapus hudsonius* are northeastern species, whereas *Blarina carolinensis* and *Oryzomys fossilis* are of southern origin.

Neofiber cannot be considered a southern genus. It was widespread when extensive marsh lands were available. In Port Kennedy Cave and the Rezabek local faunas it was found associated with *Ondatra*. Its present distribution is limited by the lack of favorable habitat in other regions.

One of the outstanding examples of sympatry in the Late Illinoian Mt. Scott local fauna is that *Sorex arcticus* and *S. palustris* (northern species) are sympatric with *Blarina carolinensis* and *Oryzomys fossilis* (southern forms). These two groups are now allopatric. Some genera, as well as many species, are found to have occurred during parts of the Pleistocene as sympatric members of a fauna, but are now allopatric. Such occurrences have a definite bearing both on the climate in which they lived as well as the microenvironments that were available in the area in which the fauna lived.

The Early Illinoian faunas indicate a more moist interval in the regions in which they lived than occurs there at the present time. This is evident from the presence of *Sorex palustris, Paradipoides stovalli,* the giant beaver *(Castoroides ohioensis),* and *Neofiber fossilis*. The giant beaver required large marsh areas joined by a large body of water. It is considered to have developed in the early northern lake regions of North America from a pre-Pleistocene *Procastoroides* stock. There is no evidence that *Paradipoides* (a large *Dipoides*) and *Castoroides* ever built dams. *Castor* was able to provide its own environment by building dams. There is no reason to assume that winters during Early and Middle Illinoian times in the region where the faunas lived were any colder than at present, but the climate lacked the summer extremes in temperature that occur there today. Such summer temperatures would have been fatal to the northern shrews and other species. Those further interested in this climate should see Smith (1954), Kapp (1965), and Miller (1966).

Milder winter temperatures than now exist in the region are indicated by the invasion of southern species into the region during the Late Illinoian and the retraction of some northern mammals from the region. The biota found in the deposits that yielded the Mt. Scott mammals indicates a more effective moisture for the region than now occurs there, as well as the lack of summer extremes in temperature. The climate during this interval in Meade County is not considered to have had winter temperatures as low as those proposed by Hibbard and Taylor (1960) for the Butler Spring local fauna or by Stephens (1960) for the Doby Springs local fauna. I consider the winter temperatures to have been like those at the present time in southern New Jersey, with cool summers as found at present in northern New Jersey and southern Wisconsin.

SANGAMON MAMMALIAN LOCAL FAUNAS OF THE HIGH PLAINS REGION

Fifty-two different kinds of mammals are recognized from three faunas (Table 7). The only new genus occurring in this period of time is the South American armadillo *(Dasypus)*. No mammalian genera are known to have become extinct. I have not included faunas from southern Texas assigned to this interval of time because they were based on the concept that the Sangamon began with the carbon-14 date of 25,000 to 40,000 years BP. A younger age is assigned to the Slaton local fauna than that which Dalquest (1967) mentioned, because the mollusks that I collected some years ago (Taylor, 1965) are considered as Sangamon. This fauna is considered tentatively to be an Early Sangamon fauna. The fauna is different from any reported from the Great Plains Province. One of the abundant mammals is *Neofiber leonardi*, which occurs with *Cynomys vetus, Dasypus* (cf. *novemcinctus*), and *Equus (Hemionus) calobatus*.

The Cragin Quarry local fauna occurs above the Mt. Scott local fauna on the Big Springs Ranch in the Kingsdown Formation, which is capped by the Sangamon caliche. The fauna is typical of an interglacial interval. A number of the species are southern and southwestern in distribution. Nineteen of the species are referred to Recent taxa, and none shows a northern relationship. Vertebrates occurring in the area that are southern and do not occur there today are *Geochelone, Gopherus, Terrapene carolina llanensis* Oelrich, *Phrynosoma modestum* Girard, *Holbrookia texana* (Troschel) (see Ethridge, 1958), *Agkistrodon contortrix* (Linnaeus) (see Brattstrom, 1967), *Notiosorex crawfordi* (Coues), and *Lasiurus (Dasypterus) golliheri* Hibbard. For a detailed account of the environment and climate, see Hibbard and Taylor (1960, pp. 33-39), who stated: "The former climate differed radically from that of the present by being much less continental: winter temperatures may never have reached freezing and summers were slightly less hot than those of today. Mean annual rainfall may have been the same."

On the XI Ranch in Meade County, Kansas, Sangamon deposits occur above the Late Illinoian deposits that contain the Butler Spring local fauna. The Sangamon deposits are capped by a caliche. In these deposits a tortoise *(Gopherus)* was found in its old burrow, as well as parts of other *Gopherus* in the Sangamon deposits. Twenty-eight taxa of mammals were recovered from deposits associated with remains of *Gopherus* (see Schultz, 1967, p. 331).

The Jinglebob biota occurs in a valley fill in the Rexroad Formation and is assigned to its time position on the basis of geomorphology of the area and the morphological grade (systematic position) of the mammalia. Some have suggested that it might be of Early Wisconsin age, others wish to assign it to a Kansan age because *Bison latifrons* occurs in the fauna. The fauna is known from 23 different mammals; 10 are referred to living species. The detailed work by Semken (1966) on *Ondatra* certainly places the fauna in the Late Sangamon or Early Wisconsin.

For a detailed description of the environment and climate, see Hibbard (1955, pp. 199-204) and Kapp (1965). In summary, at the time this fauna lived, a moist, temperate, equable climate must have existed in that part of the plains region. Moist conditions along the stream valleys allowed a coastal-plain biota comparable, in part, to the present coastal-plain fauna to live as far north as southwestern Kansas. "For pines, and possibly either *Myrica* or *Comptonia,* to spread into the southern High Plains from the southeast and Gulf Coast, a moist climate is postulated" (Kapp, 1965, p. 238). Because of the presence of the large molluscan fauna (Hibbard and Taylor, 1960, pp. 59-60), the box turtle (*Terrapene carolina llanensis*), the short-tailed shrew, the rice rat, and the meadow vole, it is concluded that the rainfall approximated 40 to 45 inches a year, that the winters were as warm or slightly warmer than they are now in southeastern Arkansas, and that the summers were cooler and more moist in comparison with those of the present.

WISCONSIN LOCAL FAUNAS FROM THE GREAT PLAINS AND CENTRAL LOWLAND PROVINCES

The Wisconsin mammalian faunas from Texas are far better known than those from the rest of the region under discussion. I have listed two faunas from Kansas, one from Oklahoma, one from New Mexico, and nine from Texas. The list and bibliographies were prepared before I received a copy of the book *Pleistocene Extinctions.* Lundelius (1967) has listed 16 Wisconsin and Holocene faunal sites from Texas, and has given an excellent analysis of these faunas. I consider Wisconsin time to have begun more than 75,000 years before the present. The oldest fauna included in the list is the Clear Creek local fauna from Denton County, Texas. Students interested in the occurrence of Wisconsin man with associated vertebrates are referred to Sellards (1952).

These Wisconsin faunas include 134 different taxa of mammals, of which 78 are referred to living species. These faunas contain 14 extinct genera, not including the horses, although all major groups are still living outside of the Western Hemisphere.

Seven genera of mammals are new to the region: *Eumops,* the mastiff bat; *Conepatus,* the hognose skunk; *Pappogeomys* (*Cratogeomys*), a pocket gopher; *Erethizon,* the porcupine; *Antilocapra,* the pronghorn; *Euceratherium,* a shrub-oxen; and *Ovis,* the bighorn sheep. Most of the small mammals are representatives of living species. The geographic distribution of some taxa during the Late Wisconsin was much different in Texas (Lundelius, 1967) than now. Genera and species in Texas that were sympatric and are now allopatric definitely indicate a difference in climatic conditions between Late Wisconsin time and the present.

In 1960 I published a figure (1960, fig. 1) of a large *Geochelone* from Dallas County, Texas, and assigned it a Sangamon age. Bob Slaughter has rightly questioned the age assignment, because it appears that these large land tortoises lived well into Wisconsin time in both Florida and Texas.

If these large land tortoises of Texas prove to be of Wisconsin age, it simply means that the climate was warm enough at that time in that region to allow them to live there. Auffenberg and Milstead (1965, p. 560) stated: "The available record of Pleistocene reptiles in Florida indicates that this area possessed a sufficiently equable climate so that many cold-sensitive types existed there throughout the Pleistocene until the Wisconsin, perhaps even until near its closing."

In 1960 I wrote: "The extreme winter temperatures, as now known, are considered to have developed during the Wisconsin." I emphasize that the continental climate now characteristic of the Great Plains and Central Lowland provinces and the present sharp zonation of climate is a product of the Late Wisconsin.

SUMMARY

In the interpretation of past climates it is natural to think of the association of present temperatures and precipitation. These two elements must be considered separately in the reconstruction of Pleistocene climates in the Interior Plains.

The Nebraskan interval of time is poorly represented by vertebrate remains. None of the Early Pleistocene mammals thus far known can be considered as indicators of a truly cold climate. Faunal evidence shows that the vertebrates in the plains region lived under less severe climatic conditions then than existed during later glacial times. Faunas indicate that the climate during each of the later glaciations was slightly cooler than the preceding glacial climate.

During each interglacial period, subtropical climates extended much farther north than now. Large land tortoises *(Geochelone)* are known from the Aftonian of Brown County, Nebraska; Meade County, Kansas; and Crosby and Randall counties, Texas. The climate during much of the interglacial periods indicates that the Arctic Ocean was not frozen over.

The Late Kansan Cudahy fauna, from below the Pearlette ash that has been taken in Texas, Kansas, and Iowa, does not indicate the marked continental climate of the present as having existed at that time. Below the Cudahy fauna in Baylor and Knox counties, Texas, remains of large land tortoises (two meters in length) are known from deposits laid down during Kansan time.

The Borchers mammalian fauna is the only one known from the Yarmouth. Its composition is in marked contrast to the Cudahy fauna from the same area in Kansas. A large land tortoise *(Geochelone)* is a member of this fauna.

The Illinoian fauna takes on the aspect of a more northern Recent fauna than either the Nebraskan or Kansan faunas. The climate differed from that of the present in being generally cooler in the summers but not as cold in the winters. The rainfall was the same as now, or slightly greater, but more effective.

During Sangamon time, large *Gopherus* (tortoises) made their way north to Meade County, Kansas, at an elevation of 2500 to 2600 feet, as did a medium-sized *Geochelone*. The climate differed radically from that of the present by being much less continental.

Small vertebrates recovered from Wisconsin deposits are chiefly the same species found living at the present time, although many of the species of both vertebrates and invertebrates that occurred sympatrically in these faunas are now allopatric species.

The marked extinction of many of the large vertebrates during the Late Wisconsin may have been because the patterns of North American climates had become more complex and were composed of more sharply contrasting conditions. The climatic zoning as known at present, with the extreme winter temperatures of the Interior Plain, is in a large part due to the strong continentality of the climate (Hibbard, 1960).

ACKNOWLEDGMENTS

I am indebted to the following students, who, in the spring of 1967, helped to prepare the bibliographies on faunas for a seminar on the interpretation of Pleistocene climates: Kraig Adler, Philip Bjork, Alan Feduccia, Lynn Fichter, Timothy Lawlor, John Lundberg, Gerald Paulson, George Zug, and Richard Zakrzewski.

I am grateful to J. Knox Jones, Jr., William Akersten, and my wife, Faye Ganfield Hibbard, for help with the manuscript.

Data accumulated under National Science Foundation projects G-5635, G-19458, GB 1526, and GB-5450 were used both in the text and the preparation of the faunal tables.

LITERATURE CITED

Auffenberg, W., and W. W. Milstead
  1965. Reptiles in the Quaternary of North America. Pp. 557-568, *in* The Quaternary of the United States (H. E. Wright, Jr., and D. G. Frey, eds.), Princeton Univ. Press, Princeton, New Jersey, x+922 pp.
Berry, E. G., and B. B. Miller
  1966. A new Pleistocene fauna and a new species of *Biomphalaria* (Basommatophora: Planorbidae) from southwestern Kansas, U.S.A. Malacologia, 4:261-267.
Bjork, P. R.
  1970. The Carnivora of the Hagerman local fauna (Late Pliocene) of southwestern Idaho. Trans. Amer. Philos. Soc., in press.
Blair, W. F.
  1950. The biotic provinces of Texas. Texas Jour. Sci., 2:93-177.
  1954. Mammals of the Mesquite Plains biotic district in Texas and Oklahoma, and speciation in the central grasslands. Texas Jour. Sci., 6:235-264.
Blair, W. F., and T. H. Hubbell
  1938. The biotic districts of Oklahoma. Amer. Midland Nat., 20:425-454.
Brattstrom, B. H.
  1967. A succession of Pliocene and Pleistocene snake faunas from the High Plains of the United States. Copeia, 1967:188-202.
Brodkorb, P.
  1964. Notes on fossil turkeys. Quart. Jour. Florida Acad. Sci., 27:223-229.
Collins, C. T.
  1964. Fossil ibises from the Rexroad fauna of the Upper Pliocene of Kansas. Wilson Bull., 76:43-49.
Dalquest, W. W.
  1967. Mammals of the Pleistocene Slaton local fauna of Texas. Southwestern Nat., 12:1-30.
Dice, L. R.
  1943. The biotic provinces of North America. Univ. Michigan Press, Ann Arbor, viii+78 pp.
Etheridge, R.
  1958. Pleistocene lizards of the Cragin Quarry fauna of Meade County, Kansas. Copeia, 1958:94-101.

Evernden, J. F., D. E. Savage, G. H. Curtis, and G. T. James.
  1964. Potassium-argon dates and the Cenozoic mammalian chronology of North America. Amer. Jour. Sci., 262:145-198.
Fenneman, N. M.
  1931. Physiography of western United States. McGraw-Hill Book Co., New York, xiii +534 pp.
Hibbard, C. W.
  1951. Vertebrate fossils from the Pleistocene Stump Arroyo member, Meade County, Kansas. Contrib. Mus. Paleontol., Univ. Michigan, 9:227-245.
  1953. *Equus (Asinus) calobatus* Troxell and associated vertebrates from the Pleistocene of Kansas. Trans. Kansas Acad. Sci., 56: 111-126.
  1955. The Jinglebob interglacial (Sangamon?) fauna from Kansas and its climatic significance. Contrib. Mus. Paleontol., Univ. Michigan, 12:179-228.
  1956. Vertebrate fossils from the Meade Formation of southwestern Kansas. Papers Michigan Acad. Sci., Arts, Letters, 41:145-203.
  1958a. New stratigraphic names for Early Pleistocene deposits in southwestern Kansas. Amer. Jour. Sci., 256:54-59.
  1958b. Summary of North American Pleistocene mammalian local faunas. Papers Michigan Acad. Sci., Arts, Letters, 43:3-32.
  1960. An interpretation of Pliocene and Pleistocene climates in North America. Ann. Rep. Michigan Acad. Sci., Arts, Letters, 62:5-30.
  1963. The origin of the P_3 pattern of *Sylvilagus, Caprolagus, Oryctolagus* and *Lepus*. Jour. Mamm., 44:1-15.
Hibbard, C. W., and W. W. Dalquest
  1966. Fossils from the Seymour Formation of Knox and Baylor counties, Texas, and their bearing on the Late Kansan climate of that region. Contrib. Mus. Paleontol., Univ. Michigan, 21:1-66.
Hibbard, C. W., C. E. Ray, D. E. Savage, D. W. Taylor, and J. E. Guilday
  1965. Quaternary mammals of North Amer-

ica. Pp. 509-525, *in* The Quaternary of the United States (H. E. Wright, Jr., and D. G. Frey, eds.), Princeton Univ. Press, Princeton, New Jersey, x+922 pp.

Hibbard, C. W., and D. W. Taylor
1960. Two Late Pleistocene faunas from southwestern Kansas. Contrib. Mus. Paleontol., Univ. Michigan, 16:1-223.

Hibbard, C. W., and R. J. Zakrzewski
1967. Phyletic trends in the Late Cenozoic microtine *Ophiomys* gen. nov. from Idaho. Contrib. Mus. Paleontol., Univ. Michigan, 21:255-271.

Kapp, R. O.
1965. Illinoian and Sangamon vegetation in southwestern Kansas and adjacent Oklahoma. Contrib. Mus. Paleontol., Univ. Michigan, 19:167-255.

Lundelius, E. L., Jr.
1967. Late-Pleistocene and Holocene faunal history of central Texas. Pp. 287-319, *in* Pleistocene extinctions: the search for a cause (P. S. Martin and H. E. Wright, Jr., eds.), Yale Univ. Press, New Haven, Connecticut, x+453 pp.

McGrew, P. O.
1944. An Early Pleistocene (Blancan) fauna from Nebraska. Field Mus. Nat. Hist., Geol. Ser., 9:33-66.

Miller, B. B.
1966. Five Illinoian molluscan faunas from the Southern Great Plains. Malacologia, 4:173-260.

Savage, D. E.
1951. Late Cenozoic vertebrates of the San Francisco Bay region. Univ. California Publ. Geol. Sci., 28:215-314.

Schultz, G. E.
1967. Four superimposed Late-Pleistocene vertebrate faunas from southwest Kansas. Pp. 321-336, *in* Pleistocene extinctions: the search for a cause (P. S. Martin and H. E. Wright, Jr., eds.), Yale Univ. Press, New Haven, Connecticut, x+453 pp.

Sellards, E. H.
1952. Early man in America. Univ. Texas Press, Austin, 211 pp.

Semken, H. A., Jr.
1966. Stratigraphy and paleontology of the McPherson *Equus* beds (Sandahl local fauna), McPherson County, Kansas. Contrib. Mus. Paleontol., Univ. Michigan, 20:121-178.

Smith, C. L.
1954. Pleistocene fishes of the Berends fauna of Beaver County, Oklahoma. Copeia, 1954:282-289.

Stephens, J. J.
1960. Stratigraphy and paleontology of a Late Pleistocene basin, Harper County, Oklahoma. Bull. Geol. Soc. Amer., 71: 1675-1702.

Taylor, D. W.
1960. Late Cenozoic molluscan faunas from the High Plains. U.S. Geol. Surv., Prof. Paper, 337:iv+1-94.
1965. The study of Pleistocene nonmarine mollusks in North America. Pp. 597-611, *in* The Quaternary of the United States (H. E. Wright, Jr., and D. G. Frey, eds.), Princeton Univ. Press, Princeton, New Jersey, x+922 pp.
1966. Summary of North American Blancan nonmarine mollusks. Malacologia, 4:1-172.

Thornthwaite, C. W.
1948. An approach toward a rational classification of climate. Geogr. Rev., 38:55-94.

Wetmore, A.
1944. Remains of birds from the Rexroad fauna of the Upper Pliocene of Kansas. Univ. Kansas Sci. Bull., 30:89-105.

Zakrzewski, R. J.
1969. The rodents from the Hagerman local fauna, Upper Pliocene of Idaho. Contrib. Mus. Paleontol., Univ. Michigan, 23:1-36.

APPENDIX

Tables have been prepared for the mammalian faunas from the Great Plains and Central Lowland provinces that are mapped in Figure 1. Bibliographies (complete through 1967 and with some entries beyond that date) have been prepared for each biota or fauna. Readers should refer to *The Mammals of North America* (2 vols., Ronald Press, New York, 1959) by E. R. Hall and K. R. Kelson for maps of the present distribution of species and subspecies of mammals.

In the tables (1-8) that follow, a plus sign (+) denotes extinct genera and an asterisk (*) indicates extinct species; a bar (—) records the presence of a taxon in a fauna and an "x" indicates a holotype.

TABLE 1. Mammals from the Rexroad Local Fauna (Upper Pliocene), KU Localities 2, 2a, and 3 from the Great Plains Province of Meade County, Kansas, elevation 2500 feet

Species	KU 2	KU 2a	KU 3	Species	KU 2	KU 2a	KU 3
Order Insectivora				*Liomys centralis* Hibbard			x
Family Soricidae				Family Cricetidae			
Sorex taylori Hibbard	x	—	—	*Onchomys gidleyi* Hibbard		x	—
Family Talpidae				+Symmetrodontomys simplicidens Hibbard		—	x
+Hesperoscalops rexroadi Hibbard	—	x	—	+Bensonomys eliasi (Hibbard)	x	—	—
Order Chiroptera				*Peromyscus baumgartneri* Hibbard			x
Family Vespertilionidae				*Peromyscus kansasensis* Hibbard			x
Lasiurus fossilis Hibbard			—	*Baiomys rexroadi* Hibbard			x
Order Carnivora				*Neotoma quadriplicatus* (Hibbard)			x
Family Canidae				*Sigmodon intermedius* Hibbard	x	—	—
Canis sp.			—	+Ogmodontomys poaphagus poaphagus Hibbard		—	x
Canis lepophagus Johnston			—	+Pliophenacomys primaevus Hibbard		x	
+Borophagus sp.			—	+Nebraskomys sp.			—
Urocyon progressus Stevens			x	Order Edentata			
Family Procyonidae				Family Megalonychidae			
Bassariscus casei Hibbard			x	+Megalonyx sp.			—
Procyon rexroadensis Hibbard			x	Order Proboscidea			
Family Mustelidae				Family Gomphotheriidae			
+Satherium piscinaria (Leidy)			—	+Stegomastodon rexroadensis Woodburne			x
+Satherium sp. (large)			—	Order Lagomorpha			
+Trigonictis kansasensis Hibbard			x	Family Leporidae			
Taxidea taxus (Schreber)			—	+Hypolagus regalis Hibbard			x
+Buisnictis breviramus (Hibbard)			x	+Pratilepus kansasensis Hibbard			x
Spilogale rexroadi Hibbard			x	+Nekrolagus progressus (Hibbard)			x
Family Felidae				+Notolagus lepusculus (Hibbard)		—	x
Felis sp. (large)			—	Order Artiodactyla			
Felis lacustris Gazin		—		Family Tayassuidae			
Felis rexroadensis Stephens			x	+Platygonus bicalcaratus Cope	—	—	—
+Homotheriini (genus and species indet.)			—	Family Camelidae			
Order Rodentia				+Tanupolama blancoensis Meade			—
Family Sciuridae				Family Cervidae			
Spermophilus howelli (Hibbard)			x	*Odocoileus brachyodontus* Oelrich			—
Spermophilus rexroadensis (Hibbard)			x	*Odocoileus sp. (large)			—
Family Castoridae				Family Antilocapridae (genus and species indet.)			
+Dipoides rexroadensis Hibbard			—	Order Perissodactyla			
+Procastoroides sweeti Barbour and Schultz				Family Equidae			
Family Geomyidae				+Nannippus phlegon (Hay)		—	—
Geomys minor Gidley		—	—	+Plesippus sp.		—	—
Geomys jacobi Hibbard			x				
Family Heteromyidae							
Perognathus gidleyi Hibbard			x				
+Prodipodomys rexroadensis Hibbard			x				

BIBLIOGRAPHY OF THE REXROAD LOCAL FAUNA

Baker, F. C.
  1938. New land and freshwater Mollusca from the Upper Pliocene of Kansas and a new species of *Gyraulus* from Early Pleistocene strata. Nautilus, 51:126-131.

Brattstrom, B. H.
  1967. A succession of Pliocene and Pleistocene snake faunas from the High Plains of the United States. Copeia, 1967:188-202.

Brodkorb, P.
1963. Catalogue of fossil birds: part 1 (Archaeopterygiformes through Ardeiformes). Bull. Florida State Mus., Biol. Ser., 7:179-293.
1964. Catalogue of fossil birds: part 2 (Anseriformes through Galliformes). Bull. Florida State Mus., Biol. Ser., 8:195-335.
1964. Notes on fossil turkeys. Quart. Jour. Florida Acad. Sci., 27:223-229.

Collins, T. C.
1964. Fossil ibises from the Rexroad fauna of the Upper Pliocene of Kansas. Wilson Bull., 76:43-49.

Dawson, M. D.
1958. Later Tertiary Leporidae of North America. Univ. Kansas Paleontol. Contrib., Vertebrata, 6:1-75.

Elias, M. K., and others
1945. Blancan as a time term in the Central Great Plains. Science, 101:270-271.

Etheridge, R.
1960. The Pliocene lizard genus *Eumecoides* Taylor. Bull. Southern California Acad. Sci., 59:62-69.

Feduccia, J. A.
1968. The Pliocene rails of North America. Auk, 85:441-453.

Ford, N. L.
1966. Fossil owls from the Rexroad fauna of the Upper Pliocene of Kansas. Condor, 68:472-475.

Franzen, D. S.
1947. The pocket gopher, *Geomys quinni* McGrew, in the Rexroad fauna, Blancan age, of southwestern Kansas. Trans. Kansas Acad. Sci., 50:55-59.

Franzen, D. S., and A. B. Leonard
1947. Fossil and living Pupillidae (Gastropoda-Pulmonata) in Kansas. Univ. Kansas Sci. Bull., 31:311-411.

Frye, J. C.
1940. A preliminary report on the water supply of the Meade Artesian Basin, Meade County, Kansas. Bull. Kansas Geol. Surv., 35:1-39.
1942. Geology and ground water resources of Meade County, Kansas. Bull. Kansas Geol. Surv., 45:3-152.

Frye, J. C., and C. W. Hibbard
1941. Pliocene and Pleistocene stratigraphy and paleontology of the Meade Basin, southwestern Kansas. Bull. Kansas Geol. Surv., 38:389-424.

Frye, J. C., and A. B. Leonard
1952. Pleistocene Geology of Kansas. Bull. Kansas Geol. Surv., 99:1-230. [All Rexroad faunal sites with mollusks are assigned a Nebraskan age.]

Gehlbach, F. R.
1965. Amphibians and reptiles from the Pliocene and Pleistocene of North America: a chronological summary and selected bibliography. Texas Jour. Sci., 17:56-70.

Hazard, E. B.
1961. The subgeneric status and distribution in time of *Citellus rexroadensis*. Jour. Mamm., 42:477-483.

Herrington, H. B., and D. W. Taylor
1958. Pliocene and Pleistocene Sphaeriidae (Pelecypoda) from the central United States. Occas. Papers Mus. Zool., Univ. Michigan, 596:1-28.

Hibbard, C. W.
1938. An Upper Pliocene fauna from Meade County, Kansas. Trans. Kansas Acad. Sci., 40:239-265.
1939. Four new rabbits from the Upper Pliocene of Kansas. Amer. Midland Nat., 21:506-513.
1939. Nekrolagus, a new name for Pediolagus Hibbard, not Marelli. Amer. Midland Nat., 21: table of contents.
1941. New mammals from the Rexroad fauna, Upper Pliocene of Kansas. Amer. Midland Nat., 26:337-368.
1941. Paleoecology and correlation of the Rexroad fauna from the Upper Pliocene of southwestern Kansas, as indicated by the mammals. Univ. Kansas Sci. Bull., 27:79-104.
1941. Mammals of the Rexroad fauna from the Upper Pliocene of southwestern Kansas. Trans. Kansas Acad. Sci., 44:265-313.
1944. Abnormal tooth pattern in the lower dentition of the jackrabbit, *Lepus californicus deserticola* (Mearns). Jour. Mamm., 25:64-66.
1948. Late Cenozoic climatic conditions in the High Plains of western Kansas. Bull. Geol. Soc. Amer., 59:592-597.
1950. Mammals of the Rexroad Formation from Fox Canyon, Meade County, Kansas. Contrib. Mus. Paleontol., Univ. Michigan, 8:113-192.
1952. A new *Bassariscus* from the Upper Pliocene of Kansas. Jour. Mamm., 33:379-381.
1953. The insectivores of the Rexroad fauna, Upper Pliocene of Kansas. Jour. Paleontol., 27:21-32.
1954. Second contribution to the Rexroad fauna. Trans. Kansas Acad. Sci., 57:221-237.
1956. Vertebrate fossils from the Meade Formation of southwestern Kansas. Papers Michigan Acad. Sci., Arts, Letters, 41:145-203. [Discussion of *Bensonomys eliasi* (Hibbard).]
1960. An interpretation of Pliocene and Pleistocene climates in North America. Ann. Rep. Michigan Acad. Sci., Arts, Letters, 62:5-30.
1963. The origin of the P_3 pattern of *Sylvilagus, Caprolagus, Oryctolagus* and *Lepus*. Jour. Mamm., 44:1-15.
1967. New rodents from the Late Cenozoic of Kansas. Papers Michigan Acad. Sci., Arts, Letters, 52:115-131.

Leonard, A. B.
1948. Invertebrates of the Blancan. Bull. Geol. Soc. Amer., 59:589-591.
1950. A Yarmouthian molluscan fauna in the midcontinent region of the United States. Univ. Kansas Paleontol. Contrib., Mollusca, 3:1-48. [Rexroad fauna considered as Aftonian.]

Murray, B. G. Jr.
1967. Grebes from the Late Pliocene of North America. Condor, 69:277-288.

Oelrich, T. M.
1952. A new *Testudo* from the Upper Pliocene of Kansas with additional notes on associated Rexroad mammals. Trans. Kansas Acad. Sci., 55:300-311.
1953. Additional mammals from the Rexroad fauna. Jour. Mamm., 34:373-378.
1954. A horned toad, *Phrynosoma cornutum*, from the Upper Pliocene of Kansas. Copeia, 1954:262-263.

Packard, R. L.
1960. Speciation and evolution of the pygmy mice, genus Baiomys. Univ. Kansas Publ., Mus. Nat. Hist., 9:579-670.

Smith, C. L.
1962. Some Pliocene fishes from Kansas, Oklahoma, and Nebraska. Copeia, 1962:505-520.

Smith, H. T. U.
  1940. Geological studies in southwestern Kansas. Bull. Kansas Geol. Surv., 34:1-212. [Description of type Rexroad Formation.]

Stephens, J. J.
  1959. A new Pliocene cat from Kansas. Papers Michigan Acad. Sci., Arts, Letters, 44:41-46.

Stevens, J. B.
  1965. Geology of the Meade County State Park area, Kansas. Papers Michigan Acad. Sci., Arts, Letters, 50:215-233.

Stevens, M. S.
  1965. A new species of *Urocyon* from the Upper Pliocene of Kansas. Jour. Mamm., 46:265-269.

Taylor, D. W.
  1960. Late Cenozoic molluscan faunas from the High Plains. U.S. Geol. Surv., Prof. Paper, 337:iv+1-94.
  1966. Summary of North American Blancan nonmarine mollusks. Malacologia, 4:1-172.

Taylor, E. H.
  1941. Extinct lizards from Upper Pliocene deposits of Kansas. Bull. Kansas Geol. Surv., 38:165-176.
  1942. Extinct toads and frogs from the Upper Pliocene deposits of Meade County, Kansas. Univ. Kansas Sci. Bull., 28:199-235.

Tihen, J. A.
  1960. On *Neoscaphiopus* and other Pliocene pelobatid frogs. Copeia, 1960:89-94.
  1962. A review of New World fossil bufonids. Amer. Midland Nat., 68:1-50.

Tordoff, H. B.
  1951. Osteology of Colinus hibbardi, a Pliocene quail. Condor, 53:23-30.
  1959. A condor from the Upper Pliocene of Kansas. Condor, 61:338-343.

Wetmore, A.
  1944. Remains of birds from the Rexroad fauna of the Upper Pliocene of Kansas. Univ. Kansas Sci. Bull., 30:89-105.

Woodburne, M. O.
  1961. Upper Pliocene geology and vertebrate paleontology of part of the Meade Basin, Kansas. Papers Michigan Acad. Sci., Arts, Letters, 46:61-101.

Zakrzewski, R. J.
  1967. The primitive vole, *Ogmodontomys,* from the Late Cenozoic of Kansas and Nebraska. Papers Michigan Acad. Sci., Arts, Letters, 52:133-150.

TABLE 2. Nebraskan Mammalian Local Fauna from Dixon Farm Quarries, Kingman County, Kansas, elevation 1570 feet

Species		Species	
Order Insectivora		Family Geomyidae	
Family Soricidae		*Geomys* sp.	—
Sorex dixonensis Hibbard	x	Family Castoridae	
Sorex leahyi Hibbard	x	?*Procastoroides*	—
Sorex sp.	—	Family Cricetidae	
Blarina sp.	—	*Peromyscus* sp.	—
Order Carnivora		+*Pliopotamys meadensis* Hibbard	—
Family Mustelidae		+*Nebraskomys mcgrewi* Hibbard	—
Mustelid (genus and species indet.)	—	+*Ophiomys meadensis* (Hibbard)	—
Order Rodentia		+*Pliolemmus antiquus* Hibbard	—
Family Sciuridae		*Synaptomys (Synaptomys) rinkeri* Hibbard	x
Spermophilus sp.	—		

BIBLIOGRAPHY OF NEBRASKAN LOCAL FAUNA

Dixon Local Fauna

Brattstrom, B. H.
 1967. A succession of Pliocene and Pleistocene snake faunas from the High Plains of the United States. Copeia, 1967:188-202.
Frye, J. C., and A. B. Leonard
 1952. Pleistocene geology of Kansas. Bull. Kansas Geol. Surv., 99:1-230.
Harrell, B. E.
 1959. Notes on fossil birds from the Pleistocene of Kansas and Oklahoma. Proc. South Dakota Acad. Sci., 38:103-106.
Herrington, H. B., and D. W. Taylor
 1958. Pliocene and Pleistocene Sphaeriidae (Pelecypoda) from the central United States. Occas. Papers Mus. Zool., Univ. Michigan, 596:1-28.
Hibbard, C. W.
 1956. Vertebrate fossils from the Meade Formation of southwestern Kansas. Papers Michigan Acad. Sci., Arts, Letters, 41:145-203.
 1957. Notes on Late Cenozoic shrews. Trans. Kansas Acad. Sci., 60:327-336.
 1958. New stratigraphic names for Early Pleistocene deposits in southwestern Kansas. Amer. Jour. Sci., 256:54-59.

Hibbard, C. W., C. E. Ray, D. E. Savage, D. W. Taylor, and J. E. Guilday
 1965. Quaternary mammals of North America. Pp. 509-525, *in* The Quaternary of the United States (H. E. Wright, Jr., and D. G. Frey, eds.), Princeton Univ. Press, Princeton, New Jersey, x+922 pp.
Taylor, D. W.
 1958. Geologic range and relationship of the freshwater snail *Anisus pattersoni*. Jour. Paleontol., 32:1149-1153.
 1960. Late Cenozoic molluscan faunas from the High Plains. U.S. Geol. Surv. Prof. Paper, 337:iv+1-94.
 1965. The study of Pleistocene nonmarine mollusks in North America. Pp. 597-611, *in* The Quaternary of the United States (H. E. Wright, Jr., and D. G. Frey, eds.), Princeton Univ. Press, Princeton, New Jersey, x+922 pp.
 1966. Summary of North American Blancan nonmarine mollusks. Malacologia, 4:1-172.
Tihen, J. A.
 1955. A new Pliocene species of *Ambystoma*, with remarks on other fossil ambystomids. Contrib. Mus. Paleontol., Univ. Michigan, 12:229-244.

TABLE 3. Aftonian Mammalian Local Faunas[1] of the Great Plains Province: BL (Broadwater-Lisco, Morrill County, Nebraska); SD (Sand Draw, restricted to *Geochelone* horizon, Brown County, South Dakota, elevation 2425 feet); DP and S (Deer Park and Sanders, respectively, Meade County, Kansas, approximate elevation 2570 feet); B (Blanco, Crosby County, Texas); CC (Cita Canyon, Randall County, Texas)

Species	BL	SD	DP	S	B	CC
Order Insectivora						
Family Soricidae						
Sorex sandersi Hibbard		—		x		
Sorex sp.		—				
Order Carnivora						
Family Canidae						
Canis lepophagus Johnston						x
Canis sp.		—				
+*Borophagus diversidens* Cope					x	—
Borophagus sp.	—					
Family Ursidae						
?*Ursus* sp.						—
Family Procyonidae						
Procyon sp.						—
Family Mustelidae						
+*Satherium priscinaria middleswarti* Barbour and Schultz	x					
+*Canimartes cumminsi* Cope					x	
+*Trigonictis kansasensis* Hibbard		—	—	—		
+*Buisnictis* sp.		—	—			
Mephitis sp.		—				
Taxidea cf. *taxus* (Schreber)		—	—	—		—
Family Hyaenidae						
+*Chasmaporthetes johnstoni* (Stirton and Christian)						x
Family Felidae						
Felis aff. *issiodorensis* (Croizet and Jobert)						—
Felis studeri Savage						x
Lynx cf. *rufus* (Schreber)						—
Panthera palaeoonca Meade					x	
+*Ischyrosmilus johnstoni* Mawby						x
+*Smilodon* sp.		—	—			
Order Rodentia						
Family Sciuridae						
+*Paenamarmota barbouri* Hibbard and Schultz	—					
Cynomys meadensis Hibbard				x		
Cynomys sp.		—				
Spermophilus sp.		—	—	—		
Family Heteromyidae						
Perognathus cf. *pearlettensis* Hibbard				—		
+*Prodipodomys* sp.				—		
Dipodomys sp.		—				
Family Geomyidae						
Geomys quinni McGrew		—	—			
Geomys tobinensis (Hibbard)				—		
Geomys sp.		—				—
Family Castoridae						
+*Procastoroides sweeti* Barbour and Schultz	x	—	—			
Family Cricetidae						
Peromyscus sp.		—				
Neotoma sp.		—				
+*Bensonomys meadensis* Hibbard		—		x		
Sigmodon cf. *intermedius* Hibbard		—				
Sigmodon sp.					—	
Pliophenacomys sp.		—				
+*Pliopotamys meadensis* Hibbard		—	x			
+*Nebraskomys mcgrewi* Hibbard		x				
+*Ogmodontomys poaphagus poaphagus* Hibbard		—	—			
+*Ophiomys meadensis* (Hibbard)				x		
+*Pliolemmus antiquus* Hibbard		—	x	—		
Family Dipodidae						
Zapus sandersi sandersi Hibbard				x		
Zapus sp.		—				

[1] These faunas, except the Broadwater-Lisco, are considered by M. F. Skinner and C. W. Hibbard (in manuscript) as being Pleistocene, but older than the first continental glaciation and thus older than the Dixon local fauna.

TABLE 3. (Continued)

Taxon				
Order Edentata				
Family Megalonychidae				
+*Megalonyx leptostomus* Cope	x	—		
Megalonyx sp.	—			
Family Mylodontidae				
+*Paramylodon* sp.	—		—	
Family Glyptodontidae				
+*Glyptotherium texanum* Osborn	x	—		
Order Proboscidea				
Family Gomphotheriidae				
+*Stegomastodon primitivus* Osborn	—	—		
+*Stegomastodon mirificus* (Leidy)	—	—	—	
+*Rhynchotherium falconeri* Osborn		x		
+*Rhynchotherium* sp.	—			
+*Serbelodon? praecursor* (Cope)		x		
Family Mammutidae				
+*Mammut* sp.	—	—		
Order Lagomorpha				
Family Leporidae				
+*Hypolagus* sp. (large)	—	—	—	
+*Hypolagus* sp. (small)	—			
?*Sylvilagus* sp.	—			
Order Artiodactyla				
Family Tayassuidae				
+*Platygonus bicalcaratus* Cope	—	x	—	
Platygonus sp.	—			
Family Camelidae				
+*Titanotylopus spatulus* (Cope)	—	x		
+*Tanupolama blancoensis* Meade		x		
+*Tanupolama* sp.		—		
+*Camelops* sp. (large)		—		
+*Camelops* sp. (small)		—		
+*Pliauchenia* sp.		—		
Family Cervidae				
Odocoileus sp.		—		
Family Antilocapridae				
+*Capromeryx arizonensis schultzi* Skinner	x			
+*Capromeryx* sp.	—		—	
Antilocaprid (genus and species indet.)			—	
Order Perissodactyla				
Family Equidae				
+*Nannippus phlegon* (Hay)	—	—	x	—
+*Plesippus simplicidens* (Cope)	—	—	—	x	—
Equus (Asinus) excelsus Leidy	—	—		
Equus (Asinus) cumminsii Cope		—	x	
+*Pliohippus (Astrohippus)* sp.			—	
Equid (small)	—			

BIBLIOGRAPHY OF AFTONIAN LOCAL FAUNAS

Blanco Local Fauna

Cope, E. C.
  1892. Report on the paleontology of the Vertebrata. Ann. Rep. Geol. Surv. Texas, 3:251-259.
  1892. A contribution to a knowledge of the fauna of the Blanco beds of Texas. Proc. Acad. Nat. Sci. Philadelphia, 44:226-229.
  1892. A hyaena and other Carnivora from Texas. Proc. Acad. Nat. Sci. Philadelphia, 44:326-327.
  1892. A contribution to the vertebrate paleontology of Texas. Proc. Amer. Philos. Soc., 30:123-131.
  1892. The age of the Staked Plains of Texas. Amer. Nat., 26:49-50.
  1892. A hyena and other Carnivora from Texas. Amer. Nat., 26:1028-1029.
  1892. The fauna of the Blanco epoch. Amer. Nat., 26:1058-1059.
  1893. A preliminary report on the vertebrate paleontology of the Llano Estacado. Ann. Rep. Geol. Surv. Texas, 4:1-136.

Evans, G. L., and G. E. Meade
  1945. Quaternary of the Texas High Plains. Univ. Texas Publ., 4401:485-507.

Gidley, J. W.
  1903. On two species of *Platygonus* from the Pliocene of Texas. Bull. Amer. Mus. Nat. Hist., 19:477-481.

Hibbard, C. W., C. E. Ray, D. E. Savage, D. W. Taylor, and J. E. Guilday
  1965. Quaternary mammals of North America. Pp. 509-525, in The Quaternary of the United States (H. E. Wright, Jr., and D. G. Frey, eds.), Princeton Univ. Press, Princeton, New Jersey, x+922 pp.

Johnston, C. S., and D. E. Savage
  1955. A survey of various Late Cenozoic vertebrate faunas of the Panhandle of Texas. Part

I: introduction, description of localities, preliminary faunal lists. Univ. California Publ. Geol. Sci., 31:27-50.
Matthew, W. D., and R. A. Stirton
1930. Osteology and affinities of *Borophagus*. Univ. California Publ. Geol. Sci., 19:171-216.
Meade, G. E.
1945. The Blanco fauna. Univ. Texas Publ., 4401:509-556.
Osborn, H. F.
1903. *Glyptotherium texanum*, a new glyptodont, from the Lower Pleistocene of Texas. Bull. Amer. Mus. Nat. Hist., 19:491-494.
1923. New subfamily, generic and specific stages in the evolution of the Proboscidea. Amer. Mus. Novit., 99:1-4.
1936. Proboscidea. Amer. Mus. Nat. Hist., New York, vol. 1, 802 pp.
Simpson, G. G.
1941. Large Pleistocene felines of North America. Amer. Mus. Novit., 1136:1-27.
Stirton, R. A.
1940. Phylogeny of North American Equidae. Univ. California Publ. Geol. Sci., 25:165-198.
Stirton, R. A., and V. L. VanderHoof
1933. *Osteoborus*, a new genus of dogs, and its relations to *Borophagus* Cope. Univ. California Publ. Geol. Sci., 23:175-182.
VanderHoof, V. L.
1936. Notes on the type of *Borophagus diversidens* Cope. Jour. Mamm., 17:415-416.
1937. Critical observations on the Canidae in Cope's original collection from the Blanco of Texas. Proc. Geol. Soc. Amer., for 1936, p. 389 (abstract).
Wood, H. E., and others
1941. Nomenclature and correlation of the North American continental Tertiary. Bull. Geol. Soc. Amer., 52:1-48.

Broadwater-Lisco Fauna

Barbour, E. H., and C. B. Schultz
1936. Notice of a new bone bed in the Early Pleistocene of Morrill County, Nebraska. Bull. Univ. Nebraska State Mus., 1:450.
1937. An Early Pleistocene fauna from Nebraska. Amer. Mus. Novit., 942:1-10.
1939. A new giant camel *Gigantocamelus fricki*, gen. et sp. nov. Bull. Univ. Nebraska State Mus., 2:17-27.
Hibbard, C. W., and E. S. Riggs
1949. Upper Pliocene vertebrates from Keefe Canyon, Meade County, Kansas. Bull. Geol. Soc. Amer., 60:829-860.
Hibbard, C. W., and C. B. Schultz
1948. A new sciurid of Blancan age from Kansas and Nebraska. Bull. Univ. Nebraska State Mus., 3:19-29.
Schultz, C. B., and T. M. Stout
1948. Pleistocene mammals and terraces in the Great Plains. Bull. Geol. Soc. Amer., 59:553-588.
Skinner, M. F.
1942. The fauna of Papago Springs Cave, Arizona, and a study of *Stockoceros*; with three new antilocaprines from Nebraska and Arizona. Bull. Amer. Mus. Nat. Hist., 80:143-220.
Webb, S. D.
1966. The osteology of *Camelops*. Bull. Los Angeles County Mus. Sci., 1:1-54.

Cita Canyon Local Fauna

Auffenberg, W.
1962. A new species of *Geochelone* from the Pleistocene of Texas. Copeia, 1962:627-636.
Brodkorb, P.
1964. Catalogue of fossil birds: part 2 (Anseriformes through Galliformes). Bull. Florida State Mus., Biol. Ser., 8:195-335.
Hibbard, C. W.
1958. Summary of North American Pleistocene mammalian local faunas. Papers Michigan Acad. Sci., Arts, Letters, 43:3-32.
Johnston, C. S.
1937. Osteology of *Bysmachelys canyonensis*, a new turtle from the Pliocene of Texas. Jour. Geol., 45:439-447.
1938. Preliminary report on the vertebrate type locality of Cita Canyon, and the description of an ancestral coyote. Amer. Jour. Sci., 35:383-390.
Johnston, C. S., and D. E. Savage
1955. A survey of various Late Cenozoic vertebrate faunas of the Panhandle of Texas. Part I. Univ. California Publ. Geol. Sci., 31:27-50.
Kurtén, B.
1963. Notes on some Pleistocene mammal migrations from the Palaearctic to the Nearctic. Eiszeitalter u. Gegenwart, 14:96-103.
Mawby, J. E.
1965. Machairodonts from the Late Cenozoic of the Panhandle of Texas. Jour. Mamm., 46:573-587.
Miller, A. H., and R. I. Bowman
1956. Fossil birds of the Late Pliocene of Cita Canyon, Texas. Wilson Bull., 68:38-46.
Miller, L., and C. S. Johnston
1937. A Pliocene record of Parapavo from Texas. Condor, 39:229.
Savage, D. E.
1955. A survey of various Late Cenozoic vertebrate faunas of the Panhandle of Texas. Part II. Proboscidea. Univ. California Publ. Geol. Sci., 31:51-74.
1960. A survey of various Late Cenozoic vertebrate faunas of the Panhandle of Texas. Part III. Felidae. Univ. California Publ. Geol. Sci., 36:317-344.
Stirton, R. A., and W. G. Christian
1940. A member of the Hyaenidae from the Upper Pliocene of Texas. Jour. Mamm., 21:445-448.
1941. *Ailurdaena* Stirton and Christian referred to *Chasmaporthetes* Hay. Jour. Mamm., 22:198.
Williams, E.
1950. *Testudo cubensis* and the evolution of the Western Hemisphere tortoises. Bull. Amer. Mus. Nat. Hist., 95:1-36.

Deer Park Local Fauna

Brattstrom, B. H.
1967. A succession of Pliocene and Pleistocene snake faunas from the High Plains of the United States. Copeia, 1967:188-202.
Franzen, D. S.
1947. The pocket gopher, *Geomys quinni* McGrew, in the Rexroad fauna, Blancan age, of southwestern Kansas. Trans. Kansas Acad. Sci., 50:55-59.
Frye, J. C., and C. W. Hibbard
1941. Pliocene and Pleistocene stratigraphy and

paleontology of the Meade Basin, southwestern Kansas. Bull. Kansas Geol. Surv., 38:389-424.

Hibbard, C. W.
1938. An Upper Pliocene fauna from Meade County, Kansas. Trans. Kansas Acad. Sci., 40: 239-265.
1941. Paleoecology and correlation of the Rexroad fauna from the Upper Pliocene of southwestern Kansas, as indicated by the mammals. Univ. Kansas Sci. Bull., 27:79-104.
1941. Mammals of the Rexroad fauna from the Upper Pliocene of southwestern Kansas. Trans. Kansas Acad. Sci., 44:265-313.
1948. Late Cenozoic climatic conditions in the High Plains of western Kansas. Bull. Geol. Soc. Amer., 59:592-597.
1949. Pleistocene stratigraphy and paleontology of Meade County, Kansas. Contrib. Mus. Paleontol., Univ. Michigan, 7:63-90.
1949. Pleistocene vertebrate paleontology in North America. Bull. Geol. Soc. Amer., 60: 1417-1428.
1956. Vertebrate fossils from the Meade Formation of southwestern Kansas. Papers Michigan Acad. Sci., Arts, Letters, 41:145-203.
1957. Tentative list of Pleistocene fossil mammals in North America, with their stratigraphic occurrence. Pp. 458-467, in Glacial and Pleistocene geology (R. F. Flint), John Wiley and Sons, Inc., New York, xiii+553 pp.
1958. New stratigraphic names for Early Pleistocene deposits in southwestern Kansas. Amer. Jour. Sci., 256:54-59.
1958. Summary of North American Pleistocene mammalian local faunas. Papers Michigan Acad. Sci., Arts, Letters, 43:3-32.
1960. An interpretation of Pliocene and Pleistocene climates in North America. Ann. Rep. Michigan Acad. Sci., Arts, Letters, 62:5-30.
1967. New rodents from the Late Cenozoic of Kansas. Papers Michigan Acad. Sci., Arts, Letters, 52:115-131.

Hibbard, C. W., C. E. Ray, D. E. Savage, D. W. Taylor, and J. E. Guilday
1965. Quaternary mammals of North America. Pp. 509-525, in The Quaternary of the United States (H. E. Wright, Jr., and D. G. Frey, eds.), Princeton Univ. Press, Princeton, New Jersey, x+922 pp.

Stirton, R. A.
1940. Phylogeny of North American Equidae. Univ. California Publ. Geol. Sci., 25:165-198. [Figures 38 and 39, page 137, of *Nannippus phlegon* are from the Deer Park local fauna, not the Rexroad fauna as stated.]

Taylor, D. W.
1960. Late Cenozoic molluscan faunas from the High Plains. U.S. Geol. Surv., Prof. Paper, 337:iv+1-94.
1966. Summary of North American Blancan nonmarine mollusks. Malacologia, 4:1-172.

Zakrzewski, R. J.
1967. The primitive vole, *Ogmodontomys*, from the Late Cenozoic of Kansas and Nebraska. Papers Michigan Acad. Sci., Arts, Letters, 52: 133-150.

Sand Draw Local Fauna

Baker, F. C.
1938. New land and fresh water Mollusca from the Upper Pliocene of Kansas and a new species of *Gyraulus* from Early Pleistocene strata. Nautilus, 51:126-131.

Brattstrom, B. H.
1967. A succession of Pliocene and Pleistocene snake faunas from the High Plains of the United States. Copeia, 1967:188-202.

Franzen, D. S.
1947. The pocket gopher, *Geomys quinni* McGrew, in the Rexroad fauna, Blancan age, of southwestern Kansas. Trans. Kansas Acad. Sci., 50:55-59.

Frick, C.
1937. Horned ruminants of North America. Bull. Amer. Mus. Nat. Hist., 69:1-669.

Herrington, H. B., and D. W. Taylor
1958. Pliocene and Pleistocene Sphaeriidae (Pelecypoda) from the central United States. Occas. Papers Mus. Zool., Univ. Michigan, 596:1-28.

Hibbard, C. W.
1957. Two new Cenozoic microtine rodents. Jour. Mamm., 38:39-44.
1958. Summary of North American Pleistocene mammalian local faunas. Papers Michigan Acad. Sci., Arts, Letters, 43:3-32.
1960. An interpretation of Pliocene and Pleistocene climates in North America. Ann. Rep. Michigan Acad. Sci., Arts, Letters, 62:5-30.
1967. New rodents from the Late Cenozoic of Kansas. Papers Michigan Acad. Sci., Arts, Letters, 52:115-131.

Jehl, J. R., Jr.
1966. Fossil birds from the Sand Draw local fauna (Aftonian) of Brown County, Nebraska. Auk, 83:669-670.

McGrew, P. O.
1944. An Early Pleistocene (Blancan) fauna from Nebraska. Field Mus. Nat. Hist., Geol. Ser., 9:33-66.

Osborn, H. F.
1936. Proboscidea. Amer. Mus. Nat. Hist., New York, vol. 1, 802 pp.

Taylor, D. W.
1954. A new Pleistocene fauna and new species of fossil snails from the High Plains. Occas. Papers Mus. Zool., Univ. Michigan, 557:1-16.
1958. Geologic range and relationship of the freshwater snail *Anisus pattersoni*. Jour. Paleontol., 32:1149-1153.
1960. Late Cenozoic molluscan faunas from the High Plains. U.S. Geol. Surv., Prof. Paper, 337:iv+1-94.
1966. Summary of North American Blancan nonmarine mollusks. Malacologia, 4:1-172.

Zakrzewski, R. J.
1967. The primitive vole, *Ogmodontomys*, from the Late Cenozoic of Kansas and Nebraska. Papers Michigan Acad. Sci., Arts, Letters, 52: 133-150.

Sanders Local Fauna

Brattstrom, B. H.
1967. A succession of Pliocene and Pleistocene snake faunas from the High Plains of the United States. Copeia, 1967:188-202.

Harrell, B. E.
1959. Notes on fossil birds from the Pleistocene of Kansas and Oklahoma. Proc. South Dakota Acad. Sci., 38:103-106.

Hibbard, C. W.
1956. Vertebrate fossils from the Meade Formation of southwestern Kansas. Papers Michigan Acad. Sci., Arts, Letters, 41:145-203.

1958. Summary of North American Pleistocene mammalian local faunas. Papers Michigan Acad. Sci., Arts, Letters, 43:3-32.

1958. New stratigraphic names for Early Pleistocene deposits in southwestern Kansas. Amer. Jour. Sci., 256:54-59.

Hibbard, C. W., C. E. Ray, D. E. Savage, D. W. Taylor, and J. E. Guilday
    1965. Quaternary mammals of North America. Pp. 509-525, *in* The Quaternary of the United States (H. E. Wright, Jr., and D. G. Frey, eds.), Princeton Univ. Press, Princeton, New Jersey, x+922 pp.

Klingener, D.
    1963. Dental evolution of *Zapus*. Jour. Mamm., 44:248-260.

Russell, R. J.
    1968. Evolution and classification of the pocket gophers of the subfamily Geomyinae. Univ. Kansas Publ., Mus. Nat. Hist., 16:473-579. [Saunders local fauna and Saunders *Geomys* =Sanders.]

Taylor, D. W.
    1960. Late Cenozoic molluscan faunas from the High Plains. U.S. Geol. Surv., Prof. Paper, 337:iv+1-94.
    1966. Summary of North American Blancan nonmarine mollusks. Malacologia, 4:1-172.

Tihen, J. A.
    1955. A new Pliocene species of *Ambystoma*, with remarks on other fossil ambystomids. Contrib. Mus. Paleontol., Univ. Michigan, 12:229-244.

TABLE 4. Kansan Mammalian Local Faunas from the Great Plains and Central Lowland provinces: S (Seger Gravel Pit, Meade County, Kansas, approximate elevation 2620 feet); G (Gilliland, Baylor, and Knox counties, Texas); H (Holloman, Tillman County, Oklahoma); RC (Rock Creek, Briscoe County, Texas); C (Cudahy Fauna, from Kansas and Texas)

Species	S	G	H	RC	C
Order Insectivora					
Family Soricidae					
Sorex cinereus Kerr					
*Sorex lacustris (Hibbard)					x
*Sorex megapalustris Paulson					x
*Microsorex pratensis Hibbard					x
Cryptotis parva (Say)					
Blarina sp.					—
Family Talpidae					
Scalopus sp.					
Order Carnivora					
Family Canidae					
Canis cf. latrans Say					—
Canis cf. lupus Linnaeus					—
*Canis (Aenocyon) dirus Leidy					
*Canis texanus Troxell				x	
Canis sp.			—		
Family Ursidae					
+Arctodus sp.					—
Family Procyonidae					
Procyon sp.					—
Family Mustelidae					
Mustela cf. erminea Linnaeus					—
Mustela nigripes (Audubon and Bachman)					—
Family Felidae					
Felis cf. concolor Linnaeus					—
Felis sp.					—
+Homotherium (Dinobastis) sp.[1]					—
Order Rodentia					
Family Sciuridae					
Cynomys ludovicianus (Ord)					
Spermophilus richardsonii (Sabine)					—
Spermophilus cf. tridecemlineatus (Mitchill)					
Spermophilus cf. franklinii (Sabine)					—
Spermophilus sp.					—
Family Heteromyidae					
Perognathus sp.					
Family Geomyidae					
Thomomys sp.					
*Geomys tobinensis (Hibbard)					—
Family Cricetidae					
*Reithrodontomys moorei (Hibbard)					x
*Peromyscus cragini Hibbard					x
Onychomys sp.					—
*Ondatra annectens (Brown)					—
Ondatra sp. (large)					—
Phenacomys sp.					—
*Microtus paroperarius Hibbard					x
*Microtus (Pedomys) llanensis (Hibbard)					x
*Pitymys meadensis Hibbard					x
*Synaptomys (Mictomys) meltoni Paulson					x
Family Dipodidae					
*Zapus sanderi sanderi Hibbard					
Order Edentata					
Family Megalonychidae					
+Nothrotherium cf. shastense Sinclair					—
+Megalonyx jeffersoni (Desmarest)					—
+Megalonyx sp.		—			—
Family Mylodontidae					
+Paramylodon harlani (Owen)	—	—	—		
Family Dasypodidae					
+Chlamytherium septentrionale (Leidy)					
Family Glyptodontidae					
+Glyptodon fredericensis (Meade)		—	—		
+Glyptodon sp.					—
Order Proboscidea					
Family Gomphotheriidae					
+Stegomastodon cf. mirificus (Leidy)		—	—	—	
+Cuvieronius sp.					—
Family Elephantidae					
+Mammuthus imperator haroldcooki (Hay)	—	—	—	—	
Order Lagomorpha					
Family Leporidae					
Sylvilagus sp.					—
Order Artiodactyla					
Family Tayassuidae					
+Platygonus cf. cumberlandensis Gidley					—
+Platygonus sp.	—	—	—		
Family Camelidae					
+Titanotylopus sp.					—
+Camelops hesternus (Leidy)					—
+Camelops niobrarensis (Leidy)					—
+Camelops sp.					
+Tanupolama macrocephala (Cope)	—	—	—		
+Tanupolama seymourensis Hibbard and Dalquest		x			
+Tanupolama cf. blancoensis Meade					—
Family Cervidae					
Odocoileus sp.					—
Cervid (genus and species indet.)					—
Family Antilocapridae					
+Tetrameryx knoxensis Hibbard and Dalquest		x			
+Capromeryx sp.					—
Antilocaprid (genus and species indet.)					—
Family Bovidae					
+Preptoceras? mayfieldi Troxell				x	
Order Perissodactyla					
Family Equidae					
+Nannippus phlegon (Hay)		—			
+Plesippus cf. simplicidens (Cope)					—
*Equus giganteus Gidley	—	—			
*Equus achates Hay and Cook	—	—			
*Equus scotti Gidley	—	—	—	x	—
*Equus sp.					
*Equus (Hemionus) calobatus Troxell				x	
Family Tapiridae					
Tapirus copei Simpson					

[1] Fide Kurtén, Pleistocene mammals of Europe, p. 76, 1968.

BIBLIOGRAPHY OF KANSAN LOCAL FAUNAS

Cudahy Fauna (Late Kansan) from the Central Lowland Province in Baylor and Knox counties, Texas, and the Great Plains Province in Randall County, Texas, and Meade County, Kansas

Brattstrom, B. H.
1967. A succession of Pliocene and Pleistocene snake faunas from the High Plains of the United States. Copeia, 1967:188-202.

Chantell, C. J.
1966. Late Cenozoic hylids from the Great Plains. Herpetologica, 22:259-264.

Franzen, D. S.
1946. A new fossil pupillid. Nautilus, 60:24-25.

Franzen, D. S., and A. B. Leonard
1947. Fossil and living Pupillidae (Gastropoda-Pulmonata) in Kansas. Univ. Kansas Sci. Bull., 30:311-411.

Frye, J. C., and A. B. Leonard
1952. Pleistocene geology of Kansas. Bull. Kansas Geol. Surv., 99:1-230.

Frye, J. C., A. B. Leonard, and C. W. Hibbard
1943. Westward extension of the Kansas "Equus beds." Jour. Geol., 51:33-47.

Frye, J. C., A. Swineford, and A. B. Leonard
1948. Correlation of Pleistocene deposits of the Central Great Plains with the glaciated section. Jour. Geol., 56:501-525.

Getz, L. L.
1960. Middle Pleistocene carnivores from southwestern Kansas. Jour. Mamm., 41:361-365.

Getz, L. L., and C. W. Hibbard
1965. A molluscan faunule from the Seymour Formation of Baylor and Knox counties, Texas. Papers Michigan Acad. Sci., Arts, Letters, 50:275-297.

Herrington, H. B., and D. W. Taylor
1958. Pliocene and Pleistocene Sphaeriidae (Pelecypoda) from the central United States. Occas. Papers Mus. Zool., Univ. Michigan, 596:1-28.

Hibbard, C. W.
1944. Stratigraphy and vertebrate paleontology of Pleistocene deposits of southwestern Kansas. Bull. Geol. Soc. Amer., 55:707-754.
1949. Pleistocene vertebrate paleontology in North America. Bull. Geol. Soc. Amer., 60:1417-1428.
1949. Pleistocene stratigraphy of Meade County, Kansas. Contrib. Mus. Paleontol., Univ. Michigan, 7:63-90.
1953. *Equus (Asinus) calobatus* Troxel and associated vertebrates from the Pleistocene of Kansas. Trans. Kansas Acad. Sci., 56:111-126.
1954. A new *Synaptomys*, an addition to the Borchers interglacial (Yarmouth?) fauna. Jour. Mamm., 35:249-252.
1958. Summary of North American Pleistocene mammalian local faunas. Papers Michigan Acad. Sci., Arts, Letters, 43:3-32.

Hibbard, C. W., and W. W. Dalquest
1966. Fossils from the Seymour Formation of Knox and Baylor counties, Texas, and their bearing on the Late Kansan climate of that region. Contrib. Mus. Paleontol., Univ. Michigan, 21:1-66.

Holman, J. A.
1965. Pleistocene snakes from the Seymour Formation of Texas. Copeia, 1965:102-104.

Johnston, C. S., and D. E. Savage
1955. A survey of various Late Cenozoic vertebrate faunas of the Panhandle of Texas. Univ. California Publ. Geol. Sci., 31:27-50.

Klingener, D.
1963. Dental evolution of *Zapus*. Jour. Mamm., 44:248-260.

Leonard, A. B.
1946. Three new pupillids from the Lower Pleistocene of central and southwestern Kansas. Nautilus, 60:20-24.
1948. Five new Yarmouthian planorbid snails. Nautilus, 62:41-47.
1950. A Yarmouthian molluscan fauna in the midcontinent region of the United States. Univ. Kansas Contrib. Paleontol., Mollusca, 3:1-48.

Paulson, G. R.
1961. The mammals of the Cudahy fauna. Papers Michigan Acad. Sci., Arts, Letters, 46:127-153.

Taylor, D. W.
1958. Geologic range and relationship of the freshwater snail *Anisus pattersoni*. Jour. Paleontol., 32:1149-1153.
1960. Late Cenozoic molluscan faunas from the High Plains. U.S. Geol. Surv., Prof. Paper, 337:iv+1-94.
1965. The study of Pleistocene nonmarine mollusks in North America. Pp. 597-611, *in* The Quaternary of the United States (H. E. Wright, Jr., and D. G. Frey, eds.), Princeton Univ. Press, Princeton, New Jersey, x+922.

Tihen, J. A.
1955. A new Pliocene species of *Ambystoma*, with remarks on other fossil ambystomids. Contrib. Mus. Paleontol., Univ. Michigan, 12:229-244.
1960. Notes on Late Cenozoic hylid and leptodactylid frogs from Kansas, Oklahoma and Texas. Southwestern Nat., 5:66-70.
1962. A review of New World fossil bufonids. Amer. Midland Nat., 68:1-50.

Gilliland Local Fauna (Middle or Late Kansan, pre-Cudahy) from the Central Lowland Province in Baylor and Knox counties, Texas

Brodkorb, P.
1964. Catalogue of fossil birds: part 2 (Anseriformes through Galliformes). Bull. Florida State Mus., Biol. Ser., 8:195-335.
1964. Notes on fossil turkeys. Quart. Jour. Florida Acad. Sci., 27:223-229.

Getz, L. L., and C. W. Hibbard
1965. A molluscan faunule from the Seymour Formation of Baylor and Knox counties, Texas. Papers Michigan Acad. Sci., Arts, Letters, 50:275-297.

Hibbard, C. W.
1960. An interpretation of Pliocene and Pleistocene climates in North America. Ann. Rep. Michigan Acad. Sci., Arts, Letters, 62:5-30.

Hibbard, C. W., and W. W. Dalquest
1960. A new antilocaprid from the Pleistocene of Knox County, Texas. Jour. Mamm., 41:20-23.
1962. Artiodactyls from the Seymour Formation of Knox County, Texas. Papers Michigan Acad. Sci., Arts, Letters, 47:83-89.

1966. Fossils from the Seymour Formation of Knox and Baylor counties, Texas, and their bearing on the Late Kansan climate of that region. Contrib. Mus. Paleontol., Univ. Michigan, 21:1-66.

Holman, J. A.
1965. Pleistocene snakes from the Seymour Formation of Texas. Copeia, 1965:102-104.

Melton, W. G., Jr.
1964. *Glyptodon fredericensis* (Meade) from the Seymour Formation of Knox County, Texas. Papers Mich. Acad. Sci., Arts, Letters, 49:129-146.

Preston, R. E.
1966. Turtles of the Gilliland faunule from the Pleistocene of Knox County, Texas. Papers Michigan Acad. Sci., Arts, Letters, 51:221-239.

Stricklin, F. L., Jr.
1961. Degradational stream deposits of the Brazos River, central Texas. Bull. Geol. Soc. Amer., 72:19-36.

Holloman Local Fauna (Middle or Late Kansan, pre-Cudahy) from the Central Lowland Province in Tillman County, Oklahoma

Hay, O. P.
1916. Description of two extinct mammals of the order Xenarthra from the Pleistocene of Texas. Proc. U.S. Nat. Mus., 51:107-123.

Hay, O. P., and H. J. Cook
1928. Preliminary descriptions of fossil mammals recently discovered in Oklahoma, Texas, and New Mexico. Proc. Colorado Mus. Nat. Hist., 8:33.
1930. Fossil vertebrates collected near, or in association with, human artifacts at localities near Colorado, Texas; Frederick, Oklahoma; and Folsom, New Mexico. Proc. Colorado Mus. Nat. Hist., 9:4-40.

Hibbard, C. W., and W. W. Dalquest
1962. Artiodactyls from the Seymour Formation of Knox County, Texas. Michigan Acad. Sci., Arts, Letters, 47:83-99.
1966. Fossils from the Seymour Formation of Knox and Baylor counties, Texas, and their bearing on the Late Kansan climate of that region. Contrib. Mus. Paleontol., Univ. Michigan, 21:1-66.

Meade, G. E.
1953. An Early Pleistocene vertebrate fauna from Frederick, Oklahoma. Jour. Geol., 61:452-460.

Osborn, H. F.
1936. Proboscidea. Amer. Mus. Nat. Hist., New York, vol. 1, 804 pp.

Tihen, J. A.
1960. Notes on Late Cenozoic hylid and leptodactylid frogs from Kansas, Oklahoma and Texas. Southwestern Nat., 5:66-70.

Wood, A. E.
1953. Pleistocene prairie-dog from Frederick, Oklahoma. Jour. Mamm., 14:160.

Rock Creek Local Fauna (Kansan, pre- or post-Cudahy) from the Great Plains Province in Briscoe County, Texas

Auffenberg, W.
1962. A redescription of *Testudo hexagonata* Cope. Herpetologica, 18:25-34.

Cope, E. D.
1893. A preliminary report on the vertebrate paleontology of the Llano Estacado. Ann. Rep. Geol. Surv. Texas, 4:1-137.

Gidley, J. W.
1900. A new species of Pleistocene horse from the Staked Plains of Texas. Bull. Amer. Mus. Nat. Hist., 13:111-116.

Hay, O. P.
1924. The Pleistocene of the middle region of North America and its vertebrated animals. Publ. Carnegie Inst. Washington, 322A:1-385.

Hibbard, C. W., and W. W. Dalquest
1966. Fossils from the Seymour Formation of Knox and Baylor counties, Texas, and their bearing on the Late Kansan climate of that region. Contrib. Mus. Paleontol., Univ. Michigan, 21:1-66.

Matthew, W. D.
1920. New specimen of the Pleistocene bear, *Arctotherium*, from Texas. Bull. Geol. Soc. Amer., 31:224-225.

Troxell, E. L.
1915. The vertebrate fossils of Rock Creek, Texas. Amer. Jour. Sci., 39:613-638.
1915. A fossil ruminant from Rock Creek, Texas, *Preptoceras mayfieldi* sp. nov. Amer. Jour. Sci., 40:479-482.

Seger Gravel Pit Fauna (Early Kansan) from the Great Plains Province in Meade County, Kansas

Hibbard, C. W.
1951. Vertebrate fossils from the Pleistocene Stump Arroyo Member, Meade County, Kansas. Contrib. Mus. Paleontol., Univ. Michigan, 9:227-245.
1958. Summary of North American Pleistocene mammalian local faunas. Papers Michigan Acad. Sci., Arts, Letters, 43:3-32.

TABLE 5. Yarmouth (Borchers) Mammalian Local Fauna from the Great Plains Province in Meade County, Kansas, approximate elevation 2470 feet

Species		Species	
Order Insectivora		Family Cricetidae	
Family Soricidae		*Onychomys fossilis* Hibbard	x
Sorex taylori Hibard	—	*Reithrodontomys pratincola* Hibbard	x
Order Carnivora		*Neotoma (Paraneotoma) taylori* Hibbard	x
Family Canidae		*Sigmodon hilli* Hibbard	x
Canis cf. *latrans* Say	—	*Ondatra* cf. *idahoensis* Wilson	—
Urocyon atwaterensis Getz	x	*Synaptomys (Synaptomys) landesi* Hibbard	x
Family Mustelidae		Family Dipodidae	
Mustela cf. *frenata* Lichtenstein	—	*Zapus burti* Hibbard	x
Spilogale cf. *putorius ambarvalis* Bangs	—	Order Lagomorpha	
Family Felidae		Family Leporidae	
Felis sp.	—	+*Hypolagus* sp. (small)	—
		Lepus cf. *californicus* Gray	—
Order Rodentia		Order Artiodactyla	
Family Sciuridae		Family Tayassuidae	
Spermophilus meadensis (Hibbard)	x	+*Platygonus* sp.	—
Spermophilus cragini (Hibbard)	x	Family Camelidae	
Family Geomyidae		+*Camelops* sp.	—
Geomys sp.	—	Family Antilocapridae (genus and species indet.)	—
Family Heteromyidae		Order Perissodactyla	
Perognathus pearlettensis Hibbard	x	Family Equidae	
Perognathus gidleyi Hibbard	—	*Equus* cf. *giganteus* Gidley	—
+*Etadonomys tiheni* Hibbard	x		

BIBLIOGRAPHY OF THE BORCHERS LOCAL FAUNA

Brattstrom, B. H.
  1967. A succession of Pliocene and Pleistocene snake faunas from the High Plains of the United States. Copeia, 1967:188-202.

Getz, L. L.
  1960. Middle Pleistocene carnivores from southwestern Kansas. Jour. Mamm., 41:361-365.

Hibbard, C. W.
  1941. The Borchers fauna, a new Pleistocene interglacial fauna from Meade County, Kansas. Bull. Kansas Geol. Surv., 38:197-220.
  1942. Pleistocene mammals from Kansas. Bull. Kansas Geol. Surv., 41:261-269.
  1943. *Etadonomys*, a new Pleistocene heteromyid rodent, and notes on other Kansas mammals. Trans. Kansas Acad. Sci., 46:185-191.
  1949. Pleistocene vertebrate paleontology in North America. Bull. Geol. Soc. Amer., 60:1417-1428.
  1954. A new *Synaptomys*, an addition to the Borchers interglacial (Yarmouth?) fauna. Jour. Mamm., 35:249-252.
  1959. Late Cenozoic microtine rodents from Wyoming and Idaho. Papers Michigan Acad. Sci., Arts, Letters, 44:3-40.

Klingener, D.
  1963. Dental evolution of *Zapus*. Jour. Mamm., 44:248-260.

Tihen, J. A.
  1955. A new Pliocene species of *Ambystoma*, with remarks on other fossils ambystomids. Contrib. Mus. Paleontol., Univ. Michigan, 12:229-244.
  1962. A review of New World fossil bufonids. Amer. Midland Nat., 68:1-50.

TABLE 6. Illinoian Mammalian Local Faunas of the Great Plains Province: E (Early Illinoian of Kansas—Adams, Rezabek, and Sandahl in part—and Oklahoma—Berends and Doby Springs); L (Late Illinoian of Kansas—Butler Spring, Mt. Scott, and Sandahl Gravel Pit)

Species	E	L	Species	E	L
Order Insectivora			*Onychomys* cf. *leucogaster* (Wied-Neuwied)	—	—
Family Soricidae			*Reithrodontomys* cf. *megalotis* (Baird)	—	
Sorex cinereus Kerr	—	—	*Reithrodontomys* cf. *montanus* (Baird)	—	
Sorex arcticus Kerr	—	—	*Reithrodontomys* sp.	—	
Sorex palustris Richardson	—	—	*Neotoma* cf. *floridana* (Ord)	—	
Blarina brevicauda fossilis Hibbard	x		*Neotoma* sp.	—	
Blarina carolinensis (Bachman)	—		*Oryzomys palustris fossilis* Hibbard	—	
Cryptotis parva (Say)	—		*Synaptomys australis* Simpson	—	
Family Talpidae			*Microtus pennsylvanicus* (Ord)	—	—
Scalopus aquaticus (Linnaeus)	—		*Neofiber leonardi* Hibbard		x
Order Chiroptera			*Ondatra nebracensis* (Hollister)	—	
Family Vespertilionidae			Family Dipodidae		
Lasiurus cinereus (Palisot de Beauvois)	—		*Zapus hudsonius transitionalis* Klingener	—	x
Order Carnivora			Order Edentata		
Family Canidae			Family Megalonychidae		
Canis latrans Say	—		+*Megalonyx jeffersoni* (Desmarset)	—	—
Vulpes velox (Say)	—		Family Mylodontidae		
Family Mustelidae			+*Paramylodon harlani* (Owen)	—	—
Mustela vison Schreber	—		Order Proboscidea		
Family Felidae			Family Mammutidae		
+*Homotherium (Dinobastis) serus* (Cope)	—		+*Mammut americanus* (Kerr)	—	
Order Rodentia			Family Elephantidae		
Family Sciuridae			+*Mammuthus columbi* (Falconer)	—	
Spermophilus richardsonii (Sabine)	—	—	+*Mammuthus imperator* (Leidy)	—	
Spermophilus tridecemlineatus (Mitchill)	—	—	+*Mammuthus* sp.	—	
Spermophilus sp.	—	—	Order Lagomorpha		
Cynomys ludovicianus (Ord)	—	—	Family Leporidae		
Cynomys vetus Hibbard	—	—	*Sylvilagus* sp.		—
+*Burosor effossorius* Starrett	x		*Lepus* sp.	—	—
Family Castoridae			Order Artiodactyla		
+*Paradipoides stovalli* Rinker and Hibbard		x	Family Tayassuidae		
+*Castoroides ohioensis* Foster	—	—	+*Mylohyus masutus* (Leidy)		
Castor canadensis Kuhl	—		Family Camelidae		
Family Geomyidae			+*Camelops kansanus* Leidy		—
Geomys bursarius (Shaw)	—	—	*Camelops* sp.		—
Geomys sp.	—	—	Family Cervidae		
Thomomys sp.	—	—	*Odocoileus* sp.		—
Family Heteromyidae			Family Antilocapridae (genus and species indet.)		—
Perognathus hispidus Baird	—		Family Bovidae		
Perognathus sp.	—	—	*Bison* cf. *latifrons* (Harlan)	—	—
Dipodomys cf. *ordii* Woodhouse	—		Order Perissodactyla		
Family Cricetidae			Family Equidae		
Peromyscus berendsensis Starrett	x		*Equus scotti* Gidley	—	
Peromyscus oklahomensis Stephens	x		*Equus* cf. *niobrarensis* Hay	—	—
Peromyscus cf. *cochrani* Hibbard	—		*Equus (Asinus)* cf. *conversidens* Owen	—	—
Peromyscus progressus Hibbard	—		*Equus (Hemionus)* sp.	—	—
Peromyscus sp.	—	—	*Equus* sp.	—	—

BIBLIOGRAPHY OF ILLINOIAN LOCAL BIOTAS

Adams Local Biota (Early Illinoian) from the Great Plains Province in Meade County, Kansas

Hibbard, C. W., and D. W. Taylor
  1960. Two Late Pleistocene faunas from southwestern Kansas. Contrib. Mus. Paleontol., Univ. Michigan, 16:1-223.
Kapp, R. O.
  1965. Illinoian and Sangamon vegetation in southwestern Kansas and adjacent Oklahoma. Contrib. Mus. Paleontol., Univ. Michigan, 19:167-255.
Miller, B. B.
  1966. Five Illinoian molluscan faunas from the Southern Great Plains. Malacologia, 4:173-260.
Schultz, G. E.
  1965. Pleistocene vertebrates from the Butler

Spring local fauna, Meade County, Kansas. Papers Michigan Acad. Sci., Arts, Letters, 50: 235-265.

1967. Four superimposed Late-Pleistocene vertebrate faunas from southwest Kansas. Pp. 321-336, *in* Pleistocene extinctions: the search for a cause (P. S. Martin and H. E. Wright, Jr., eds.), Yale Univ. Press, New Haven, Connecticut, x+453 pp.

1969. Geology and paleontology of a Late Pleistocene basin in southwest Kansas. Geol. Soc. Amer. Spec. Paper, 105:viii+1-85.

Berends Local Biota (Early Illinoian) from the Great Plains Province in Beaver County, Oklahoma

Brattstrom, B. H.
    1967. A succession of Pliocene and Pleistocene snake faunas from the High Plains of the United States. Copeia, 1967:188-202.

Herrington, H. B., and D. W. Taylor
    1958. Pliocene and Pleistocene Sphaeriidae (Pelecypoda) from the central United States. Occas. Papers Mus. Zool., Univ. Michigan, 596:1-28.

Hibbard, C. W.
    1963. A Late Illinoian fauna from Kansas and its climatic significance. Papers Michigan Acad. Sci., Arts, Letters, 48:187-221.

Hibbard, C. W., and D. W. Taylor
    1960. Two Late Pleistocene faunas from southwestern Kansas. Contrib. Mus. Paleontol., Univ. Michigan, 16:1-223.

Kapp, R. O.
    1965. Illinoian and Sangamon vegetation in southwestern Kansas and adjacent Oklahoma. Contrib. Mus. Paleontol., Univ. Michigan, 19: 167-255.

Mengel, R. M.
    1952. White pelican from the Pleistocene of Oklahoma. Auk, 69:81-82.

Miller, B. B.
    1966. Five Illinoian molluscan faunas from the Southern Great Plains. Malacologia, 4:173-260.

Rinker, G. C., and C. W. Hibbard
    1952. A new beaver, and associated vertebrates, from the Pleistocene of Oklahoma. Jour. Mamm., 33:98-101.

Semken, H. A., Jr.
    1966. Stratigraphy and paleontology of the McPherson Equus beds (Sandahl local fauna), McPherson County, Kansas. Contrib. Mus. Paleontol., Univ. Michigan, 20:121-178.

Smith, C. L.
    1954. Pleistocene fishes of the Berends fauna of Beaver County, Oklahoma. Copeia, 1954:282-289.
    1958. Additional Pleistocene fishes from Kansas and Oklahoma. Copeia, 1958:176-180.

Starrett, A.
    1956. Pleistocene mammals of the Berends fauna of Oklahoma. Jour. Paleontol., 30:1187-1192.

Taylor, D. W.
    1954. A new Pleistocene fauna and new species of fossil snails from the High Plains. Occas. Papers Mus. Zool., Univ. Michigan, 557:1-16.

Taylor, D. W., and C. W. Hibbard
    1955. A new Pleistocene fauna from Harper County, Oklahoma. Oklahoma Geol. Surv. Circ., 37:1-23.

Butler Spring Local Biota (Late Illinoian) from the Great Plains Province in Meade County, Kansas

Brattstrom, B. H.
    1967. A succession of Pliocene and Pleistocene snake faunas from the High Plains of the United States. Copeia, 1967:188-202.

Herrington, H. B., and D. W. Taylor
    1958. Pliocene and Pleistocene Sphaeriidae (Pelecypoda) from the central United States. Occas. Papers Mus. Zool., Univ. Michigan, 596:1-28.

Hibbard, C. W., and D. W. Taylor
    1960. Two Late Pleistocene faunas from southwestern Kansas. Contrib. Mus. Paleontol., Univ. Michigan, 16:1-223.

Kapp, R. O.
    1965. Illinoian and Sangamon vegetation in southwestern Kansas and adjacent Oklahoma. Contrib. Mus. Paleontol., Univ. Michigan, 19: 167-255.

Miller, B. B.
    1966. Five Illinoian molluscan faunas from the Southern Great Plains. Malacologia, 4:173-260.

Schultz, G. E.
    1965. Pleistocene vertebrates from the Butler Spring local fauna, Meade County, Kansas. Papers Michigan Acad. Sci., Arts, Letters, 50: 235-265.
    1967. Four superimposed Late-Pleistocene vertebrate faunas from southwest Kansas. Pp. 321-336, *in* Pleistocene extinctions: the search for a cause (P. S. Martin and H. E. Wright, Jr., eds.), Yale Univ. Press, New Haven, Connecticut, x+453 pp.
    1969. Geology and paleontology of a Late Pleistocene basin in southwest Kansas. Geol. Soc. Amer. Spec. Paper, 105:viii+1-85.

Smith, C. L.
    1958. Additional Pleistocene fishes from Kansas and Oklahoma. Copeia, 1958:176-180.

Doby Springs Local Biota (Early Illinoian) from the Great Plains Province in Harper County, Oklahoma

Brattstrom, B. H.
    1967. A succession of Pliocene and Pleistocene snake faunas from the High Plains of the United States. Copeia, 1967:188-202.

Etheridge, R.
    1960. The slender glass lizard, *Ophisaurus attenuatus*, from the Pleistocene (Illinoian glacial) of Oklahoma. Copeia, 1960:46-47.
    1961. Late Cenozoic glass lizards *(Ophisaurus)* from the Southern Great Plains. Herpetologica, 17:179-186.

Gutentag, E. D., and R. H. Benson
    1962. Neogene (Plio-Pleistocene) fresh-water ostracodes from the Central High Plains. Bull. Kansas Geol. Surv., 157:1-60.

Herrington, H. B., and D. W. Taylor
    1958. Pliocene and Pleistocene Sphaeriidae (Pelecypoda) from the central United States. Occas. Papers Mus. Zool., Univ. Michigan, 596:1-28.

Hibbard, C. W., and D. W. Taylor
    1960. Two Late Pleistocene faunas from south-

western Kansas. Contrib. Mus. Paleontol., Univ. Michigan, 16:1-223.

Kapp, R. O.
1965. Illinoian and Sangamon vegetation in southwestern Kansas and adjacent Oklahoma. Contrib. Mus. Paleontol., Univ. Michigan, 19: 167-255.

Klingener, D.
1963. Dental evolution of *Zapus*. Jour. Mamm., 44:248-260.

Miller, B. B.
1966. Five Illinoian molluscan faunas from the Southern Great Plains. Malacologia, 4:173-260.

Smith, C. L.
1958. Additional Pleistocene fishes from Kansas and Oklahoma. Copeia, 1958:176-180.

Stephens, J. J.
1960. Stratigraphy and paleontology of a Late Pleistocene basin, Harper County, Oklahoma. Bull. Geol. Soc. Amer., 71:1675-1702.

Tihen, J. A.
1962. A review of New World fossil bufonids. Amer. Midland Nat., 68:1-50.

Mt. Scott Local Biota (Late Illinoian) from the Great Plains Province in Meade County, Kansas

Brattstrom, B. H.
1967. A succession of Pliocene and Pleistocene snake faunas from the High Plains of the United States. Copeia, 1967:188-202.

Etheridge, R.
1961. Late Cenozoic glass lizards *(Ophisaurus)* from the Southern Great Plains. Herpetologica, 17:179-186.

Hibbard, C. W.
1963. A Late Illinoian fauna from Kansas and its climatic significance. Papers Michigan Acad. Sci., Arts, Letters, 48:187-221.

Kapp, R. O.
1965. Illinoian and Sangamon vegetation in southwestern Kansas and adjacent Oklahoma. Contrib. Mus. Paleontol., Univ. Michigan, 19: 167-255.

Klingener, D.
1963. Dental evolution of *Zapus*. Jour. Mamm., 44:248-260.

Miller, B. B.
1961. A Late Pleistocene molluscan faunule from Meade County, Kansas. Papers Mich. Acad. Sci., Arts, Letters, 46:103-125.
1966. Five Illinoian molluscan faunas from the Southern Great Plains. Malacologia, 4:173-260.

Milstead, W. W.
1967. Fossil box turtles *(Terrapene)* from central North America, and box turtles of eastern Mexico. Copeia, 1967:168-179.

Smith, G. R.
1963. A Late Illinoian fish fauna from southwestern Kansas and its climatic significance. Copeia, 1963:278-285.

Rezabek Local Fauna (Illinoian) from the Great Plains Province in Lincoln County, Kansas

Brattstrom, B. H.
1967. A succession of Pliocene and Pleistocene snake faunas from the High Plains of the United States. Copeia, 1967:188-202.

Frye, J. C., A. B. Leonard, and C. W. Hibbard
1943. Westward extension of the Kansas "Equus beds." Jour. Geol., 51:33-47.

Hibbard, C. W.
1943. The Rezabek fauna, a new Pleistocene fauna from Lincoln County, Kansas. Univ. Kansas Sci. Bull., 24:235-247.

Sandahl Local Fauna (Illinoian) from the border of the Great Plains and Central Lowland provinces in McPherson County, Kansas, elevation 1420 feet

Cope, E. D.
1889. The Edentata of North America. Amer. Nat., 23:651-664.

Harnly, H. J.
1934. Vertebrate fossils from McPherson Equus beds. Trans. Kansas Acad. Sci., 37:151.

Hibbard, C. W.
1952. Vertebrate fossils from Late Cenozoic deposits of central Kansas. Univ. Kansas Paleontol. Contrib., Vertebrata, 2:1-14.

Lillegraven, J. A.
1966. *Bison crassicornis* and the ground sloth *Megalonyx jeffersoni* in the Kansas Pleistocene. Trans. Kansas Acad Sci., 69:294-300.

Lindahl, J.
1891. Description of a skull of *Megalonyx leidyi* n. sp. Trans. Amer. Philos. Soc., 17:1-10.

Miller, B. B.
1970. The Sandahl molluscan fauna (Illinoian) from McPherson County, Kansas. Ohio Jour. Sci., 70:39-50.

Nininger, H. H.
1928. Pleistocene fossils from McPherson County, Kansas. Trans. Kansas Acad. Sci., 31:96-97.

Semken, H. A., Jr.
1966. Stratigraphy and paleontology of the McPherson Equus beds (Sandahl local fauna), McPherson County, Kansas. Contrib. Mus. Paleontol., Univ. Michigan, 20:121-178.

Semken, H. A., Jr., and C. D. Griggs
1965. The long-nosed peccary, *Mylohyus nasutus*, from McPherson County, Kansas. Papers Michigan Acad. Sci., Arts, Letters, 50:267-274.

TABLE 7. Sangamon Mammalian Local Faunas of the Great Plains Province: S (Slaton, Lubbock County, Texas); C (Cragin Quarry, including Butler Spring area, Meade County, Kansas); J (Jinglebob, Meade County, Kansas)

Species	S	C	J	Species	S	C	J
Order Insectivora				*Reithrodontomys* cf. *fulvescens* J. A. Allen		—	
Family Soricidae				*Peromyscus progressus* Hibbard		x	
Sorex cf. *vagrans* Baird	—			*Peromyscus* cf. *progressus* Hibbard		—	
Sorex cf. *cinereus* Kerr		—		*Peromyscus cochrani* Hibbard		?	x
Blarina cf. *brevicauda* (Say)		—		*Peromyscus* sp.			
Notiosorex crawfordi (Coues)	—			*Neotoma albigula* Hartley		—	
Order Chiroptera				*Neotoma micropus* Baird		—	
Family Vespertilionidae				*Neotoma* sp.			—
Lasiurus cinereus (Palisot de Beauvois)		—		*Oryzomys fossilis* Hibbard			x
**Lasiurus (Dasypterus) golliheri* (Hibbard)		x		*Ondatra zibethicus* (Linnaeus)		—	
Near *Lasiurus* sp.		—		*Microtus pennsylvanicus* (Ord)		—	—
Order Carnivora				*Microtus (Pedomys) ochrogaster* (Wagner)		—	—
Family Ursidae				**Neofiber leonardi* Hibbard		—	
+*Arctodus simus* (Cope)		—		**Synaptomys australis* Simpson			—
Family Canidae				Family Dipodidae			
Canis latrans Say		—	—	*Zapus hudsonius adamsi* Hibbard			x
**Canis (Aenocyon) dirus* Leidy		—	—	Order Edentata			
Vulpes velox (Say)		—		Family Mylodontidae			
Family Mustelidae				+*Paramylodon harlani* (Owen)			
Mustela cf. *frenata* Lichtenstein		—		Family Dasypodidae			
Taxidea taxus (Schreber)		—		*Dasypus* cf. *novemcinctus* Linnaeus		—	
Spilogale putorius (Linnaeus)				Order Proboscidea			
Spilogale putorius interrupta (Rafinesque)		—		Family Elephantidae			
Mephitis sp.		—		+*Mammuthus* cf. *imperator* (Leidy)		—	
Family Felidae				+*Mammuthus columbi* (Falconer)		—	—
Felis concolor Linnaeus		—		Order Lagomorpha			
**Panthera atrox* (Leidy)		—		Family Leporidae			
+*Homotherium* (*Dinobastis*?)		—		*Sylvilagus* sp.		—	—
Order Rodentia				*Lepus californicus* Gray		—	
Family Sciuridae				*Lepus* sp.		—	
Cynomys ludovicianus (Ord)		—		Order Artiodactyla			
**Cynomys vetus* Hibbard		—		Family Tayassuidae			
Spermophilus cf. *tridecemlineatus* (Mitchill)		—		+*Platygonus* cf. *compressus* Le Conte			
Spermophilus cf. *spilosoma* Bennett				+*Platygonus* sp.		—	
Spermophilus sp.		—		Family Camelidae			
Family Castoridae				+*Camelops kansanus* Leidy		—	
Castor canadensis Kuhl		—		+*Camelops sulcatus* (Cope)		—	
Family Geomyidae				+*Tanupolama macrocephala* (Cope)		—	—
Geomys bursarius (Shaw)		—	—	Family Cervidae			
Geomys sp.		—	—	*Odocoileus virginianus* (Zimmermann)		—	—
Family Heteromyidae				Family Antilocapridae			
Perognathus hispidus Baird	—	—	—	+*Capromeryx furcifer* Matthew		—	
Perognathus sp.	—	—	—	+*Capromeryx minimus* Meade	x		
Dipodomys ordii Woodhouse	—	—	—	Family Bovidae			
Family Cricetidae				**Bison latifrons* (Harlan)			—
**Onychomys jinglebobensis* Hibbard			x	Order Perissodactyla			
**Onychomys* cf. *jinglebobensis* Hibbard		—		Family Equidae			
Onychomys cf. *leucogaster* (Wied-Neuwied)		—		**Equus niobrarensis* Hay		—	
Reithrodontomys cf. *montanus* (Baird)		—		**Equus* cf. *scotti* Gidley		—	
Reithrodontomys cf. *megalotis* (Baird)		—		**Equus (Asinus) conversidens* Owen		—	—
				**Equus (Hemionus) calobatus* Troxell		—	
				**Equus (Hemionus)* sp. small		—	

BIBLIOGRAPHY OF SANGAMON LOCAL BIOTAS

Cragin Quarry Local Biota (including Butler Spring)

Auffenberg, W.
    1966. A new species of Pliocene tortoise, genus *Geochelone*, from Florida. Jour. Paleontol., 40:877-882.

Brattstrom, B. H.
    1967. A succession of Pliocene and Pleistocene snake faunas from the High Plains of the United States. Copeia, 1967:188-202.

Etheridge, R.
    1958. Pleistocene lizards of the Cragin Quarry fauna of Meade County, Kansas. Copeia, 1958: 94-101.
    1960. Additional notes on the lizards of the Cragin Quarry fauna. Papers Michigan Acad. Sci., Arts, Letters, 45:113-117.

Frye, J. C.
    1942. Geology and groundwater resources of Meade County, Kansas. Bull. Kansas Geol. Surv., 45:1-152.

Frye, J. C., and C. W. Hibbard
    1941. Pliocene and Pleistocene stratigraphy and paleontology of the Meade Basin, southwestern Kansas. Bull. Kansas Geol. Surv., 38:389-424.

Hay, O. P.
    1917. On a collection of fossil vertebrates made by Dr. F. W. Cragin in the Equus beds of Kansas. Univ. Kansas Sci. Bull., 10:39-51.

Herrington, H. B., and D. W. Taylor
    1958. Pliocene and Pleistocene Sphaeriidae (Pelecypoda) from the central United States. Occas. Papers Mus. Zool., Univ. Michigan, 596:1-28.

Hibbard, C. W.
    1937. Notes on some vertebrates from the Pleistocene of Kansas. Trans. Kansas Acad. Sci., 40:223-237.
    1939. Notes on some mammals from the Pleistocene of Kansas. Trans. Kansas Acad. Sci., 42: 463-470.
    1949. Pleistocene stratigraphy and paleontology of Meade County, Kansas. Contrib. Mus. Paleontol., Univ. Michigan, 7:63-90.
    1949. Pleistocene vertebrate paleontology in North America. Bull. Geol. Soc. Amer., 60: 1417-1428.
    1958. Summary of North American Pleistocene mammalian local faunas. Papers Michigan Acad. Sci., Arts, Letters, 43:3-32.
    1958. New stratigraphic names for Early Pleistocene deposits in southwestern Kansas. Amer. Jour. Sci., 256:54-59.

Hibbard, C. W., and D. W. Taylor
    1960. Two Late Pleistocene faunas from southwestern Kansas. Contrib. Mus. Paleontol., Univ. Michigan, 16:1-223.

Kapp, R. O.
    1965. Illinoian and Sangamon vegetation in southwestern Kansas and adjacent Oklahoma. Contrib. Mus. Paleontol., Univ. Michigan, 19: 167-255.

Milstead, W. W.
    1967. Fossil box turtles *(Terrapene)* from central North America, and box turtles of eastern Mexico. Copeia, 1967:168-179.

Schultz, G. E.
    1967. Four superimpsed Late-Pleistocene vertebrate faunas from southwest Kansas. Pp. 321-336, *in* Pleistocene extinctions: the search for a cause (P. S. Martin and H. E. Wright, Jr., eds.), Yale Univ. Press, New Haven, Connecticut, x+453 pp.
    1969. Geology and paleontology of a Late Pleistocene basin in southwest Kansas. Geol. Soc. Amer. Spec. Paper, 105:viii+1-85.

St. John, O.
    1887. Notes on the geology of southwestern Kansas. Rep. Kansas State Bd. Agric., 5:132-152.

Tihen, J. A.
    1960. On *Neoscaphiopus* and other Pliocene pelobatid frogs. Copeia, 1960:89-94.
    1960. Notes on Late Cenozoic hylid and leptodactylid frogs from Kansas, Oklahoma and Texas. Southwestern Nat., 5:66-70.
    1962. A review of New World fossil bufonids. Amer. Midland Nat., 68:1-50.

Jinglebob Local Biota (Late Sangamon)

Brattstrom, B. H.
    1967. A succession of Pliocene and Pleistocene snake faunas from the High Plains of the United States. Copeia, 1967:188-202.

Chantell, C. J.
    1966. Late Cenozoic hylids from the Great Plains. Herpetologica, 22:259-264.

Herrington, H. B., and D. W. Taylor
    1958. Pliocene and Pleistocene Sphaeriidae (Pelecypoda) from the central United States. Occas. Papers Mus. Zool., Univ. Michigan, 596:1-28.

Hibbard, C. W.
    1955. The Jinglebob interglacial (Sangamon?) fauna from Kansas and its climatic significance. Contrib. Mus. Paleontol., Univ. Michigan, 12: 47-96.

Hibbard, C. W., and D. W. Taylor
    1960. Two Late Pleistocene faunas from southwestern Kansas. Contrib. Mus. Paleontol., Univ. Michigan, 16:1-223.

Klingener, D.
    1963. Dental evolution of *Zapus*. Jour. Mamm., 44:248-260.

Milstead, W. W.
    1967. Fossil box turtles *(Terrapene)* from central North America, and box turtles of eastern Mexico. Copeia, 1967:168-179.

Oelrich, T. M.
    1953. A new box turtle from the Pleistocene of southwestern Kansas. Copeia, 1953:33-38.

Rinker, G. C.
    1949. *Tremarctotherium* from the Pleistocene of Meade County, Kansas. Contrib. Mus. Paleontol., Univ. Michigan, 7:107-112.

Stout, M.
    1965. Explanation of figure 9-41. P. 69, *in* Guidebook for [INQUA] Field Conference D, Central Great Plains (C. B. Schultz and H. T. U. Smith, eds.), Nebraska Acad. Sci., Lincoln, 123 pp.

Taylor, D. W., and C. W. Hibbard
    1955. A new Pleistocene fauna from Harper County, Oklahoma. Oklahoma Geol. Surv. Circ., 37:1-23.

Tihen, J. A.
    1954. A Kansas herpetofauna. Copeia, 1954:217-221.

1960. Notes on Late Cenozoic hylid and leptodactylid frogs from Kansas, Oklahoma and Texas. Southwestern Nat., 5:66-70.

van der Schalie, H.
1953. Mollusks from an interglacial deposit (Sangamon? age) in Meade County, Kansas. Nautilus, 66:80-90.

Slaton Local Fauna (Early Sangamon)

Dalquest, W. W.
1967. Mammals of the Pleistocene Slaton local fauna of Texas. Southwestern Nat., 12:1-30.

Meade, G. E.
1952. The water rat in the Pleistocene of Texas. Jour. Mamm., 33:87-89.

Milstead, W. W.
1967. Fossil box turtles *(Terrapene)* from central North America, and box turtles of eastern Mexico. Copeia, 1967:168-179.

Taylor, D. W.
1965. The study of Pleistocene nonmarine mollusks in North America. Pp. 597-611, *in* The Quaternary of the United States (H. E. Wright, Jr., and D. G. Frey, eds.), Princeton Univ. Press, Princeton, New Jersey, x+922 pp.

TABLE 8. Wisconsin Mammalian Local Faunas of the Great Plains and Central Lowlands provinces; NM (New Mexico—Burnet Cave); T (Texas—Centipede Cave; Clear Creek; Damp Cave; Easley Ranch; Friesenhahn Cave; Howard Ranch; Longhorn Cavern; Miller's Cave; Scharbaur); O (Oklahoma—Bar M, locality 1); K (Kansas—Jones Ranch; Robert)

Species	NM	T	O	K
Order Marsupialia				
Family Didelphidae				
Didelphis marsupialis Linnaeus		–		
Didelphis marsupialis virginiana Kerr		–		
Order Insectivora				
Family Soricidae				
Sorex cinereus Kerr		–	–	
Sorex palustris Richardson		–	–	
Blarina brevicauda (Say)		–	–	
Blarina brevicauda fossilis Hibbard				–
Cryptotis parva (Say)		–		
Notiosorex crawfordi (Coues)		–		
Family Talpidae				
Scalopus aquaticus Linnaeus		–		
Order Chiroptera				
Family Vespertilionidae				
Myotis velifer (J. A. Allen)		–		
Myotis sp.	–	–		
Eptesicus fuscus (Palisot de Beauvois)	–	–		
Family Molossidae				
Eumops perotis (Schinz)		–		
Order Carnivora				
Family Ursidae				
+*Arctodus simus* (Cope)		–		
+*Arctodus* sp.	–			
Ursus americanus Pallus		–		
Family Canidae				
Canis latrans Say		–		
Canis lupus nubilus Say		–		
Canis (Aenocyon) dirus (Leidy)		–		
Vulpes vulpes macroura Baird		–		
Vulpes velox (Say)		–	–	
Urocyon cinereoargenteus (Schreber)		–		
Family Procyonidae				
Bassariscus astutus (Lichtenstein)		–		
Procyon lotor (Linnaeus)		–		
Family Mustelidae				
Mustela cf. *vison* Schreber		–		
Mustela nigripes (Audubon and Bachman)		–		
Taxida taxus (Schreber)		–	–	
Spilogale putorius (Linnaeus)		–		
Mephitis mephitis (Schreber)		–		
Conepatus mesoleucus (Lichtenstein)		–		
Family Felidae				
Felis onca Linnaeus		–		
Felis concolor Linnaeus		–		
Lynx rufus (Schreber)		–		
+*Homotherium (Dinobastis) serus* (Cope)		–		
+*Smilodon* sp.		–		
Order Rodentia				
Family Sciuridae				
Marmota flaviventris (Audubon and Bachman)	–			
Cynomys ludovicianus (Ord)	–	–		–
Spermophilus richardsonii (Sabine)				–
Spermophilus tridecemlineatus (Mitchill)				–
Peromyscus cf. *difficilis* (J. A. Allen)				
Peromyscus sp.		–		
Neotoma floridana (Ord)		–		
Neotoma cf. *micropus* Baird		–		
Neotoma albigula Hartley		–		
Neotoma lepida Thomas		–	–	
Neotoma mexicana Baird		–	–	
Neotoma cinerea (Ord)		–	–	
Neotoma sp.				–
Oryzomys palustris fossilis Hibbard				–
Sigmodon hispidus Say and Ord		–		
Ondatra zibethicus (Linnaeus)		–		
Microtus pennsylvanicus (Ord)		–	–	–
Microtus longicaudus (Merriam)		–		
Microtus mexicanus (Saussure)	–	–		
Microtus (Pedomys) ochrogaster (Wagner)		–		
Microtus (Pitymys) pinetorum (Le Conte)		–		
Microtus ochrogaster and/or *M. pinetorum*		–		
Synaptomys cooperi Baird				–
Synaptomys cooperi paludis Hibbard and Rinker				–
Synaptomys australis Simpson		–		
Family Dipodidae				
Zapus hudsonius (Zimmermann)				–
Family Erethizontidae				
Erethizon dorsatum (Linnaeus)		–		
Order Edentata				
Sloth		–		
Family Dasypodidae				
Dasypus bellus (Simpson)		–	–	
Order Proboscidea				
+Family Mammutidae				
+*Mammut* sp.		–		
Family Elephantidae				
+*Mammuthus columbi* (Falconer)		–		
+*Mammuthus* cf. *columbi* (Falconer)				–
+*Mammuthus* sp.		–		
Order Lagomorpha				
Family Leporidae				
Sylvilagus floridanus (J. A. Allen)		–		
Sylvilagus audubonii (Baird)	–	–		
Sylvilagus sp.				–
Lepus townsendii Bachman	–			
Lepus alleni Mearns		–		
Lepus californicus Gray		–		
Lepus sp.		–		
Order Artiodactyla				
Family Tayassuidae				
+*Platygonus compressus* Le Conte		–		
+*Platygonus* sp.		–		
+*Mylohyus nasutus* (Leidy)		–		
+*Mylohyus* sp.		–		
Family Camelidae				
+*Camelops* cf. *hesternus* (Leidy)		–		
+*Camelops* sp.		–	–	

TABLE 8. (Continued)

Species	NM	T	O	K	Species	NM	T	O	K
Spermophilus cf. *tridecemlineatus* (Mitchill)		—			+*Tanupolama* sp.	—			
Spermophilus mexicanus (Erxleben)		—			Family Cervidae				
Spermophilus spilosoma Bennett		—			*Odocoileus hemionus* (Rafinesque)	—			
Spermophilus franklinii (Sabine)		—			*Odocoileus virginianus* (Zimmermann)	—	—		
Spermophilus variegatus (Erxleben)	—	—			*Odocoileus* cf. *virginianus* (Zimmermann)	—			
Ammospermophilus leucurus (Merriam)		—			*Odocoileus* sp.		—		—
Sciurus cf. *niger* Linnaeus	—				+*Sangamonia* sp.	—			
Family Castoridae					*Rangifer? fricki* Schultz and Howard				x
Castor canadensis Kuhl	—				Family Antilocapridae				
Family Geomyidae					+*Stockoceros onusrosagris* (Roosevelt and Burden)	—			
Thomomys bottae fulvus (Woodhouse)	—				+*Capromeryx* sp.				
Thomomys cf. *talpoides* (Richardson)		—			*Antilocapra americana* (Ord)	—			
Geomys bursarius (Shaw)	—				*Antilocapra* sp.	—			
Geomys arenarius Merriam	—				Family Bovidae				
Geomys sp.		—	—		*Bison* cf. *latifrons* (Harlan)	—			
Pappogeomys (*Cratogeomys*) *castanops* (Baird)	—				*Bison antiquus* Leidy	—			
Family Heteromyidae					*Bison* cf. *antiquus* Leidy	—			
Perognathus merriami J. A. Allen	—				*Bison* sp.	—	—	—	
Perognathus hispidus Baird	—				+*Eucheratherium collinum* Furlong and Sinclair	—			
Perognathus sp.		—	—		+*Preptoceras sinclairi* Furlong	—			
Dipodomys ordii Woodhouse	—	—			*Ovis canadensis auduboni* Merriam	—			
Family Cricetidae					*Ovis canadensis* Shaw	—			
Onychomys leucogaster (Wied-Neuwied)	—	—			Order Perissodactyla				
Reithrodontomys montanus (Baird)	—				Family Tapiridae				
Reithrodontomys megalotis (Baird)	—				*Tapirus* sp.	—			
Reithrodontomys fulvescens J. A. Allen	—				Family Equidae				
Reithrodontomys sp.		—			*Equus midlandensis* Quinn	—			
Peromyscus cochrani Hibbard	—				*Equus* cf. *scotti* Gidley	—			
Peromyscus maniculatus (Wagner)	—	—			*Equus* (*Asinus*) *excelsus* Leidy	—			
Peromyscus leucopus (Rafinesque)	—				*Equus* (*Asinus*) cf. *excelsus* Leidy	—			
					Equus (*Asinus*) *conversidens* Owen	—			
					Equus (*Asinus*) cf. *conversidens* Owen	—			
					Equus (*Asinus*) sp. small	—			
					Equus sp. large	—			
					Equus sp. medium	—			

BIBLIOGRAPHY OF WISCONSIN LOCAL FAUNAS

Bar M Local Fauna (Late Wisconsin, carbon-14 date, locality 1, 21,360±1250 BP) from the Great Plains Province in Harper County, Oklahoma

Myers, A. J.
   1965. Late Wisconsin date for the Bar M local fauna. Oklahoma Geol. Notes, 25:168-170.

Taylor, D. W., and C. W. Hibbard
   1955. A new Pleistocene fauna from Harper County, Oklahoma. Oklahoma Geol. Surv. Circ., 37:1-23.

   1965. The study of Pleistocene nonmarine mollusks in North America. Pp. 597-611, *in* The Quaternary of the United States (H. E. Wright, Jr., and D. G. Frey, eds.), Princeton Univ. Press, Princeton, New Jersey, x+922 pp.

Burnet Cave Local Fauna (Late Wisconsin, carbon-14 date, 7432±300 BP), from the Great Plains Province in Eddy County, New Mexico

Howard, E. B.
   1930. Archaeological research in the Guadalupe Mountains. Jour. Univ. Pennsylvania Mus., 21:189-202.
   1931. Field work in the Southwest. Bull. Univ. Pennsylvania Mus., 3:11-14.

   1932. Caves along the slopes of the Guadalupe Mountains. Bull. Texas Archaeol. Paleontol. Soc., 4:7-19.
   1935. Evidence of early man in North America, based on geological and archaeological work in New Mexico. Jour. Univ. Pennsylvania Mus., 24:55-171.

Libby, W. F.
1954. Chicago radiocarbon dates, IV. Science, 119:135-140.
Murray, K. F.
1957. Pleistocene climate and the fauna of Burnet Cave, New Mexico. Ecology, 38:129-132.
Schultz, C. B., and E. B. Howard
1935. The fauna of Burnet Cave, Guadalupe Mountains, New Mexico. Proc. Acad. Nat. Sci. Philadelphia, 87:273-298.

Skinner, M. F.
1942. The fauna of Papago Springs Cave, Arizona, and a study of *Stockoceros;* with three new antilocaprines from Nebraska and Arizona. Bull. Amer. Mus. Nat. Hist., 80:143-220.
Stearns, C. E.
1942. A fossil marmot from New Mexico and its climatic significance. Amer. Jour. Sci., 240:867-878.

Centipede Cave Local Fauna (Late Wisconsin) from the Great Plains Province in Val Verde County, Texas

Lundelius, E. L., Jr.
1963. Non-human skeletal material [*in* Centipede and Damp caves: excavation in Val Verde County, Texas, 1958]. Bull. Texas Archaeol. Soc., 33:127-129.

1967. Late-Pleistocene and Holocene faunal history of central Texas. Pp. 287-319, *in* Pleistocene extinctions: the search for a cause (P. S. Martin and H. E. Wright, Jr., eds.), Yale Univ. Press, New Haven, Connecticut, x+453 pp.

Clear Creek Local Fauna (Late Wisconsin, carbon-14 date, 28,840±4740 BP) from the eastern edge of the Central Lowland Province in Denton County, Texas

Cheatum, E. P., and D. Allen
1963. An ecological local molluscan fauna in Texas. Jour. Grad. Res. Center, Southern Methodist Univ., 31:174-179.
Holman, J. A.
1963. Late Pleistocene amphibians and reptiles of the Clear Creek and Ben Franklin local faunas of Texas. Jour. Grad. Res. Center, Southern Methodist Univ., 31:152-167.
Schlichting, H. E., Jr.
1963. Charophytes of Pleistocene age from Delta and Denton counties, Texas. Jour. Grad. Res. Center, Southern Methodist Univ., 31:180-181.
Slaughter, B. H., and R. Ritchie
1963. Pleistocene mammals of the Clear Creek local fauna, Denton County, Texas. Jour. Grad. Res. Center, Southern Methodist Univ., 31:117-131.
Taylor, D. W.
1965. The study of Pleistocene nonmarine mollusks in North America. Pp. 597-611, *in* The Quaternary of the United States (H. E. Wright, Jr., and D. G. Frey, eds.), Princeton Univ. Press, Princeton, New Jersey, x+922 pp.
Uyeno, T.
1963. Late Pleistocene fishes of the Clear Creek and Ben Franklin local faunas of Texas. Jour. Grad. Res. Center, Southern Methodist Univ., 31:168-173.

Damp Cave Local Fauna (Late Wisconsin) from the Great Plains Province in Val Verde County, Texas

Lundelius, E. L., Jr.
1963. Non-human skeletal material [*in* Centipede and Damp caves: excavation in Val Verde County, Texas, 1958]. Bull. Texas Archaeol. Soc., 33:127-129.

1967. Late-Pleistocene and Holocene faunal history of central Texas. Pp. 287-319, *in* Pleistocene extinctions: the search for a cause (P. S. Martin and H. E. Wright, Jr., eds.), Yale Univ. Press, New Haven, Connecticut, x+453 pp.

Easley Ranch Local Fauna (Early Wisconsin) from the Central Lowland Province in Foard County, Texas

Dalquest, W. W.
1962. The Good Creek Formation, Pleistocene of Texas, and its fauna. Jour. Paleontol., 36:568-582.
Holman, J. A.
1962. A Texas Pleistocene herpetofauna. Copeia, 1962:255-261.

Taylor, D. W.
1965. The study of Pleistocene nonmarine mollusks in North America. Pp. 597-611, *in* The Quaternary of the United States (H. E. Wright, Jr., and D. G. Frey, eds.), Princeton Univ. Press, Princeton, New Jersey, x+922 pp.

Friesenhahn Cave Local Fauna (Middle to Late Wisconsin) from near the southern edge of the Great Plains Province in Bexar County, Texas

Auffenberg, W.
1962. A new species of *Geochelone* from the Pleistocene of Texas. Copeia, 1962:627-636.
Evans, G. L.
1961. The Friesenhahn Cave. Bull. Texas Mem. Mus., 2:1-22.
Hay, O. P.
1920. Description of some Pleistocene vertebrates found in the United States. Proc. U.S. Nat. Mus., 58:83-146.
Kennerly, T. E., Jr.
1956. Comparisons between fossil and Recent species of the genus *Perognathus*. Texas Jour. Sci., 8:74-86.

Lundelius, E. L., Jr.
1960. *Mylohyus nasutus*, long-nosed peccary of the Texas Pleistocene. Bull. Texas Mem. Mus., 1:1-40.
1967. Late Pleistocene and Holocene faunal history of central Texas. Pp. 287-319, *in* Pleistocene extinctions: the search for a cause (P. S. Martin and H. E. Wright, Jr., eds.), Yale Univ. Press, New Haven, Connecticut, x+453 pp.
Meade, G. E.
1961. The sabre-toothed cat, *Dinobastis serus*. Bull. Texas Mem. Mus., 2:23-60.
Mecham, J. S.
1959. Some Pleistocene amphibians and reptiles

from Friesenhahn Cave, Texas. Southwestern Nat., 3:17-27.

Milstead, W. W.
1956. Fossil turtles of Friesenhahn Cave, Texas, with the description of a new species of *Testudo*. Copeia, 1956:162-171.
1967. Fossil box turtles *(Terrapene)* from central North America, and box turtles of eastern Mexico. Copeia, 1967:168-187.

Pettus, D.
1956. Fossil rabbits (Lagomorpha) of the Friesenhahn Cave deposits, Texas. Southwestern Nat., 1:109-115.

Tamsitt, J. R.
1957. *Peromyscus* from the Late Pleistocene of Texas. Texas Jour. Sci., 9:355-363.

Tihen, J. A.
1962. A review of New World fossil bufonids. Amer. Midland Nat., 68:1-50.

Howard Ranch Local Fauna (Late Wisconsin, carbon-14 dates, 16,775±565 to 19,098±1047 BP) from the Central Lowland Province in Hardeman County, Texas

Dalquest, W. W.
1965. New Pleistocene formation and local fauna from Hardeman County, Texas. Jour. Paleontol., 39:63-79.

Frye, J. C., and A. B. Leonard
1963. Pleistocene geology of Red River Basin in Texas. Texas Bur. Econ. Geol. Rep. Inv., 49:1-48. [The molluscan fauna from the Howard Ranch local fauna is listed and considered as Late Kansan age.]

Holman, J. A.
1964. Pleistocene amphibians and reptiles from Texas. Herpetologica, 20:73-83.

Taylor, D. W.
1965. The study of Pleistocene nonmarine mollusks in North America. Pp. 597-611, in The Quaternary of the United States (H. E. Wright, Jr., and D. G. Frey, eds.), Princeton Univ. Press, Princeton, New Jersey, x+922 pp.

Jones Local Fauna (Late Wisconsin, carbon-14 date of *Ambystoma tigrinum*, neotenic, bed on shell, I-3461, 26,700±1500 and 29,000±1300 BP) from the Great Plains Province in Meade County, Kansas, elevation 2420 feet

Brattstrom, B. H.
1967. A succession of Pliocene and Pleistocene snake faunas from the High Plains of the United States. Copeia, 1967:188-202.

Chantell, C. J.
1966. Late Cenozoic hylids from the Great Plains. Herpetologica, 22:259-264.

Crane, H. R., and J. B. Griffin
1961. University of Michigan radiocarbon dates VI. Radiocarbon, 3:105-125. [M-1103, Jones local fauna>30,000 BP.]

Downs, T.
1954. Pleistocene birds from the Jones fauna of Kansas. Condor, 56:207-221.

Franzen, D. S., and A. B. Leonard
1947. Fossil and living Pupillidae (Gastropoda-Pulmonata) in Kansas. Univ. Kansas Sci. Bull., 31:311-411.

Frye, J. C.
1942. Geology and ground-water resources of Meade County, Kansas. Bull. Kansas Geol. Surv., 45:1-152.

Frye, J. C., and C. W. Hibbard
1941. Pliocene and Pleistocene stratigraphy and paleontology of the Meade Basin, southwestern Kansas. Bull. Kansas Geol. Surv., 38:389-424.

Frye, J. C., and A. B. Leonard
1951. Stratigraphy of the Late Pleistocene loesses of Kansas. Jour. Geol., 59:287-305.
1952. Pleistocene geology of Kansas. Bull. Kansas Geol. Surv., 99:1-230.

Goodrich, C.
1940. Mollusks of a Kansas Pleistocene deposit. Nautilus, 53:77-79. [Jones fauna comes from 15 feet below top of exposure.]

Gutentag, E. D., and R. H. Benson
1962. Neogene (Plio-Pleistocene) fresh-water ostracodes from the Central High Plains. Bull. Kansas Geol. Surv., 157:1-60. [Locality 7 is Jones fauna.]

Herrington, H. B., and D. W. Taylor
1958. Pliocene and Pleistocene Sphaeriidae (Pelecypoda) from the central United States.

Occas. Papers Mus. Zool., Univ. Michigan, 596:1-28.

Hibbard, C. W.
1940. A new Pleistocene fauna from Meade County, Kansas. Trans. Kansas Acad. Sci., 43:417-425.
1942. Pleistocene mammals from Kansas. Bull. Kansas Geol. Surv., 41:261-269.
1949. Pleistocene stratigraphy and paleontology of Meade County, Kansas. Contrib. Mus. Paleontol., Univ. Michigan, 7:63-90.
1949. Pleistocene vertebrate paleontology in North America. Bull. Geol. Soc. Amer., 60:1417-1428.
1958. Summary of North American Pleistocene mammalian local faunas. Papers Michigan Acad. Sci., Arts, Letters, 43:3-32.

Hibbard, C. W., C. E. Ray, D. E. Savage, D. W. Taylor, and J. E. Guilday
1965. Quaternary mammals of North America. Pp. 509-525, in The Quaternary of the United States (H. E. Wright, Jr., and D. G. Frey, eds.), Princeton Univ. Press, Princeton, New Jersey, x+922 pp.

Hibbard, C. W., and D. W. Taylor
1960. Two Late Pleistocene faunas from southwestern Kansas. Contrib. Mus. Paleontol., Univ. Michigan, 16:1-223.

Taylor, D. W., and C. W. Hibbard
1955. A new Pleistocene fauna from Harper County, Oklahoma. Oklahoma Geol. Surv. Circ., 37:1-23.

Tihen, J. A.
1942. A colony of fossil neotenic *Ambystoma tigrinum*. Univ. Kansas Sci. Bull., 28:189-198.
1955. A new Pliocene species of *Ambystoma*, with remarks on other fossil ambystomids. Contrib. Mus. Paleontol., Univ. Michigan, 12:229-244.
1960. Notes on Late Cenozoic hylid and leptodactylid frogs from Kansas, Oklahoma and Texas. Southwestern Nat., 5:66-70.
1962. A review of New World fossil bufonids. Amer. Midland Nat., 68:1-50.

Longhorn Cavern Local Fauna (Late Wisconsin) from the Great Plains Province in Burnet County, Texas

Patton, T. H.
   1963. Fossil remains of southern bog lemming in Pleistocene deposits of Texas. Jour. Mamm., 44:275-277.

Semken, H. A., Jr.
   1961. Fossil vertebrates from Longhorn Cavern, Burnet County, Texas. Texas Jour. Sci., 13: 290-310.

Miller's Cave Local Fauna (Late Wisconsin to Recent) from the Great Plains Province in Llano County, Texas

Lundelius, E. L., Jr.
   1967. Late Pleistocene and Holocene faunal history of central Texas. Pp. 287-319, *in* Pleistocene extinctions: the search for a cause (P. S. Martin and H. E. Wright, Jr., eds.), Yale Univ. Press, New Haven, Connecticut, x+453 pp.

Patton, T. H.
   1963. Fossil vertebrates from Miller's Cave, Llano County, Texas. Bull. Texas Mem. Mus., 7:1-41.

Robert Local Fauna (Late Wisconsin, carbon-14 date, 11,100±390 BP) from the Great Plains Province in Meade County, Kansas

Schultz, G. E.
   1967. Four superimposed Late-Pleistocene vertebrate faunas from southwest Kansas. Pp. 321-336, *in* Pleistocene extinctions: the search for a cause (P. S. Martin and H. E. Wright, Jr., eds.), Yale Univ. Press, New Haven, Connecticut, x+453 pp.
   1969. Geology and paleontology of a Late Pleistocene basin in southwest Kansas. Geol. Soc. Amer. Spec. Paper, 105:viii+1-85.

Scharbauer Local Fauna (Middle to Late Wisconsin) from the Great Plains Province in Midland County, Texas

Gazin, C. L.
   1955. Identification of some vertebrate fossil material from the Scharbauer site, Midland, Texas. Appendix 2, p. 119, *in* The Midland discovery (F. Wendorf, A. D. Krieger, and C. C. Albritton), Univ. Texas Press, Austin.

Quinn, J. H.
   1955. Report on the horse remains from the Scharbauer site. Appendix 1, pp. 117-118, *in* The Midland discovery (F. Wendorf, A. D. Krieger, and C. C. Albritton), Univ. Texas Press, Austin.
   1957. Pleistocene Equidae of Texas. Texas Bur. Econ. Geol. Rep. Inv., 33:1-51.